The Papovaviridae

Volume 2
THE PAPILLOMAVIRUSES

THE VIRUSES

Series Editors
HEINZ FRAENKEL-CONRAT, *University of California*
Berkeley, California

ROBERT R. WAGNER, *University of Virginia School of Medicine*
Charlottesville, Virginia

THE VIRUSES: Catalogue, Characterization, and Classification
Heinz Fraenkel-Conrat

THE ADENOVIRUSES
Edited by Harold S. Ginsberg

THE HERPESVIRUSES
Volumes 1–3 • Edited by Bernard Roizman
Volume 4 • Edited by Bernard Roizman and Carlos Lopez

THE PAPOVAVIRIDAE
Volume 1 • Edited by Norman P. Salzman
Volume 2 • Edited by Norman P. Salzman and Peter M. Howley

THE PARVOVIRUSES
Edited by Kenneth I. Berns

THE PLANT VIRUSES
Volume 1 • Edited by R. I. B. Francki
Volume 2 • Edited by M. H. V. Van Regenmortel and Heinz Fraenkel-Conrat

THE REOVIRIDAE
Edited by Wolfgang K. Joklik

THE RHABDOVIRUSES
Edited by Robert R. Wagner

THE TOGAVIRIDAE AND FLAVIVIRIDAE
Edited by Sondra Schlesinger and Milton J. Schlesinger

THE VIROIDS
Edited by T. O. Diener

The Papovaviridae

Volume 2

THE PAPILLOMAVIRUSES

Edited by
NORMAN P. SALZMAN
and
PETER M. HOWLEY
National Institutes of Health
Bethesda, Maryland

PLENUM PRESS • NEW YORK AND LONDON

Library of Congress Cataloging in Publication Data

(Revised for volume 2)

The papovaviridae.

(The Viruses)
Volume 2 edited by Norman P. Salzman and Peter M. Howley.
Includes bibliographies and index.
Contents: v. 1. The polyomaviruses—The papillomaviruses.
1. Papovaviruses—Collected works. I. Salzman, Norman P. II. Howley, Peter M.
III. Series: The Viruses. [DNLM: 1. Papovaviridae. QW 165.5.P2 P218]
QR406.P36 1986 576′.6484 86-15160
ISBN 0-306-42452-5 (v. 2)

A Division of Plenum Publishing Corporation
233 Spring Street, New York, N.Y. 10013

Printed in the United States of America

Contributors

Eberhard Amtmann, Institute for Virus Research, German Cancer Research Center, 6900 Heidelberg, Federal Republic of Germany
Harri Ahola, Department of Medical Genetics, University of Uppsala, Biomedical Center, S-751 23 Uppsala, Sweden
Carl C. Baker, Laboratory of Tumor Virus Biology, National Cancer Institute, Bethesda, Maryland 20892
Daniel DiMaio, Department of Human Genetics, Yale University School of Medicine, New Haven, Connecticut 06510
Peter M. Howley, Laboratory of Tumor Virus Biology, National Cancer Institute, Bethesda, Maryland 20892
Robert F. LaPorta, Department of Oral Biology and Pathology, School of Dental Medicine, State University of New York, Stony Brook, New York 11794
Jorge Moreno-Lopez, Department of Veterinary Microbiology (Virology), Swedish University of Agricultural Sciences, Biomedical Center, S-751 23 Uppsala, Sweden
Carl Olson, Department of Veterinary Science, University of Wisconsin, Madison, Wisconsin 53706
Gérard Orth, Unité des Papillomavirus, Unité INSERM 190, Institut Pasteur, 75724 Paris Cedex 15, France
Ulf Pettersson, Department of Medical Genetics, University of Uppsala, Biomedical Center, S-751 23 Uppsala, Sweden
Herbert Pfister, Institut für Klinische Virologie, Universität Erlangen-Nürnberg, D-8520 Erlangen, Federal Republic of Germany
Richard Schlegel, Laboratory of Tumor Virus Biology, National Cancer Institute, Bethesda, Maryland 20892
Achim Schneider, Sektion für Gynäkologische Zytologie und Histologie, Universitäts-Frauenklinik, 7900 Ulm, Federal Republic of Germany
Bettie M. Steinberg, Department of Otolaryngology, Long Island Jewish Medical Center, New Hyde Park, New York 11042

Arne Stenlund, Department of Medical Genetics, University of Uppsala, Biomedical Center, S-751 23 Uppsala, Sweden; *Present address:* Department of Molecular Biology, University of California, Berkeley, California 94720

Lorne B. Taichman, Department of Oral Biology and Pathology, School of Dental Medicine, State University of New York, Stony Brook, New York 11794

Klaus Wayss, Institute for Pathology, German Cancer Research Center, 6900 Heidelberg, Federal Republic of Germany

Felix O. Wettstein, Department of Microbiology and Immunology, UCLA School of Medicine, Los Angeles, California 90024

Harald zur Hausen, Deutsches Krebsforschungszentrum, 6900 Heidelberg, Federal Republic of Germany

Preface

In recent years there has been an explosion in research on the papillomaviruses. The viral nature of human warts was first suggested 80 years ago by Ciuffo, who demonstrated transmission using cell-free filtrates. Shope described the first papillomavirus over 50 years ago as the etiologic agent in infectious papillomatosis in rabbits. Subsequent studies by Rous established that benign rabbit papillomas induced by this virus could progress to carcinomas when treated with specific nonviral cofactors. Despite these rich beginnings, the papillomavirus field lay virtually dormant until the late 1970s because no one was able to propagate these viruses in culture successfully. In the late 1970s the molecular cloning of the papillomavirus genomes permitted investigators to partially circumvent this obstacle to their progress. The cloning of the viral genomes permitted the standardization of viral reagents and provided sufficient material to begin a systematic evaluation of the biology of this group of viruses.

This volume contains a series of chapters designed to provide a historical perspective on this field and to review the current state of research involving the papillomaviruses. Chapters 1 and 2 serve as introductory chapters, providing a general description and overview of the papillomaviruses and a historical perspective. Chapters 3–5 cover the molecular biology and cell biology of the papillomaviruses, focusing on transcription, gene expression in epithelial cells, and viral transformation. The rabbit and the *Mastomys natalensis* provide animal models for studying carcinogenic progression and are covered in Chapters 6 and 7. In humans, HPVs have been associated with cutaneous carcinomas in patients with epidermodysplasia verruciformis and with anogenital carcinomas; these topics are reviewed in Chapters 8 and 9. Laryngeal papillomas are associated with the same HPV types seen in the genital tract and are also capable of malignant progression, as discussed in Chapter 10. The viral DNA of the bovine papillomavirus type 1 has been developed into a mammalian cell cloning vector based on its ability to remain as a stable plasmid in transformed cells, and its use as a cloning vector is reviewed in Chapter 11. Finally, an appendix is included presenting an analysis of sequences of the viral genomes available at the time this book was completed.

Norman P. Salzman
Peter M. Howley

Contents

Chapter 1

Papillomaviruses: General Description, Taxonomy, and Classification

Herbert Pfister

Chapter 2

Animal Papillomas: Historical Perspectives

Carl Olson

Chapter 3

Organization and Expression of Papillomavirus Genomes

Ulf Pettersson, Harri Ahola, Arne Stenlund, and Jorge Moreno-Lopez

Chapter 4

The Expression of Papillomaviruses in Epithelial Cells

Lorne B. Taichman and Robert F. LaPorta

Chapter 5

Papillomavirus Transformation

Peter M. Howley and Richard Schlegel

Chapter 6

Papillomaviruses and Carcinogenic Progression I: Cottontail Rabbit (Shope) Papillomavirus

Felix O. Wettstein

Chapter 7

Papillomaviruses and Carcinogenic Progression II: The *Mastomys natalensis* Papillomavirus

Eberhard Amtmann and Klaus Wayss

Chapter 8

Epidermodysplasia Verruciformis

Gérard Orth

Chapter 9

The Role of Papillomaviruses in Human Anogenital Cancer

Harald zur Hausen and Achim Schneider

Chapter 10

Laryngeal Papillomas: Clinical Aspects and *in Vitro* Studies

Bettie M. Steinberg

Chapter 11

Papillomavirus Cloning Vectors

Daniel DiMaio

Appendix

Sequence Analysis of Papillomavirus Genomes

Carl C. Baker

CHAPTER 1

Papillomaviruses
General Description, Taxonomy, and Classification

HERBERT PFISTER

I. INTRODUCTION

Papillomaviruses are classified as the genus *Papillomavirus* of the Papovaviridae family by virtue of their capsid structure and biochemical composition (Matthews, 1982). The icosahedral particles contain only DNA and protein, and show a buoyant density in CsCl of 1.34 g/cm^3. The single DNA molecule in each virion is double-stranded and circular. Polyomalike viruses form the second Papovaviridae genus, which is distinguished from papillomaviruses by a smaller capsid (55 nm versus 45 nm) and a shorter DNA (7.25 kbp to 8.4 kbp versus 5.25 kbp).

Papillomaviruses induce tumors in skin and mucosa, thus explaining the name of this virus group (from the Latin *papilla*, nipple, pustule and the Greek suffix *oma* to denote tumor). In some tumors the viruses occur in large amounts (up to 10^{13} physical particles per gram tumor) and are easy to isolate. It was therefore possible to unequivocally demonstrate the viral etiology by transmission experiments with cell-free filtrates of tumor extracts (Rowson and Mahy, 1967). Other tumors contain very few virus particles, and the underlying virus types were not accessible to further analysis until recently. This was due to the lack of suitable cell culture systems for *in vitro* propagation of papillomaviruses. Molecular cloning of viral DNAs helped to overcome that problem. At the moment,

HERBERT PFISTER • Institut für Klinische Virologie, Universität Erlangen-Nürnberg, D-8520 Erlangen, Federal Republic of Germany.

we face an almost exponential increase in papillomavirus DNA isolates, which turn out to be very heterogeneous. Sequencing of the DNAs and their availability as labeled probes provided much information on the biochemistry of papillomaviruses, but our knowledge about molecular mechanisms of replication and transformation is still far behind that of the polyomaviruses. On the other hand, there is growing interest in papillomaviruses because of their association with human malignancies (see Chapters 8 and 9). Viral DNA was shown to persist in carcinomas and a possible role of the virus during malignant conversion warrants continued research efforts.

II. PROPERTIES OF THE VIRIONS

A. Particle Structure and Protein Composition

Papillomavirus particles have a sedimentation coefficient, $s_{20,w}$, of 300. They are composed of 72 capsomers in a skew structure, either left-handed (cottontail rabbit papillomavirus) or right-handed (human papillomavirus) (Finch and Klug, 1965; Klug and Finch, 1965). After negative staining, the capsomers appear as hollow cylinders of equal height and width, which are connected at their base by fibrous bridgelike structures (Yabe *et al.*, 1979). Filamentous forms occur, which are regarded to result from aberrant maturation. They are made up of hexagonal capsomers and show a buoyant density of "empty," i.e., DNA-free capsids (Noyes, 1965).

The amount of virus particles that can be isolated from biopsy material is often not sufficient for protein analysis. Studies on viruses from humans and animals revealed major structural proteins with molecular weights in the range of 53,000–59,000 (Favre *et al.*, 1975a; Gissmann *et al.*, 1977; Pfister *et al.*, 1977, 1979c; Lancaster and Olson, 1978; Müller and Gissmann, 1978). DNA sequences coding for the major capsid protein of bovine papillomavirus 1 were identified by hybrid selected *in vitro* translation (Engel *et al.*, 1983). The region harbors a large open reading frame (L1), which codes for 495 amino acids starting from the first ATG codon. A comparison of nucleotide sequences from several papillomaviruses showed L1 to be highly conserved. Amino acid homologies were in the range of 40% to 54%.

It is not surprising from these data that the capsid proteins share common antigenic determinants. Most of them appear to be masked, however, in intact particles. Immunization with native virions triggers a type-specific antibody response (Le Bouvier *et al.*, 1966; Gissmann *et al.*, 1977; Orth *et al.*, 1978b; Pfister *et al.*, 1979b) and cross-reacting antisera were only obtained with detergent-disrupted particles (Jensen *et al.*, 1980). Sera of rabbits, which bear a transplantable, papillomavirus-induced carcinoma, were also shown to react with the major structural polypeptide of disrupted human papillomavirus 1 (Orth *et al.*, 1978a). This may be explained by the fact that carcinoma cells do not regularly produce virus particles but only unassembled structural proteins. A few sera from carcinoma carriers reacted with native heterologous particles. This indicates additional, probably weak, genus-specific antigens on the virion surface, which do not elicit antibody formation during routine immunization protocols.

Cross-reacting antisera readily disclose papillomavirus capsid antigens in thin sections of warts from humans and animals (Jenson *et al.*, 1980; Lack *et al.*, 1980; Pfister and Meszaros, 1980; Woodruff *et al.*, 1980). The sera do not react with capsid proteins of polyomalike viruses. They were extensively used to screen tumors for papillomavirus fingerprints whenever specific probes were not as yet available.

Apart from the major capsid protein, additional proteins were described in papillomavirus particles. Some of them seem to result from conversion of the major polypeptide (Pfister *et al.*, 1977). Only one additional large open reading frame (L2) exists in the so-called "late" part of the bovine papillomavirus 1 genome, which is specifically transcribed in productive cells of a bovine wart (Amtmann and Sauer, 1982; Engel *et al.*, 1983). A corresponding gene product has not been identified and its localization within the viral capsid remains to be established. One protein with a molecular weight of about 70,000 was frequently observed within purified capsids (Favre *et al.*, 1975a; Gissmann *et al.*, 1977; Lancaster and Olson, 1978). It could theoretically be explained as a fusion product of reading frames L2 and L1.

The DNA of bovine papillomaviruses 1 and 2 was shown to be associated with histone-related proteins in a chromatinlike complex (Favre *et al.*, 1977). The cellular histones H2a, H2b, H3, and H4 were also detected in human papillomavirus 1 capsids. The electrophoretic mobility of the viral proteins in an acetic acid/urea system differed from the mobility of cellular histones (Pfister and zur Hausen, 1978a) which might be due to acetylation, as described for polyoma virus (Schaffhausen and Benjamin, 1976). Histonelike proteins were generally not found in empty capsids or in DNA-containing particles of human papillomavirus 4 or bovine papillomavirus 3 (Gissmann *et al.*, 1977; Pfister *et al.*, 1979c). It must be questioned if this reflects a basic difference in DNA packaging, or a different stability of the virions which could lead to a partial loss of low-molecular-weight histones.

B. Nucleic Acid

The DNAs of most papillomaviruses have an average molecular weight of 5.0×10^6, corresponding to about 7.8 kbp. The DNAs of some viruses are smaller (7.2 kbp) (Pfister *et al.*, 1979c; Campo *et al.*, 1980; Jarrett *et al.*, 1984a; Kremsdorf *et al.*, 1984) but this is not likely to indicate a subgroup, because all human viruses with a shorter DNA are closely related to "full-size" DNA viruses according to DNA–DNA hybridization. Only a virus from oral papillomas of the dog harbors significantly larger DNA (molecular weight of about 5.5×10^6) (Pfister and Meszaros, 1980). The G + C content varies between 41% and 50% (Crawford and Crawford, 1963; Müller and Gissmann, 1978; Kremsdorf *et al.*, 1983).

The complete nucleotide sequences of four papillomavirus DNAs have been determined and reveal a strikingly similar genome organization (Chen *et al.*, 1982; Danos *et al.*, 1982; Schwarz *et al.*, 1983; Giri *et al.*, 1985). All major open reading frames are on one strand, thus indicating that there is only one coding DNA strand. The large reading frames are of comparable size and in similar positions, which allows the construction of a consensus genome (Fig. 1). A 1-kbp noncoding region harbors transcription control signals and the origin of replication, which was mapped on bovine papillomavirus 1 DNA by electron microscopy of replicative intermediates (Waldeck *et al.*, 1984). A 4.5-kbp area follows, which contains two major (E1 and E2) and a number of minor reading frames. Genetic experiments with bovine papillomavirus 1 identified a number of early gene functions within this region that play a part in episomal persistence of viral DNA in high copy number and in oncogenic transformation of host cells. No gene products have yet been identified, but some functions were shown to act in *trans* (see Chapter 3). Reading frames L2 and L1 lie within 2.3 kbp.

The sequences are highly homologous, especially within E1 and L1, which leads to cross-hybridization between the DNAs of different papillomaviruses under conditions of reduced stringency (Law *et al.*, 1979; Heilman *et al.*, 1980). The degree of relationship varies between the var-

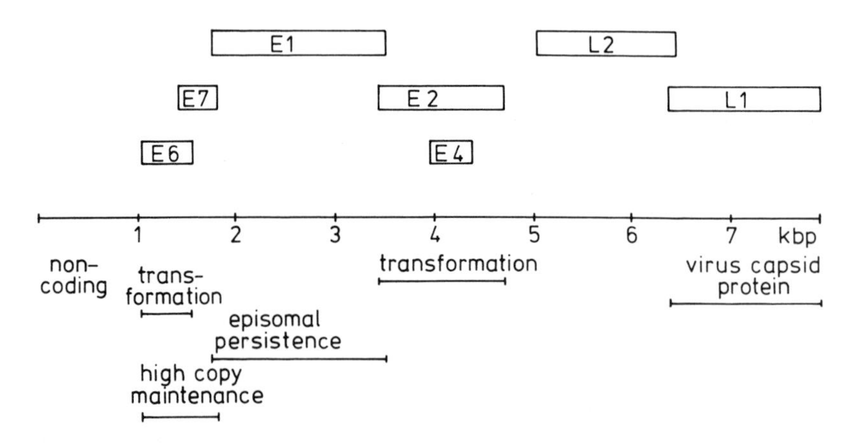

FIGURE 1. Consensus genome organization of papillomaviruses. The open bars represent open reading frames, which are labeled E or L depending on their position in the "early" or "late" region of the genome (see text). The numbers follow current nomenclature. Reading frame E3 without an ATG start codon at the beginning and reading frame E5, the position of which varies considerably between different papillomaviruses, are not shown. Gene functions as characterized for bovine papillomavirus 1 are mapped below.

ious viruses but most of them show significant hybridization at 50°C below the melting temperature of the DNA.

III. BIOLOGICAL PROPERTIES

A. Characteristics of the.Benign Tumors

Papillomaviruses induce epithelial or fibroepithelial tumors of skin or mucosa. All human papillomaviruses cause pure epithelial proliferations. The tumors are basically benign, show limited growth, and usually regress spontaneously (Rulison, 1942; Massing and Epstein, 1963).

1. Pure Epithelial Lesions

The host range of most papillomaviruses is confined to epithelial cells and the tumors are characterized by hyperplasia of cells in the spinous layer (Fig. 2). After *in situ* hybridization with viral DNA probes, cell

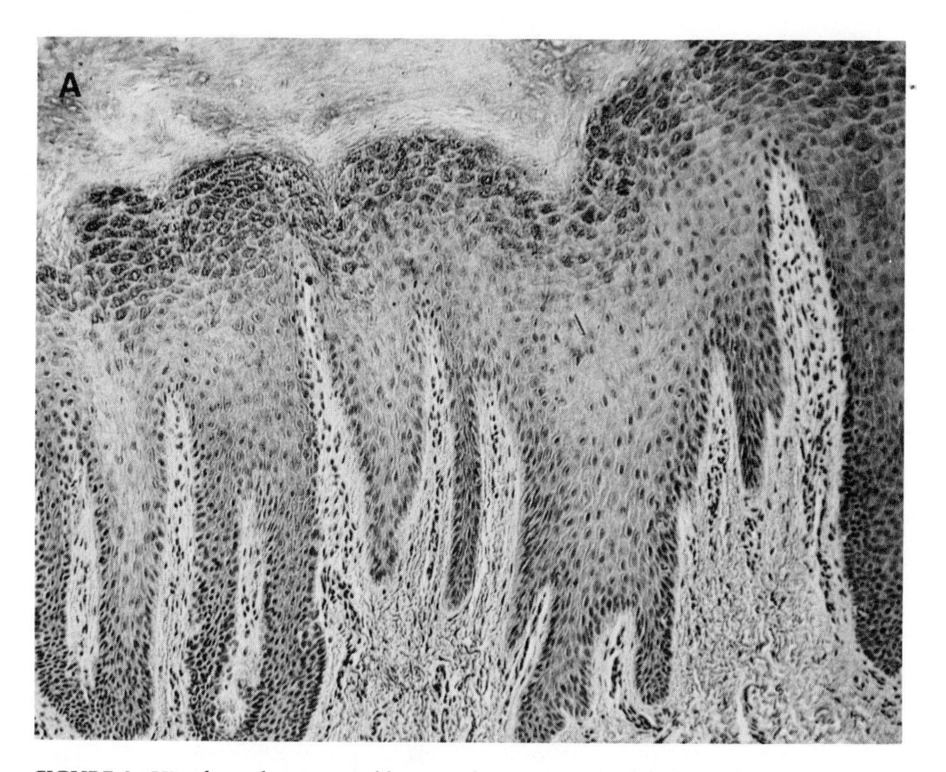

FIGURE 2. Histological sections of human skin warts induced by human papillomavirus 1 (A), 2 (B), or 3 (C). The epithelial proliferations show acanthosis and hyperkeratosis as common features. Note different virus-specific cytopathogenic effects in the stratum granulosum. (Photographed by A. Gassenmaier; H & E, × 100.)

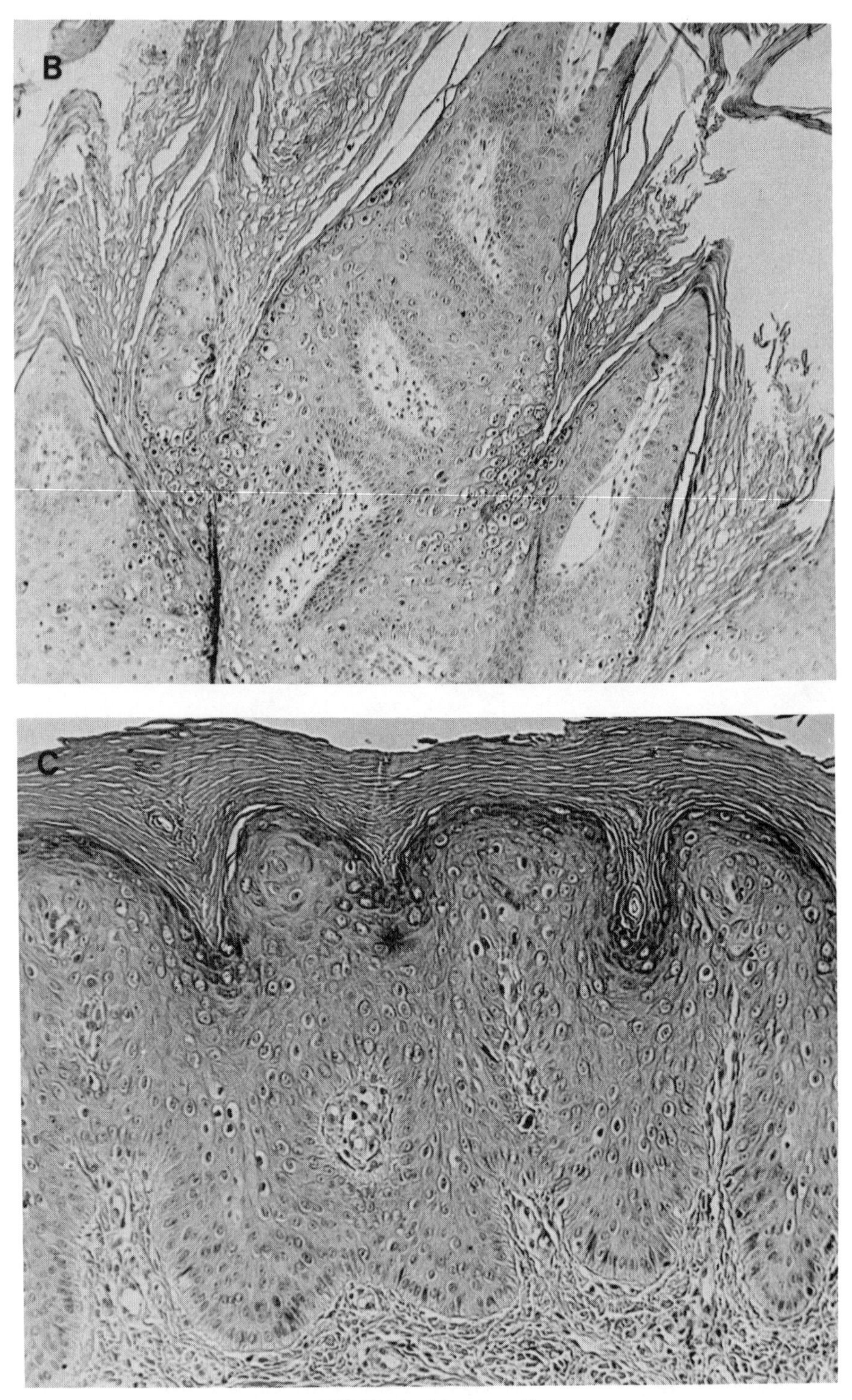

FIGURE 2. (*Continued*)

nuclei of the stratum spinosum and granulosum become heavily labeled, which indicates extensive viral DNA replication (Orth *et al.*, 1971; Grussendorf and zur Hausen, 1979). With some virus types, labeling starts in the first or the second suprabasal cell layer, and with others, only a few nuclei at the tumor periphery are positive. The negative results with many cells, especially with basal cells, are easily explained by the relatively low sensitivity of the *in situ* hybridization technique.

DNA synthesis could be demonstrated in keratinizing cells of the tumors, which is in contrast to normal epidermis and chemically induced papillomas, where DNA synthesis is restricted to basal and parabasal cells (Croissant and Orth, 1967; Rashad, 1969). Papillomaviruses are obviously able to raise the blockade of DNA replication in differentiating cells, which indicates an interference with the corresponding cellular control mechanism.

The amount of mature virus particles in the tumors varies considerably depending on the virus type and the host organism. For example, the particle yield from human papillomavirus-1-induced warts differs by several orders of magnitude and the maximum levels of 10^{12}–10^{13} particles/wart are never reached by human papillomavirus 4 (Dreger and zur Hausen, personal communication). Mature virions are first seen in association with the nucleoli of cells in the stratum spinosum (Almeida *et al.*, 1962). The capsids are spread throughout the nuclei of the stratum granulosum and appear embedded in keratin in the stratum corneum, partially in paracrystalline arrays.

The analysis of the glucose-6-phosphate dehydrogenase phenotype in common warts of heterozygous women suggested a monoclonal origin (Murray *et al.*, 1971) but both isozymes were found in four genital warts (Friedmann and Fialkow, 1976), thus indicating a multicellular origin.

The data are consistent with the following model for wart induction. The virus probably enters the skin through microtraumas and infects a basal cell(s). This should result in increased proliferation accompanied by cell differentiation similar to that in normal epidermis. Cells of the stratum germinativum are nonpermissive for papillomavirus replication but become increasingly permissive with ongoing differentiation, which results in viral DNA replication, synthesis of structural proteins, and induction of virus-specific cytopathogenic effects. Mature viruses may initiate new infection cycles thus accounting for a tumor of multicellular origin.

The incubation period varies between 3 and 18 months in humans (Rowson and Mahy, 1967) and between 3 and 8 weeks in rabbits (Ito, 1975). The efficiency of papilloma induction is rather low as estimated from experimental systems (ID_{50} 9×10^7; Bryan and Beard, 1940). There are no clear-cut ideas about the events wich take place during the long incubation period and which are apparently rare or several in number. Viral gene expression could turn out to be a crucial step in tumor formation. Papillomavirus transcription is extremely restricted and only a few RNA molecules per cell are usually detected in contrast to more than

100 DNA copies (see Chapter 3). Skin tissue of the multimammate mouse harbors up to several thousand copies of episomal papillomavirus genomes without detectable RNA and without pathological alterations. As soon as tumors appear, viral transcripts can also be demonstrated (Amtmann *et al.*, 1984).

2. Fibropapillomas and Fibromas

Fibropapillomas start to develop soon after papillomavirus infection in contrast to epithelial warts. A fibroblastic reaction can be observed within the first week after experimental infection (Olson *et al.*, 1969) and progresses to massive fibroplasia during the next few weeks. It lasts 4–6 weeks before acanthosis becomes visible in the epithelium overlying the fibroma (Fig. 3) and in some cases (e.g., with the deer fibroma virus) the

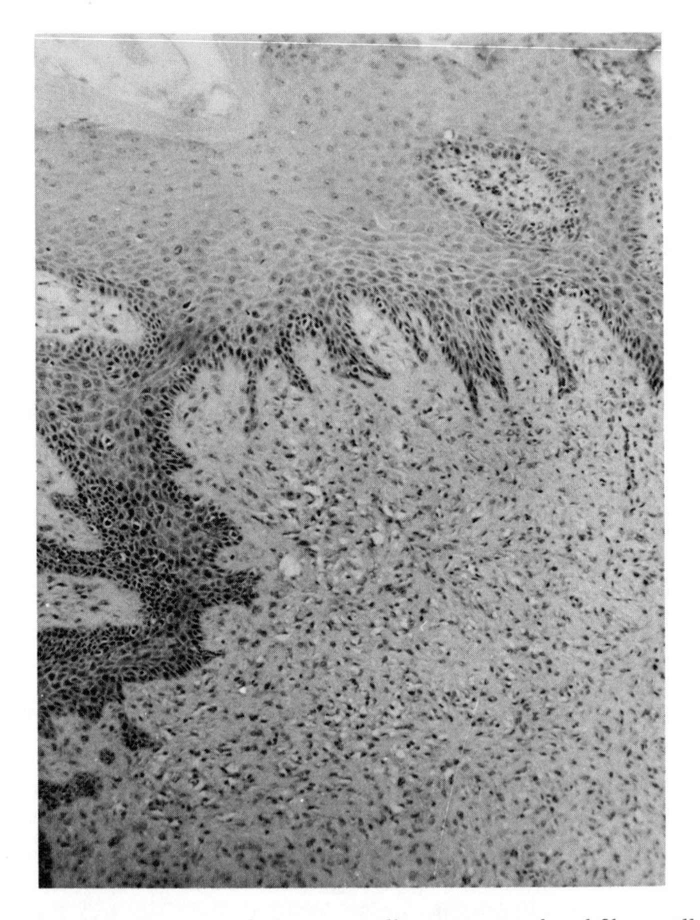

FIGURE 3. Histological section of a bovine papillomavirus-1-induced fibropapilloma from the neck of a cow. Fibroblasts form the principal mass of the tumor. The overlying epithelium shows moderate acanthosis. (Photographed by A. Gassenmaier; H & E, × 100.)

epithelium is only minimally affected. No virus particles can be detected in the fibroblasts, i.e., they are nonpermissive. The distribution of mature viruses in the epithelial moiety of the fibropapilloma is comparable to that in the epithelial warts.

The transformation of fibroblasts can be reproduced *in vitro* and was studied in depth with bovine papillomavirus 1 (see Chapter 5). Primary fibroblasts from cattle, mouse, and hamster are susceptible to transformation, which results in altered cell morphology, increased growth rate, and immortalization (Black *et al.*, 1963; Thomas *et al.*, 1963, 1964; Boiron *et al.*, 1964). Permanently growing mouse cell lines NIH 3T3 and C127 are transformed to focus formation, growth in soft agar, and to tumorigenicity in the nude mouse (Dvoretzky *et al.*, 1980). Transformation follows single-hit kinetics. Transfection studies with deletion mutants (Nakabayashi *et al.*, 1983; Schiller *et al.*, 1984) and cDNA clones (Yang *et al.*, 1985) of bovine papillomavirus 1 showed that transforming ability resides within two independent parts of the viral genome, which cover the reading frames E6 and E2, respectively (see Fig. 1 and Chapter 5). Integration of viral DNA does not appear to be essential for transformation. No integrated DNA was detected in cloned bovine papillomavirus 1-transformed C127 mouse cells, i.e., only less than 0.1 virus genome equivalent per cell could have been integrated (Law *et al.*, 1981). Circular viral DNA persists extrachromosomally with a high copy number of 20–120 genome equivalents per cell and usually without any major deletions or rearrangements. Congruent data were obtained with bovine cells from the conjunctiva and the palate (Moar *et al.*, 1981b) and with cells from connective tissue tumors of cattle, horses (Amtmann *et al.*, 1980; Lancaster, 1981), and hamsters (Breitburd *et al.*, 1981; Moar *et al.*, 1981a; Pfister *et al.*, 1981a).

B. Immune Response of the Host

The host reacts to papillomavirus infection with humoral and cellular immune mechanisms against virus particles and tumor tissue (for review see Jablonska *et al.*, 1982; Pfister, 1984). Both IgM and IgG antibodies against viral capsid proteins were detected in sera from infected humans, rabbits, and calves. The antibody response is slow and titers usually remain low. Sera from rabbits carrying papillomas were shown to neutralize the virus *in vitro*. No evidence was obtained, however, that antibodies had any effect on tumor development, once the virus had infected the epidermis (Kidd *et al.*, 1936). A protective role of serum antibodies against naturally occurring reinfection must thus be questioned, because the virus is very likely to meet its target cell without antibody contact.

Cell-mediated immune reactions seem to play the key role both in wart regression and in immunity. In contrast to humoral immune defects, cell-mediated immune deficiencies usually imply an increased susceptibility to reinfection, and the tumors hardly regress. Histological examination of regressing tumors showed inflammatory infiltration of macrophages and lymphocytes, thus demonstrating the activity of cellular immune mechanisms. This type of immunity is likely to be directed against tumor-specific antigens although this has not been directly demonstrated. Vaccination of rabbits with tumor tissue increased the frequency of regression from 25% to 50–90% (Evans *et al.*, 1962) and formalized suspensions of wart tissues were shown to immunize calves against challenge with infectious wart suspensions (Olson *et al.*, 1962). To obtain a vaccine that is applicable to man, it will be essential to define the active principle of such crude preparations, i.e., the virus-specific early protein(s) or a virally induced cellular protein.

C. Malignant Conversion—General Aspects

A number of papillomaviruses induce tumors that may eventually progress to carcinomas (Table I; Chapters 6–11). Most of these systems will be discussed in detail in the second half of this volume and only some general aspects will be covered here.

One general characteristic of malignant conversion is that carcinomas are not an early consequence of papillomavirus infection but occur on the basis of long-persisting papillomas with "latency periods" between 1 and 40 years. This indicates that additional events must come along to induce malignant growth. Synergistic effects of external cofactors were indeed recognized both in experimental and in naturally occurring systems. Chemical carcinogens, tumor promoters, UV light, and X-irradiation considerably increase the risk of malignant conversion and reduce the latency period. The development of malignant tumors also turned out to be significantly influenced by the genetic disposition of the host.

An important etiological role of papillomaviruses in the progression from warts to carcinomas is suggested by the following findings:

1. Epidemiology clearly indicates a correlation between specific papillomavirus infections and carcinoma development (zur Hausen, 1977; Jarrett *et al.*, 1978; Orth *et al.*, 1979; Spradbrow and Hoffmann, 1980; Meisels *et al.*, 1982).
2. Carcinoma cells are not permissive for papillomavirus replication and no mature particles are produced but the viral DNA persists usually in high copy number. It is of special interest that distinct virus types prevail in the malignant tumors whereas the host may be infected by a large panel of papillomaviruses leading to benign tumors (Orth *et al.*, 1980; Ostrow *et al.*, 1982; Pfister *et al.*, 1983a; Dürst *et al.*, 1983; Boshart *et al.*, 1984). This could be indicative of an increased carcinogenic potential of those very types.

TABLE I. Malignant Tumors Associated with Papillomavirus Infection

Virus	Host	Tumor	"Latency period"	Oncogenic cofactors	Demonstration of viral		References
					DNA	RNA	
HPV-5, 8 (14, 17, 20)	Man	Skin carcinoma in epidermodysplasia verruciformis	On average 25a	UV light	+	+	See Chapter 8
HPV-16, 18 (10, 11)	Man	Cervical carcinoma	10–30a		+	+	See Chapter 9
?	Man	Laryngeal carcinoma	5–40a	X rays, tobacco smoke	+		See Chapter 10
CRPV	Rabbit	Skin carcinoma	1a	Chemical carcinogens; however, not necessary	+	+	See Chapter 6
BPV-4	Cattle	Esophageal carcinoma		Bracken fern diet	–		
?	Cattle (Hereford)	Ocular carcinoma	3 months–2a	UV light	+		Spradbrow and Hoffmann (1980)
	Cattle (Friesian)	Skin carcinoma	Several months	UV light			Spradbrow (personal communication)
?	Sheep	Skin carcinoma	Several months	UV light	?		Vanselow and Spradbrow (1982)
MnPV	Multimammate mouse	Keratoacanthomas, skin carcinomas	Several months	Tumor promoter TPA	+	+	See Chapter 7

3. Virus-specific transcripts were demonstrated in rabbit skin carcinomas (Nasseri *et al.*, 1982), human skin carcinomas (Yutsudo *et al.*, personal communication), and human cervical carcinomas (Schwarz and Freese, personal communication). Fine-structure analysis of RNAs from rabbit tumors showed that only reading frames E6 and E7 could be translated from these messages (Danos *et al.*, Wettstein and Nasseri, personal communications). It is remarkable that viral transcripts are still present in a cottontail rabbit papillomavirus-induced carcinoma, which was kept in animals by repeated transplantation for over 30 years (McVay *et al.*, 1982). The observation that the viral DNA within this transplantable carcinoma is highly methylated except for the early promoter region (Wettstein and Stevens, 1983) makes continued transcription even more significant. It points to a selective pressure to keep the promoter undermethylated—i.e., active—and strongly suggests that the expression of specific viral genes is essential for the maintenance of the neoplastic state.

Little is known at the moment about molecular mechanisms that might be involved in malignant conversion of papillomas. Comparative analysis of virus types which were found to be associated with carcinomas and those which were only detected in benign lesions has not as yet disclosed relevant differences.

Integration of viral DNA into host chromosomes was observed in a number of carcinomas (Wettstein and Stevens, 1980, 1982; Boshart *et al.*, 1984) but only extrachromosomal viral DNA was detected in others (Orth *et al.*, 1980; Green *et al.*, 1982; Ostrow *et al.*, 1982; Pfister *et al.*, 1983a). Integration of viral DNA theoretically allows special mechanisms of carcinogenesis. Insertion of viral transcription control signals could activate cellular (onco)genes either via their own promoter or in the form of fusion transcripts from viral and cellular DNA. On the other hand, integration may disrupt transcription units of the virus as observed in human cervical carcinoma cells (Schwarz *et al.*, personal communication). This could imply qualitative and quantitative changes in viral gene expression and an altered oncogenic potential of the virus. With the cottontail rabbit papillomavirus, however, it was not possible to demonstrate an increased cancerogenic potential of the viral DNA from carcinomas. Transfection of the skin of domestic rabbits with DNA extracts from carcinomas led to papillomas and not directly to carcinomas (Ito, 1963).

Quantitative differences were noted for virus-specific transcripts in papillomas and carcinomas of rabbits (Nasseri *et al.*, 1982). This might be due to differences in methylation of the viral genome in benign and malignant tumors (Wettstein and Stevens, 1983). Further characterization of the RNAs and their protein products will be necessary to evaluate the significance of the quantitative shift.

Neoplasms of man and animals are characterized by abnormal karyotypes, which are supposed to play an important role in carcinogenesis. It will be worthwhile to determine if some papillomaviruses preferentially induce chromosomal aberrations. Bovine papillomavirus-1-induced fibrosarcomas of Syrian hamsters all showed abnormalities, most frequently with chromosomes 1, 4, and 15 (Gamperl *et al.*, 1984). It is of special interest that cervical dysplasias, which are distinguished by abnormal mitotic figures, are preferentially induced by human papillomavirus 16, which persists in cervical carcinomas (Crum *et al.*, 1984). This would provide an attractive model for carcinogenesis but it could not explain a necessity for continued DNA persistence and transcription.

Genetic experiments are underway to assess the biological activity of cottontail rabbit papillomavirus DNA, which was mutagenized *in vitro* and transfected into domestic rabbits. It will be very interesting to see if viral functions can be affected, which are essential specifically for the progression from papillomas to carcinomas. The rabbit system looks most promising to address this question experimentally.

IV. CLASSIFICATION AND NOMENCLATURE

Classification of papillomaviruses is presently based on the host range and the relatedness of the nucleic acids (Coggin and zur Hausen, 1979). Biological properties are not regarded as a reliable criterion at the moment, because it is sometimes hard to assess the biological role of individual papillomaviruses. A number of specificities will be established, however: preferences for target cells (e.g., keratinocytes, fibroblasts), tissues (e.g., skin, mucosa), and malignant tumors. They may provide useful criteria for future classification (see Section VI).

It is generally accepted that each virus is named first after its natural host. This is usually unequivocal because most papillomaviruses have a very restricted host range, which excludes even experimental infections of other species (Rowson and Mahy, 1967; Koller and Olson, 1972). Keratinocyte-specific viruses from different species show no DNA cross-hybridization under stringent conditions, either. Several viruses, which are able to transform fibroblasts, reveal a less restricted host range, but there is no problem in identifying the natural host, which is the only one where keratinocytes are affected. It was suggested that the host should be called by its English name (Coggin and zur Hausen, 1979), but in common practice the terms refer to host family, subfamily, or species in the English or Latin form. The virus names are abbreviated by one or two letters for host designation, followed by "PV" for papillomavirus. Examples are HPV and BPV for human and bovine papillomaviruses, respectively, CRPV for cottontail rabbit papillomavirus, and MnPV for *Mastomys natalensis* papillomavirus, the virus of the multimammate mouse.

For historical reasons the virus from canine oral papillomas is referred to as COPV (Cheville and Olson, 1964). To standardize nomenclature this virus should be called CaPV.

Viruses from a given species are presently classified as types, subtypes, and variant strains according to their DNA homologies (Coggin and zur Hausen, 1979). To compare viral DNAs, hybridization is carried out in liquid phase and single strands are finally removed by hydroxylapatite chromatography or S1 nuclease digestion. In the case of less than 50% cross-hybridization, two isolates are considered as independent types; they are defined as subtypes if cross-hybridization exceeds the 50% value but is obviously incomplete. According to common practice types are labeled 1, 2, 3 . . . and subtypes 1a, 1b, 1c . . . .

The 50% borderline is certainly rather arbitrary and should be open for discussion as more information becomes available on biological properties and evolutionary aspects. When compared with other classification systems in virology, for example serotypes of adenoviruses, less than 50% cross-hybridization represents a very stringent criterion. The 50% value as derived from liquid-phase hybridization must not be confused, however, with 50% homology on the nucleotide sequence level. Those viruses that have been sequenced show a considerable number of patchy homologies, but no cross-hybridization at all under conditions of high stringency. Within conserved regions such as reading frames E1 or L1, nucleotide homology amounts to about 50%! This indicates that different viruses are actually much more related than anticipated from the liquid-phase hybridization test.

If two isolates differ only in a few restriction enzyme cleavage sites but show homology close to 100%, they should be regarded as variant strains rather than subtypes. There is no borderline defined, however, nor are exact data on cross-hybridization available in many of these cases. It was suggested to label strain variants a_1, a_2, a_3 . . . .

A number of papillomavirus DNAs were directly cloned from biopsy material and not from purified virus particles. They are accepted as papillomavirus-specific on the basis of the following criteria: (1) size in the range of 7 to 8 kbp; (2) persistence as extrachromosomal, circular DNA; (3) sequence homology with other papillomaviruses and/or cross-hybridization under relaxed conditions; (4) genome organization comparable to other papillomaviruses. For HPV6, the viral origin of the cloned DNA was confirmed more directly. Extracts from biopsy tissue were centrifuged in CsCl density gradients, DNA was prepared from individual fractions, and annealing with the cloned, radioactively labeled DNA was observed at the density of papillomavirus particles (Gissmann and zur Hausen, 1980).

Classification of papillomaviruses on the DNA level was sometimes supported by serological data. In most cases, however, it was impossible because of limiting amounts of antigen.

V. DESCRIPTION OF VIRUS TYPES

A. Human Papillomaviruses

1. Virus Types

On the basis of less than 50% cross-hybridization, 29 HPV types have been differentiated (for references see Table II). They can be divided into several groups, which show no significant sequence homology when tested under stringent hybridization conditions. The members of individual groups cross-hybridize from less than 1% to 40%. Six groups have as yet only one representative.

HPV-1 was regarded as the only prototype of HPV for a long time due to its prevalence in plantar warts and the abundant particle synthesis.

TABLE II. Human Papillomaviruses

Group	Type	Reference	Reference for DNA cloning
A	HPV-1	Favre *et al.* (1975b)	Danos *et al.* (1980), Heilman *et al.* (1980)
B	HPV-2	Orth *et al.* (1977)	Heilman *et al.* (1980)
	HPV-3	Orth *et al.* (1978b)	Kremsdorf *et al.* (1983)
	HPV-10	Green *et al.* (1982)	Kremsdorf *et al.* (1983)
	HPV-27	Ostrow *et al.* (personal communication)	
	HPV-28	Favre *et al.* (personal communication)	
	HPV-29	Favre *et al.* (personal communication)	
C	HPV-4	Gissmann *et al.* (1977)	Heilman *et al.* (1980)
D D1	HPV-5	Orth *et al.* (1980)	Kremsdorf *et al.* (1982)
	HPV-8	Orth *et al.* (1980)	Pfister *et al.* (1981b)
	HPV-12	Kremsdorf *et al.* (1983)	Kremsdorf *et al.* (1983)
	HPV-14	Tsumori *et al.* (1983)	Tsumori *et al.* (1983)
	HPV-19		
	HPV-20	Kremsdorf *et al.* (1984)	Kremsdorf *et al.* (1984)
	HPV-21	Gassenmaier *et al.* (1984)	Gassenmaier *et al.* (1984)
	HPV-22		
	HPV-23		
	HPV-25		
D2	HPV-9	Kremsdorf *et al.* (1982)	Kremsdorf *et al.* (1982)
	HPV-15		
	HPV-17	Kremsdorf *et al.* (1984)	Kremsdorf *et al.* (1984)
D3	HPV-24		
E	HPV-6	Gissmann and zur Hausen (1980)	de Villiers *et al.* (1981)
	HPV-11	Gissmann *et al.* (1982)	Gissmann *et al.* (1982)
	HPV-13	Pfister *et al.* (1983b)	Pfister *et al.* (1983b)
	HPV-26	Ostrow *et al.* (1984)	Ostrow *et al.* (1984)
F	HPV-7	Orth *et al.* (1981) Ostrow *et al.* (1981)	
G	HPV-16	Dürst *et al.* (1983)	Dürst *et al.* (1983)
H	HPV-18	Boshart *et al.* (1984)	Boshart *et al.* (1984)

A few strain variants have been described (Gissmann *et al.*, 1977; Pfister, 1980) but the fairly large number of isolates are remarkably homogeneous and one variant, which has been cloned and sequenced (Danos *et al.*, 1980, 1982), forms an overwhelming majority.

HPV-4 and 7 are less frequent. Different cleavage patterns have been described for HPV-4 (Heilman *et al.*, 1980; Pfister, 1980).

HPV-16 and 18 DNAs were cloned from cervical carcinomas. It was not possible to demonstrate the group-specific viral antigen in these tumors but this is basically not surprising as malignant tumors are generally not permissive for papillomavirus replication. The cloned nucleic acids were classified as papillomavirus DNA on the basis of their typical size, their cross-hybridization with other papillomavirus DNAs under relaxed conditions, and their colinearity with these DNAs as revealed by hybridization with subgenomic DNA fragments.

Group B comprises six virus types. HPV-3a, 10a, and 28 are closely related (with 20%–36% cross-reactivity in liquid-phase hybridization experiments) and HPV-3 and 10 share a number of restriction enzyme cleavage sites (Kremsdorf *et al.*, 1983; Favre, personal communication). HPV-2a showed only weak sequence homology with these viruses when tested by Southern blot hybridization. HPV-2 and 10 are rather heterogeneous themselves, however. HPV-2c, for example, was found to be the most prevalent HPV-2 subtype, which showed only 55% cross-hybridization with HPV-2a (Fuchs and Pfister, 1984). Substantial differences were observed between the physical maps of HPV-2a, b, and c (Orth *et al.*, 1977, personal communication; Heilman *et al.*, 1980; Fuchs and Pfister, 1984). Two HPV-10 subtypes were recently cloned from clinically diagnosed common warts and tentatively designated HPV-P and HPV-PW (Ostrow *et al.*, 1983). They revealed about 60% homology with each other and 73 and 53% with HPV-10a, respectively. These HPV-10 subtypes showed only 24 and 13% homology with HPV-3. HPV-27 is rather close to HPV-2 (20% homology; Ostrow, personal communication) and HPV-29 represents an interesting link between HPV-2- and HPV-3-related viruses. It showed 7% homology with HPV-2 and 15% homology with HPV-3 (Favre, personal communication). It must be noted from such findings that transitions between types and subtypes become more fluid when more isolates enter the scheme.

Group D occupies a special position. The HPVs of group D were divided into three subgroups on the basis of only weak cross-hybridization among viral genomes belonging to different subgroups (Kremsdorf *et al.*, 1984). Altogether, cross-reactivity was observed in the range of 0–40%. However, heteroduplex analysis under less stringent conditions showed that group D viruses are much more related among themselves than to other HPV isolates, thus justifying their classification within one group (Croissant and Orth, personal communication). A heterogeneity in DNA sizes was observed, which crosses the subgroup classification (Kremsdorf *et al.*, 1984). All D2 viruses (9, 15, 17) have shorter DNAs with about 7.2

kbp, but so do HPV-22 and 23. The remaining D1 viruses and HPV-24 have DNAs with 7.7 kbp as usually observed with HPVs.

Two subtypes each were described for HPV-14 and 17, which shared about 90% of their sequences and a number of restriction enzyme cleavage sites (Kremsdorf *et al.*, 1984). Independent HPV DNA clones from different countries revealed nearly identical cleavage maps; e.g., three isolates of HPV-5 (Kremsdorf *et al.*, 1982; Ostrow *et al.*, 1982; Pfister *et al.*, 1983a) and two isolates each of HPV-19 and 20 (Gassenmaier *et al.*, 1984; Kremsdorf *et al.*, 1984). This indicates that group D viruses exist as defined entities although they are confined to few isolated cases which are distributed worldwide. Looking at four variant strains of HPV-8 (Pfister *et al.*, 1981b; Kremsdorf *et al.*, 1983; Gassenmaier *et al.*, 1984; Pfister, unpublished findings), a cluster of differing cleavage sites attracts attention (Fig. 4). It spans about 1.5 kbp starting within the noncoding region of HPV-8 (Fuchs *et al.*, 1986) and extending into the early region. A more detailed analysis on the sequence level will determine whether this is indicative of an increased selective pressure on a special genome region.

Group E comprises at least three viruses, which affect the mucosa. HPV-6 and 11 are closely related, showing 25% cross-hybridization and very similar biological properties. A number of subtypes were described (Gissmann *et al.*, 1983; Mounts and Kashima, 1984), many of which will probably turn out to be cleavage variants of one prototype. HPV-13 is more distantly related. It revealed only 3–4% cross-reactivity with HPV-6 and 11 (Pfister *et al.*, 1983b).

HPV-26 was isolated from skin papillomas of a patient who exhibited an immune deficiency syndrome. Southern blot analysis under stringent conditions revealed a weak sequence homology with HPV-6, and an even weaker homology with HPV-3 (Ostrow *et al.*, 1984). The distribution of

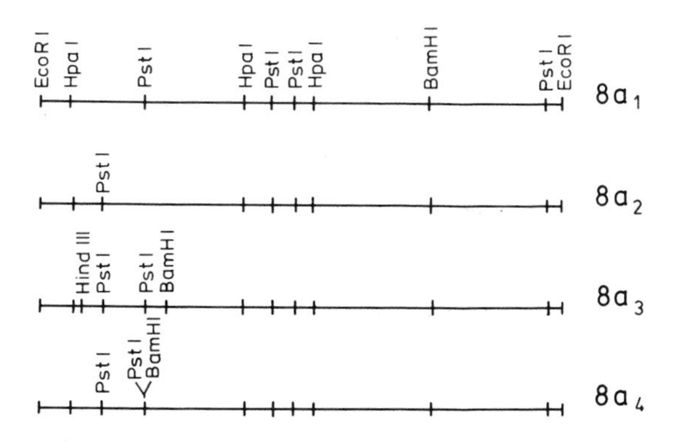

FIGURE 4. Physical maps of restriction enzyme cleavage sites on the genomes of four variant strains of HPV-8. Note a clustering of differences in the left moiety of the genomes. a_1: Pfister *et al.* (1981b); a_2: Kremsdorf *et al.* (1983); a_3: Gassenmaier *et al.* (1984); a_4: Pfister *et al.* (1983a).

homologous sequences within the genomes remains to be established. Grouping HPV-26 together with HPV-6 is only tentative at the moment.

Specific antisera were obtained for a few HPV types by immunization of rabbits or guinea pigs with purified virus particles. No cross-reaction was observed between HPV-1, 2, 3, and 4. An antiserum against pooled group D viruses did not react with these viruses either (Gissmann *et al.*, 1977; Orth *et al.*, 1977, 1978b; Pfister and zur Hausen, 1978b).

2. Biological Properties

Individual HPV types are associated with distinct clinical symptoms and it is interesting to note that viruses that are related on the molecular level show similar biological properties (Table III). The following picture will probably turn out to be rather simplified as more HPV types and subtypes are being identified and as more tumors are being examined. However, new isolates are generally not likely to annul the prevalence of known types in the different tumors and this permits some conclusions on tropism and lesion specificity of HPV prototypes.

From skin warts, seven HPV types have been isolated to date. The lesions are characterized by acanthosis, prominent keratohyalin granules in the stratum granulosum, vacuolization, and hyperkeratosis. On the basis of localization and morphology, they are differentiated into plantar, common, and flat warts. HPV-1 predominates in deep plantar warts and

TABLE III. HPV Types and Clinical Symptoms

	HPV type													
Tumor type	1	2 27 29	3 10 28	4	5, 12, 19, 21, 23,	8 14 20 22 25	9 15 17	24	6 11	7	13	16	18	26
Verruca vulgaris	+	+ + +		+ +						+				+
Verruca plantaris	+ + +	+		+										
Verruca plana			+ + +											
Epidermodysplasia verruciformis			+		+ + +		+ +	+						
Condyloma acuminatum	+	+							+ +					
Condyloma planum									+ +			+	?	
Bowenoid papulosis												+		
Laryngeal papilloma									+ +					
Focal, epithelial hyperplasia Heck	+										+ +			
Conjunctival papilloma									+					

HPV-2 both in exophytic, papillomatous common warts from the hands and in mosaic warts from the planta pedis (Orth *et al.*, 1977; Pfister and zur Hausen, 1978b; Laurent *et al.*, 1982). HPV-3 or 10 prevail in flat, juvenile warts (Orth *et al.*, 1978b; Kremsdorf *et al.*, 1983) and HPV-4 was detected in small plantar and common warts, which usually grow endophytically and hardly rise over the skin surface (Gissmann *et al.*, 1977; Pfister *et al.*, 1980). There are a lot of exceptions to the rule, however. A survey on 35 common warts revealed HPV-2 24 times, HPV-4 7 times, and HPV-1 4 times (Pfister, unpublished findings). HPV-1 and 2 have also been found in anogenital lesions (Krzyzek *et al.*, 1980) where HPV-6 and 11 usually predominate (Gissmann *et al.*, 1983). In some cases it proved to be difficult to differentiate morphologically between HPV-2-induced common warts and HPV-3- or 10-induced flat warts (Ostrow *et al.*, 1983; Fuchs and Pfister, 1984; Jablonska, personal communication).

A more specific correlation emerges from wart histology (Jablonska *et al.*, 1980; Orth *et al.*, 1981; Gross *et al.*, 1982). HPV-1-induced warts are characterized by extensive papillomatosis and hyperkeratosis and by eosinophilic, cytoplasmic keratohyalin inclusions, which may become very large in the stratum granulosum (Fig. 2A). Viral progeny particles are often produced in large amounts. A yield of 10^{12} particles per wart was estimated (Barrera-Oro *et al.*, 1962). The capsids form paracrystalline aggregates, which appear as basophilic nuclear inclusions in conventionally stained histological sections. HPV-2-induced warts reveal numerous keratohyalin granules of varying size and stainability as well as papillomatosis, acanthosis, and hyperkeratosis (Fig. 2B). HPV-3-induced flat warts show only moderate or almost no papillomatosis but extensive perinuclear vacuolization of the cells of the upper Malpighian layer (Fig. 2C). HPV-4-induced warts are characterized by papillomatosis, hyperkeratosis, and large clear cells, which show crescent-shaped, peripherally located nuclei and appear clustered in the squamous and granular layers.

HPV-7 was isolated only from common warts of people who handle meat (Orth *et al.*, 1981; Ostrow *et al.*, 1981). The lesions are characterized by papillomatosis, acanthosis, hyper- and parakeratosis. Large, sometimes binucleated, clear cells are surrounded by heavily staining cells. From the profession of the patients, it seems possible that HPV-7 represents an as yet unknown animal papillomavirus.

HPV-26 DNA was cloned from biopsies of a 53-year-old patient with a 2-year history of multiple common warts and condylomas. Multiple squamous cell carcinomas of the face and forehead were first diagnosed 12 years before (Ostrow *et al.*, 1984). Examination of immunological parameters revealed low levels of total and interactive T cells. It remains to be established to what extent HPV-26 infection depends on immune deficiency of the host.

Congenital defects of cell-mediated immunity clearly are one basis of the nosological unit epidermodysplasia verruciformis (Jablonska *et al.*,

1982). The patients develop skin warts during childhood, which persist throughout life and gradually spread to cover most of the body. Part of the tumors represent flat warts, which harbor HPV-3 or 10, and HPV-2-induced common warts occur in some patients (Orth *et al.*, 1978b, 1980). The basic role of the impaired immunity is confirmed by the increased incidence of HPV-2 and 3 infections in immunosuppressed transplant recipients (Pfister *et al.*, 1979a, unpublished). Such findings point to a correlation between the grade of cell-mediated immunity depression and the susceptibility to distinct HPV types.

Severely impaired T-cell functions of epidermodysplasia verruciformis patients are probably at least one basis for the infections with HPV-5, 8, and all other group D viruses (Glinski *et al.*, 1976, 1981; Obalek *et al.*, 1980). These types were never isolated from warts of the normal population except for HPV-5, which was detected in lesions of two renal transplant recipients (Lutzner *et al.*, 1980). More extensive screenings of transplant patients failed to detect additional cases of group D infections (Orth, personal communication; Pfister and Schell, unpublished findings), which could be explained by the existence of rather specific immune defects or by additional prerequisites in epidermodysplasia verruciformis patients.

Group D viruses induce typical lesions, which set the tone of the clinical picture of epidermodysplasia verruciformis (Lewandowsky and Lutz, 1922; Jablonska *et al.*,1979; Orth *et al.*, 1979). The extremely flat warts and reddish or pityriasis versicolorlike plaques show a very characteristic histology. Giant keratinocytes with a basophilic, pale-staining cytoplasm are found in the stratum spinosum and granulosum of the epidermis. The nuclear chromatin is marginated and both nucleus and cytoplasm appear vacuolated. Patients with epidermodysplasia verruciformis are often infected with several HPV types belonging to group D. It is therefore difficult to establish type-specific effects of these viruses but they seem to induce very similar dysplasias. Of special interest is the prevalence of HPV-5 and 8 in squamous cell skin carcinomas of the patients (Orth *et al.*, 1980; Ostrow *et al.*, 1982; Pfister *et al.*, 1983a; see also Chapter 8).

HPV types 6, 11, 13, 16, and 18 affect the mucosa. As mentioned above, HPV-1 and 2 happen to infect the mucosa from time to time as does HPV-10 (Green *et al.*, 1982) but these types are rare guests there.

HPV-6 and 11 were isolated from a broad lesion spectrum, namely from condylomas, cervical dysplasias, laryngeal papillomas, and conjunctival papillomas (Gissmann *et al.*, 1983; Lass *et al.*, 1983; Mounts and Kashima, 1984). Dysplasias of the cervix uteri were only recently realized to be associated with HPV infection (Laverty *et al.*, 1978; Della Torre *et al.*, 1978) and are now referred to as flat condylomas or condylomata plana (Meisels *et al.*, 1982). Laryngeal papillomas have a bimodal age distribution with a first peak between 2 and 5 years and a second one between 40 and 60 years. From clinical observations, people suspected that infection of children may take place during delivery from a mother with condylomas. This hypothesis is now supported by the observation that both tumors are associated with HPV types 6 and 11. Laryngeal papillomas in adults were also shown to harbor HPV-6 DNA in three cases (Mounts and Kashima, 1984). One could imagine that these tumors are partially due to reactivation of a virus, which persists after apparent or inapparent infection during childhood, since it was possible to demonstrate persisting HPV DNA in clinically normal mucosa (Steinberg *et al.*, 1983; see Chapter 10).

According to initial data, HPV-11 seemed to predominate in cervical dysplasias and laryngeal papillomas and HPV-6 in exophytic condylomas (Gissmann *et al.*, 1983), but this was rather equalized by further isolates (Mounts and Kashima, 1984; Orth, personal communication). HPV-6- and 11-induced condylomas cannot be histologically differentiated either (Gross, personal communication). They reveal a panel of distinguishable clear cells without any correlation with the virus type. In a study of 21 patients, extensive anatomical spread in the respiratory tract and severe clinical course were found to be associated with isolates showing the typical HPV-11 cleavage pattern (Mounts and Kashima, 1984). Certainly many more cases need to be analyzed to substantiate this observation.

One recent report noted HPV DNA in a conjunctival papilloma, which hybridized with HPV-11 under stringent conditions (Lass *et al.*, 1983). A more detailed characterization of the virus remains to be done.

HPV-16 and 18 DNAs are prevalent in cervical carcinomas from where they were cloned directly (see Chapter 9). Whereas HPV-18 has never been detected in benign tumors, HPV-16 is occasionally observed in condylomata acuminata and more frequently in cervical dysplasias (Gissmann *et al.*, 1984).

The macroscopically rather inconspicuous bowenoid papulosis histologically resembles a carcinoma *in situ*. It can be distinguished by the different age distribution of the patients and the multicentric origin of the lesions. The tumors usually regress spontaneously. HPV-16 DNA was detected in several cases and HPV-6 DNA once (Zachow *et al.*, 1982; Ikenberg *et al.*, 1983).

HPV-13 probably represents the predominant etiological agent of focal epithelial hyperplasia Heck (Pfister *et al.*, 1983b). This disease was first described in American Indians (Archard *et al.*, 1965). The condition has also been reported in other races, but is extremely rare in Caucasians (Orfanos *et al.*, 1974; van Wyk, 1977). The reason for this specificity is unknown. Heck's disease occurs mainly in children and young adults and appears clustered in certain families, thus underlining a genetic component. Soft nodules are observed on the labial and buccal mucosae. Histological examination shows acanthosis with elongation and anastomosis of the rete ridges. Some cells are binuclear and the nuclei reveal abnormally dense granules (Praetorius-Clausen, 1969). The tumors may persist for several years but do not become malignant and finally tend to undergo spontaneous remission.

B. Bovine Papillomaviruses

Six types of BPV have been identified, which clearly form two groups. The first one is composed of BPV-1, 2, and 5 and the second one of BPV-3, 4, and 6 (Table IV).

BPV-1 and 2 show cross-hybridization of about 45% when tested by DNA–DNA reassociation kinetics (Lancaster and Olson, 1978). A number of restriction enzyme cleavage sites map at identical positions when the genomes are aligned at their single *Hin*dIII sites (Lancaster, 1979). Heteroduplex analysis revealed almost complete homology within the transforming region but the segments coding for structural proteins showed only partial homology (Campo and Coggins, 1982). This explains the limited serological cross-reactivity between BPV-1 and 2, which is in the range of 10%, and the different efficiency of mouse erythrocyte agglutination (Lancaster and Olson, 1978). BPV-5 is more distantly related, showing only 5% DNA homology with BPV-2 and poor cross-reactivity in immunohistochemical tests (Campo *et al.*, 1981). Common sequences of BPV-2 and 5 are equally distributed over the whole genome. The genomes of the three viruses are of the same size, i.e., 7946 bp in the case of BPV-1 as derived from the DNA sequence (Chen *et al.*, 1982).

BPV-1, 2, and 5 cause fibropapillomas. BPV-1 and 2 induce cutaneous fibropapillomas and paragenital lesions (Lancaster and Olson, 1978; Pfister *et al.*, 1979c; Jarrett *et al.*, 1980). BPV-2-specific DNA was also detected in fibropapillomas of the alimentary tract (Jarrett *et al.*, 1984a) but no mature virus could be isolated from these lesions. This may imply that even epithelial cells of the alimentary tract mucosa are nonpermissive for a complete productive cycle of BPV-2. Experimental infection of calves with BPV-1 or 2 results in urinary bladder tumors or in meningiomas depending on the route of inoculation. BPV-1- or 2-specific DNA in high copy number was repeatedly demonstrated in naturally occurring equine sarcoids (Lancaster *et al.*, 1979; Amtmann *et al.*, 1980). Experimental infection of horses led to histologically similar tumors, thus incriminating BPV-1 and 2 as etiological agents of this tumor in the alien, nonpermissive host (Olson and Cook, 1951). The two viruses have a remarkably broad experimental host range. Fibromas or fibrosarcomas develop upon infection of hamsters (Friedman *et al.*, 1963), mice (Boiron *et al.*, 1964), and pikas (Puget *et al.*, 1975), and fibroblasts of mice (Thomas *et al.*, 1964) and hamsters (Black *et al.*, 1963) are transformed *in vitro*. The viruses persist on the DNA level in these nonproductive cells (Pfister, 1984). By means of such experimental systems, which are rather easy to handle,

TABLE IV. Animal Papillomavirus Types

Group	Type	Natural host	Reference	Reference for DNA cloning
A	BPV-1	Cattle	Lancaster and Olson (1978)	Howley *et al.* (1980)
	BPV-2	Cattle	Lancaster and Olson (1978)	Howley *et al.* (1980)
	BPV-5	Cattle	Campo *et al.* (1981)	
	DFV	Deer	Lancaster and Sundberg (1982)	
	EEPV	European elk	Moreno-Lopez *et al.* (1981)	Stenlund *et al.* (1983)
	RPV	Reindeer	Moreno-Lopez (personal communication)	
B	BPV-3	Cattle	Pfister *et al.* (1979c)	Coggins *et al.* (1983)
	BPV-4	Cattle	Campo *et al.* (1980)	Campo and Coggins (1982)
	BPV-6	Cattle	Jarrett *et al.* (1984b)	Jarrett *et al.* (1984b)
C	CRPV	Cottontail rabbit	Shope (1933)	Wettstein and Stevens (1980)
D	MnPV	Multimammate mouse	Müller and Gissmann (1978)	
E	CaPV	Dog	Pfister and Meszaros (1980)	
F	FPV	Chaffinch	Lina *et al.* (1973), Osterhaus *et al.* (1977)	Moreno-Lopez *et al.* (1984)

BPV-1 became one of the most extensively studied papillomaviruses (see Chapters 3 and 5).

BPV-5 was recovered from "rice-grain"-like papillomas of the teats (Campo *et al.*, 1981). Also with this lesion tumor development begins with transformation of fibroblasts in the dermis. The overlying keratinocytes finally form a thick plaque, the rice grain, followed by development of papillomatosis (Jarrett *et al.*, 1984b).

A recent virus isolate from bovine fibropapillomas showed restriction enzyme cleavage patterns completely different from those of BPV-1, 2, or 5 (Spradbrow *et al.*, 1983). The DNA had a molecular weight of 4.9×10^6. Classification must await cross-hybridization experiments.

BPV-3, 4, and 6 fundamentally differ from the first group of BPVs. The DNAs of these viruses are significantly smaller and do not cross-hybridize with DNAs of BPV-1, 2, or 5 under stringent conditions. The genome size was determined to be 6.7 kbp by gel electrophoresis and 7.2 kbp by electron microscopy (Pfister *et al.*, 1979c; Campo *et al.*, 1980; Coggins *et al.*, 1983; Jarrett *et al.*, 1984b). BPV-3 and 4 share about 12% of their DNA sequences as determined by reassociation kinetics under stringent conditions (Coggins *et al.*, 1983). Heteroduplex analysis and Southern blot hybridization showed that the homologous sequences are distributed over the entire genome length. BPV-6 is more distantly related. Its DNA hybridized only under conditions of low stringency to DNAs of BPV-3 and 4 (Jarrett *et al.*, 1984b).

The major structural protein of BPV-3 is somewhat larger than those of BPV-1 and 2 (59,000 versus 53,000) (Pfister *et al.*, 1979c). No cross-reactivity was observed with monospecific antisera against BPV-1 and 3, respectively. An antiserum against disrupted BPV-2, which is known to detect group-specific papillomavirus antigens and which reacted with BPV-5 at a dilution of 1 : 100, failed to detect BPV-4 and 6 capsid proteins (Jarrett *et al.*, 1984b).

BPV-3, 4, and 6 are associated with purely epithelial tumors. BPV-3 was isolated from skin papillomas of Australian cattle (Pfister *et al.*, 1979c) and BPV-4 induces epithelial proliferations of the alimentary tract (Campo *et al.*, 1980), some of which undergo transformation to carcinomas (Jarrett *et al.*, 1978). BPV-6 leads to frond epithelial papillomas on the skin of the udder (Jarrett *et al.*, 1984b).

Three different papillomavirus-specific DNAs were recently cloned as *Bam*HI fragments from flat (pFL1 and pFL2) and filiform (pFI1) papillomas of the teats (Olson and Lancaster, personal communication). They were 7 kbp (pFL1 + 2) and 6.2 kbp (pFI1) in size. Hybridization analysis under stringent conditions revealed limited sequence homology between pFL1 and 2 and BPV-1 whereas pFI1 turned out to be unrelated to BPV-1. A possible relationship with BPV-3, 4, and especially 6 remains to be established.

Papillomavirus particles were demonstrated in negatively stained preparations from various ocular lesions such as conjunctival plaque, conjunctival papilloma, and papilloma of the eyelid (Ford *et al.*, 1982). These viruses have not yet been classified. They deserve interest, however, as the lesions are regarded as precursors to bovine ocular squamous cell carcinoma (Spradbrow and Hoffmann, 1980).

C. Other Papillomaviruses from Animals

Papillomaviruses are widespread among mammals. They have been described in sheep (Gibbs *et al.*, 1975), goats (Davis and Kemper, 1936), deer (Shope *et al.*, 1958), reindeer (Moreno-Lopez *et al.*, personal communication), European elks (Moreno-Lopez *et al.*, 1981), horses (Fulton *et al.*, 1970), rabbits (Shope, 1933; Parsons and Kidd, 1942), beavers (Carlson *et al.*, 1983), multimammate mice (Müller and Gissmann, 1978), dogs (Chambers and Evans, 1959; Watrach, 1969), coyotes (Greig and Charlton, 1973), monkeys (Rangan *et al.*, 1980), pigs (Lancaster and Olson, 1982), opossums (Koller, 1972), and elephants (Sundberg *et al.*, 1981). There are two avian viruses infecting the chaffinch (Osterhaus *et al.*, 1977) and the red parrot (Jacobson *et al.*, 1983). Many of these viruses have not been well characterized to date. They affect the skin or the mucosa and lead to papillomas, fibropapillomas, or fibromas (Table V). The viruses that were tested revealed the group-specific antigen and/or DNA cross-hybridization with other papillomaviruses under relaxed conditions (Law *et al.*, 1979; Jenson *et al.*, 1980; Pfister and Meszaros, 1980; Sundberg *et al.*, 1981; Carlson *et al.*, 1983; Jacobson *et al.*, 1983; Stenlund *et al.*, 1983).

The cottontail rabbit papillomavirus (CRPV) was studied in great detail. The viral DNA was cloned, mapped, and sequenced (Favre *et al.*, 1982; Giri *et al.*, 1985) and the protein composition of viral particles was investigated (Favre *et al.*, 1975a). CRPV induces skin papillomas in its natural host. Usually they regress spontaneously but sometimes persist and undergo malignant conversion. The domestic rabbit is also suscep-

TABLE V. Lesion and Site Specificity of Animal Papillomaviruses

	Lesion		
Site	Papilloma	Fibropapilloma	Fibroma
Skin	BPV-3, 6	BPV-1, 2, 5	DFV
	CRPV	EEPV	
	Equine PV	RPV	
	MnPV	Sheep PV	
	Elephant PV		
	FPV		
Mucosa	BPV-4	BPV-2	
	CaPV		
	RaPV		
	Elephant PV		

tible to CRPV. The lesions tend to persist on this animal and develop into carcinomas in 75% of the cases (see Chapter 6).

The oral mucosa of rabbits proved refractory to CRPV. It can be infected by a rabbit oral papillomavirus (RaPV), however, which seems to be rather unrelated to CRPV as effective immunity to CRPV did not protect from RaPV infection (Parsons and Kidd, 1942). The virus is latent in the mouth. Occasional spontaneous virus production was observed with completely healthy animals. When the mucosa is injured, RaPV becomes activated and leads to epithelial papillomas around fibrovascular stalks, as it does spontaneously, as well.

The *Mastomys natalensis* papillomavirus (MnPV) represents another example for persistence of latent papillomavirus genomes (Amtmann *et al.*, 1984). Viral DNA could already be detected in embryos of multimammate mice but the first tumors were not noted before the age of 1 year. The MnPV-induced papillomas may progress to keratoacanthomas and squamous cell carcinomas (Reinacher *et al.*, 1978). This is discussed in detail in Chapter 7. The viral DNA shows a relatively high GC content of 50% (Müller and Gissmann, 1978) and has some homology with DNA of a miopapovavirus, which induces papillomas in hamsters (Scherneck *et al.*, 1984).

The DNA of the canine oral papillomavirus (CaPV) is significantly larger than that of any other papillomavirus (5.4×10^6; Pfister and Meszaros, 1980). The nucleotide sequence is being determined (Lancaster, personal communication) and it will be interesting to see how the genome organization of CaPV differs from that of the other types.

Several papillomaviruses, which lead to fibropapillomas or fibromas in their natural host (Table V), reveal limited DNA homology under stringent conditions in spite of infecting different host species. Deer fibroma virus (DFV) DNA hybridized to about 9% with BPV-1 DNA and 3% with BPV-2 DNA (Lancaster and Sundberg, 1982). Similar to the bovine viruses, DFV agglutinated mouse erythrocytes and a hemagglutination inhibition test with monospecific antisera indicated a cross-reactivity of roughly 1% between DFV and BPV-1 surface antigens. Extensive homology could also be detected between the genomes of DFV, European elk papillomavirus (EEPV), and reindeer papillomavirus (RPV), but the latter two revealed no sequence homology with BPV-1 under stringent hybridization conditions (Moreno-Lopez *et al.*, 1981, personal communication). DFV, EEPV, RPV, and sheep papillomavirus induce tumors in the hamster as do BPV-1 and 2 (Koller and Olson, 1972; Gibbs *et al.*, 1975; Stenlund *et al.*, 1983).

One representative of the avian papillomaviruses was studied in more detail. It was isolated from skin papillomas of the chaffinch and called FPV (Moreno-Lopez *et al.*, 1984). Partial nucleotide sequence analysis permitted aligning the genomes of FPV and BPV-1 by means of extensive homologies within reading frames E1 and L1. This comparison suggests that avian and mammalian papillomaviruses have a similar genome organization.

VI. EVOLUTIONARY ASPECTS

A. Papillomaviruses and Miopapovaviruses

Papillomaviruses can be clearly defined as one genus apart from the miopapovaviruses. No DNA cross-hybridization was observed even under conditions of very low stringency (Law *et al.*, 1979) and the genome organization within the two genera fundamentally differs in that miopapovaviruses have two coding DNA strands whereas papillomaviruses apparently have only one (Danos *et al.*, 1983; Engel *et al.*, 1983). Genus-specific antigens on viral capsid proteins are not shared by members of the respective other genus (Jenson *et al.*, 1980).

A papovavirus isolated from skin tumors of the Syrian hamster (Graffi *et al.*, 1967, 1968) shares some properties of papillomaviruses and miopapovaviruses. The viral capsid is about 40 nm in diameter and contains circular DNA with 39.2% GC and a molecular weight of 3.8×10^6 (Scherneck *et al.*, 1979). The DNAs of the hamster papovavirus (HaPV) and polyoma virus cross-hybridize under stringent conditions (Scherneck *et al.*, 1984). Homologous sequences were localized at the 3′ end of the early region of polyoma and in the entire late region, especially within sequences coding for VP2/3 and the amino-terminal part of VP1. On this basis, HaPV is classified as a miopapovavirus. Like papillomaviruses, HaPV induces cutaneous papillomatosis, is produced in the nuclei of the keratinized cell layer, and does not grow in cell culture systems. Preliminary hybridization experiments under relaxed conditions indicated sequence homology between HaPV and MnPV but not between HaPV and HPV-1 or 4 (Scherneck *et al.*, 1984). This led to speculations that HaPV emerged by recombination between polyoma- and papillomavirus-related sequences. Nucleotide sequence data will be necessary to substantiate this assumption.

On the amino acid level, regions of homology were found between the large T antigens of SV40 and polyoma and the E1-derived protein of papillomaviruses (Danos *et al.*, 1984), but these homologies are likely to be dictated by functional aspects. For example, the analogy around residue 450 of the E1 protein is shared by prokaryotic ATP synthases and by the p21 protein, which is coded by the Harvey murine sarcoma virus *ras* gene. The respective amino acids form the nucleotide binding sites of the enzymes.

B. Evolution of Papillomaviruses

Host range was discussed before as an accepted and so far unequivocal criterion for schematic classification of papillomaviruses. There can be little doubt, however, that host range does not reflect the natural relationship of individual virus types. This becomes clear with BPV-1. It shares little DNA homology with many bovine papillomaviruses (BPV-

3, 4, 6) and induces histologically different tumors. On the other hand, BPV-1 shows biological properties very similar to DFV, EEPV, and RPV, which infect other species. This is complemented by considerable cross-hybridization between the DNAs of these viruses even under stringent conditions. As a matter of fact, BPV-1 and colleagues nicely crystallize as a well-defined papillomavirus species mainly characterized by the ability to induce fibropapillomas or fibromas *in vivo.*

Human papillomaviruses do not form a homogeneous group either. When DNAs of HPV-1, 2, 4, BPV-1, and CRPV were allowed to cross-hybridize with decreasing stringency, the first signals were observed at 36° below the melting temperature of the DNA. Sequences with more mismatch reacted only at 43° or further below the melting temperature (Law *et al.*, 1979; Heilman *et al.*, 1980). According to this criterion, HPV-1 is more closely related to BPV-1 and CRPV than to HPV-2 or 4.

HPV-1, HPV-6, BPV-1, and CRPV were compared on the level of amino acid sequences within reading frames E1 and L1. The protein sequences were optimally aligned and the percentage of identical amino acids for pairwise alignments was determined (Danos *et al.*, 1984). HPV-1 again turned out to be closer to CRPV than to HPV-6, the other human virus.

Homologies across species barriers are even more striking within reading frame L2. A comparison of HPV-1, 6, 8, BPV-1, and CRPV revealed highly conserved amino- and carboxy-termini in all five types. Only HPV-1, 8, and CRPV, however, showed homology over the entire length of L2 (Danos *et al.*, 1984; Fuchs *et al.*, 1986). HPV-8 and CRPV also have related sequences close to the conserved amino-terminus, which are partially absent from the HPV-1 genome and completely missing in HPV-6 and BPV-1 DNA. The reading frames L2 thus show a relationship in the succession CRPV/HPV-8–HPV-1–HPV-6–BPV-1. This correlates with the tissue tropism of the respective viruses. HPV-1, 8, and CRPV infect skin keratinocytes, HPV-6 affects the mucosa, and BPV-1 has a dual tropism for fibroblasts and keratinocytes.

The close relationship of viruses from quite unrelated species might be explained either by a convergent evolution under selective pressure exerted by the host cell or by a separation of viruses with different tropism early in evolution prior to the separation of the host organisms. In this connection it will be interesting to evaluate the relationship between HPV-6 and the canine papillomavirus, which affects the oral mucosa.

Tissue tropism cannot account for the great distance between HPV-1, 2, and 4, which all induce skin warts. More data on genome organization and biological properties are needed to evaluate their evolutionary relationship.

C. Papillomaviruses from Epidermodysplasia Verruciformis

The rare disease epidermodysplasia verruciformis is distinguished by an amazing plurality of HPV, which led to speculations about the origin of this heterogeneity. First of all, it is probably a question of systematic cloning of viral DNAs with the intention to correlate virus type and cancerogenic potential, which led to the characterization of so many types. Second, the patients select for a number of HPV types by their impaired cellular immunity (Jablonska *et al.*, 1982). There is no evidence for an increased mutation rate in the course of the lifelong infection of an individual patient, however. Isolates from different countries show very similar cleavage maps on the one hand and different types are too far apart on the other hand to explain divergence just by mutation.

Data on cross-hybridization in liquid phase point to a hybrid nature of HPV-14 and 25 (Kremsdorf *et al.*, 1984; Gassenmaier *et al.*, 1984). They show a roughly equal distance to HPV-19 and 20 or HPV-19 and 8, respectively, all three of which are more distantly related (Fig. 5). The hybrid nature of HPV-25 was underlined by the demonstration of separate genome regions, which hybridized more efficiently with HPV-8 or 19, respectively (Gassenmaier *et al.*, 1984). These findings are suggestive of recombination but the limited overall homology gives no evidence for a recent event. The results could also be explained by separate evolution with conservation of different genome regions.

D. Reservoir of Papillomaviruses

Papillomaviruses are usually widespread among young individuals of the respective host species. Some virus types, however, are only observed under very special circumstances as for example in hosts with impaired immunity (HPVs of group D), within certain ethnic groups (HPV-13), or in association with nonproductive malignant tumors (HPV-18). This raises questions as to their origin and reservoir.

Although papillomaviruses can be expected to be rather stable, it seems extremely unlikely that HPVs of group D, for example, exist without a reservoir. There is no evidence for endogenous papillomavirus sequences. Inapparent infection is therefore the most likely explanation for

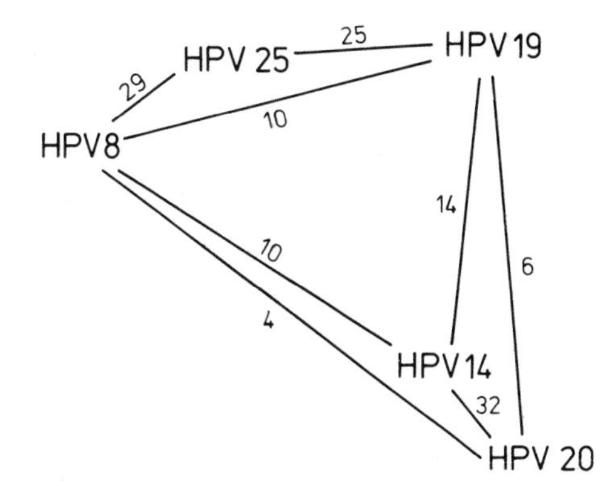

FIGURE 5. Relationship between human papillomaviruses 8, 14, 19, 20, and 25 as derived from DNA cross-hybridization in liquid phase [values in percent of homologous hybridization from Kremsdorf *et al.* (1984) and Gassenmaier *et al.* (1984)].

the spread of such viruses. Clinically inapparent persistence of viral DNA was described for MnPV in multimammate mice (Amtmann *et al.*, 1984) and for HPV-11 in human laryngeal mucosa (Steinberg *et al.*, 1983). As observed with RaPV, infectious particles can be shed by completely healthy animals (Parsons and Kidd, 1942), which guarantees the distribution of the virus without disease. This may also happen with the "epidermodysplasia-specific" HPV-8 for preliminary studies demonstrated agglutinating antibodies in ten sera from 100 healthy adults (Pfister *et al.*, 1981b).

The above-mentioned examples are likely to represent a group of extremely well-adapted papillomaviruses, which exist as primarily harmless passengers. Only extreme situations of one kind or the other will provide a chance to identify these viruses and study their role in disease.

ACKNOWLEDGMENTS. Original work described herein was supported by the Deutsche Forschungsgemeinschaft (SFB 118). I deeply appreciate the secretarial assistance of D. Brückner.

REFERENCES

Almeida, J. D., Howatson, A. F., and Williams, M. G., 1962, Electron microscope study of human warts: Sites of virus production and nature of the inclusion bodies, *J. Invest. Dermatol.* **38:**337–345.

Amtmann, E., and Sauer, G., 1982, Bovine papilloma virus transcription: Polyadenylated RNA species and assessment of the direction of transcription, *J. Virol.* **43:**59–66.

Amtmann, E., Müller, H., and Sauer, G., 1980, Equine connective tissue tumors contain unintegrated bovine papilloma virus DNA, *J. Virol.* **35:**962–964.

Amtmann, E., Volm, M., and Wayss, K., 1984, Tumour induction in the rodent *Mastomys natalensis* by activation of endogenous papilloma virus genomes, *Nature* **308:**291–292.

Archard, H. O., Heck, J. W., and Stanley, H. R., 1965, Focal epithelial hyperplasia: An unusual oral mucosal lesion found in Indian children, *Oral Surg.* **20:**201–212.

Barrera-Oro, J. G., Smith, K. O., and Melnick, J. L., 1962, Quantitation of papovavirus particles in human warts, *J. Natl. Cancer Inst.* **29:**583–595.

Black, P. H., Hartley, J. W., Rowe, W. P., and Huebner, R. J., 1963, Transformation of bovine tissue culture cells by bovine papilloma virus, *Nature* **199:**1016–1018.

Boiron, M., Levy, J. P., Thomas, M., Friedman, J. C., and Bernard, J., 1964, Some properties of bovine papilloma virus, *Nature* **201:**423–424.

Boshart, M., Gissmann, L., Ikenberg, H., Kleinheinz, A., Scheurlen, W., and zur Hausen, H., 1984, A new type of papillomavirus DNA, its presence in genital cancer biopsies and in cell lines derived from cervical cancer, *EMBO J.* **3:**1151–1157.

Breitburd, F., Favre, M., Zoorob, R., Fortin, D., and Orth, G., 1981, Detection and characterization of viral genomes and search for tumoral antigens in two hamster cell lines derived from tumors induced by bovine papillomavirus type 1, *Int. J. Cancer* **27:**693–702.

Bryan, W. R., and Beard, J. W., 1940, Correlation of frequency of positive inoculations with inoculation period and concentration of purified papilloma protein, *J. Infect. Dis.* **66:**245–253.

Campo, M. S., and Coggins, L. W., 1982, Molecular cloning of bovine papillomavirus genomes and comparison of their sequence homologies by heteroduplex mapping, *J. Gen. Virol.* **63:**255–264.

Campo, M. S., Moar, M. H., Jarrett, W. F. H., and Laird, H. M., 1980, A new papillomavirus associated with alimentary tract cancer in cattle, *Nature* **286:**180–182.
Campo, M. S., Moar, M. H., Laird, H. M., and Jarrett, W. F. H., 1981, Molecular heterogeneity and lesion site specificity of cutaneous bovine papillomaviruses, *Virology* **113:**323–335.
Carlson, B. L., Hill, D., and Nielsen, S. W., 1983, Cutaneous papillomatosis in a beaver, *J. Am. Vet. Med. Assoc.* **183:**1283–1284.
Chambers, V. C., and Evans, C. A., 1959, Canine oral papillomatosis. 1. Virus assay and observations on the various stages of the experimental infection, *Cancer Res.* **19:**1188–1195.
Chen, E. Y., Howley, P. M., Levinson, A. D., and Seeburg, P. H., 1982, The primary structure and genetic organization of the bovine papillomavirus type 1 genome, *Nature* **299:**529–534.
Cheville, N. F., and Olson, C., 1964, Cytology of the canine oral papilloma, *Am. J. Pathol.* **45:**849–872.
Coggin, J. R., Jr., and zur Hausen, H., 1979, Workshop on papillomaviruses and cancer, *Cancer Res.* **39:**545–546.
Coggins, L. W., Hettich, I., Smith, K. T., Slater, A. A., Roe, F. A., Pfister, H., and Campo, M. S., 1983, The genomes of bovine papillomaviruses types 3 and 4 are colinear, *J. Gen. Virol.* **64:**2771–2776.
Crawford, L. V., and Crawford, E. M., 1963, A comparative study of polyoma and papilloma viruses, *Virology* **21:**258–263.
Croissant, O., and Orth, G., 1967, Mise en êvidence par autoradiographie d'une synthèse d'ADN dans les cellules en voie de kératinisation des papillomes provoquès par le virus de Shope chez le lapin domestique, *C.R. Acad. Sci.* **265:**1341–1344.
Crum, C. P., Ikenberg, H., Richart, R. M., and Gissmann, L., 1984, Human papillomavirus type 16 and early cervical neoplasia, *N. Engl. J. Med.* **310:**880–883.
Danos, O., Katinka, M., and Yaniv, M., 1980, Molecular cloning, refined physical map and heterogeneity of methylation sites of papilloma virus type 1a DNA, *Eur. J. Biochem.* **109:**457–461.
Danos, O., Katinka, M., and Yaniv, M., 1982, Human papillomavirus 1a complete DNA sequence: A novel type of genome organization among papovaviridae, *EMBO J.* **1:**231–236.
Danos, O., Engel, L. W., Chen, E. Y., Yaniv, M., and Howley, P. M., 1983, Comparative analysis of the human type 1a and bovine type 1 papillomavirus genomes, *J. Virol.* **46:**557–566.
Danos, O., Giri, I., Thierry, F., and Yaniv, M., 1984, Papillomavirus genomes: Sequences and consequences, *J. Invest. Dermatol.* **83:**7s–11s.
Davis, C. L., and Kemper, H. E., 1936, Common warts (papillomata) in goats, *J. Am. Vet. Med. Assoc.* **41:**175–179.
Della Torre, G., Pilotti, S., De Palo, G., and Rilke, F., 1978, Viral particles in cervical condylomatous lesions, *Tumori* **64:**549–553.
deVilliers, E.-M., Gissmann, L., and zur Hausen, H., 1981, Molecular cloning of viral DNA from human genital warts, *J. Virol.* **40:**932–935.
Dürst, M., Gissmann, L., Ikenberg, H., and zur Hausen, H., 1983, A new type of papillomavirus DNA from a cervical carcinoma and its prevalence in cancer biopsies from different geographic regions, *Proc. Natl. Acad. Sci. USA* **80:**3812–3815.
Dvoretzky, I., Shober, R., Chattopadhyay, S. K., and Lowy, D. R., 1980, A quantitative *in vitro* focus forming assay for bovine papilloma virus, *Virology* **103:**369–375.
Engel, L. W., Heilman, C. A., and Howley, P. M., 1983, Transcriptional organization of bovine papillomavirus type 1, *J. Virol.* **47:**516–528.
Evans, C. A., Gormann, L. R., Ito, Y., and Weiser, R. S., 1962, Antitumor immunity in the SHOPE papilloma–carcinoma complex of rabbits. I. Papilloma regression induced by homologous and autologous tissue vaccines, *J. Natl. Cancer Inst.* **29:**277–285.
Favre, M., Breitburd, F., Croissant, O., and Orth, G., 1975a, Structural polypeptides of rabbit, bovine, and human papilloma viruses, *J. Virol.* **15:**1239–1247.
Favre, M., Orth, G., Croissant, O., and Yaniv, M., 1975b, Human papillomavirus DNA: Physical map, *Proc. Natl. Acad. Sci. USA* **72:**4810–4814.

Favre, M., Breitburd, F., Croissant, O., and Orth, G., 1977, Chromatin-like structures obtained after alkaline disruption of bovine and human papillomaviruses, *J. Virol.* **21:**1205–1209.

Favre, M., Jibard, N., and Orth, G., 1982, Restriction mapping and physical characterization of the cottontail rabbit papillomavirus genome in transplantable VX2 and VX7 domestic rabbit carcinomas, *Virology* **119:**298–309.

Finch, J. I., and Klug, A., 1965, The structure of viruses of the papilloma–polyoma type. III. Structure of rabbit papilloma virus, *J. Mol. Biol.* **13:**1–12.

Ford, J. N., Jennings, P. A., Spradbrow, P. B., and Francis, J., 1982, Evidence for papillomaviruses in ocular lesions in cattle, *Res. Vet. Sci.* **32:**257–259.

Friedman, J. C., Levy, J. P., Lasneret, J., Thomas, M., Boiron, M., and Bernard, J., 1963, Induction de fibromes sous-cutanés chez le hamster doré par inoculation d'extraits à cellulaires de papillomes bovins, *C.R. Acad. Sci. III* **257:**2328–2331.

Friedmann, J. M., and Fialkow, P. J., 1976, Viral "tumorigenesis" in man: Cell markers in condylomata acuminata, *Int. J. Cancer* **17:**57–61.

Fuchs, P. G., and Pfister, H., 1984, Cloning and characterization of papillomavirus type 2c DNA, *Intervirology* **22:**177–180.

Fuchs, P. G., Iftner, Th., Weninger, J. and Pfister, H., 1986, Epidermodysplasia verruciformis-associated human papillomavirus 8: Genomic sequence and comparative analysis, *J. Virol.* **58:**626–634.

Fulton, R. E., Doane, F. W., and Macpherson, L. W., 1970, The fine structure of equine papilloma and the equine papilloma virus, *J. Ultrastruct. Res.* **30:**328–343.

Gamperl, R., Amtmann, E., and Pfister, H., 1984, Chromosomal changes in bovine papillomavirus type 1-induced Syrian hamster tumors, *Cancer Genet. Cytogenet.* **12:**151–162.

Gassenmaier, A., Lammel, M., and Pfister, H., 1984, Molecular cloning and characterization of the DNAs of human papillomaviruses 19, 20 and 25 from a patient with epidermodysplasia verruciformis, *J. Virol.* **52:**1019–1023.

Gibbs, E. P. J., Smale, C. J., and Lauman, M. J. P., 1975, Warts in sheep, *J. Comp. Pathol.* **85:**327–334.

Giri, I., Danos, O., and Yaniv, M. 1985, Genomic structure of the cottontail rabbit (Shope) papillomavirus, *Proc. Natl. Acad. Sci. USA* **82:**1580–1584.

Gissmann, L., and zur Hausen, H., 1980, Partial characterization of viral DNA from human genital warts (condylomata acuminata), *Int. J. Cancer* **25:**605–609.

Gissmann, L., Pfister, H., and zur Hausen, H., 1977, Human papilloma virus (HPV): Characterization of 4 different isolates, *Virology* **76:**569–580.

Gissmann, L., Diehl, V., Schultz-Coulon, H. J., and zur Hausen, H., 1982, Molecular cloning and characterization of human papilloma virus DNA derived from a laryngeal papilloma, *J. Virol.* **44:**393–400.

Gissmann, L., Wolnik, L., Ikenberg, H., Koldovsky, U., Schnürch, H. G., and zur Hausen, H., 1983, Human papillomavirus types 6 and 11 DNA sequences in genital and laryngeal papillomas and in some cervical cancers, *Proc. Natl. Acad. Sci. USA* **80:**560–563.

Gissmann, L., Boshart, M., Dürst, M., Ikenberg, H., Wagner, D., and zur Hausen, H., 1984, Presence of human papillomavirus (HPV) DNA in genital tumors, *J. Invest. Dermatol.* **83:**26s–28s.

Glinski, W., Jablonska, S., Langner, A., Obalek, S., Haftek, M., and Proniewska, M., 1976, Cell-mediated immunity in epidermodysplasia verruciformis, *Dermatologica* **153:**218–227.

Glinski, W., Obalek, S., Jablonska, S., and Orth, G., 1981, T cell defect in patients with epidermodysplasia verruciformis due to human papillomavirus type 3 and 5, *Dermatologica* **162:**141–147.

Graffi, A., Schramm, T., Bender, E., Bierwolf, D., and Graffi, I., 1967, Über einen neuen virushaltigen Hauttumor beim Goldhamster, *Arch. Geschwulstforsch.* **30:**277–283.

Graffi, A., Schramm, T., Graffi, I., Bierwolf, D., and Bender, E., 1968, Virus-associated skin-tumors of the Syrian hamster: Preliminary note, *J. Natl. Cancer Inst.* **40:**867–873.

Green, M., Brackmann, K. H., Sanders, P. R., Loewenstein, P. M., Freel, J. H., Eisinger, M., and Switlyk, S. A., 1982, Isolation of a human papillomavirus from a patient with epidermodysplasia verruciformis: Presence of related viral DNA genomes in human urogenital tumors, *Proc. Natl. Acad. Sci. USA* **79:**4437–4441.

Greig, A. S., and Charlton, K. M., 1973, Electron microscopy of the virus of oral papillomatosis in the coyote, *J. Wildl. Dis.* **9:**359–361.

Gross, G., Pfister, H., Gissmann., L., and Hagedorn, M., 1982, Correlation between human papillomavirus type (HPV) and histology of warts, *J. Invest. Dermatol.* **78:**160–164.

Grussendorf, E. I., and zur Hausen, H., 1979, Localization of viral DNA replication in sections of human warts by nucleic acid hybridization with complementary RNA of human papilloma virus type 1, *Arch. Dermatol. Res.* **264:**55–63.

Heilman, C. A., Law, M. F., Israel, M. A., and Howley, P. M., 1980, Cloning of human papilloma virus genomic DNAs and analysis of homologous polynucleotide sequences, *J. Virol.* **36:**395–407.

Howley, P. M., Law, M. F., Heilman, C., Engel, L., Alonso, M. C., Israel, M. A., Lowy, D. R., and Lancaster, W. D., 1980, Molecular characterization of papillomavirus genomes, *Cold Spring Harbor Conf. Cell Prolif.* **7:**233–247.

Ikenberg, H., Gissmann, L., Gross, G., Grussendorf-Conen, E. I., and zur Hausen, H., 1983, Human papillomavirus type-16-related DNA in genital Bowen's disease and in Bowenoid papulosis, *Int. J. Cancer* **32:**563–565.

Ito, Y., 1963, Studies on subviral tumorigenesis: Carcinoma derived from nucleic acid induced papillomas of rabbit skin, *Acta. Unio. Int. Contra Cancrum* **19:**280–283.

Ito, Y., 1975, Papilloma–myxoma viruses, in: *Cancer*, Vol. 2 (F. F. Becker, ed.), pp. 323–341, Plenum Press, New York.

Jablonska, S., Orth, G., Jarzabek-Chorzelska, M., Glinski, W., Obalek, S., Rzesa, G., Croissant, O., and Favre, M., 1979, Twenty-one years of follow-up studies of familial epidermodysplasia verruciformis, *Dermatologica* **158:**309–327.

Jablonska, S., Orth, G., Glinski, W., Obalek, S., Jarzabek-Chorzelska, M., Croissant, O., Favre, M., and Rzesa, G., 1980, Morphology and immunology of human warts and familial warts, in: *Leukaemias, Lymphomas and Papillomas: Comparative Aspects* (P. A. Bachmann, ed.), pp. 107–131, Taylor & Francis, London.

Jablonska, S., Orth, G., and Lutzner, M. A., 1982, Immunopathology of papillomavirus-induced tumors in different tissues, *Springer Semin. Immunopathol.* **5:**33–62.

Jacobson, E. R., Mladinich, C. R., Clubb, S., Sundberg, J. P., and Lancaster, W. D., 1983, Papilloma-like infection in an African gray parrot, *J. Am. Vet. Med. Assoc.* **183:**1307–1308.

Jarrett, W. F. H., McNeil, P. E., Grimshaw, W. T. R., Selman, I. E., and McIntyre, W. I. M., 1978, High incidence area of cattle cancer with a possible interaction between an environmental carcinogen and a papilloma virus, *Nature* **274:**215–217.

Jarrett, W. F. H., McNeil, P. E., Laird, H. M., O'Neil, B. W., Murphy, J., Campo, M. S., and Moar, M. H., 1980, Papilloma viruses in benign and malignant tumors of cattle, in: *Viruses in Naturally Occurring Cancers* (M. Essex, G. Todaro, and H. zur Hausen, eds.), pp. 215–222, Cold Spring Harbor Laboratory, Cold Spring Harbor, N.Y.

Jarrett, W. F. H., Campo, M. S., O'Neil, B. W., Laird, H. M., and Coggins, L. W., 1984a, A novel bovine papillomavirus (BPV-6) causing true epithelial papillomas of the mammary gland skin: A member of a proposed new BPV group, *Virology* **136:**255–264.

Jarrett, W. F. H., Campo, M. S., Blaxter, M. L., O'Neil, B. W., Laird, H. M., Moar, M. H., and Sartirana, M. L., 1984b, Alimentary fibropapilloma in cattle: A spontaneous tumour, non-permissive for papillomavirus replication, *J. Natl. Cancer Inst.* **73:**449–504.

Jenson, A. B., Rosenthal, J. R.. Olson, C., Pass, F., Lancaster, W. D., and Shah, K., 1980, Immunological relatedness of papillomaviruses from different species, *J. Natl. Cancer Inst.* **64:**495–500.

Kidd, J. G., Beard, J. W., and Rous, P., 1936, Serological reactions with a virus causing a rabbit papilloma which becomes cancerous. II. Tests of the blood of animal carrying various tumors, *J. Exp. Med.* **64:**63–78.

Klug, A., and Finch, J. T., 1965, Structure of virus of the papilloma–polyoma type. I. Human wart virus, *J. Mol. Biol.* **11**:403–423.

Koller, L. D., 1972, Cutaneous papillomatosis in an opossum, *J. Natl. Cancer Inst.* **49**:309–313.

Koller, L. D., and Olson, C., 1972, Attempted transmission of warts from man, cattle, and horses, and of deer fibroma, to selected hosts, *J. Invest. Dermatol.* **58**:366–368.

Kremsdorf, D., Jablonska, S., Favre, M., and Orth, G., 1982, Biochemical characterization of two types of human papillomaviruses associated with epidermodysplasia verruciformis, *J. Virol.* **43**:436–447.

Kremsdorf, D., Jablonska, S., Favre, M., and Orth, G., 1983, Human papillomaviruses associated with epidermodysplasia verruciformis. II. Molecular cloning and biochemical characterization of human papillomavirus 3a, 8, 10, and 12 genomes, *J. Virol.* **48**:340–351.

Kremsdorf, D., Favre, M., Jablonska, S., Obalek, S., Rueda, L. A., Lutzner, M. A., Blanchet-Bardon, C., van Voorst Vader, P. C., and Orth, G., 1984, Molecular cloning and characterization of the genomes of nine newly recognized human papillomavirus types associated with epidermodysplasia verruciformis, *J. Virol.* **52**:1013–1018.

Krzyzek, R. A., Watts, S. L., Anderson, D. L., Faras, A. J., and Pass, F., 1980, Anogenital warts contain several distinct species of human papillomavirus, *J. Virol.* **36**:236–244.

Lack, E. E., Jenson, A. B., Smith, H. G., Healy, G. B., Pass, F., and Vawter, G. F., 1980, Immunoperoxidase localization of human papillomavirus in laryngeal papillomas, *Intervirology* **14**:148–154.

Lancaster, W. D., 1979, Physical maps of bovine papillomavirus type 1 and type 2 genomes, *J. Virol.* **32**:684–687.

Lancaster, W. D., 1981, Apparent lack of integration of bovine papillomavirus DNA in virus-induced equine and bovine tumors and virus-transformed mouse cells, *Virology* **108**:251–255.

Lancaster, W. D., and Olson, C., 1978, Demonstration of two distinct classes of bovine papilloma virus, *Virology* **89**:371–379.

Lancaster, W. D., and Olson, C., 1982, Animal papillomaviruses, *Microbiol. Rev.* **46**:191–207.

Lancaster, W. D., and Sundberg. J. P., 1982, Characterization of papilloma viruses isolated from cutaneous fibromas of white-tailed deer and mule deer, *Virology* **123**:212–216.

Lancaster, W. D., Theilen, G. H., and Olson, C., 1979, Hybridization of bovine papilloma virus type 1 and type 2 DNA to DNA from virus-induced hamster tumors and naturally occurring equine connective tissue tumors, *Intervirology* **11**:227–233.

Lass, J. H., Grove, A. S., Papale, J. J., Albert, D. M., Jenson, A. B., and Lancaster, W. D., 1983, Detection of human papillomavirus DNA sequences in conjunctival papilloma, *Am. J. Ophthalmol.* **96**:670–674.

Laurent, R., Kienzler, J. L., Croissant, O., and Orth, G., 1982, Two anatomoclinical types of warts with plantar localization: Specific cytopathogenic effects of papillomavirus, *Arch. Dermatol. Res.* **274**:101–111.

Laverty, C. R., Russell, P., Hills, E., and Booth, N., 1978, The significance of noncondylomatous wart virus infection of the cervical transformation zone: A review with discussion of two illustrative cases, *Acta Cytol.* **22**:195–201.

Law, M.-F., Lancaster, W. D., and Howley, P. M., 1979, Conserved sequences among the genomes of papillomaviruses, *J. Virol.* **32**:199–207.

Law, M.-F., Lowy, D. R., Dvoretzky, I., and Howley, P. M., 1981, Mouse cells transformed by bovine papillomavirus contain only extrachromosomal viral DNA sequences, *Proc. Natl. Acad. Sci. USA* **78**:2727–2731.

Le Bouvier, G. L., Sussmann, M., and Crawford, L. V., 1966, Antigenic diversity of mammalian papillomaviruses, *J. Gen. Microbiol.* **45**:497–501.

Lewandowsky, F., and Lutz, W., 1922, Ein Fall einer bisher nicht beschriebenen Hauterkrankung (epidermodysplasia verruciformis), *Arch. Dermatol. Syph.* **141**:193–203.

Lina, P. H. C., van Noord, M. J., and de Groot, F. G., 1973, Detection of virus in squamous papillomas of the wild bird species *Fringilla coelebs*, *J. Natl. Cancer Inst.* **50**:567–571.

Lutzner, M., Croissant, O., Ducasse, M.-F., Kreis, H., Crosnier, J., and Orth, G., 1980, A potentially oncogenic human papillomavirus (HPV-5) found in two renal allograft recipients, *J. Invest. Dermatol.* **75:**353–356.

McVay, P., Fretz, M., Wettstein, F., Stevens, J., and Ito, Y., 1982, Integrated Shope virus DNA is present and transcribed in the transplantable rabbit tumour Vx-7, *J. Gen. Virol.* **60:**271–278.

Massing, A. M., and Epstein, W. L., 1963, Natural history of warts: A two-year study, *Arch. Dermatol.* **87:**306.

Matthews, R. E. F., 1982, Classification and nomenclature of viruses, *Intervirology* **17:**1–199.

Meisels, A., Morin, C., and Casas-Cordero, M., 1982, Human papillomavirus infection of the uterine cervix, *Int. J. Gynecol. Pathol.* **1:**75–94.

Moar, M. H., Campo, M. S., Laird, H. M., and Jarrett, W. F. H., 1981a, Unintegrated viral DNA sequences in hamster tumor induced by bovine papilloma virus, *J. Virol.* **39:**945–949.

Moar, M. H., Campo, M. S., Laird, H., and Jarrett, W. F. H., 1981b, Persistence of nonintegrated viral DNA in bovine cells transformed *in vitro* by bovine papillomavirus type 2, *Nature* **293:**749–751.

Moreno-Lopez, J., Pettersson, U., Dinter, Z., and Philipson, L., 1981, Characterization of a papilloma virus from the European elk (EEPV), *Virology* **112:**589–595.

Moreno-Lopez, J., Ahola, H., Stenlund, A., Osterhaus, A., and Pettersson, U., 1984, Genome of an avian papillomavirus, *J. Virol.* **51:**872–875.

Mounts, P., and Kashima, H., 1984, Association of human papillomavirus subtype and clinical course in respiratory papillomatosis, *Laryngoscope* **94:**28–33.

Müller, H., and Gissmann, L., 1978, Mastomys natalensis papilloma virus (MnPV), the causative agent of epithelial proliferations: Characterization of the virus particle, *J. Gen. Virol.* **41:**315–323.

Murray, R. F., Hobbs, J., and Payne, B., 1971, Possible clonal origin of common warts (verruca vulgaris), *Nature* **232:**51–52.

Nakabayashi, Y., Chattopadhyay, S. K., and Lowy, D. R., 1983, The transforming function of bovine papillomavirus DNA, *Proc. Natl. Acad. Sci. USA* **80:**5832–5836.

Nasseri, M., Wettstein, F. O., and Stevens, J. G., 1982, Two colinear and spliced viral transcripts are present in non-virus-producing benign and malignant neoplasms induced by the Shope (rabbit) papillomavirus, *J. Virol.* **44:**263–268.

Noyes, W. F., 1965, Structure of the human wart virus, *Virology* **23:**65–72.

Obalek, S., Glinski, W., Haftek, M., Orth, G., and Jablonska, S., 1980, Comparative studies on cell-mediated immunity in patients with different warts, *Dermatologica* **161:**73–83.

Olson, C., and Cook, R. H., 1951, Cutaneous sarcoma-like lesions of the horse caused by the agent of bovine papilloma, *Proc. Soc. Exp. Biol. Med.* **77:**281–284.

Olson, C., Luedke, A. J., and Brobst, D. F., 1962, Induced immunity of skin, vagina, and urinary bladder to bovine papillomatosis, *Cancer Res.* **22:**463–468.

Olson, C., Gordon, D. E., Robl, M. G., and Lee, K. P., 1969, Oncogenicity of bovine papilloma virus, *Arch. Environ. Health* **19:**827–837.

Orfanos, C. E., Strunk, W., and Gartmann, H., 1974, Fokale epitheliale Hyperplasie der Mundschleimhaut: Hecksche Krankheit, *Dermatologica* **149:**163–175.

Orth, G., Jeanteur, P., and Croissant, O., 1971, Evidence for and localization of vegetative viral DNA replication by autoradiographic detection of RNA–DNA hybrids in sections of tumors induced by Shope papilloma virus, *Proc. Natl. Acad. Sci. USA* **68:**1876–1880.

Orth, G., Favre, M., and Croissant, O., 1977, Characterization of a new type of human papilloma virus that causes skin warts, *J. Virol.* **24:**108–120.

Orth, G., Breitburd, F., and Favre, M., 1978a, Evidence for antigenic determination shared by the structural polypeptides of (Shope) rabbit papillomavirus and human papillomavirus type 1, *Virology* **91:**243–255.

Orth, G., Jablonska, S., Favre, M., Croissant, O., Jarzabek-Chorzelska, M., and Rzesa, G., 1978b, Characterization of two new types of human papilloma viruses in lesions of epidermodysplasis verruciformis, *Proc. Natl. Acad. Sci. USA* **75:**1537–1541.

Orth, G., Jablonska, S., Jarzabek-Chorzelska, M., Obalek, S., Rzesa, G., Favre, M., and Croissant, O., 1979, Characteristics of the lesions and risk of a malignant conversion associated with the type of human papillomavirus involved in epidermodysplasia verruciformis, *Cancer Res.* **39:**1074–1082.

Orth, G., Favre, M., Breitburd, F., Croissant, O., Jablonska, S., Obalek, S., Jarzabek-Chorzelska, M., and Rzesa, G., 1980, Epidermodysplasia verruciformis: A model for the role of papilloma viruses in human cancer, *Cold Spring Harbor Conf. Cell Prolif.* **7:**259–282.

Orth, G., Jablonska, S., Favre, M., Croissant, O., Obalek, S., Jarzabek-Chorzelska, M., and Jibard, N., 1981, Identification of papillomaviruses in butcher's warts, *J. Invest. Dermatol.* **76:**97–102.

Osterhaus, A. M. D. E., Ellens, D. J., and Horzinek, M. C., 1977, Identification and characterization of a papillomavirus from birds (Fringillidae), *Intervirology* **8:**351–359.

Ostrow, R. S., Krzyzek, R., Pass, F., and Faras, A. J., 1981, Identification of a novel human papilloma virus in cutaneous warts of meat handlers, *Virology* **108:**21–27.

Ostrow, R. S., Bender, M., Niimura, M., Seki, T., Kawashima, M., Pass, F., and Faras, A. J., 1982, Human papillomavirus DNA in cutaneous primary and metastasized squamous cell carcinomas from patients with epidermodysplasia verruciformis, *Proc. Natl. Acad. Sci. USA* **79:**1634–1638.

Ostrow, R., Zachow, K., Watts, S., Bender, M., Pass, F., and Faras, A., 1983, Characterization of two HPV-3 related papillomaviruses from common warts that are distinct clinically from flat warts or epidermodysplasia verruciformis, *J. Invest. Dermatol.* **80:**436–440.

Ostrow, R. S., Zachow, K. R., Thompson, O., and Faras, A. J., 1984, Molecular cloning and characterization of a unique type of human papillomavirus from an immune deficient patient, *J. Invest. Dermatol.* **82:**362–366.

Parsons, R. J., and Kidd, J. G., 1942, Oral papillomatosis of rabbits: A virus disease, *J. Exp. Med.* **77:**233–250.

Pfister, H., 1980, Comparative aspects of papillomatosis, in: *Leukaemias, Lymphomas and Papillomas: Comparative Aspects* (P. A. Bachmann, ed.), pp. 93–106, Taylor & Francis, London.

Pfister, H., 1984, Biology and biochemistry of papillomaviruses, *Rev. Physiol. Biochem. Pharmacol.* **99:**111–181.

Pfister, H., and Meszaros, J., 1980, Partial characterization of a canine oral papillomavirus, *Virology* **104:**243–246.

Pfister, H., and zur Hausen, H., 1978a, Characterization of proteins of human papilloma virus (HPV) and antibody response to HPV 1, *Med. Microbiol. Immunol.* **166:**13–19.

Pfister, H., and zur Hausen, H., 1978b, Seroepidemiological studies of human papilloma virus (HPV 1) infections, *Int. J. Cancer* **21:**161–165.

Pfister, H., Gissmann, L., and zur Hausen, H., 1977, Partial characterization of proteins of human papilloma viruses (HPV) 1–3, *Virology* **83:**131–137.

Pfister, H., Gross, G., and Hagedorn, M., 1979a, Characterization of human papillomavirus 3 in warts of a renal allograft patient, *J. Invest. Dermatol.* **73:**349–353.

Pfister, H., Huchthausen, B., Gross, G., and zur Hausen, H., 1979b, Seroepidemiological studies of bovine papillomavirus infections, *J. Natl. Cancer Inst.* **62:**1423–1425.

Pfister, H., Linz, U., Gissmann, L., Huchthausen, B., Hoffmann, D., and zur Hausen, H., 1979c, Partial characterization of a new type of bovine papillomaviruses, *Virology* **96:**1–8.

Pfister, H., Gissmann, L., zur Hausen, H., and Gross, G., 1980, Characterization of human and bovine papilloma viruses and of the humoral immune response to papilloma virus infection, *Cold Spring Harbor Conf. Cell Prolif.* **7:**249–258.

Pfister, H., Fink, B., and Thomas, C., 1981a, Extrachromosomal bovine papillomavirus type 1 DNA in hamster fibromas and fibrosarcomas, *Virology* **115:**414–418.

Pfister, H., Nürnberger, F., Gissmann, L., and zur Hausen, H., 1981b, Characterization of a human papillomavirus from epidermodysplasia verruciformis lesions of a patient from Upper Volta, *Int. J. Cancer* **27:**645–650.

Pfister, H., Gassenmaier, A., Nürnberger, F., and Stüttgen, G., 1983a, HPV 5 DNA in a carcinoma of epidermodysplasia verruciformis patient infected with various human papillomavirus types, *Cancer Res.* **43:**1436–1441.

Pfister, H., Hettich, I., Runne, U., Gissmann, L., and Chilf, G. N., 1983b, Characterization of human papillomavirus type 13 from lesions of focal epithelial hyperplasia Heck, *J. Virol.* **47:**363–366.

Praetorius-Clausen, F., 1969, Histopathology of focal epithelial hyperplasia: Evidence of viral infection, *Tandlaegebladet* **73:**1013–1022.

Puget, A., Favre, M., and Orth, G., 1975, Induction de tumeurs fibroblastiques cutanées où sous-cutanées chez l'Ochotone afghan (Ochotono rufescens rufescens) par inoculation du virus du papillome bovine, *C. R. Acad. Sci. III* **280:**2813–2816.

Rangan, S. R. S., Gutter, A., Baskin, G. B., and Anderson, D., 1980, Virus-associated papillomas in colobus monkeys (Colobus guereza), *Lab. Anim. Sci.* **30:**885–889.

Rashad, A. L., 1969, Radioautographic evidence of DNA synthesis in well differentiated cells of human skin papillomas, *J. Invest. Dermatol.* **53:**356–362.

Reinacher, M., Müller, H., Thiel, W., and Rudolph, R. L., 1978, Localization of papillomavirus and virus-specific antigens in the skin of tumor-bearing Mastomys natalensis (GRA Giessen), *Med. Microbiol. Immunol.* **165:**93–99.

Rowson, K. E. K., and Mahy, B. W. J., 1967, Human papova (wart) virus, *Bacteriol. Rev.* **313:**110–131.

Rulison, R. H., 1942, Warts: A statistical study of nine hundred twenty-one cases, *Arch. Dermatol. Syphilol.* **46:**66–81.

Schaffhausen, B. S., and Benjamin, T. L., 1976, Deficiency in histone acetylation in nontransforming host range mutants of polyoma virus, *Proc. Natl. Acad. Sci. USA* **73:**1092–1096.

Scherneck, S., Böttger, M., and Feunteun, J., 1979, Studies on the DNA of an oncogenic papovavirus of the Syrian hamster, *Virology* **96:**100–107.

Scherneck, S., Vogel, F., Nguyen, H. L., and Feunteun, J., 1984, Sequence homology between polyoma virus, simian virus 40 and a papilloma producing virus from a Syrian hamster: Evidences for highly conserved sequences, *Virology* **137:**41–48.

Schiller, J. T., Vass, W. C., and Lowy, D. R., 1984, Identification of a second transforming region in bovine papillomavirus DNA, *Proc. Natl. Acad. Sci. USA* **81:**7880–7884.

Schwarz, E., Dürst, M., Demankowski, C., Lattermann, O., Zech, R., Wolfsperger, E., Suhai, S., and zur Hausen, H., 1983, DNA sequence and genome organization of genital human papillomavirus type 6b, *EMBO J.* **2:**2341–2348.

Shope, R. E., 1933, Infectious papillomatosis of rabbits; with a note on the histopathology, *J. Exp. Med.* **58:**607–624.

Shope, R. E., Mangold, R., McNamara, L. G., and Dumbell, K. R., 1958, An infectious cutaneous fibroma of the Virginia white-tailed deer (*Odocoileus virginianus*), *J. Exp. Med.* **108:**797–802.

Spradbrow, P. B., and Hoffmann, D., 1980, Bovine ocular squamous cell carcinoma, *Vet. Bull.* **50:**449–459.

Spradbrow, P. B., Ford, J. N., and Samuel, J. L., 1983, Papillomaviruses with unusual restriction endonuclease profiles from bovine skin warts. *J. Aust. Vet.* **60:**259.

Steinberg, B. M., Topp, W. C., Schneider, P. S., and Abramson, A. L., 1983, Laryngeal papillomavirus infection during clinical remission, *N. Engl. J. Med.* **308:**1261–1264.

Stenlund, A., Moreno-Lopez, J., Ahola, H., and Pettersson, U., 1983, European elk papillomavirus: Characterization of the genome, induction of tumors in animals, and transformation *in vitro*, *J. Virol.* **48:**370–376.

Sundberg, J. P., Russell, W. C., and Lancaster, W., 1981, Papillomatosis in Indian elephants, *J. Am. Med. Assoc.* **179:**1247–1248.

Thomas, M., Levy, J. P., Tanzer, J., Boiron, M., and Bernard J., 1963, Transformation in vitro de cellules de peau de veau embryonnaire sous l'action d'extrants a cellulaires de papillomes bovins, *C. R. Acad. Sci. III* **257:**2155–2158.

Thomas, M., Boiron, M., Tanzer, J., Levy, J. P., and Bernard, J., 1964, *In vitro* transformation of mice cells by bovine papilloma virus, *Nature* **202:**709–710.

Tsumori, T., Yutsudo, M., Nakano, Y., Tanigaki, T., Kitamura, H., and Hakura, A., 1983, Molecular cloning of a new human papilloma virus isolated from epidermodysplasia verruciformis lesions, *J. Gen. Virol.* **64:**967–969.

Vanselow, A. A., and Spradbrow, P. B., 1982, Papillomaviruses, papillomas and squamous cell carcinomas in sheep, *Vet. Rec.* **110:**561–562.

Waldeck, W., Rösl, F., and Zentgraf, H., 1984, Origin of replication in episomal bovine papilloma virus type 1: DNA isolated from transformed cells, *EMBO J.* **3:**2173–2178.

Watrach, A. M., 1969, The ultrastructure of canine cutaneous papilloma, *Cancer Res.* **29:**2079–2084.

Wettstein, F. O., and Stevens, J. G., 1980, Distribution and state of viral nucleic acid in tumors induced by Shope papilloma virus, *Cold Spring Harbor Conf. Cell Prolif.* **7:**301–307.

Wettstein, F. O., and Stevens, J. G., 1982, Variable-sized free episomes of Shope papilloma virus DNA are present in all non-virus-producing neoplasms and integrated episomes are detected in some, *Proc. Natl. Acad. Sci. USA* **79:** 790–794.

Wettstein, F. O., and Stevens, J. G., 1983, Shope papilloma virus DNA is extensively methylated in non-virus-producing neoplasms, *Virology* **126:**493–504.

Woodruff, J. D., Braun, L., Cavallieri, R., Gupta, P., Pass, F., and Shah, K. V., 1980, Immunological identification of papillomavirus antigen in paraffin-processed condyloma tissues from the female genital tract, *Obstet. Gynecol.* **56:**727–732.

Yabe, Y., Sadakane, H., and Isono, H., 1979, Connection between capsomeres in human papilloma virus, *Virology* **96:**547–552.

Yang, Y.-C., Okayama, H., and Howley, P. M. 1985, Bovine papilloma-virus contains multiple transforming genes, *Proc. Natl. Acad. Sci. USA* **82:**1030–1034.

Zachow, K. R., Ostrow, R. S., Bender, M., Watts, S., Okagaki, T., Pass, F., and Faras, A. J., 1982, Detection of human papillomavirus DNA in anogenital neoplasias, *Nature* **300:**771–773.

zur Hausen, H., 1977, Human papilloma viruses and their possible role in squamous cell carcinomas, *Curr. Top. Microbiol. Immunol.* **78:**1–30.

CHAPTER 2

Animal Papillomas
Historical Perspectives

CARL OLSON

Because cutaneous papillomas can be readily recognized, they have long been known in animals. Their occurrence in the horse was described in the 9th century by a stablemaster of the Caliph of Bagdad (Erk, 1976).

Experimental exploration of animal papillomas began first with transmission, then demonstration of a viral etiology and host specificity of the viruses. Transmission of canine oral papillomatosis was done in 1898 (McFadyean and Hobday) and skin papilloma from horse to horse was experimentally transmitted in 1901 (Cadeac). However, experimental transplantation of a canine invasive carcinoma of the nose was reported in 1876 by Novinsky (cited by Shimkin, 1955). This may have been canine venereal sarcoma or possibly a sequela of canine oral papillomavirus such as was observed by Watrach *et al.* (1970).

Papillomas have now been shown to be transmissible in twenty species of animals (Table I), and the naturally occurring papillomavirus of cattle, deer, sheep, and cottontail rabbit will cause tumors in other species.

I. CANINE ORAL PAPILLOMATOSIS

Canine oral papillomas (COP) occur and are readily transmitted in susceptible dogs (Fig. 1). It can be a problem in kennels where young dogs must be protected by a prophylactic vaccine prepared from a suspension

CARL OLSON • Department of Veterinary Science, University of Wisconsin, Madison, Wisconsin 53706.

TABLE I. Transmissible Animal Papillomas

Species	Location	Transmission: Within species	Transmission: To other species	Lesion produced
Cattle type 1, 2	Skin fibropapilloma	Skin and brain, fibroma	Horse	Fibroma
			Hamster	Fibrosarcoma
	Fibropapilloma	Genital epithelium	Pika	Fibroma
	Papilloma carcinoma	Urinary bladder		
Cattle atyp.	Skin	Papilloma	NR[a]	
Cattle 4	Alimentary adenocarcinoma	Papilloma	ND[b]	
Cattle 5, 6	Teat	Papilloma	ND	
Cattle 3	Teat	Papilloma	ND	
Deer				
Whitetail	Skin	Fibropapilloma	Hamster	Fibroma
Blacktail	Skin	Fibropapilloma[c]	ND	
Mule	Skin	Fibropapilloma[c]	ND	
Reindeer	Skin	Fibropapilloma[c]		
Elk	Skin	Fibropapilloma[c]	Hamster	Fibroma
Other cervids	Skin	Fibropapilloma[c]	ND	
Sheep	Skin	Fibropapilloma	Hamster	Fibroma
Goat	Skin	Fibropapilloma[c]	ND	
Horse	Skin	Papilloma	NR	
Rabbit				
Cottontail	Skin	Carcinoma	Domestic	Carcinoma
Domestic	Oral	Papilloma	NR	
Dog	Oral	Papilloma	NR	
Coyote	Oral	Papilloma[c]	Dog	Papilloma
Wolf	Oral	Papilloma[c]	ND	
Pig	Genital	Condyloma	ND	
Primates				
Cebus	Skin	Papilloma	ND	
Chimpanzee	Skin	Papilloma[c]	ND	
Human	Skin	Papilloma[c]	NR	
Chaffinch	Leg	Papilloma	ND	
Pronghorn Antelope	Skin	ND	ND	
Opossum	Skin	NR	ND	
Elephant	Oral	ND	ND	

[a] NR, no response.
[b] ND, not done.
[c] Naturally occurring.

of COP. The incubation period for oral papillomas can vary from 4 to 10 weeks and regression usually follows in 3 to 14 weeks with essentially the same experience in England (McFadyean and Hobday, 1898), the United States (Chambers and Evans, 1959; Cheville and Olson, 1964a), and Japan (Konishi *et al.*, 1972; Tokita and Konishi, 1975). The first reaction to COP virus is hyperplasia of keratin-producing cells in which no viral particles

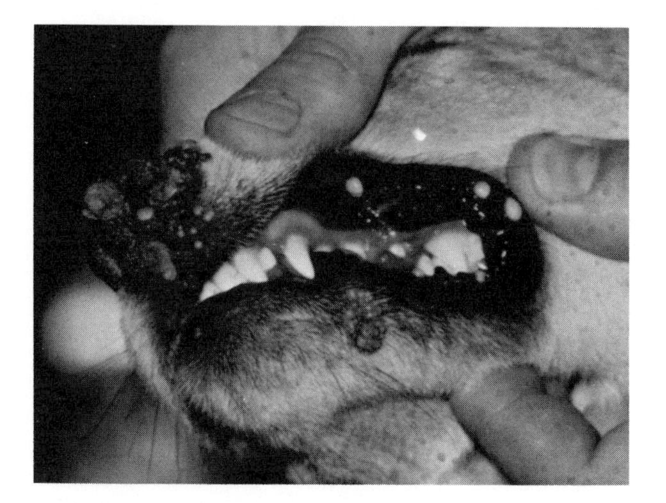

FIGURE 1. Naturally occurring canine oral papillomas. The occurrence of papillomas on skin of the lower jaw and about the nose is unusual.

can be detected. Later, larger vesicular cells appear containing Feulgen-positive nuclear inclusions. Intranuclear COP virus appeared first in these cells in the stratum spinosum, and later in linear and mosaic patterns in the stratum granulosum and corneum (Cheville and Olson, 1964a). Watrach *et al.* (1970) reported a squamous cell carcinoma in association with extensive oral papillomas that had persisted for 8 months in a $1\frac{1}{2}$-year-old dog. COP virions were present in the papillomas. With COP, most workers were not able to produce papillomas on the skin of dogs. However, Watrach (1969) found abundant papillomavirus in two cases of canine cutaneous papillomas. One was a 9-month-old German shepherd with numerous small gray-white "protuberands" on posterior abdomen and adjacent thigh areas. The other was on the neck of a 1-year-old poodle. Transmission was not attempted since the biopsies had been fixed in formalin. Davis and Sabine (1975–76) reported dermal papillomas but no oral papillomas in racing greyhounds of Australia. Virions were demonstratable by electron microscopy and cell-free filtrates produced experimental papillomas in 27 to 49 days. Untreated lesions regressed by 59 days.

Tokita and Konishi (1975) produced papillomas on the scarified surface of the cornea and sclera in 8 of 17 dogs and at the mucocutaneous junction of the eyelid in 7 of 17 dogs. Only 1 of 5 puppies inoculated on the skin of the abdomen, back, and face developed papilloma. Inoculation of the vagina, penis, stomach, urinary bladder, and brain produced negative results. The duration of COP was prolonged 117 to 337 days in 3 dogs given a carcinogen (*N*-methyl-*N*-nitro-*N*-nitrosoguanidine) orally for 64 days after the papillomas appeared. Papillomas of the eyelid and lips with characteristic papilloma virions occurred in a 5-

year-old dog with natural oral papillomas. In another natural case, bilateral papillomas of the conjunctival–corneal areas were found in a 9-year-old dog (Hare and Howard, 1977) suggesting exposure to an oncogenic agent but no attempts were made by electron microscopy or transmission to identify a papillomavirus.

DeMonbreun and Goodpasture (1932) reported oral papillomas to have been extensive in a dog with numerous warts on the posterior pharyngeal wall and a few on the epiglottis but none below the level of the glottis.

In another study, COP material failed to induce COP lesions in the larynxes of four dogs whose buccal mucosa responded to the same material with production of typical COP. Specimens from four cases of human laryngeal papillomas were inoculated into two young dogs. Scarified inoculations were on buccal and laryngeal mucosa as well as intradermally on abdomen and vagina. No lesions developed or were found on the dogs. One dog of each pair underwent necropsy at 8 weeks and the four remaining dogs were inoculated with COP virus at 1 year and found susceptible to COPV (Olson, unpublished data).

The COP virus did not cause warts in kittens, mice, rats, guinea pigs, rabbits, or monkeys (*M. rhesus*) (DeMonbreun and Goodpasture, 1932). The demonstration of papillomalike virions in a natural case of oral papillomatosis in a coyote (Greig and Charlton, 1973) should prompt study of the action of COP virus in this animal. Samuel *et al.* (1978) produced oral papillomas in two beagles with oral papilloma material from a coyote.

Tokita *et al.* (1977) cultivated trypsinized COP cells most readily when the papillomas were 15 to 22 days old. Both epithelial-like cells and fibroblasts grew in the cultures which together with the culture media were inoculated on the mucosa of young dogs with negative results.

II. SHOPE RABBIT PAPILLOMA

The history of the Shope rabbit papilloma and cottontail rabbit papillomavirus (CRPV) has been well reviewed by Gross (1970) which should be consulted since there are many interesting details of personal comments of Richard E. Shope and Jerome T. Syverton not available elsewhere.

Paraphrased comments of Shope to Gross (1970): A visitor to Princeton viewing a rabbit with experimental fibroma said it was nothing compared to rabbits around Cherokee, Iowa, with "horns" coming out of their heads (Fig. 2); almost every other rabbit had them. He returned to Cherokee with the visitor but failed to get a specimen in 4 days. Shope left a five-dollar bill and a jar of glycerol for the "rabbit horns" and promised an additional five-dollar bill upon delivery of the glycerinated "horns." A week later, specimens arrived and transmission experiments began.

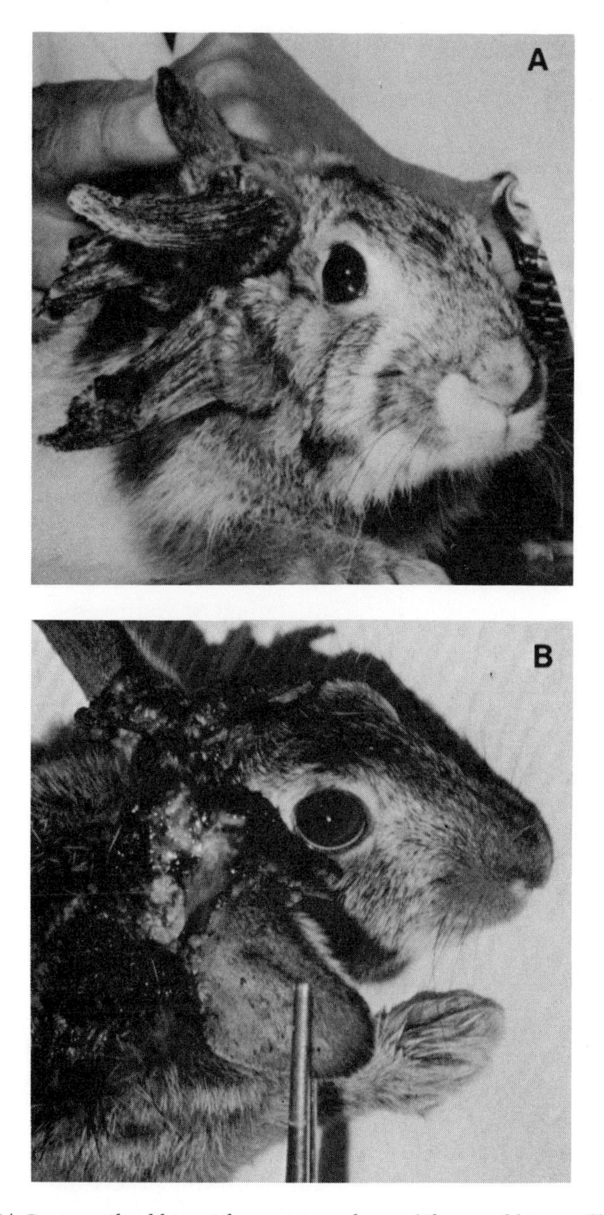

FIGURE 2. (A) Cottontail rabbit with cutaneous horns (Shope rabbit papilloma) captured near the Mississippi River in Minnesota by James T. Prince. (B) Cutaneous horns and adjacent carcinoma in a wild cottontail rabbit, courtesy of W. C. Phelps and the Department of Microbiology, University of Minnesota, Minneapolis.

CRPV appears to be native to cottontails in certain geographical areas. It will produce papillomas in both cottontail and domestic rabbits (Shope, 1933), but infectious virus is regularly produced only in the papillomas of the cottontail. Rous and Beard (1935) found that visceral transplants of early or nonregressing cutaneous papillomas could become squamous carcinomas. When kept for more than 6 months, the papillomas in 25% of cottontail and 75% of domestic rabbits developed cancerous skin lesions (Syverton, 1952). In these carcinomas, the virus particles could not be recovered. Cell culture lines have been developed from papillomas of cottontail rabbits. In these cells, a cytoplasmic Shope papillomavirus antigen could be detected by immunofluorescence (Osato and Ito, 1968) but CRVP virions were not seen and no papillomas or tumors were produced upon inoculation in any rabbits (Ishimoto *et al.*, 1970). Viral DNA is associated in carcinomas (Stevens and Wettstein, 1979), can persist in a free episomal form (Wettstein and Stevens, 1982), and can still be detected in the transplantable cell lines after 30 years (McVay *et al.*, 1982; Favre *et al.*, 1982).

The early experiments of Shope (1937) demonstrated that preparations either infectious or noninfectious on the skin would effectively immunize when administered intraperitoneally. From these results he stated "that this virus was effectively masked, perhaps by combination with a neutralizing substance or by alteration to a noninfective phase, was indicated by the failure of the noninfectious suspension to infect when applied to scarified rabbit skin." The noninfectious material produced virus-neutralizing antibodies. Again to quote, "it was, in effect, a biologically inactivated virus vaccine."

III. EQUINE PAPILLOMATOSIS

This condition usually affects young horses and ponies with small warts usually about the mouth, nose, and head (Fig. 3). The warts cause no discomfort and regress in a few weeks to 4 months after they are first noticed. They may occur as an epizootic in groups of horses or ponies as young susceptible animals become available for exposure. There may be antigenic differences with different equine papillomaviruses (EPV). It was noted that a horse developed a natural infection while papillomas from two previous experimental exposures were present and then the animal was resistant to a third experimental infection (Cook and Olson, 1951). It may be significant that the equine papillomas on the skin of the breast and neck illustrated by Cadeac (1901) were large and confluent. Such large papillomas are not seen today and might indicate a change in the oncogenicity of the virus d-ıring the past 80 years. Cadeac (1901) also mentioned the occurrence of papillomas at birth in both colts and calves. Such congenital cutaneous papillomas are seen today (Njoku and Burwash, 1972).

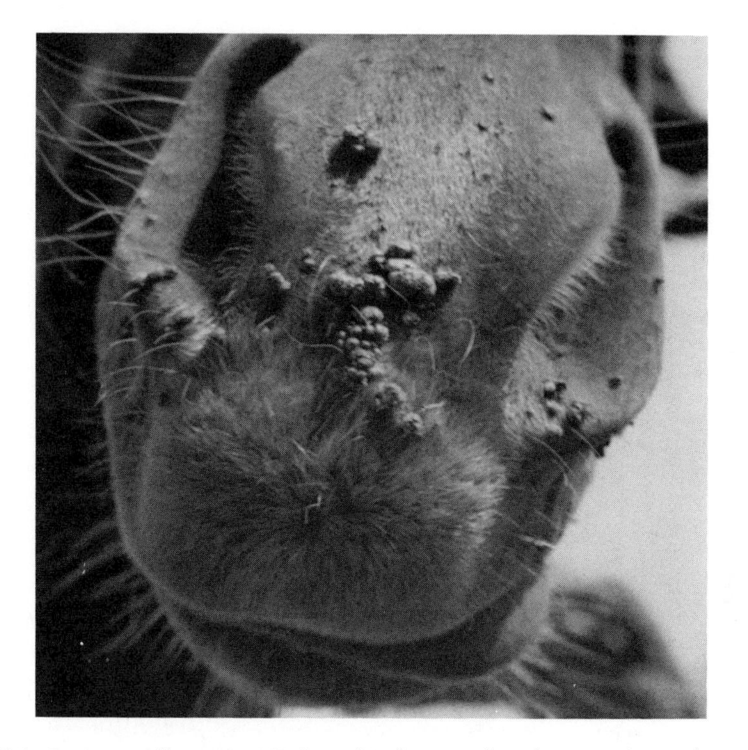

FIGURE 3. Equine papillomatosis 60 days after first noted on the nose of a yearling quarter horse. The warts regressed a month later.

Calves, lambs, dogs, rabbits, guinea pigs (Cook and Olson, 1951) as well as *Macaca mulatta, M. speciosus,* and *Saimiri sciurea* monkeys (Koller and Olson, 1972) were not susceptible to EPV.

IV. BOVINE PAPILLOMATOSIS

The contagiousness of bovine papillomatosis has been recognized by veterinarians for many years as Cadeac documented in 1901.

The first unequivocal transmission of bovine papillomavirus (BPV) was reported by Creech (1929) and the fibroplasia with both naturally occurring and experimentally produced cutaneous warts was noted. The cutaneous warts on cattle are more properly termed fibropapillomas (Fig. 4) because of the fibroplasia which develops before the epithelial portion of the fibropapilloma. The significance of the BPV-induced fibroplasia was overlooked for years. Experimental production of a fibroblastic tumor in the horse with BPV (Olson and Cook, 1951) aroused curiosity about the role of BPV in fibroplasia, which was found to be the initial reaction following intradermal infection of BPV in calves (Cheville and Olson, 1964b). Such BPV-induced fibromas have been produced in the skin, brain

FIGURE 4. Section of an experimental fibropapilloma 72 days after inoculation. Note the early epithelial hyperplasia over the fibroma replacing the normal corium of the skin.

(Fig. 5), urinary bladder (Fig. 6), and genital mucosa of cattle but fibroblasts in other locations seem to be less susceptible to stimulation (Gordon and Olson, 1968). BPV introduced into the dermis of calves remains infective for only 48 hr but the locally exposed fibroblasts have been transformed and respond with fibroplasia (Barthold and Olson, unpublished data). This was shown by first inoculating BPV with carbon marker into the corium, then removing the site and using it as an autograft to a new site. Sites grafted after 48 hr had growth confined to the graft only; those grafted earlier developed fibroma in the adjacent graft bed.

Opportunity to study the epizootiology of warts was presented in a group of 110 susceptible cattle confined in lots for 2½ years (Bagdonas and Olson, 1953). The 12 original cases occurred in 4 foci, then spread to affect three-fourths of the cattle. In the first episode, lasting 30 weeks, 57 animals were affected while in the lots. Two years later, in a second episode, 22 new cases developed while in pasture and 32 cattle previously affected became reinfected. Warts developed after 3½ to 4 months from direct contact with affected animals in 29 cases and in 14 other cases, infection spread through a woven wire fence. The more frequent sites of papillomas were the chin, neck, shoulder, and dewlap, areas subject to abrasions and wounds providing exposure to the infection.

An early report by Magalhaes (1920) described development of mul-

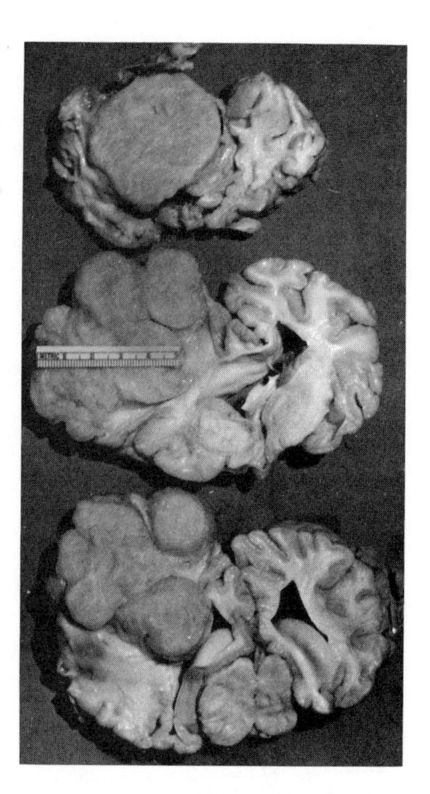

FIGURE 5. Cross section of a calf brain with fibroma compressing and replacing cerebrum 105 days after local injection of bovine papillomavirus.

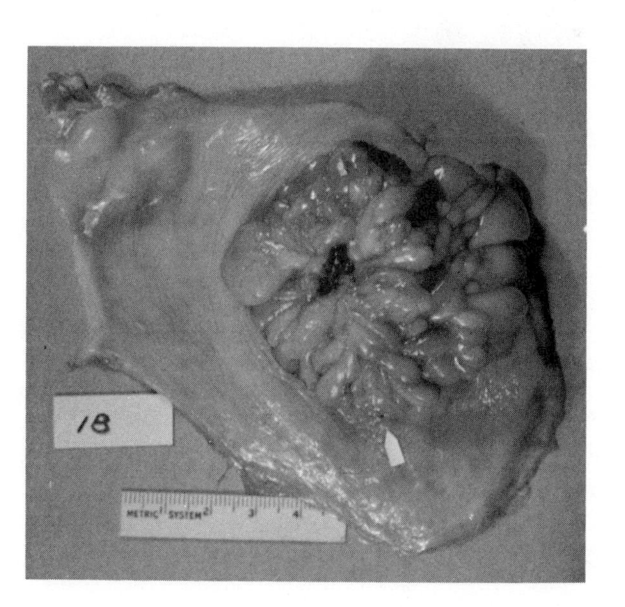

FIGURE 6. Mucosal aspect of bovine urinary bladder with polyploid tumor overlying fibroma 374 days after injection of bovine papillomavirus in the submucosa.

tiple papillomas in the skin of two calves 20 days after intravenous injection of bovine papilloma suspension. This suggested the existence of a BPV viremia. In attempts to repeat this experiment, 12 calves were each given intravenously up to 150 ml of bovine papilloma suspension and only 5 calves developed papillomas at skin wounds made up to 10 min afterwards even though similar wounds were made at eight successive intervals up to 16 weeks (Lee and Olson, 1968). The wounds were made to expose the epithelium to hemorrhage which should cause infection from viremia which apparently was extremely brief.

Following exposure of the skin to BPV (Fig. 7), or even formalin-inactivated BPV, cattle usually become resistant to reinfection. There can be papillomas secondary (Fig. 8) to the initial growth of fibropapillomas suggestive of a poor immune response of the host or exposure to an immunogenically different BPV. Humoral antibodies to BPV can be demonstrated following exposure of both cattle and horses by neutralization (Segre *et al.*, 1955) (Fig. 9) or precipitation (Lee and Olson, 1969). Hemagglutination inhibition (Favre *et al.*, 1974) and complement fixation (Lancaster *et al.*, 1977) tests showed antigenic differences in two classes of BPV.

The existence of two classes of BPV was clearly established by Lancaster and Olson (1978) by DNA–DNA reassociation kinetics as well as

FIGURE 7. Papillomas experimentally produced on a calf at ten sites where bovine papillomavirus was applied to scarified skin 5 months earlier.

FIGURE 8. Secondary spread of papillomas on lower neck and shoulder from localized sites of experimental papillomas in midneck region.

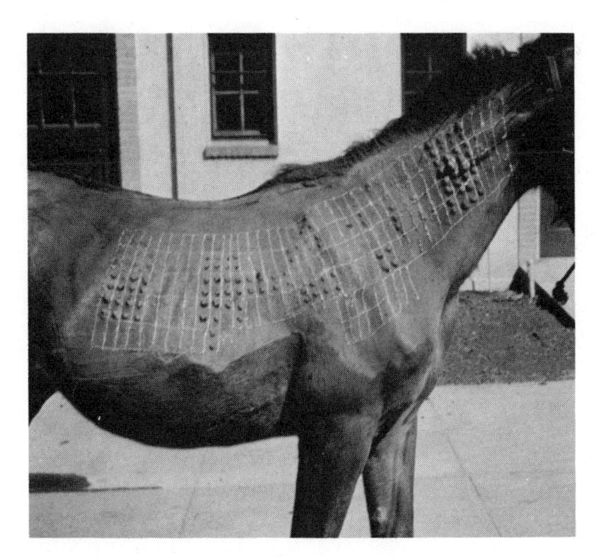

FIGURE 9. Skin tests on susceptible horse for neutralization of bovine papillomavirus mixed with serum from horses and cattle with experimental BPV infection. Each test site inoculated in duplicate. The fibromas, indicating no neutralization, were evident at 30 days when photo was taken.

polynucleotide sequences assayed by hybridization. Another BPV-like virus not yet well understood occurs in atypical bovine papillomatosis (Fig. 10) (Barthold *et al.*, 1974). This apparently infrequent papilloma does not have a fibromatous element (Fig. 11), tends to persist and recur on the same animal, is contagious but experimentally very poorly or nontransmissible, and the virions seem to be immunologically unrelated to the two classes of typical BPV. Atypical bovine papilloma suspension does not cause tumors in BPV-susceptible horses or hamsters (Koller and Olson, 1972). Pfister *et al.* (1979) suggested that this is similar to BPV-3 isolated by them from two cases of skin papilloma. BPV-4 has been found in epithelial tumors of the alimentary tract (Campo *et al.*, 1980) and BPV-5 from teat papillomas (Campo *et al.*, 1981).

The BPV DNA sequences of BPV-1 and 2 could be recognized by molecular hybridization in DNA of tumors in hamsters by the respective type 1 or 2 BPV used to produce the tumor (Lancaster *et al.*, 1979). The DNA sequences of BPV-1 and 2 were found (Lancaster *et al.*, 1977) in naturally occurring equine sarcoid tumors (Jackson, 1936; Olson, 1948) whose viral etiology had not been established even though morphologically similar tumors had been previously produced in the horse (Olson and Cook, 1951). Partial hybridization in other equine sarcoids suggested another type of papillomavirus (Lancaster *et al.*, 1979).

Meischke (1978) working in Jarrett's laboratory investigated in considerable detail the papillomas on the teats of cows. He found teat pap-

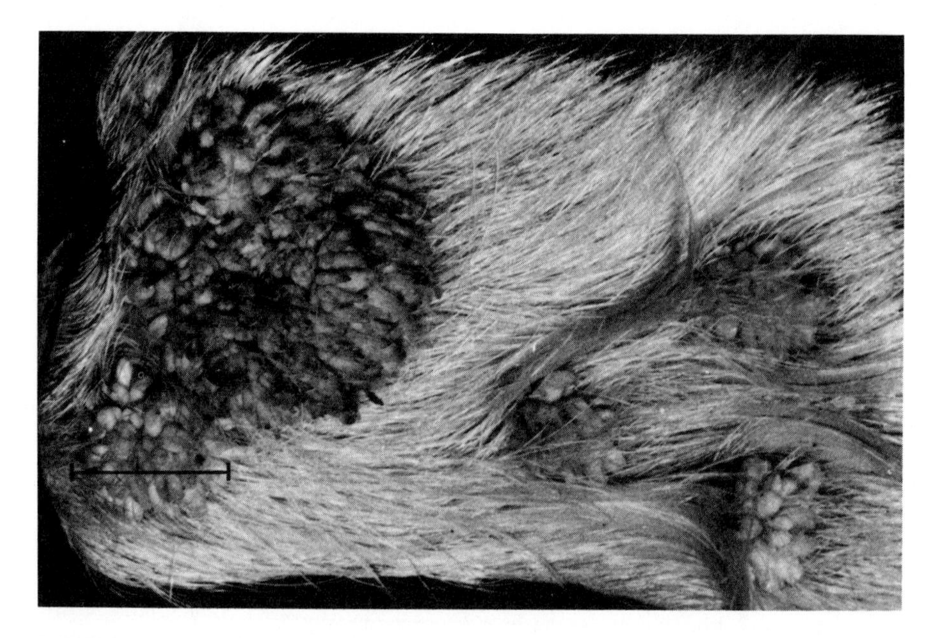

FIGURE 10. Atypical papillomas on skin of a cow. These do not have a fibromatous component, are contagious, but very poorly or nontransmissible.

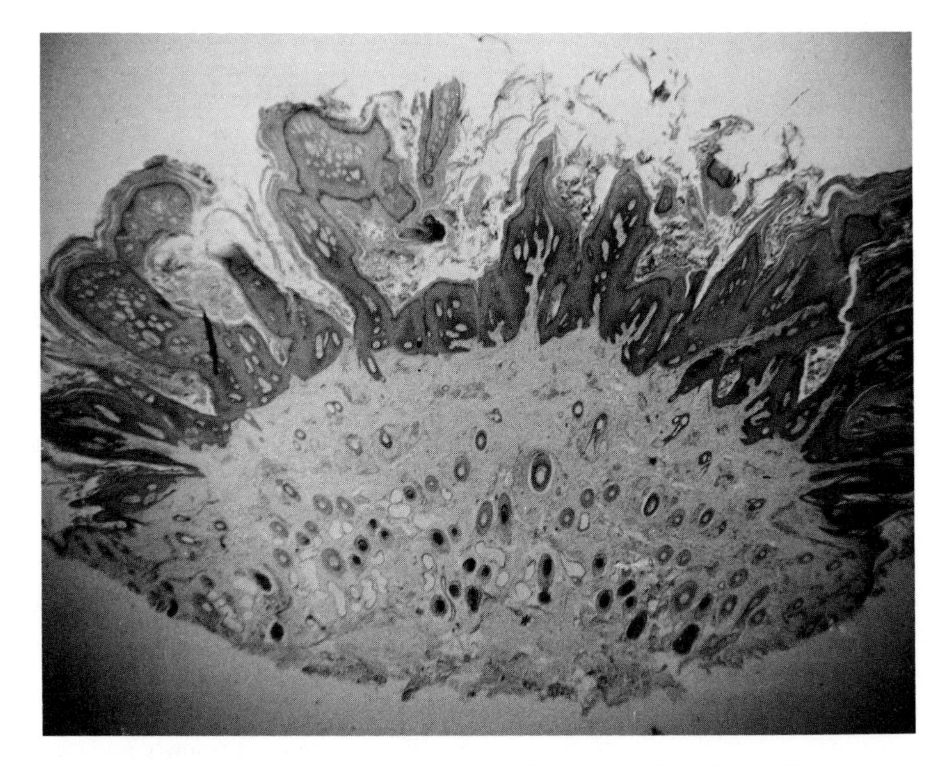

FIGURE 11. Histological reaction of atypical papilloma from skin of cows. Note normal dermis with hair follicles and lack of fibroma.

illomas in 260 of 721 (36%) cattle in a Glasgow abattoir. Papilloma virions could be demonstrated in frond, flat round, and rice grain types of papillomas. The rice grain type lacked the fibroma seen in the other lesions (Meischke, 1979c) and resembled atypical bovine papillomatosis (Barthold *et al.*, 1974).

Bovine fetal cell lines from the skin, palate, conjunctiva, and vascular meninges were transformed by BPV from skin warts (fibropapillomas) but not by the BPV of teat papillomas or rice grain lesions which he designated as "focal epithelial hyperplasia BPV" (Meischke, 1979c). The BPV from these produced similar lesions on the skin of calves which did not immunize against a subsequent challenge of fibropapilloma material (Meischke, 1979a,b). The transformed cells from tissue culture did not produce lesions in test calves but would produce tumors in athymic nude mice (Meischke, 1979a). Sera from calves with regressing fibropapillomas when used with complement would inhibit transformation of fetal cell lines. This could be absorbed from such sera with BPV-induced tumor cells or BPV-transformed fetal cell lines but not with BPV suspensions (Meischke, 1979a). Three of the ten calves used by Meischke had papillomas of the buccal mucosa, soft palate, tongue, and esophagus the etiology of which is somewhat obscure because of the variety of treatments (Meischke, 1978). The finding (which should be confirmed) of BPV-like virions in epithelial cells in milk suggests the shedding of BPV from teat lesions during milking and a potential source of infection for young calves (Meischke, 1979a).

A study of teat papillomas in a Wisconsin abattoir revealed a 25% frequency of similar papillomas (Olson *et al.*, 1982). Atypical filiform (Fig. 12) and flat as well as fibropapillomas had BPV capsid antigen. The atypical papillomas were experimentally reproduced on teats but haired skin was less susceptible. Molecular studies of the viruses are under way.

It is interesting to speculate that there are probably two immune mechanisms involved with bovine fibropapillomatosis. One is the humoral response to BPV which was initially a 19 S precipitin antibody followed by 7 S antibody (Lee and Olson, 1969). These precipitin antibodies appeared earlier than development of resistance to reinfection with BPV but were not correlated with growth or regression of the bovine fibropapilloma (Lee and Olson, 1969). Antibodies in the serum will neutralize the ability of BPV in bovine wart suspensions to produce fibroblastic tumors in the horse (Fig. 9) (Segre *et al.*, 1955) and to transform cells in tissue culture (Black *et al.*, 1963). The other immune mechanism involves the BPV-induced fibroma cells on which a membrane antigen can be demonstrated (Barthold and Olson, 1974a). Regression of such fibromas is apparently due to a cell-mediated immune response and can occur in the fibromas which develop before the epithelial proliferation of papillomas (Barthold and Olson, 1974b). The BPV-induced fibromas of cattle and horses have a common membrane neoantigen as demonstrated by immunofluorescence (Barthold and Olson, 1978).

A curious lesion called subcutaneous papillomatous cysts (Fig. 13) can be produced in a calf susceptible to BPV (Koller and Olson, 1971a).

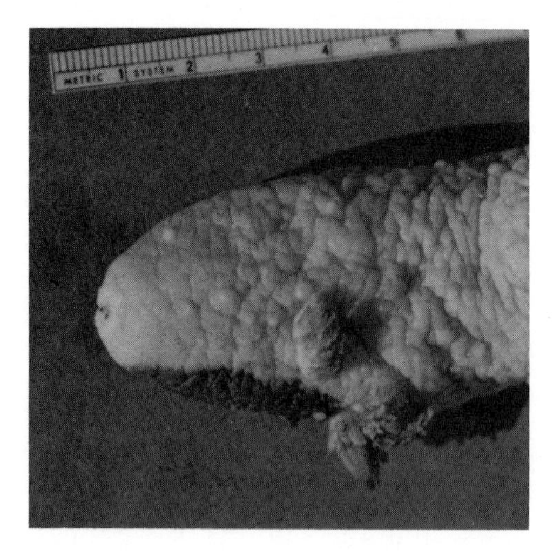

FIGURE 12. Atypical papillomas on the teat of a cow. These do not have a fibromatous component and are transmissible.

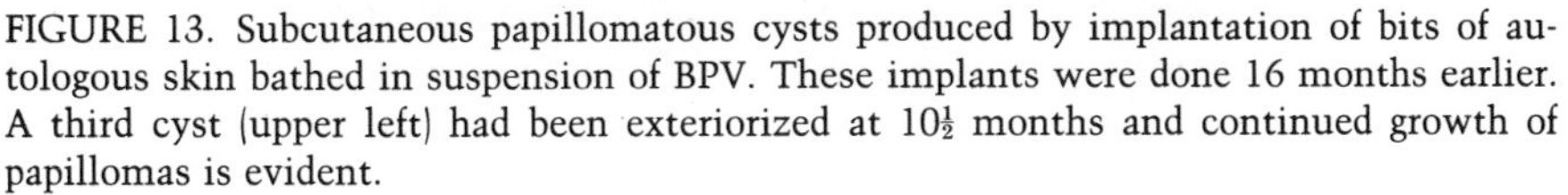

FIGURE 13. Subcutaneous papillomatous cysts produced by implantation of bits of autologous skin bathed in suspension of BPV. These implants were done 16 months earlier. A third cyst (upper left) had been exteriorized at $10\frac{1}{2}$ months and continued growth of papillomas is evident.

Bits of skin bathed in a suspension of BPV and then implanted subcutaneously as autologous implants developed into multiple cysts lined with papillomas. There was remarkable growth of the original small amount of epithelium. At about 4 months there was fibroblastic proliferation around the small cyst. As the cyst became larger with time, the fibroblastic element was reduced perhaps because of a cell-mediated immune reaction. The papillomatous lining persisted for several months and by about 20 months had disappeared leaving a cyst filled with thick caseous material representing desquamated keratinized epithelium. The cysts contained virions of BPV and also their antibodies.

Formalin-treated suspensions of bovine wart tissue have been commercially available as a vaccine for many years. It has been recommended and used in treatment of existing bovine fibropapillomatosis with varied claims of success. A controlled trial failed to give evidence of a therapeutic effect (Olson and Skidmore, 1959) but such vaccine is effective in prevention of warts when used prophylactically prior to exposure to BPV (Olson *et al.*, 1968).

In addition to the horse, hamsters (Friedman *et al.*, 1963), C3H/eB mice (Boiron *et al.*, 1964; Friedman *et al.*, 1965), and the pika (*Ochotona rufescens rufescens*) (Puget *et al.*, 1975) are susceptible to BPV and respond by developing fibroblastic tumors. Such tumors are malignant in the hamster and may metastasize (Fig. 14) from the subcutis to the lungs,

FIGURE 14. Hamster with large fibroma on left shoulder area, site of inoculation with BPV isolate 247, 315 days earlier. Smaller tumors are evident on tail and right rear leg. The latter are regarded as metastases since they developed later and were not likely to have been implanted from bits of original tumor.

tail, and extremities of the legs (Robl and Olson, 1968). The lag between inoculation of any age of hamster and manifestation of tumor may be 6 months to 1 year. Growth is progressive indicating either a lack of neoantigen on the cell membrane to which the host can react or failure of the immune mechanism to control the new growth. Chondroblasts are stimulated to grow when BPV is placed near the ear cartilage in hamsters (Robl and Olson, 1968).

Bovine fetal cell lines have been transformed by several laboratories (Black *et al.*, 1963; Thomas *et al.*, 1963; Meischke, 1979a,b). Hamster embryo cells are altered by BPV (Black *et al.*, 1963; Morgan and Meinke, 1980) and produce tumors in hamsters (Geraldes, 1969). The transformed bovine fetal cells were not oncogenic in calves (Black *et al.*, 1963; Meischke, 1979b); however, hamster fibroma cell lines were oncogenic in hamsters (Breitburd *et al.*, 1981).

Mouse cell lines NIH 3T3 and C127 can be transformed by BPV (Dvoretzky *et al.*, 1980) and by a fragment of its DNA (Lowry *et al.*, 1980). The virus is transcribed in such cells (Heilman *et al.*, 1982).

Enzootic bovine hematuria (Kalkus, 1913; Fujimoto *et al.*, 1951; Pamukcu, 1955) is an entity of naturally occurring tumors of the bovine urinary bladder associated with ingestion (Fig. 15) of bracken fern (*Pteridium aquilina*). In some geographical areas (Fig. 16), BPV has been found in such tumors (Olson *et al.*, 1965). BPV has readily produced fibromatous tumors with cystitis glandularis (Fig. 17) in the urinary bladder of exper-

FIGURE 15. Cow eating bracken fern in an area of Brazil where enzootic hematuria occurs. Note the sparse growth of pasture grass in background which forces the cow to eat bracken.

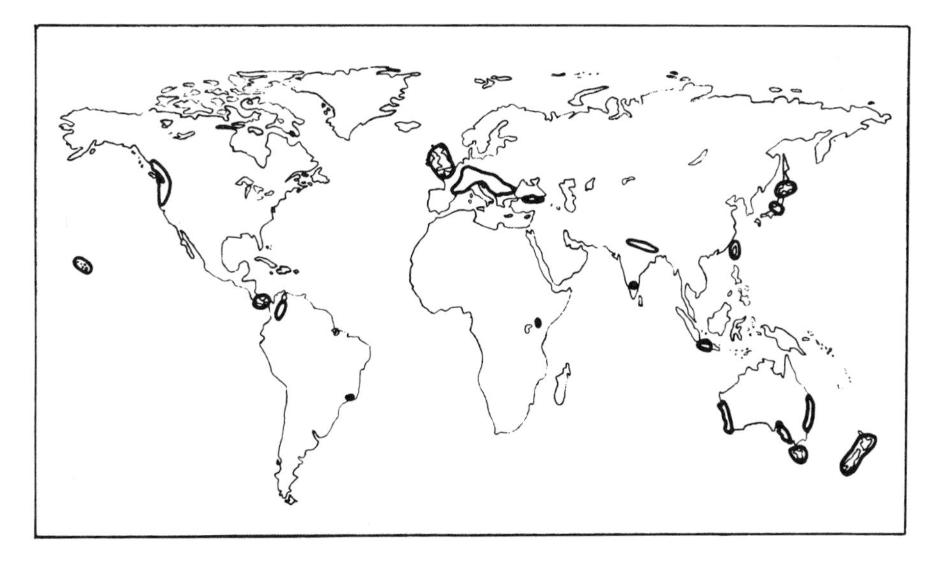

FIGURE 16. Enzootic hematuria of cattle has been reported from the areas encircled. Although bracken fern exists almost worldwide, fern with carcinogen seems to be restricted to certain soil and growth conditions.

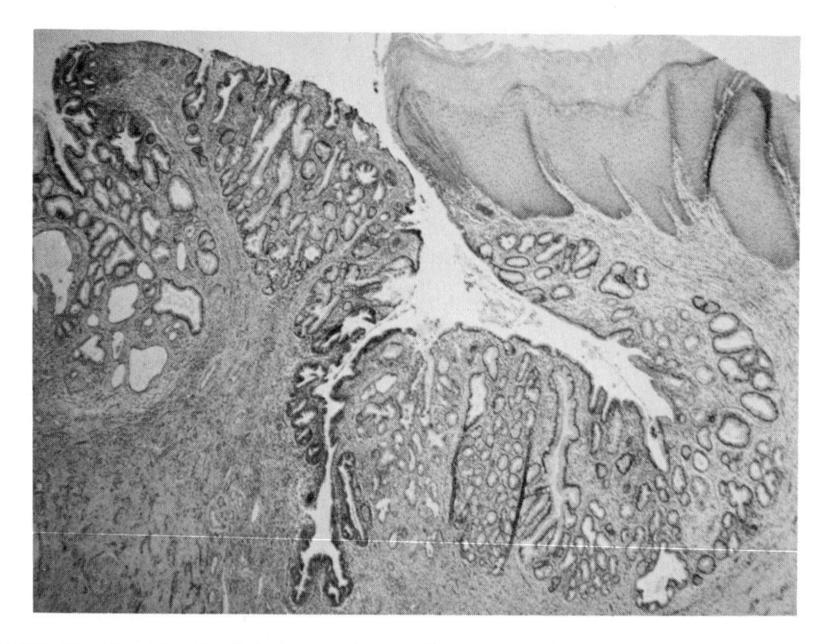

FIGURE 17. Cystitis glandularis together with an area of squamous metaplasia of the urinary bladder epithelium. BPV was isolated from this case of enzootic hematuria.

imental calves (Fig. 18) (Brobst and Olson, 1965). Yoshikawa and Oyamada (1971) studied a series of 116 cases of "bovine vesical fibromatosis" with a frequency of 16.07% in cattle coming to the abattoir in Towada, Japan. Enzootic bovine hematuria with epithelial lesions as well as fibromatosis is also common in this part of Japan. Epidermoid carcinoma of the pharynx, esophagus, and rumen was found in a series of 44 cases of enzootic hematuria by Tokarnia *et al.* (1969) in Brazil. Jarrett *et al.* (1978a,b) have demonstrated papilloma virions in these lesions which were found in 19% of 7746 cattle examined in a Glasgow abattoir. The squamous carcinomas of the alimentary tract occurred almost exclusively in cattle from the upland where bracken fern grows. Affected cattle had been there for several years. Study of the relation between BPV and a plant carcinogen should yield basic information on carcinogenesis. The carcinogen in bracken fern is still unknown but apparently develops only with certain soil conditions. For example, the bracken fern growing naturally in Wisconsin does not seem to cause bladder lesions either in cattle on farms or when experimentally fed for more than 1 year (Dobereiner *et al.*, 1966). Enzootic bovine hematuria is reported from rather restricted areas (Fig. 16) of the world (Dobereiner *et al.*, 1966) and has ceased to occur when improved pastures provide sufficient forage and cattle no longer eat bracken fern.

BPV can be titrated on the skin of calves (Fig. 19) as well as by its transforming actions on certain mouse cell lines.

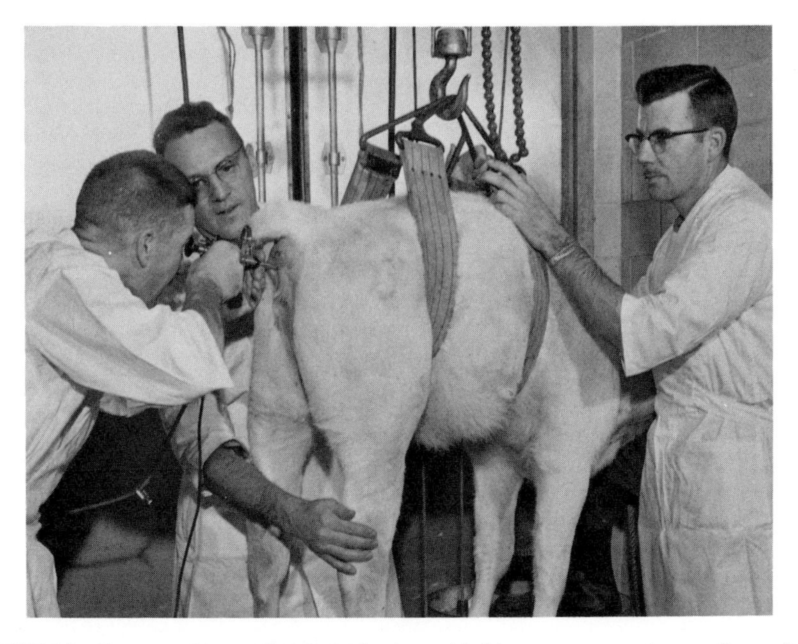

FIGURE 18. Cystoscopic examination of urinary bladder tumor experimentally produced with BPV. Canvas bands immobilize the calf and the rear band also displaces abdominal viscera forward.

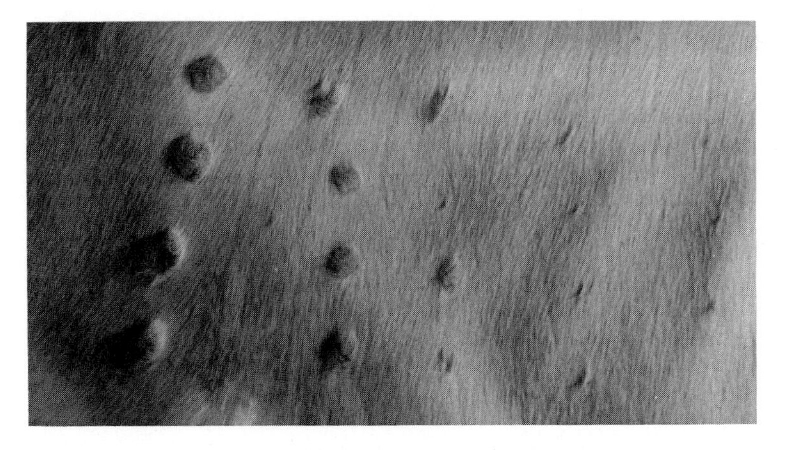

FIGURE 19. Titration of bovine papilloma isolate 207 in 15% suspension on skin of a calf 80 days after inoculation. Dilutions from left to right of: 1:100; 1:1000; 1:10,000; 1:50,000; and 1:100,000 were inoculated in four vertical sites. Very slight growth also occurred at one site each of 1:500,000 and 1:1,000,000 dilutions.

V. PAPILLOMATOSIS IN SHEEP

Warts either occur rarely in sheep or are overlooked because of the wool covering. Gibbs *et al.* (1975) described four cases with a transmissible papillomavirus. The papillomas were found on the legs and muzzle. In one case, the wart was a fibropapilloma, and in three other cases, as well as the experimentally produced warts, the dermal fibroplasia was not as marked. Warts appeared 45 to 50 days after inoculation, became keratinized, and then regressed by day 90. Hamsters developed fibromas with the sheep papilloma material but cattle and goats were refractory.

Recently, Vanselow and Spradbrow (1982) found papilloma virions by electron microscopy in squamous carcinomas and papillomas of Merino sheep in Australia. Such squamous carcinomas in nonpigmented areas poorly covered by wool (ear, muzzle, eye, and vulva) had been recognized as an economic problem for many years. It is believed that papillomavirus and sunlight are involved in a complex etiology of these tumors. Virions have not been demonstrated in papillomas of the udder and vulva (Spradbrow, 1982).

VI. ORAL PAPILLOMATOSIS OF RABBITS

These small papillomas are usually found under the tongue and perhaps often overlooked for they are seldom mentioned or studied. In the original report of Parsons and Kidd (1936), 17% of 385 domestic rabbits from various sources were affected. The virus would produce papillomas in the mouth of domestic and wild rabbits in 2 to 4 weeks which lasted for several months. The virus did not produce papillomas on the lips or abdominal skin. Rabbits were immune to the oral virus after regression but not to Shope papillomavirus. Since the virus could be washed from the mouth of affected animals and even prior to development of clinical papillomas, infection was probably spread by oral secretions.

VII. DEER FIBROMATOSIS

The deer fibroma virus discovered by Shope (1955) grows in the rather thin layer of epithelial cells (Tajima *et al.*, 1968) covering the more massive growth of fibroma (Fig. 20). The description of natural and experimental fibroma in white-tailed deer (*Odocoileus virginianus*) by Shope *et al.* (1958) is a classic report covering 6 years of study. Fibromas could be readily produced by inoculation on scarified skin. They were first evident in 7 to 8 weeks as very small elevations and in another 7 weeks became 2 to 3 mm in diameter. Later the fibromas regressed in most of

FIGURE 20. Extensive growth of fibromas on the face of a deer. The heavy growth on upper eyelids blinded the deer and it was found near a fence with a broken neck. Such extensive growth is probably associated with an impaired immune system.

the deer. In one deer, the fibromas continued to grow for 10 months. It was speculated that while most deer are susceptible, the fibromas soon regress except for the rare individual in which they become extensive via uncontrolled growth. In such individuals, pulmonary metastases have occurred (Fay, 1970; Koller and Olson, 1971b). The deer fibroma virus is not infectious for cattle (Shope *et al.*, 1958), sheep, rabbits, guinea pigs, the horse, or three species of monkeys (Koller and Olson, 1972) but will produce fibromas subcutaneously but not in the brain of hamsters (Robl, Gordon, and Olson, unpublished data). Cases of deer fibromatosis have been reported from Vermont to California and Alaska.

The fibromatosis which occurs in both white-tailed and mule deer is caused by papillomaviruses which are slightly different (Groff and Lancaster, 1982). The virions occur in the moderately hyperplastic epithelium covering the fibromas and the viral genome can be demonstrated in fibroma. In nature and experimentally, the lesions are usually small and regress. Deer are not susceptible to BPV. Warts, papillomas, and fibropapillomas have been reported in a variety of other deer, caribou, and moose (Sundberg and Nielsen, 1981).

VIII. EUROPEAN ELK PAPILLOMAVIRUSES

Cutaneous fibropapillomas have been found to be rather common in the European elk of Sweden. The European elk papillomavirus produced fibrosarcomas in hamsters and is unrelated to BPV based on molecular hybridization studies (Moreno-Lopez *et al.*, 1981).

IX. VIRAL PAPILLOMATOSIS IN OTHER ANIMALS

Viral cutaneous papillomas developed in a 1-year-old opossum born and raised in captivity (Koller, 1972). The papillomas continued to grow for 6 months and were gone at 1 year. Characteristic papilloma virions were found. Attempts were made to transmit the opossum wart and human wart material to other adult opossums but no papillomas developed.

Parish (1961) studied a transmissible genital papilloma of boar pigs resembling condyloma acuminatum of man. Cell-free filtrates produced papillomas of the prepuce and vulva by scarification or intradermal injection with an incubation period of 4 to 10 weeks. Following regression there was immunity to reinoculation but serum antibodies to the agent were not manifested until after two more injections of infective material (Parish, 1962). Lesions could not be produced in man (the author), pig, calf, rabbit, guinea pig, or mouse skin (Parish, 1961).

Several years ago, cutaneous papillomas were reported from 50 of 150 lactating female goats (Moulton, 1954). Growths were on only the udders and teats though previously they had occurred on the head, neck, shoulders, and foreleg of a goat in New Mexico. Of five biopsies, three were papillomas and two were regarded as squamous cell carcinoma with metastasis to a lymph node in one. Transmission trials were not done. A current study indicates that multiple cutaneous papillomas of the head and neck are squamous in character and mammary and genital warts are fibropapillomas (Theilen *et al.*, 1982). Thus far, a suspected virus has not been demonstrated.

A spontaneous papilloma in a brown *Cebus* monkey was reported to be transmissible to the same and other species of Old World monkeys (Lucke *et al.*, 1950). Hyperemic patches were noted 2 weeks after inoculation on scarified skin which later became confluent projections up to 15 mm high. Most growths regressed in 4 to 6 months. The paucity of reports on transmissible papillomas in primates seems unusual with the large numbers carefully supervised in various laboratories. Virus-associated papillomas have been reported in colobus monkeys (Rangan *et al.*, 1980). Three species of monkeys inoculated with human, bovine, equine papilloma and deer fibroma material did not develop papillomas (Koller and Olson, 1972).

An infectious squamous cell tumor in a colony of *Mastomys natalensis* (GRA Giessen are inbred rodents) was transmissible on scarified skin to about 50% of yellow mutant GRA Giessen animals 48 to 53 days old (Rudolph and Muller, 1976). In 6 to 8 weeks, progressive growth was evident and in two animals extended into the musculature. The histology of the natural and experimental tumors was that of a keratoacanthoma in which a papillomavirus was demonstrated.

An avian papillomalike virus has been reported in the chaffinch (*Frin-*

gilla coelebs) and brambling (*F. montifringilla*) with lesions on the legs of these birds (Osterhaus *et al.*, 1977).

Orth *et al.* (1978) demonstrated antigens in common on the virions of the Shope papilloma- and human papillomaviruses. This has been extended to show the existence of a common capsid antigen in human virus from plantar warts, hand warts, flat warts, and lesions of epidermodysplasia verruciformis, the two classes of BPV, and COP virus (Jenson *et al.*, 1980).

Sundberg *et al.* (1984) collected neoplastic lesions from a variety of animal species and found a group-specific common papillomavirus capsid antigen in many. These were examined with rabbit antisera against BPV-1 in a peroxidase–antiperoxidase (PAP) technique developed by Jenson *et al.* (1980). In the horse, 11 of 35 papillomas of the skin, penis, and vulva were positive, while 31 squamous cell carcinomas and 14 equine sarcoids were negative. In dogs, 21 of 55 cutaneous and oral papillomas, 1 of 6 on the penis, and 5 of 9 cutaneous squamous cell carcinomas were positive. In cattle, 13 of 49 cutaneous and 1 of 25 penile fibropapillomas were positive. In the domestic rabbit, 6 of 6 oral papillomas were positive. Cutaneous papillomas in white-tailed, black-tailed, and mule deer were positive for the capsid antigen which correlated with demonstration of virions by electron microscopy. A fibropapilloma from a pronghorn antelope also stained for the common capsid antigen (Sundberg *et al.*, 1983).

Virions resembling papillomavirus have been found by electron microscopy in bovine conjunctival plaques, and papillomas, as well as a papilloma of the eyelid (Ford *et al.*, 1982). They suggest such virions may be involved in the etiology of bovine ocular squamous carcinoma.

Mention should be made that Holinger *et al.* (1962) inoculated 68 animals (dogs, horses, monkeys, rabbits, mice, and guinea pigs) with human laryngeal papilloma material and all remained negative.

Atanasiu (1948) attempted transmission of a common verruca from a human finger to the eye, prepuce, and testicular skin of laboratory rabbits, mice, and skin of the thigh of an elderly porter with no success after 5 months. However, one chimpanzee (*Papio papio*) receiving the same material on the prepuce developed soft verrucae of the penile mucosa in 2 months which 3 months later became grossly and histologically similar to soft genital verrucae. A second chimpanzee remained negative.

The question of infection of man with animal papillomavirus has received some attention. A prevalence of warts in butchers and veterinarians (Bosse and Christofers, 1964; DePeuter *et al.*, 1977) suggested a relation to bovine papillomatosis. Pfister *et al.* (1979) found no antibodies to BPV-1 or 2 in man. Orth *et al.* (1981) and Ostrow *et al.* (1981) found no BPV-1 or 2 in their search for viruses in warts of butchers in Europe and the United States. However, a human virus type was found only in warts of meat handlers and possibly represents an unknown animal type. Biochemical studies have yet to be done with such human papillomaviruses and the viruses of various animal species. Although there may be only a slight possibility of infectivity between species, the genealogical origins of various papillomaviruses might become evident.

A comprehensive review by Pfister (1983) presents the biochemical features of the animal as well as human papillomaviruses. From his review, the paramount role of molecular biology is evident in leading to understanding of the pathogenesis of papillomas and associated carcinomas. Attempts at experimental transmission in animals together with the techniques of molecular biochemistry will provide further model systems for study of neoplastic disease.

REFERENCES

Atanasiu, P., 1948, Transmission de la verrue commune au single cynocephale (*Papio papio*), *Ann. Inst. Pasteur (Paris)* **74**:246–248.

Bagdonas, V., and Olson, C., 1953, Observations on the epizootiology of cutaneous papillomatosis (warts) of cattle, *J. Am. Vet. Med. Assoc.* **122**:393–397.

Barthold, S. W., and Olson, C., 1974a, Membrane antigen of bovine papilloma virus-induced fibroma cells, *J. Natl. Cancer Inst.* **52**:737–741.

Barthold, S. W., and Olson, C., 1974b, Fibroma regression in relation to antibody and challenge immunity to bovine papilloma virus, *Cancer Res.* **34**:2436–2431.

Barthold, S. W., and Olson, C., 1978, Common membrane neoantigens on bovine papilloma virus-induced fibroma cells from cattle and horses, *Am. J. Vet. Res.* **39**:1643–1645.

Barthold, S. W., Koller, L. D., Olson, C., Studer, E., and Holtan, A., 1974, Atypical warts in cattle, *J. Am. Vet. Med. Assoc.* **165**:276–280.

Black, P. H., Hartley, J. W., Rowe, W. P., and Huebner, R. J., 1963, Transformation of bovine tissue culture cells by bovine papilloma virus, *Nature* **199**:1016–1018.

Boiron, M., Levy, J. P., Thomas, M., Friedman, J. C., and Bernard, J. C., 1964, Some properties of bovine papilloma virus, *Nature* **201**:423.

Bosse, K., and Christofers, E., 1964, Beitrag zur Epidemiologie der Warzen, *Hautarzt* **15**:80–86.

Breitburd, F., Favre, M., Zoorob, R., Fortin, D., and Orth, G., 1981, Detection and characterization of viral genomes and search for tumoral antigens in two hamster cell lines derived from tumors induced by bovine papillomavirus type 1, *Int. J. Cancer* **27**:693–702.

Brobst, D. F., and Olson, C., 1965, Histopathology of urinary bladder tumors induced by bovine cutaneous papilloma agent, *Cancer Res.* **25**:12–19.

Cadeac, M., 1901, Sur la transmission experimentale des papillomes des diverses especes, *Bull. Soc. Sci. Vet. Med. Comp. Lyon* **4**:280–286.

Campo, M. S., Moar, M. H., Jarrett, W. F. H., and Laird, H. M., 1980, A new papillomavirus associated with alimentary tract cancer in cattle, *Nature* **286**:180–182.

Campo, M. S., Moar, M. H., Laird, H. M., and Jarrett, W. F. H., 1981, Molecular heterogeneity and lesion site specificity of cutaneous bovine papillomaviruses, *Virology* **113**:323–335.

Chambers, V. C., and Evans, C. A., 1959, Canine oral papillomatosis. I. Virus assay and observations on the various stages of the experimental infection, *Cancer Res.* **19**:1188–1195.

Cheville, N. F., and Olson, C., 1964a, Cytology of the canine oral papilloma, *Am. J. Pathol.* **45**:849–872.

Cheville, N. F., and Olson, C., 1964b, Epithelial and fibroblastic proliferation in bovine cutaneous papillomatosis, *Pathol. Vet.* **1**:248–257.

Cook, R. H., and Olson, C., 1951, Experimental transmission of cutaneous papilloma of the horse, *Am. J. Pathol.* **27**:1087–1097.

Creech, C. T., 1929, Experimental studies of the etiology of common warts in cattle, *J. Agric. Res. (Washington, D.C.)* **39**:723–737.

Davis, P. E., and Sabine, M., 1975–76, Virus papilloma in the racing greyhound, in: *The Racing Greyhound* (O. G. Jones, ed.), Vol. I, pp. 32–35, World Racing Federation, London.

DeMonbreun, W. A., and Goodpasture, E. W., 1932, Infectious oral papillomatosis of dogs, *Am. J. Pathol.* **8**:43–56.

DePeuter, M., DeClercq, B., and Minette, A., 1977, An epidemiological survey of virus warts of the hands among butchers, *Br. J. Dermatol.* **96**:427–431.

Dobereiner, J., Olson, C., Brown, R. R., Price, J. M., and Yess, N., 1966, Metabolites in urine of cattle with experimental bladder lesions and fed bracken fern, *Pesqui. Agropecu. Bras.* **1**:189–199.

Dvoretzky, I. G., Shober, R., Chattopadhyay, S. K., and Lowy, D. R., 1980, A quantitative *in vitro* focus forming assay for bovine papilloma virus, *Virology* **103**:369–375.

Erk, N., 1976, A study of Kitab al-Hail wal-Baitara by Muhammed Ibu ahi Hazam, *Hist. Med. Vet.* **1**:101–104.

Favre, M., Breitburd, F., Croissant, O., and Orth, G., 1974, Hemagglutinating activity of bovine papilloma virus, *Virology* **60**:572–578.

Favre, M., Jibard, N., and Orth, G., 1982, Restriction mapping and physical characterization of the cottontail rabbit papillomavirus genome in transplantable VX2 and VX7 domestic rabbit carcinomas, *Virology* **119**:298–309.

Fay, L. D., 1970, Skin tumors of Cervidae, in: *Infectious Diseases of Wild Animals* (J. W. Davis, L. H. Karstad, and D. O. Trainer, eds.), pp. 385–392, Iowa State University Press, Ames.

Ford, J. N., Jennings, P. A., Spradbrow, P. B., and Francis, J., 1982, Evidence of papillomaviruses in ocular lesions in cattle, *Res. Vet. Sci.* **32**:257–259.

Friedman, J. C., Levy, J. P., Lasneret, J., Thomas, M., Boiron, M., and Bernard, J., 1963, Induction de fibromes sous-cutanés chez le hamster doré par inoculation d'extraits acellulaires de papillomes bovins, *C.R. Seances Acad. Sci. Ser. III* **257**:2328–2331.

Friedman, J. C., Lasneret, J., Gibeaux, L., and Boiron, M., 1965, Developpement de fibromes proliferatifs chez la souris à l'aide d'extraits acellulaires de papillomes bovins et leur transformation maligne greffes isologues, *Rev. Med. Vet.* **141**:115–122.

Fujimoto, Y., Sugano, S., and Kobayashi, T., 1951, Histopathological studies on bovine hematuria in Nemura, Hokkaido, I, *Mem. Fac. Agric. Hokkaido Univ.* **1**:159–166.

Geraldes, A., 1969, Malignant transformations of hamster cells by cell-free extracts of bovine papillomas (*in vitro*), *Nature* **222**:1283–1284.

Gibbs, E. P. J., Smale, C. J., and Lauman, M. J. P., 1975, Warts in sheep, *J. Comp. Pathol.* **85**:327–334.

Gordon, D. E., and Olson, C., 1968, Meningiomas and fibroblastic neoplasia in calves induced with the bovine papilloma virus, *Cancer Res.* **28**:2423–2431.

Greig, A. S., and Charlton, K. M., 1973, Electron microscopy of the virus of oral papillomatosis in the coyote, *J. Wildl. Dis.* **9**:359–361.

Groff, D. E., and Lancaster, W. D., 1982, Two subtypes of deer fibroma virus show specific regions of rapid nucleotide divergence, Paper presented at the Cold Spring Harbor Conference on Papilloma Viruses, Sept. 14–18, p. 72.

Gross, L., 1970, *Oncogenic Viruses,* 2nd ed., pp. 49–81, Pergamon Press, Elmsford, N.Y.

Hare, C. L., and Howard, E. B., 1977, Canine conjunctivocorneal papillomatosis: A case report, *J. Am. Anim. Hosp. Assoc.* **13**:688–690.

Heilman, C. A., Engel, L., Lowy, D. R., and Howley, P. M., 1982, Virus-specific transcription in bovine papillomavirus transformed mouse cells, *Virology* **119**:22–34.

Holinger, P. H., Johnston, K. E., Conner, G. H., Conner, B. R., and Holper, J., 1962, Studies of papilloma of the larynx, *Ann. Otol. Rhinol. Laryngol.* **71**:443–447.

Ishimoto, A., Ooto, S., Kimura, I., Miyake, T., and Ito, Y., 1970, In vitro cultivation and antigenicity of cottontail rabbit papilloma cells induced by the Shope papilloma virus, *Cancer Res.* **30**:2598–2605.

Jackson, C., 1936, The incidence and pathology of tumours of domesticated animals in South Africa, *Onderstepoort J. Vet. Sci. Anim. Ind.* **6:**278–385.

Jarrett, W. F. H., McNeil, P. E., Grimshaw, W. T. R., Selman, I. E., and McIntyre, W. I. M., 1978a, High incidence area of cattle cancer with a possible interaction between an environmental carcinogen and a papilloma virus, *Nature* **274:**215–217.

Jarrett, W. F. H., Murphy, J., O'Neil, B. W., and Laird, H. M., 1978b, Virus-induced papillomas of the alimentary tract of cattle, *Int. J. Cancer* **22:**323–328.

Jenson, A. B., Rosenthal, J. R., Olson, C., Pass, F., Lancaster, W. D., and Shah, K., 1980, Immunological relatedness of papillomaviruses from different species, *J. Natl. Cancer Inst.* **64:**495–500.

Kalkus, J. W., 1913, A preliminary report of bovine red warter (cystic hematuria), *Wash. State Univ. Agric. Exp. Stn. Bull.* **112:**1–27.

Koller, L. D., 1972, Brief communication: Cutaneous papillomatosis in an opossum, *J. Natl. Cancer Inst.* **49:**309–313.

Koller, L. D., and Olson, C., 1971a, Subcutaneous papillomatous cysts produced by papilloma virus, *J. Natl. Cancer Inst.* **47:**891–898.

Koller, L. D., and Olson, C., 1971b, Pulmonary fibroblastomas in deer with cutaneous fibromatosis, *Cancer Res.* **31:**1373–1375.

Koller, L. D., and Olson, C., 1972, Attempted transmission of warts from man, cattle, and horses, and of deer fibroma, to selected hosts, *J. Invest. Dermatol.* **58:**366–368.

Konishi, S., Tokita, H., and Ogata, H., 1972, Studies on canine oral papillomatosis. I. Transmission and characterization of the virus, *Jpn. J. Vet. Sci.* **34:**263–268.

Lancaster, W. D., and Olson, C., 1978, Demonstration of two distinct classes of bovine papilloma virus, *Virology* **89:**371–379.

Lancaster, W. D., Olson, C., and Meinke, W., 1977, Bovine papilloma virus: Presence of a virus-specific DNA sequences in naturally occurring equine tumors, *Proc. Natl. Acad. Sci. USA* **74:**524–528.

Lancaster, W. D., Theilen, G. H., and Olson, C., 1979, Hybridization of bovine papilloma virus type 1 and type 2 DNA to DNA from virus-induced hamster tumors and naturally occurring equine connective tissue tumors, *Intervirology* **11:**227–233.

Lee, K. P., and Olson, C., 1968, Response of calves to intravenous and repeated intradermal inoculation of bovine papilloma virus, *Am. J. Vet. Res.* **16:**517–520.

Lee, K. P., and Olson, C., 1969, Precipitin response of cattle to bovine papilloma virus, *Cancer Res.* **29:**1393–1397.

Lowy, D. R., Dvoretzky, I., Shober, R., Law, M. F., Engel, L., and Howley, P. M., 1980, *In vitro* tumourigenic transformation by a defined subgenomic fragment of bovine papilloma virus DNA, *Nature* **287:**72–74.

Lucke, B., Ratcliffe, H., and Breedis, C., 1950, Transmissible papilloma in monkeys, *Fed. Proc.* **9:**337.

McFadyean, J., and Hobday, F., 1898, Note on the experimental transmission of warts in the dog, *J. Comp. Pathol. Ther.* **11:**341–344.

McVay, P., Fretz, M., Wettstein, F., Stevens, J., and Ito, Y., 1982, Integrated Shope virus DNA is present and transcribed in the transplantable rabbit tumour VX-7, *J. Gen. Virol.* **60:**271–278.

Magalhaes, O., 1920, Verruga dos bovideos, *Bras. Med.* **34:**430–431.

Meischke, H. R. C., 1978, *In vivo* and *in vitro* studies on bovine papilloma virus, thesis, Department of Veterinary Pathology, University of Glasgow (printed by Cloeck ein Moedigh, B. V. Amsterdam).

Meischke, H. R. C., 1979a, *In vitro* transformation by bovine papilloma virus, *J. Gen. Virol.* **43:**471–487.

Meischke, H. R. C., 1979b, Experimental transmission of bovine papilloma virus (BPV) extracted from morphologically distinct teat and cutaneous lesions and the effects of inoculation of BPV transformed fetal bovine cells, *Vet. Rec.* **104:** 360–366.

Meischke, H. R. C., 1979c, A survey of bovine teat papillomatosis, *Vet. Rec.* **104:**28–31.

Moreno-Lopez, J., Pettersson, U., Dinter, Z., and Philipson, L., 1981, Characterization of a papilloma virus from the European elk (EEPV), *Virology* **112:**589–595.

Morgan, D. M., and Meinke, W., 1980, Isolation of clones of hamster embryo cells transformed by the bovine papilloma virus, *Curr. Microbiol.* :247–251.

Moulton, J. E., 1954, Cutaneous papillomas on the udders of milk goats, *North Am. Vet.* **35:**29–33.

Njoku, C. O., and Burwash, W. A., 1972, Congenital cutaneous papilloma in a foal, *Cornell Vet.* **62:**54–57.

Olson, C., 1948, Equine sarcoid, a cutaneous neoplasm, *Am J. Vet. Res.* **9:**333–341.

Olson, C., and Cook, R. H., 1951, Cutaneous sarcoma-like lesions of the horse caused by the agent of bovine papilloma, *Proc. Soc. Exp. Biol. Med.* **77:**281–284.

Olson, C., and Skidmore, L. V., 1959, Therapy of experimentally produced bovine cutaneous papillomatosis with vaccines and excision, *J. Am. Vet. Med. Assoc.* **135:**339–343.

Olson, C., Pamukcu, A. M., and Brobst, D. F., 1965, Papilloma-like virus from bovine urinary bladder tumors, *Cancer Res.* **25:**840–849.

Olson, C., Robl, M. G., and Larson, L. L., 1968, Cutaneous and penile bovine fibropapillomatosis, *J. Am. Vet. Med. Assoc.* **153:**1189–1194.

Olson, R. O., Olson, C., and Easterday, B. C., 1982, Papillomatosis of the bovine teat (mammary papilla), *Am. J. Vet. Res.* **43:**2250–2252.

Orth, G., Breitburd, F., and Favre, M., 1978, Evidence for antigenic determinants shared by the structural polypeptides of (Shope) rabbit papillomavirus and human papillomavirus type 1, *Virology* **91:**243–255.

Orth, G., Jablonska, S., Favre, M., Croissant, O., Obalek, S., Jarzabek-Chorzelska, M., and Jibard, N., 1981, Identification of papillomaviruses in butcher's warts, *J. Invest. Dermatol.* **76:**97–102.

Osato, T., and Ito, Y., 1968, Immunofluorescence studies of Shope papilloma virus in cottontail rabbit kidney tissue cultures, *Proc. Soc. Exp. Biol. Med.* **128:**1205–1209.

Osterhaus, A. M. D. E., Ellens, D. J., and Horzinek, M. C., 1977, Identification and characterization of a papillomavirus from birds (Fringillidae), *Intervirology* **8:**351–359.

Ostrow, R. S., Krzyzek, R., Pass, F., and Faras, A. J., 1981, Identification of a novel human papilloma virus in cutaneous warts of meat handlers, *Virology* **108:**21–27.

Pamukcu, A. M., 1955, Investigation on the pathology of enzootic bovine hematuria in turkey, *Zentralbl. Veterinaermed.* **2:**409–429.

Parish, W. E., 1961, A transmissible genital papilloma of the pig resembling condyloma acuminatum of man, *J. Pathol. Bacteriol.* **81:**331–345.

Parish, W. E., 1962, An immunological study of the transmissible genital papilloma of the pig, *J. Pathol. Bacteriol.* **83:**429–442.

Parsons, R. J., and Kidd, J. G., 1936, A virus causing oral papillomatosis in rabbits, *Proc. Soc. Exp. Biol. Med.* **35:**441–443.

Pfister, H., 1983, Biology and biochemistry of papillomaviruses, *Rev. Physiol. Biochem. Pharmacol.* **99:**111–181.

Pfister, H., Huchthausen, B., Gross, G., and zur Hausen, H., 1979, Seroepidemiological studies of bovine papillomavirus infections, *J. Natl. Cancer Inst.* **62:**1423–1425.

Puget, A., Favre, M., and Orth, G., 1975, Induction de tumeurs fibroblastiques cutanées où sous-cutanées chez l'Ochotone afghan (Ochotono rufescens rufescens) par inoculation du virus du papillome bovin, *C.R. Acad Sci.* **280:**2813–2816.

Rangan, S. R. S., Gutter, A., Baskin, G. B., and Anderson, D., 1980, Virus associated papillomas in colobus monkeys (*Colobus guereza*), *Lab. Anim. Sci.* **30:**885–889.

Robl, M. G., and Olson, C., 1968, Oncogenic action of bovine papilloma virus in hamsters, *Cancer Res.* **28:**1596–1604.

Rous, P., and Beard, J. W., 1935, The progression to carcinoma of virus-induced rabbit papillomas (Shope), *J. Exp. Med.* **62:**523–548.

Rudolph, R., and Muller, H., 1976, Induktion von epidarmalem Tumorwachstum in der Haut von Mastomys natalensis durch Ubertragung virushaltigen Tumorgewebes eines Plattenspithelkarzinoms, *Zentralbl. Veterinaermed. Reihe B* **23:**143–150.

Samuel, W. M., Chalmers, G. A., and Gunson, J. R., 1978, Oral papillomatosis of coyotes (*Canis latrans*) and wolves (*Canis lupus*) of Alberta, *J. Wildl. Dis.* **14:**165–169.

Segre, D., Olson, C., and Hoerlein, A. B., 1955, Neutralization of bovine papillomavirus with serums from cattle and horses with experimental papillomas, *Am. J. Vet. Res.* **16:**517–520.

Shimkin, M. B., 1955, A note on the history of transplantation of tumors, *Cancer* **8:**653–655.

Shope, R. E., 1933, Infectious papillomatosis of rabbits; with a note on the histopathology, *J. Exp. Med.* **58:**607–624.

Shope, R. E., 1937, Immunization of rabbit to infectious papillomatosis, *J. Exp. Med.* **65:**219–231.

Shope, R. E., 1955, An infectious fibroma of deer, *Proc. Soc. Exp. Biol. Med.* **88:**533–535.

Shope, R. E., Mangold, R., McNamara, L. G., and Dumbell, K. R., 1958, An infectious cutaneous fibroma of the Virginia white-tailed deer (*Odocoileus virginianus*), *J. Exp. Med.* **108:**797–802.

Spradbrow, P., 1982, Papilloma viruses, papillomas, and ocular and cutaneous carcinomas in ruminant animals, paper presented at the Cold Spring Harbor Conference on Papilloma Viruses, Sept. 14–18, p. 73.

Stevens, J. G., and Wettstein, F. O., 1979, Multiple copies of Shope virus DNA are present in cells of benign and malignant non-virus producing neoplasms, *J. Virol.* **30:**891–898.

Sundberg, J. P., and Nielsen, S. W., 1981, Deer fibroma, a review, *Can Vet. J.* **22:**385–388.

Sundberg, J. P., Williams, E., Thorne, E. T., and Lancaster, W. D., 1983, Cutaneous fibropapilloma in a pronghorn antelope, *J. Am. Vet. Med. Assoc.* **183:**1333–1334.

Sundberg, J. P., Junge, R. E., and Lancaster, W. D., 1984, Immunoperoxidase localization of papillomaviruses in hyperplastic and neoplastic epithelial lesions of animals, *Am. J. Vet. Res.* **45:**1441–1446.

Syverton, J. T., 1952, The pathogenesis of the rabbit papilloma-to-carcinoma sequence, *Ann. N.Y. Acad. Sci.* **54:**1126–1140.

Tajima, M., Gordon, D. E., and Olson, C., 1968, Electron microscopy of bovine papilloma and deer fibroma viruses, *Am. J. Vet. Res.* **29:**1185–1194.

Theilen, G. H., Wheeldon, E. B., East, N., Madewell, B. R., and Lancaster, W. D., 1982, Goat papillomatosis, paper presented at the Cold Spring Harbor Conference on Papilloma Viruses, Sept. 14–18, pp. 73a–73b.

Thomas, M., Levy, J. P., Tanzer, J., Boiron, M., and Bernard, J., 1963, Transformation in vitro de cellules de peau de veau embryonnaire sous l'action d'extrants a cellulaires de papillomes bovins, *C.R. Seances Acad. Sci. Ser. III* **257:**2155–2158.

Tokarnia, C. H., Dobereiner, J., and Cannella, C. F. C., 1969, Ocorrencia da hematuria enzootica e de carcinomas epidermoides no trato digestivo superior em bovinos no Brasil. II. Estudos complementares, *Pesqui. Agropecu. Bras.* **4:**209–224.

Tokita, H., and Konishi, S., 1975, Studies on canine oral papillomatosis. II. Oncogenicity of canine oral papilloma virus to various tissues of dog with special reference to eye tumor, *Jpn. J. Vet. Sci.* **37:**109–120.

Tokita, H., Konishi, S., and Ogata, M., 1977, Studies on canine oral papillomatosis. III. Cultivation of papilloma cells in vitro, *Jpn. J. Vet. Sci.* **39:**619–626.

Vanselow, B. A., and Spradbrow, P. B., 1982, Papillomaviruses, papillomas and squamous cell carcinomas in sheep, *Vet. Rec.* **110:**561–562.

Watrach, A. M., 1969, The ultrastructure of canine cutaneous papilloma, *Cancer Res.* **29:**2079–2084.

Watrach, A. M., Small, E., and Case, M. T., 1970, Canine papilloma: Progression of oral papilloma to carcinoma, *J. Natl. Cancer Inst.* **45:**915–920.

Wettstein, F. O., and Stevens, J. G., 1982, Variable-sized free episomes of Shope papilloma virus DNA are present in all non-virus-producing neoplasms and integrated episomes are detected in some, *Proc. Natl. Acad. Sci. USA* **79:**790–794.

Yoshikawa, T., and Oyamada, T., 1971, Primary occurrence of bovine vesical fibromatosis, *Kitasato Arch. Exp. Med.* **34:**197–204.

CHAPTER 3

Organization and Expression of Papillomavirus Genomes

ULF PETTERSSON, HARRI AHOLA, ARNE STENLUND, AND JORGE MORENO-LOPEZ

I. INTRODUCTION

Although the papillomaviruses were discovered early in this century (Ciuffo, 1907; Shope, 1933), studies of their molecular properties progressed extremely slowly until the last decade. The lack of *in vitro* systems for virus cultivation has obviously been a retarding factor. Due to their morphological and physical properties, the papillomaviruses were assumed to be closely related to the other members of the papovavirus group, SV40 and polyomavirus (Melnick *et al.*, 1974; Matthews, 1982). It was, however, recognized that both the papillomavirus particle (Strauss *et al.*, 1950; Finch and Klug, 1965) and the viral chromosome (Crawford and Crawford, 1963; Favre *et al.*, 1975; Gissmann and zur Hausen, 1976; Orth *et al.*, 1977) were significantly larger than those of other papovaviruses.

Today the papillomavirus field is one of the most rapidly progressing areas in virology. Gene technology has probably had a greater impact on this field than on almost any other branch of virology. Molecular cloning makes it possible to produce large quantities of papillomavirus DNA

ULF PETTERSSON, HARRI AHOLA, AND ARNE STENLUND • Department of Medical Genetics, University of Uppsala, Biomedical Center, S-751 23 Uppsala, Sweden. JORGE MORENO-LOPEZ • Department of Veterinary Microbiology (Virology), Swedish University of Agricultural Sciences, Biomedical Center, S-751 23 Uppsala, Sweden. *Present address of A.S.:* Department of Molecular Biology, University of California, Berkeley, California 94720.

(Heilman *et al.*, 1980; Danos *et al.*, 1980; Pfister *et al.*, 1981; de Villiers *et al.*, 1981; Moreno-López *et al.*, 1981, 1984; Gissmann *et al.*, 1982; Clad *et al.*, 1982; Kremsdorf *et al.*, 1983; Dürst *et al.*, 1983; Boshart *et al.*, 1984) which in combination with recently developed procedures for nucleic acid sequencing permits the analysis of papillomavirus genomes at the molecular level. Also the discovery of efficient *in vitro* systems for cellular transformation allows certain biological properties of the viral genome to be studied (Thomas *et al.*, 1964; Geraldes, 1969; Dvoretzky *et al.*, 1980; Lowy *et al.*, 1980; Moar *et al.*, 1981). The structural studies of papillomavirus genomes have progressed much more rapidly than the functional analysis. Complete nucleotide sequences are known for several human (Danos *et al.*, 1981, 1983; Schwarz *et al.*, 1983; Seedorf *et al.*, 1985; Fuchs *et al.*, 1986; Cole and Streeck, 1986; Dartmann *et al.*, 1986) and animal papillomaviruses (Chen *et al.*, 1982; Giri *et al.*, 1985; Groff and Lancaster, 1985). Although the human and animal papillomavirus genomes generally exhibit a low degree of sequence homology (Law *et al.*, 1979; Heilman *et al.*, 1980), they have many properties in common, particularly regarding the organization of the open translational reading frames (ORFs). General conclusions can thus be drawn from experiments with the papillomavirus that is most amenable to experimental studies, BPV-1. This virus has served as a model for other papillomaviruses, and much of the information described in this review concerns BPV-1. Our current knowledge about papillomavirus gene functions is rather scanty, and information about papillomavirus proteins is almost nonexistent. Also our current understanding of papillomavirus transcription is very limited primarily because the mRNA levels present in papillomavirus-infected or -transformed cells are remarkably low. The transcriptional maps which are discussed below are therefore incomplete and much additional work is needed before we have a complete understanding of how the viral genome is expressed and what controlling elements are present in papillomavirus chromosomes.

II. GENERAL PROPERTIES OF PAPILLOMAVIRUS GENOMES

All papillomavirus genomes extracted from virus particles appear to consist of superhelical double-stranded DNA molecules, having a length of approximately 8000 bp (Danos *et al.*, 1982, 1983; Chen *et al.*, 1982; Schwarz *et al.*, 1983; Giri *et al.*, 1985). The genomes which have been sequenced to date range in length from 7654 nucleotides for HPV-8 to 8374 nucleotides for the deer papillomavirus (DPV). The GC contents are in general low and range from 40.3% for HPV-1a to 46.4% for the CRPV genome. All papillomavirus genomes studied so far, including those from the recently discovered human papillomavirus types 16 (Dürst *et al.*, 1983) and 18 (Boshart *et al.*, 1984), which have not yet been isolated from viral particles, have many properties in common and most conclusions that can be drawn from already available data are likely to be valid for other members of the papillomavirus group. When the first papillomavirus sequences were determined, it became apparent that the papillomaviruses have a different genome organization than other members of the papovavirus genus, like SV40 and polyomavirus (Danos *et al.*, 1982; Chen *et al.*, 1982). A characteristic feature of papillomavirus genomes is that all major ORFs are located on the same DNA strand. This is in contrast to polyomavirus and SV40 whose early and late genes are transcribed in opposite directions (for a review see Tooze, 1981).

For most DNA viruses it is possible to subdivide the genes into two functional groups, namely the "early" genes which are expressed before the onset of viral DNA replications and the "late" genes which are expressed after viral DNA replication has commenced. The "late" genes of most DNA viruses are known to encode the structural proteins which form the virus particle. The "early" and "late" genes are moreover controlled by different regulatory signals in SV40 and polyomavirus (for a review see Tooze, 1981). Since no tissue culture system is available for propagation of papillomaviruses, the same functional division of the genome cannot be made. In analogy with other papovaviruses, it has, however, been assumed that the "early" genes are synonymous with genes that are expressed in transformed cells. It has been established that a fragment comprising 69% of the BPV-1 genome is sufficient for transformation of mouse cells *in vitro* (Lowy *et al.*, 1980). The ORFs present in this fragment are therefore thought to encode "early" (E) functions and the remainder of the genome is often referred to as the "late" (L) part. Some experimental support in favor of this subdivision exists; it is known that RNA extracted from wart tissue, where the virus is able to replicate, includes sequences from the so-called L-region (Amtmann and Sauer, 1982; Engel *et al.*, 1983). Certain "late" proteins, however, seem to be encoded by sequences which are located within the E-region. Moreover, promoters located upstream of the E-region are apparently used to control the expression of "late" proteins as will be discussed in Section VI.B.

The papillomaviruses have, like other papovaviruses, a very compact genome organization. The ORFs in the E-region show considerable overlaps and there are few noncoding nucleotides from the start of the E6 ORF in the beginning of the E-region of BPV-1 to the end of L1 ORF in the L-region. The coding regions cover approximately 90% of most papillomavirus genomes (Fig. 1). A region which appears to be noncoding since it lacks ORFs of significant size does, however, exist. It is located between the end of the L-region and the beginning of the E-region and its length differs slightly among the papillomavirus genomes sequenced at present; it ranges from 397 in the HPV-8 genome to 982 nucleotides in the HPV-1a genome.

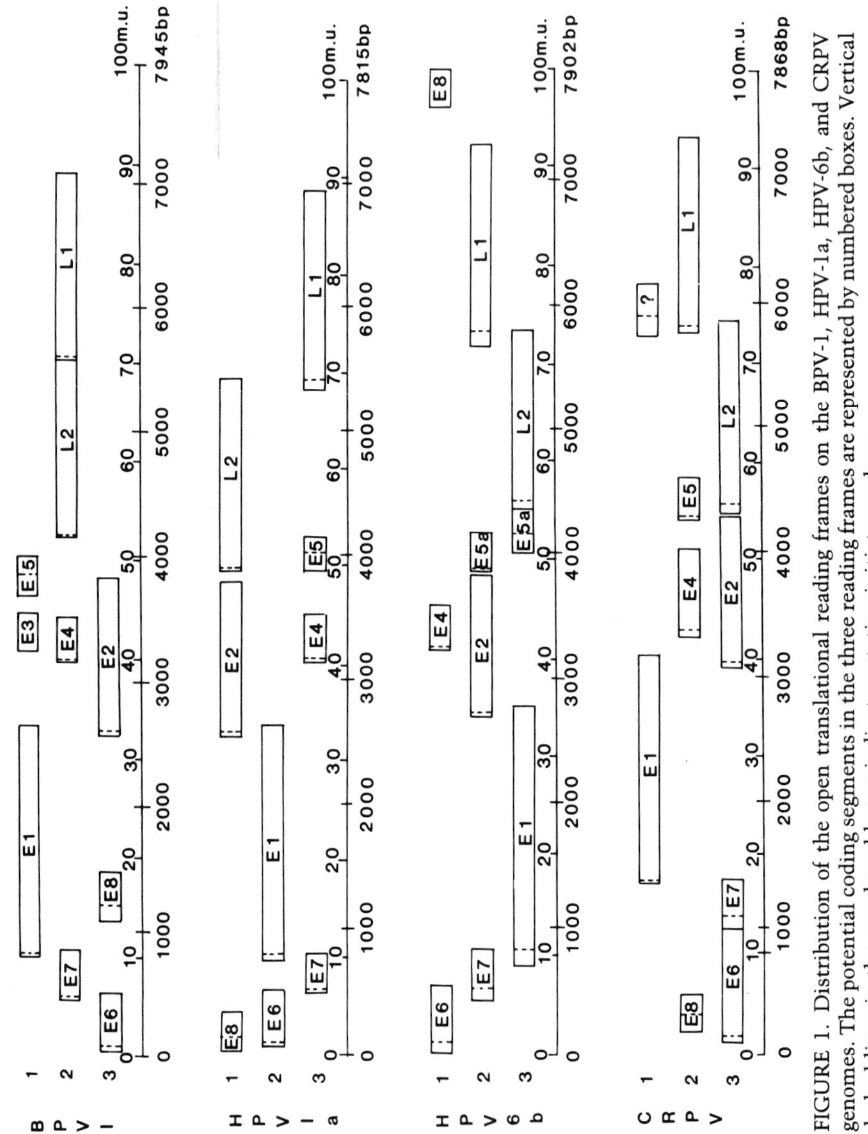

FIGURE 1. Distribution of the open translational reading frames on the BPV-1, HPV-1a, HPV-6b, and CRPV genomes. The potential coding segments in the three reading frames are represented by numbered boxes. Vertical dashed lines in the numbered boxes indicate putative initiator codons.

III. THE GENOME OF BPV-1

The BPV-1 genome consists of 7945 bp (Chen *et al.*, 1982; Danos *et al.*, 1983; revised by Stenlund *et al.*, 1985) and is usually divided into 100 map units. Position 0 is the first nucleotide in the unique recognition sequence for the *Hpa*I endonuclease, located near the 5′-end of the transforming region.

A. Organization of the E (Transforming)-Region

The transforming segment of the BPV-1 genome is located between the cleavage sites for endonucleases *Hin*dIII (coordinate 87) and *Bam*HI (coordinate 56). Eight ORFs of a significant size have been detected in this segment of the BPV-1 genome (Fig. 1) and they are designated E1 to E8 (Danos *et al.*, 1983). Five of these—E1, E2, E4, E6, and E7—appear to have equivalent counterparts in all papillomavirus genomes sequences so far and are thus likely to encode functional polypeptides. E1 which is the largest ORF extends between coordinates 10 and 34 in the BPV-1 genome. There exists a putative initiator ATG near its 5′-end and E1 has the capacity to encode a polypeptide with a molecular weight of 68K. The COOH-terminal end of the E1 ORF overlaps slightly with the second largest ORF, designated E2. This ORF, which is located between coordinates 32 and 48 in the BPV-1 genome, also contains an ATG triplet near its NH_2-terminal end and has the capacity to encode a polypeptide with a molecular weight of 45K. The ORF, designated E3, overlaps completely with E2 and is located between coordinates 41 and 45. This ORF lacks ATG triplets and must consequently be spliced to an upstream exon if expressed as a functional polypeptide. Also the E4 ORF overlaps completely with E2 and is located between coordinates 40 and 44 in the BPV-1 genome (Fig. 1). Thus, there exists a region between coordinates 41 and 44 in the BPV-1 genome which contains three overlapping ORFs. E4 contains a putative initiator ATG near its NH_2-terminal end and has the potential to encode a polypeptide with a molecular weight of 12K. Unlike the E3 ORF which has no apparent counterpart in other papillomavirus genomes, the E4 ORF seems to be present in all the sequenced papillomavirus genomes.

The E5 ORF is located at the 3′-end of the transforming region. It is short and encodes a very hydrophobic polypeptide. Although E5 is only present in certain papillomaviruses, it still seems to be functional (see Section V). The E8 ORF, which in the BPV-1 genome overlaps completely with the E1 ORF (Fig. 1), has a variable location in different papillomaviruses. It probably is not functional.

The E6 and E7 ORFs are located in the 5′-part of the transforming region and E6 spans between coordinates 0.6 and 6 in the BPV-1 genome. E7 overlaps with E6 for a short distance and is located between coordinates

6 and 11. Both E6 and E7 contain putative initiator triplets near their NH_2-terminal ends which also appears to be the case in other papillomavirus genomes. The predicted molecular weights of the E6 and E7 proteins are 16 and 14K, respectively. The arrangement of the E6 ORFs is slightly different in the CRPV genome as will be discussed below.

B. Organization of the L-Region

The L-region which covers approximately 3000 bp in the BPV-1 genome is located between coordinates 53 and 89 (Chen *et al.*, 1982). Two major ORFs, designated L1 and L2, have been identified in this region (Fig 1.) Corresponding ORFs are found in the other sequenced papillomavirus genomes and they code for proteins which are present in the papillomavirus capsid. The L1 ORF has the capacity to encode a polypeptide with a predicted molecular weight of 55.5K and the L2 ORF encodes a polypeptide with a predicted molecular weight of 50.1K (Chen *et al.*, 1982; Ahola *et al.*, 1983). Both the L1 and L2 ORFs contain ATG triplets near their NH_2-terminal ends and they are transcribed in the same direction as the E-region.

C. The Noncoding Region

The region in the BPV-1 genome between the end of the L1 ORF and the beginning of the E6 ORF appears to be noncoding. It is approximately 1000 nucleotides long and has some characteristic features; it contains several AT-rich regions, the longest being a 21-bp run. Several nearly perfect palindromes are also found in this region as well as many segments of dyad symmetry with nonhomologous loops. Noteworthy is the finding of several copies of the palindrome ACCGNNNNCGGT (Dartmann *et al.*, 1986). The structural features of the noncoding region infer that it contains control elements for DNA replication and transcription.

D. Transcriptional Signals

The nucleotide sequence of the BPV-1 genome reveals several TATA-motifs (Chen *et al.*, 1982) and the sequence TATAAA is present at nucleotides 58,* 5090, and 7109. A closely related sequence motif TATATA is present at nucleotides 4073 and 6860. Both these sequence motifs have in other studies been found to be part of the eukaryotic promoter and to be important for localizing the 5′-end of transcripts *in vivo* and *in vitro*. So far only the TATAAA sequence located at nucleotide 58 has been

*The nucleotide given is the position of the first nucleotide in the sequence motif.

demonstrated to be part of a functional promoter in BPV-1 (Ahola *et al.*, 1983; Campo *et al.*, 1983; Yang *et al.*, 1985; Stenlund *et al.*, 1985). Transcripts which have their 5′-ends near the TATA-motifs at nucleotides 6860 and 7109 have been observed in RNA from virus producing fibropapillomas. The TATATA sequence which is located at nucleotide 4073 precedes the L2 ORF and appears to be part of a promoter for transcription of a mRNA from the L-region. Available experimental data (see Section VI.B) suggest, however, that most L-region transcripts are initiated upstream of the E-region. Due to its location, it seems unlikely that the TATA-motif at nucleotide 5090 is part of a functional promoter. Additional studies are consequently required to establish which of the abovementioned TATA sequences represent functional promoters. Studies on the structure of BPV-1 mRNAs have shown that an additional less typical TATA-motif, TAATAT, which is located between nucleotides 2414 and 2419, appears to be the promoter for an abundant mRNA species, present in transformed cells (Stenlund *et al.*, 1985; Yang *et al.*, 1985).

The sequence AATAAA or a closely related sequence is usually found adjacent to 3′ polyadenylation sites (Proudfoot, 1982). This sequence is found at four positions in the BPV-1 genome, namely 4180, 6434, 7092, and 7156. The one located at nucleotide 4180 is apparently functional since a poly(A) addition site for the mRNAs in the E-region has been identified in its vicinity, i.e., at nucleotide 4203 (Yang *et al.*, 1985). It has also been demonstrated that the AATAAA sequence at nucleotide 7156 is the polyadenylation signal for mRNAs that are exclusively expressed during a productive infection (Engel *et al.*, 1983) whereas the other two hexanucleotides seem to be nonfunctional. It is, however, noteworthy that both BPV-1 and HPV-1a genomes contain two closely spaced AATAAA sequences immediately beyond the L1 ORF.

Eukaryotic transcription has been found to require so-called enhancer sequences. Their presence was first demonstrated in SV40 (Banerji *et al.*, 1981; Moreau *et al.*, 1981) and they are usually located upstream of the promoter region. One typical property of enhancers is that they function independent of their orientation and exact position relative to the promoter. Lusky *et al.* (1983) have identified a BPV-1 enhancer located at coordinate 55. It has been mapped within a 40-bp segment located between nucleotides 4394 and 4433, thus immediately beyond the common poly(A) addition site for early BPV-1 mRNAs (Weiher and Botchan, 1984). It contains a typical so-called enhancer core sequence and contains two separate functional domains (Weiher *et al.*, 1983; Weiher and Botchan, 1984). An unexpected finding was that the BPV-1 enhancer is located at the end rather than at the beginning of the early transcription unit. The enhancer sequence is of critical importance for BPV-1 gene expression under certain conditions and its function can be substituted for by other enhancers like the SV40 72-bp repeat (Lusky *et al.*, 1983). Spandidos and Wilkie (1983) demonstrated that the BPV enhancer has a preference for bovine cells. Yet another enhancer seems to be located in the noncoding bovine cells. Yet another enhancer seems to be located in the noncoding region. This enhancer is transactivated by a product from the the E2 ORF (Spalholz *et al.*, 1985).

E. The Origin of DNA Replication

Two different experimental strategies have been used in order to identify replication origins in the BPV-1 genome. Lusky and Botchan (1984) used the following approach: they assumed that a cloned subgenomic fragment of BPV-1 DNA which contained the replication origin would be able to replicate extrachromosomally in BPV-1-transformed cells. The basis for this assumption was that the resident viral genomes in the transformed cell would supply all functions that are required in *trans* and that only the replication origin would be necessary to allow a second plasmid to replicate when introduced into the transformed cell. A neomycin resistance gene (Colbère-Garapin *et al.*, 1981), linked to the various BPV-1 fragments, served as a selectable marker in these experiments. BPV-1-transformed cells were then transfected with the different recombinants and a number of cell clones were obtained which expressed the resistance marker. These were then analyzed for the presence of extrachromosomally replicating plasmids containing the marker and the results revealed surprisingly that two separate regions in the BPV-1 genome were able to support extrachromosomal replication. The two regions were designated plasmid maintenance sequences (PMS). One sequence (PMS-1) was located within a 521-bp fragment, spanning between coordinates 87 and 94, whereas the second (PMS-2) was located within a 140-bp fragment around coordinate 20, i.e., within the E1 ORF. An interesting finding was that regions within the two PMS sequences display homology although they appear to be oriented in opposite directions (Lusky and Botchan, 1984).

In order to study the structure of BPV-1 replication intermediates, Waldeck *et al.* (1984) used a different experimental approach. They isolated covalently closed circular DNA molecules from BPV-1-transformed hamster cells and used the electron microscope as an analytical tool to search for replicating genomes. Using this method a class of molecules were identified which had the properties of replication intermediates of the "Cairns' type." A single "replication eye" centering at a unique position was observed both in monomeric and in dimeric molecules. By studying molecules which had replicated to different extents it was possible to map the origin around coordinate 87 in the physical map of the BPV-1 genome. This position coincides with the PMS-1 sequence identified by Lusky and Botchan (1984). Waldeck *et al.* (1984) were unable to find replicating molecules in which PMS-2 served as the origin of replication. Their experimental procedure selected, however, for molecules which remain superhelical during replication and molecules which replicate as open circles would have been overlooked in this study.

It is noteworthy that the replication origin at coordinate 87 coincides with DNase I hypersensitive region in the viral chromatin (Rösl *et al.*, 1983).

IV. COMPARATIVE ANATOMY OF PAPILLOMAVIRUS GENOMES

A comparison between four papillomavirus genomes is given below. Ahola *et al.* (1983) have in addition sequenced more than 50% of the genome of a BPV-1 isolate, different from that originally studied by Chen *et al.* (1982). The two BPV-1 isolates were obtained from animals in Sweden and the United States, respectively, and were thus of different origin. The degree of sequence variation within a single papillomavirus serotype could consequently be assessed from a comparison between them. A total of five differences were noticed most of which were base pair transitions. Two of the changes were located in noncoding regions whereas two others were found in the L2 ORF. One of the latter results in an amino acid substitution, whereas another change was silent. The fifth difference was a silent change located in the E6 ORF. The results thus indicate that the individual papillomavirus serotypes have not been subjected to a rapid genetic drift.

A. Features of the HPV-1a Genome

An extensive comparison between the BPV-1 and HPV-1a genomes has been reported by Danos *et al.* (1983), showing that the two genomes are organized in very similar ways. The latter genome is 7815 bp long, i.e., 130 bp shorter than the BPV-1 genome. The length difference is due to the deletion of short DNA segments in two positions, rather than being caused by evenly distributed small deletions throughout the genome. It is possible to align the two genomes since most major ORFs which are present in the BPV-1 sequence are also present in the HPV-1a sequence. Furthermore, the unique *Hpa*I cleavage site which is used as a reference point in the BPV-1 genome is present in the corresponding position in HPV-1a DNA. All major ORFs are located on one DNA strand, as is the case for BPV-1. The observed similarity between the two genomes was somewhat unexpected in light of previously reported hybridization studies which revealed a low degree of homology between the HPV-1a and BPV-1 genomes (Law *et al.*, 1979; Heilman *et al.*, 1980). The E8 ORF in BPV-1 has no counterparts in the HPV-1a genome, whereas the other ORFs are found in equivalent positions. An ORF, designated E8, is present in the HPV-1a sequence, although in a different position (Fig. 1), and it shows no apparent homology.

Three main regions of homology were detected when the BPV-1 and HPV-1a sequences were compared. The first includes the COOH-terminal half of the E1 ORF and the NH_2-terminal half of the E2 ORF. A surprising observation was that the conserved part of the E2 ORF stops at the point where the overlaps with the E3 and E4 ORFs begin. This is in contrast to the SV40 and polyomavirus genomes in which the most highly conserved region is found where the VP1 and VP2 genes overlap (Ferguson and Davis, 1975). The second region of homology corresponds to the NH_2-terminal 50 amino acids of L2. Here the homology is on the order of 50% whereas the remaining part of L2 has diverged considerably except for the region near the COOH-terminal end. Here the third region of homology starts which includes most of the L1 ORF.

The regions outside of the conserved areas exhibit less homology but certain common features can nevertheless be demonstrated. An example is the E6 and E7 ORFs which show little sequence similarity, but which still have characteristics in common. The motif Cys-X-X-Cys is repeated four times in the amino acid sequence of the predicted E6 polypeptide of both genomes. In addition, the motifs are separated by similar distances. In E7 the same motif is repeated twice in both genomes with a spacing of 29 amino acid residues. It is noteworthy that a similar sequence combination is found in the small t-antigens from a variety of papovaviruses, including SV40 and polyomavirus where the motif is repeated twice (Friedman *et al.*, 1978; Seif *et al.*, 1979). The L2 ORF is poorly conserved and a number of point mutations appear to have accumulated here whereas no insertions or deletions of significant size are observed. The COOH-terminal end of the predicted protein product of the L2 ORF is highly basic in both BPV-1 and HPV-1a which is also the case for the minor capsid proteins, VP2 and VP3 of SV40 and polyomavirus. In the BPV-1 genome the L1 and L2 ORFs are located in the same translational frame and the putative initiator ATG for the L1 protein is separated by 15 bp from the stop codon of the L2 ORF. In the HPV-1a genome the L1 and L2 ORFs are located in different phases and the first ATG of the L1 ORF overlaps with the end of L2. The homology between the L1 ORFs of BPV-1 and HPV-1a begins, however, at the second ATG in the L1 ORF of HPV-1a suggesting that the translated parts of the L1 and L2 ORFs may be nonoverlapping also in HPV-1a. A noteworthy similarity is the presence of a conserved putative splice acceptor site preceding the first and the second ATGs in the L1 ORFs of BPV-1 and HPV-1a, respectively (Fig. 2).

The noncoding regions also show striking similarities with regard to both size and structure. The size of the noncoding region is 982 bp in HPV-1a and 943 in BPV-1 and the region contains AT-rich segments in both genomes and two closely spaced polyadenylation signals (nucleotides 7380 and 7426) in addition to various sequence repeats of different kinds. The AT-rich regions are longer in the HPV-1a genome and cover about

FIGURE 2. A comparison between the nucleotide sequences at the beginning of the L1 ORFs in the BPV-1, HPV-1a, HPV-6b, and CRPV genomes (data from Danos *et al.*, 1982, 1983; Chen *et al.*, 1982; Schwarz *et al.*, 1983; Giri *et al.*, 1984). The putative initiator ATG is underlined. This ATG is the second ATG in the L1 ORF of the HPV-1a genome. The indicated ATGs are in all cases preceded by a typical splice acceptor site. The asterisks indicate the postulated splice junctions.

Position	Sequence	Genome
5609	GCCTAATTTTTTTTGCAG*ATG	BPV-1
5828	CTATCTTTTTACTTGCAG*ATG	CRPV
5431	ATGTATAATGTTTTTCAG*ATG	HPV-1a
5789	TTCCCTTATTTTTTTCAG*ATG	HPV-6b

100 bp immediately beyond the L1 stop codon. A 10-bp direct repeat which is present in the BPV-1 sequence is also present in the HPV-1a genome, although organized in the opposite orientation. This implies that a function of this sequence element must be independent of the orientation. Another striking feature is that the sequences around the *Hpa*I cleavage sites, present in both genomes at coordinate 0, are very similar. This observation is interesting, since it has been shown that this region in BPV-1 contains a transcriptional control element (Ahola *et al.*, 1983; Campo *et al.*, 1983; Yang *et al.*, 1985; Stenlund *et al.*, 1985). The TATA-motif at coordinate 1 in BPV-1 which precedes the cap site and the first ATG in the E6 ORF has an equivalent in almost precisely the same position in the HPV-1a genome. The HPV-1a genome contains many additional TATA-motifs, one of which is located near the NH_2-terminal end of E7 (nucleotide 520). Another TATA-box is located in front of the L-region (nucleotide 3844). It remains, however, to be shown which of these are part of functional promoters.

The polyadenylation signal AATAAA is present at four positions in the HPV-1a genome. One is located at the end of the E-region, two are located after the L-region, and the fourth is located at nucleotide 821. The latter together with one of those located beyond the L-region are probably nonfunctional.

B. Features of the HPV-6b Genome

The HPV-6b genome, which was sequenced by Schwarz *et al.* (1983), contains 7902 bp, i.e., 43 bp smaller than the BPV-1 genome and 87 bp larger than the HPV-1a genome (Danos *et al.*, 1982, 1983; Chen *et al.*, 1982). The unique *Hpa*I restriction enzyme cleavage site GTTAAC which is found in both the BPV-1 and the HPV-1a sequence near the 5′-end of the E-region is absent from the HPV-6b sequence. A related sequence GTTAAT is, however, present in the corresponding position which can be used to align the HPV-6b genome with other papillomavirus genomes. It shares, as expected, many features with the HPV-1a and BPV-1 genomes and ORFs corresponding to E1, E2, E4, E6, and E7 as well as L1 and L2 (Fig. 1) have been recognized in the HPV-6b sequence (Schwarz *et al.*, 1983).

The E1 ORF in the HPV-6b genome has the capacity to encode a 649-amino-acid polypeptide which thus is slightly larger than the predicted E1 proteins of BPV-1 and HPV-1a. A typical property of the E1 protein is that it consists of an NH_2-terminal segment which shows a low degree of homology with E1 proteins of other papillomaviruses whereas the COOH-terminal part is very well conserved. A 57-amino-acid-long region is present in the E1 protein in which 34 residues are identical in all three viruses and only in six positions are the three sequences completely different.

The predicted E2 protein shows the same features as those of BPV-1 and HPV-1a; it displays amino acid sequence homology with E2 proteins of other papillomaviruses in the NH_2- and COOH-terminal parts, whereas the homology is reduced in the internal region when the E4 ORF overlaps with E2. The NH_2-terminal end of the E2 ORF overlaps with the COOH-terminal end of E1 for 56 nucleotides and this distance is exactly conserved in the HPV-1a and BPV-1 genomes (Danos *et al.*, 1983).

It is noteworthy that the E4 ORF which is present in approximately the same position of all papillomavirus genomes so far studied (Fig. 1) is poorly conserved.

The comparison between BPV-1 and HPV-1a revealed that the postulated E6 protein also is poorly conserved and HPV-6b follows this rule. A characteristic feature is, however, the presence of a fourfold repetition of the tetrapeptide sequence Cys-X-X-Cys, being regularly spaced as is the case in the other papillomavirus genomes.

The predicted E7 protein consists of 98 amino acids in HPV-6b and is thus 39 amino acids shorter than the predicted E6 protein of BPV-1. The E7 protein contains the tetrapeptide Cys-X-X-Cys repeated twice as the BPV-1 and HPV-1a genomes.

In the HPV-6b genome a 552-bp region is located between the end of the E2 ORF and the beginning of the L2 ORF. In BPV-1 this region contains the COOH-terminal part of the small E5 ORF, which also overlaps with E2. The HPV-6b genome has a different organization in this region (Fig. 1); it includes two ORFs designated E5a and E5b. E5a contains a 5′-terminal ATG codon and has the capacity to encode a protein of 91 amino acid residues. The predicted amino acid sequence includes a leucine-rich region which resembles the COOH-terminal region of the E5 protein in BPV-1. The E5b ORF has no apparent counterpart in the papillomavirus genomes.

The late region of HPV-6b resembles that of other papillomaviruses. The L1 and L2 ORFs are also present in HPV-6b and extensive sequence homology exists between the L1 protein of HPV-6b and L1 proteins of other papillomaviruses. This observation is consistent with the finding that antisera against denatured capsid proteins from different papillomaviruses cross-react (Jenson *et al.*, 1980). The L2 protein is less well-conserved and the homology is mainly restricted to the NH_2- and COOH-terminal portions. The COOH-terminal end of the L2 protein is rich in basic amino acids, a feature which it shares with the VP2 and VP3 proteins of polyomaviruses. It has been proposed that the conserved NH_2-terminal part of the L2 protein interacts with the highly conserved L1 polypeptide whereas the structurally diverging parts may serve some host- or tissue-specific functions (Schwarz *et al.*, 1983).

Many of the transcriptional control signals identified in BPV-1 and HPV-1a appear also to be present in the HPV-6b genome. A putative promoter region is found in front of the E6 ORF which shares significant sequence homology with the corresponding regions in HPV-1a and BPV-1. It includes a TATA-box which is followed by a putative cap site, located directly in front of the first ATG in the E6 ORF. Many additional TATA-motifs are present in the HPV-6b genome although it remains to demonstrate which of them are promoter elements. No polyadenylation signal AATAAA is present at the expected position near the 3′-end of the E-region. It may have been replaced by a related sequence, ATTAAA (nucleotide 4554), which is known to serve as a functional polyadenylation signal in other eukaryotic genes (Hagenbüchle *et al.*, 1980; Jung *et al.*, 1980). A bona fide polyadenylation signal, located downstream from the end of the L1 ORF (nucleotide 7407), is likely to be the common signal for polyadenylation of the L-region mRNAs, although it remains to be verified experimentally. The HPV-6b genome contains four additional polyadenylation signals, located at nucleotides 320, 1684, 6149, and 6422. Due to their locations it is questionable whether they are functional.

The noncoding regions in BPV-1 and HPV-1a contain, by definition, no ORFs of significant length. A small ORF, designated E8, is, however, found in the corresponding region of HPV-6b (Fig. 1). This ORF is located in a different position than in the other papillomavirus genomes. It is therefore unlikely to be functional and has no apparent relationship to the E8 ORF recognized in the BPV-1 sequence. The noncoding region of HPV-6b resembles that of other papillomavirus genomes in containing AT-rich segments with repetitions including tandem repeats, inverted repeats, and palindromic sequences.

C. Features of the CRPV Genome

The CRPV genome which is 7868 bp long was sequenced by Giri *et al.* (1985). Analysis of the ORFs reveals a genome organization which is similar to that found in BPV-1, HPV-1a, and HPV-6b (Fig. 1). The characteristic cleavage site for endonuclease *Hpa*I which is used as a reference point in both HPV-1a and BPV-1 genomes is absent from the CRPV sequence. Due to the presence of a homologous sequence in the corresponding region, it is nevertheless possible to align the CRPV genome with the other papillomavirus genomes.

ORFs corresponding to E1, E2, E4, E5, E6, and E7 as well as L1 and L2 have been identified in the CRPV sequence. An ORF designated E8 is also present in the CRPV genome (Fig. 1) although it is located in a different position than in BPV-1 (Giri *et al.*, 1985). A surprising finding was that the CRPV and HPV-1a genomes are more closely related to each other than are the two HPV genomes 1a and 6b (Giri *et al.*, 1985). The most outstanding feature of the CRPV genome is that the ORF designated E6 is considerably longer than in other papillomavirus genomes; it has the capacity to encode a polypeptide approximately twice as long, consisting of 273 amino acids and having a molecular weight of 30K, in contrast to the E6 protein of BPV-1 which has a predicted molecular weight of 16K (Giri *et al.*, 1985). The size of the E7 ORF is on the other hand approximately the same as in other papillomavirus genomes. The E1 ORF is slightly shorter than in other papillomavirus genomes, encoding a putative 602-amino-acid-long polypeptide, having a molecular weight of 68K. The E1 ORF of CRPV shares many properties with the E1 ORFs of other papillomaviruses; it contains a putative initiator ATG near the NH_2-terminal end and the COOH-terminal part shares extensive homology with other papillomavirus genomes, whereas the remainder of the E1 protein is less well conserved.

The E6 and E7 ORFs contain repetitions of the characteristic sequence motif Cys-X-X-Cys; it is repeated eight times in the E6 ORF of CRPV.

Further sequence comparisons from an evolutionary point of view reveal that the putative E2 gene product may be composed of three regions: the 100 NH_2-terminal amino acids are well conserved, showing 40–55% homology in contrast to the variable internal part which overlaps with the E4 ORF. The COOH-terminal part of the E2 protein is again well conserved. E4 is considerably larger in CRPV than in other papillomavirus genomes (Fig. 1). Two putative polyadenylation signals AATAAA are located at the end of the E-region, i.e., at nucleotides 4348 and 4468. However, only the former seems to be functional. Another typical feature of the CRPV genome is that the distance between the end of the E2 ORF and the beginning of the L2 ORF is exceptionally short. The L-region of CRPV has very similar properties as the corresponding region in other papillomavirus genomes. A well-conserved L1 ORF is present whereas the putative L2 polypeptide of CRPV exhibits little similarity with L2 proteins from other papillomaviruses except at the NH_2- and COOH-terminal ends. A third ORF, which overlaps with L1, is found in the L-region of CRPV (Fig. 1) although it remains to be determined whether it is functional.

The noncoding region in the CRPV genome, located between the L1 and E6 ORFs, covers 678 bp and is thus smaller than the noncoding regions of other sequenced papillomavirus genomes. It has probably been compressed as a consequence of the large E6 ORF that is present in CRPV. It starts with an AT-rich region of about 100 bp and is clearly homologous with the corresponding region in BPV-1. This is interesting since it contains the PMS-1 sequence, i.e., one of the two replication origins of BPV-1. It is, however, noteworthy that the two genomes appear to be inverted relative to each other in this region. A polyadenylation signal which presumably is used for the late mRNAs is found at position 7396, in the middle of an AT-rich region.

A property common to all papillomavirus genomes studied so far is the presence of numerous repeated sequences in the noncoding region. Direct repeats as long as 32 bp are present in CRPV in addition to numerous shorter direct and inverted repeats as well as palindromic sequences. Due to its complex arrangement, the noncoding region of CRPV resembles the transcriptional control regions present in SV40 and retrovirus LTRs.

A search for sequences which resemble the transcriptional control elements in the noncoding region of CRPV revealed a similar arrangement as has been found in the BPV-1 genome. A more complex organization of the transcriptional signals became, however, apparent when the mRNAs from the transforming regions were studied, and at least four initiation sites for transcription exist in the 5′-part of the E-region (Danos *et al.*, 1985). These are located at nucleotides 87, 158, 905, and 970, and TATA-motifs precede the first three of those. The results suggest that several promoters exist in the 5′-part of the E-region of CRPV in contrast to BPV-1, where only one cap site has been identified so far.

D. Features of Other Papillomavirus Genomes

Sequence information is available from several additional papillomavirus genomes. The deer papillomavirus (DPV) genome which was sequenced by Groff and Lancaster (1985) is closely related to the BPV-1 genome, although it is significantly longer, comprising 8374 bp. ORFs corresponding to E1, E2, E4, E5, E6, and E7 as well as L1 and L2 were found in the DPV genome. A short additional ORF, designated L3, is present in the L-region of DPV. However, it is questionable whether or not this is functional. Well-conserved regions were observed within the E1, E2, and L1 ORFs when the DPV sequence was compared to that of BPV-1. The unique *Hpa*I cleavage site which is used as a reference point in the BPV-1 genome is also present in the corresponding location in the DPV genome and the promoter region which is present near this position in BPV-1 has the identical counterpart in the DPV sequence.

Partial sequence information is also available from the BPV-2 genome

(Mitra and Lancaster, personal communication). It appears, as expected, to be closely related to the BPV-1 genome and no outstanding features have been recognized.

Other papillomavirus genomes which have been subjected to sequence studies include the European elk papillomavirus (EEPV) (Ahola *et al.*, 1986) and an avian papillomavirus (Moreno-López *et al.*, 1984). The EEPV genome appears to be very closely related to that of DPV. The limited sequence information that is available from the avian virus reveals well-conserved sequences within the E1 and L1 ORFs. The results demonstrate moreover that the avian papillomavirus genome has the same principal genetic organization as the mammalian papillomavirus genomes.

Sequence studies have also been performed on the genome of HPV-8 (Fuchs *et al.*, 1986), a papillomavirus which frequently is associated with the disease epidermodysplasia verruciformis. Available results reveal the same genetic organization as is typical for other papillomaviruses. Its noncoding region is very short. An interesting feature is the presence of a TATAAA-box in the COOH-terminal part of the E1 ORF which is preceded by a CAAT-box. The genomes of HPV-11 and HPV-33 have also been sequenced (Dartmann *et al.*, 1986; Cole and Streeck, 1986). Their genes are organized in a similar way as in other human papillomaviruses.

Of particular interest is the sequence information which has been collected from the HPV-16 and HPV-18 genomes, since these viruses have not yet been isolated in the form of virus particles. Their genomes are sometimes present in human genital tumors, usually in an integrated form. It may therefore be argued that these viruses are defective and never exist as virions. Available sequence information from HPV-16 and HPV-18 (Seedorf *et al.*, 1985; Gissmann *et al.*, personal communication) reveals, however, the presence of the same ORFs as are present in BPV-1 including the L1 and L2 ORFs. The results thus infer that HPV-16 and HPV-18 are nondefective papillomaviruses.

V. PREDICTED FUNCTIONS OF PAPILLOMAVIRUS PROTEINS

Presently we know very little about the proteins which are expressed in papillomavirus-transformed cells. Only the capsid proteins have to some extent been characterized. However, a number of genetic studies have been reported which make it possible to assign functions to the products of a few ORFs. Mutations introduced in the E1 ORF show that this protein is of critical importance for episomal replication of the BPV-1 genome, since cells transformed by mutants with lesions in the E1 gene exclusively contain papillomavirus DNA in the integrated form (Nakabayashi *et al.*, 1983; Lusky and Botchan, 1984; Sarver *et al.*, 1984). The region which covers the E2, E4, and E5 ORFs in BPV-1 appears to encode a transforming function, since this region has transforming properties when linked to a transcriptional control region from a retrovirus LTR (Nakabayashi *et al.*, 1983). It has also been shown that mutants with lesions in this part of the genome are defective for transformation (Sarver *et al.*, 1984; Lusky and Botchan, 1984, 1985; Schiller *et al.*, 1984; Yang *et al.*, 1985). Recent studies indicate, however, that the BPV-1 genome encodes two separate transforming functions, one encoded near the 3′ end of the E-region and the other in the E6 region (Schiller *et al.*, 1984; Yang *et al.*, 1985). The short E5 protein, which is very hydrophobic, seems to have transforming properties (Schiller *et al.*, 1986; DiMaio *et al.*, 1986). Moreover, Lusky and Botchan (1985) have discovered that cells transformed by deletion mutants which are defective in the E7 ORF contain a lower number of BPV-1 genomes, indicating that the protein product of the E7 ORF is involved in copy number control.

It is also possible to make certain predictions about the functions of some papillomavirus proteins from their postulated amino acid sequences. Clertant and Seif (1984) discovered that the predicted E1 product contains regions which show significant homology with the SV40 and polyomavirus large T-antigens at the amino acid sequence level. The homologous region is located in the COOH-terminal half of the protein and covers more than 200 amino acids. The E1 protein has moreover a similar predicted secondary structure as the T-antigens in this region and it is noteworthy that the homology was particularly pronounced in the parts of the polyoma and SV40 large T-proteins which are involved in nucleotide binding and ATPase activity. This finding lends additional support to the notion that the E1 protein functions in conjunction with DNA replication.

Comparative studies of the E2 region from different papillomaviruses suggest that the E2 proteins may be composed of three domains as discussed above. The COOH-terminal domain is strongly conserved and covers the region from which the 3′-exon of an abundant mRNA in BPV-1-transformed cells is transcribed. This domain appears to be distantly related to a portion of the protein encoded by the *mos* oncogene of Moloney murine sarcoma virus and its cellular homologue (Danos and Yaniv, 1983). Spalholz *et al.* (1985) have demonstrated that E2 encodes a protein, required for transactivation of an enhancer, located in the noncoding region.

The E6 protein of all studied papillomaviruses has a characteristic property, namely the presence of a repeated sequence motif with the structure Cys-X-X-Cys. An interesting observation was that the E6 protein of CRPV shows homology with the β chain of ATP synthase from bovine mitochondria, spinach or maize chloroplasts, and *E. coli* (Giri *et al.*, 1985). The homologies are observed along most of the E6 protein and it is only necessary to introduce one large and a few small gaps in order to align the two sequences. Amino acid residues in the enzyme which are known to be involved in the AMP binding are conserved in the E6 sequence as well as those residues which may be of importance for the binding of magnesium ions. Striking differences do, however, exist particularly with regard to the cysteine contents of the two proteins.

The E4 protein of HPV-8 shows some resemblance to the EBNA-2 protein from EBV-transformed cells (Fuchs *et al.*, 1986).

For the late proteins it appears that the bulk of the major papillomavirus capsid protein, VP1, is encoded by the L1 ORF. This conclusion is based on the finding that the amino acid composition of the VP1 protein from BPV-1 is similar to that of the predicted L1 protein (Meinke and Meinke, 1981; Chen *et al.*, 1982; Ahola *et al.*, 1983). More conclusive evidence has been obtained by expressing the L1 and L2 proteins in *E. coli* after cloning them in expression vectors (Pilacinski *et al.*, 1984). These studies show that the biosynthetically produced L1 protein, in contrast to the L2 protein, reacts well with sera prepared against bovine papillomaviruses. Antisera prepared against the biosynthetic L1 and L2 proteins, on the other hand, reacted with BPV-1 and the antisera were able to prevent the virus from transforming mouse C127 cells. These results have been interpreted to mean that the L1 protein carries the major antigenic determinants in the BPV-1 capsid. Since, however, antibodies raised against the biosynthetic L2 protein do react with BPV-1 virions, the results also suggest that the L2 protein is a minor capsid component.

VI. TRANSCRIPTIONAL ORGANIZATION OF PAPILLOMAVIRUS GENOMES

Studies on papillomavirus transcription have been severely hampered by the absence of tissue culture systems for viral replication and the low abundance of viral mRNAs in papillomavirus-transformed cells. Between 0.01 and 0.2% of the mRNA in a BPV-1-transformed cell appears to be of viral origin and some mRNA species represent a very small fraction of the total viral mRNA (Heilman *et al.*, 1983; Yang *et al.*, 1985). It has consequently been very difficult to establish a complete transcriptional map for any papillomavirus genome and several mRNA species which can be predicted to exist from the DNA sequence remain to be discovered. The concentration of certain papillomavirus mRNA species is, in fact, so low that available analytical methods are not sensitive enough to allow their detection. Several experimental systems have been used to study papillomavirus gene expression: mouse or hamster cells transformed by BPV-1, naturally occurring BPV-1 induced warts, BPV-1 induced hamster tumors and finally CRPV induced warts and tumors or established cell lines derived from these tumors (Amtmann and Sauer, 1982; Heilman *et al.*, 1982; Freese *et al.*, 1982; Nasseri *et al.*, 1982; Engel *et al.*, 1983; Ahola *et al.*, 1983; Nasseri and Wettstein, 1984a,b; Georges *et al.*, 1984; Yang *et al.*, 1985; Stenlund *et al.*, 1985; Danos *et al.*, 1985). The methods which have been used for analysis include Northern blot analysis, S1 nuclease analysis, electron microscopic heteroduplex analysis, and cDNA cloning.

A. RNAs Expressed in Rodent Cells Transformed by BPV-1

The first studies on mRNAs in BPV-1-transformed cells were reported by Amtmann and Sauer (1982), Freese *et al.* (1982), and Heilman *et al.* (1982). Northern blot and S1 nuclease analysis was used in these studies to show that BPV-1-transformed cells contain several overlapping mRNA species, ranging in size between approximately 1.1 and 4.1 kb, which has a common poly(A) addition site, now known to be located at coordinate 53 in the BPV-1 genome. The early studies of Heilman *et al.* (1982) failed to reveal spliced mRNA species. Subsequent investigations have, however, demonstrated that all major mRNAs in BPV-1-transformed cells consist of two or more exons (Ahola *et al.*, 1983; Yang *et al.*, 1985; Stenlund *et al.*, 1985). A major cap site for early BPV-1 mRNAs was mapped by Ahola *et al.* (1983) around coordinate 1 in the BPV-1 genome and later studies (Stenlund *et al.*, 1985; Yang *et al.*, 1985) have shown that it is located at nucleotide 89 in the BPV-1 sequence. This cap site defines one transcription unit which extends between nucleotides 89 and 4203 where the common polyadenylation site for all E-region mRNAs is located (Yang *et al.*, 1985). Nuclear mRNA precursors have indeed been identified (Engel *et al.*, 1982; Stenlund *et al.*, 1985) which extend between these positions in the BPV-1 genome and a variety of spliced mRNAs appear to be generated from these nuclear transcripts by differential splicing.

The positions of splice acceptor and donor sites within this transcription unit have been determined by the S1 nuclease protection technique and by sequencing of cDNA clones (Stenlund *et al.*, 1985; Yang *et al.*, 1985). The precise locations of two splice donor sites (D1, D2) and of three splice acceptor sites (A1, A2, A4) are known as well as the approximate position of a fourth splice acceptor site (A3). Figure 3 is a schematic drawing of the transcription unit and indicates the positions of the different splice junctions. The splice donor sites D1 and D2 are located at nucleotides 304 and 864, respectively, in the BPV-1 genome, whereas the acceptor sites A1, A2, and A4 are located at nucleotides 527, 557, and 3224, respectively. The splice acceptor site A3 is located at coordinate 31, although its precise position has not been determined. A set of mRNAs consisting of different exon combinations are generated by combining the splice donor and acceptor sites in different fashions. For instance, donor site D1 is in some mRNAs connected to the acceptor site A1, in others to acceptor site A4, and occasionally to the third acceptor site A2. D1 can also be connected to acceptor site A3 although this combination is very infrequent. Acceptor site A4 can be connected to either of donor sites D1 or D2, etc. Figure 3 illustrates that five exons, designated exon-

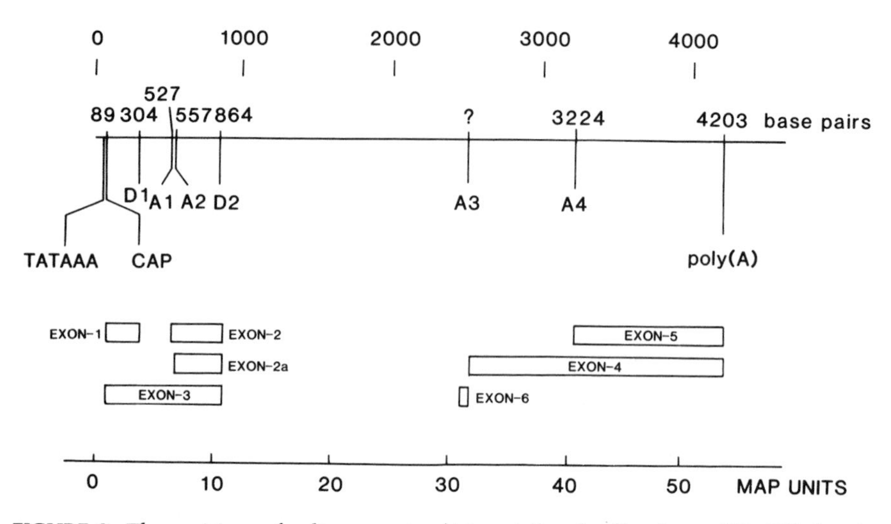

FIGURE 3. The positions of splice acceptor (A1 to A4) and splice donor (D1, D2) sites in the E-region of the BPV-1 genome. The positions of the TATA-box, the cap and the polyadenylation sites which define a transcription unit in the E-region of BPV-1 are indicted. The exons which result from differential splicing are shown by the numbered boxes.

1 to exon-5, are defined by the different splice junctions. Exon-1 is located between the cap site and donor site D1, exon-2 between acceptor site A1 and donor site D2. An overlapping and related exon, designated exon-2a, is located between acceptor site A2 and donor site D2 whereas exon-3 is located between the cap site and donor site D2. Finally, exon-4 and exon-5 which overlap are located between the polyadenylation site at nucleotide 4203 and wither of acceptor sites A3 and A4.

One additional exon has been identified in the E-region, designated exon-6 (Fig. 3), which is located between nucleotides 2443 and 2505.

Based on electron microscopic heteroduplex analysis (Stenlund *et al.*, 1985) and sequence studies of cDNA clones (Yang *et al.*, 1985), it has been possible to determine the molecular structure of several different mRNAs which are present in BPV-1-transformed cells. The two types of analyses have yielded essentially identical results. Stenlund *et al.* (1985) divided the mRNAs from the E-region of BPV-1 into five categories, designated types 1 to 5 (Fig. 4). This subdivision is followed in the discussion below.

The type 1 mRNA is the most abundant viral mRNA species present in BPV-1-transformed C127 cells. It has a length of 1.0 kb and consists of two exons, i.e., exon-5 from the 3′-end of the E-region linked to the short exon-6 at coordinate 31 (Fig. 4). Two different type 1 mRNAs appear to exist which differ with regard to the structure of their 5′-exons. One type (1A) has the structure outlined above whereas the second type (1B) contains a longer 5′-exon. The latter is also located around coordinate 31 although its precise position remains to be determined. It is possible that

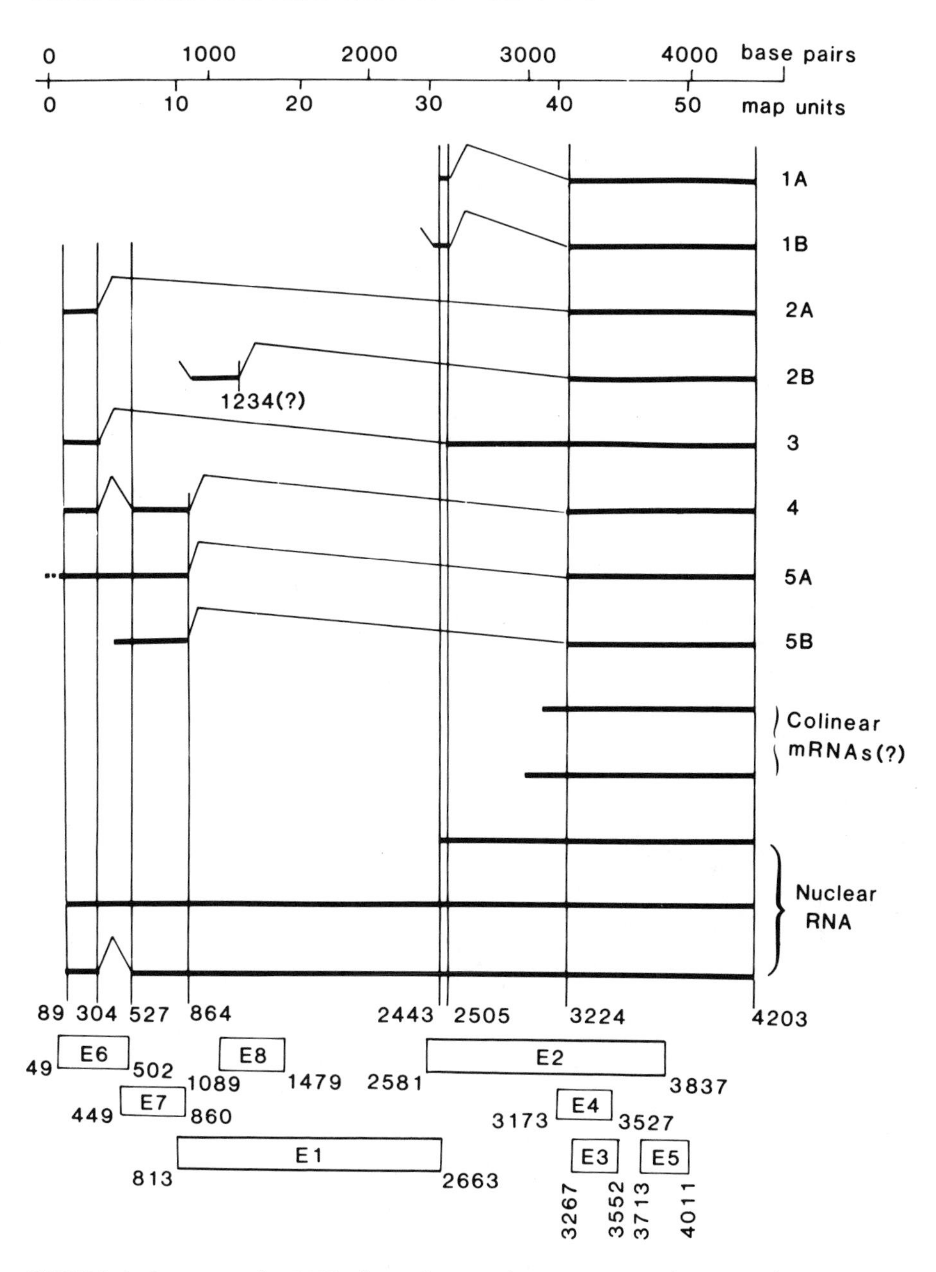

FIGURE 4. Structure of mRNAs from the transforming region of BPV-1. The exons are indicated with thick lines and the mRNAs are designated according to Stenlund *et al.* (1985). The type 5B mRNA species was described by Yang *et al.* (1985). The figure is based on information published by Stenlund *et al.* (1985) and Yang *et al.* (1985). Three nuclear RNA species, one of which is spliced, are also shown. The positions of the different splice acceptor and donor sites are indicated as well as the positions of the major open translational reading frames in the BPV-1 genome.

the 5′-exon in the type 1B mRNA is linked to yet another very short leader since heteroduplexes formed between the type 1B mRNAs and BPV-1 DNA often contain a short 5′ tail of unhybridized RNA (Stenlund *et al.* 1985; see also Fig. 5B).

The 5′-end of exon-6 at nucleotide 2443 appears to contain the capped 5′-end of the type 1A mRNA and a promoter has been mapped around coordinate 31 (H. Ahola *et al.*, unpublished information). An examination of the nucleotide sequence in this region reveals the nucleotide combination TAATAT between nucleotides 2414 and 2419. This is the expected position of a TATA-box if a cap site indeed is located at nucleotide 2443. A colinear nuclear RNA has been identified which extends between co-

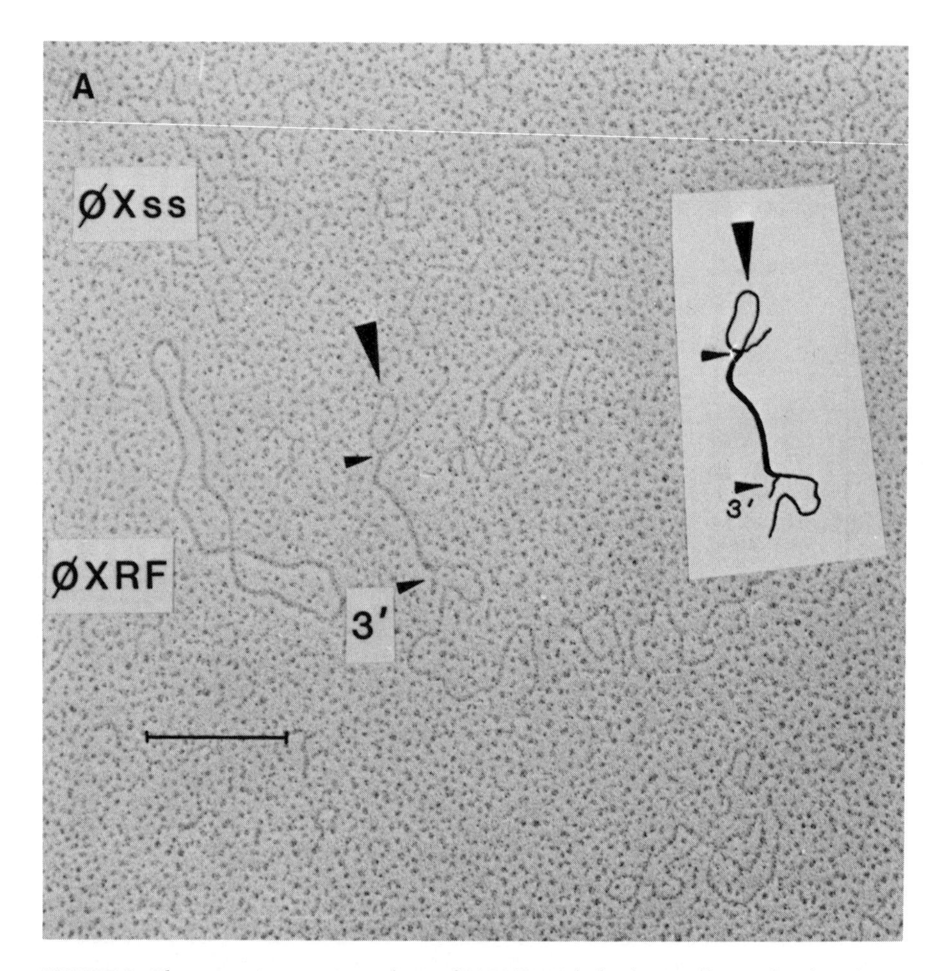

FIGURE 5. Electron microscopic analysis of RNA/DNA hybrids. Small arrowheads indicate ends of double-stranded regions, i.e., ends of transcripts; large arrowheads indicate DNA deletion loops corresponding to intervening sequences. (A) Type 1A molecule; (B) Type 1B molecule; (C) Type 2A molecule; (D) Type 4 molecule. Scale bar: 1 kb.

ordinate 31 and the poly(A) addition site at coordinate 53 (Stenlund *et al.* 1985); it might thus be the unspliced nuclear RNA precursor for the type 1A mRNAs (Fig. 4).

The 3′-exon (exon-5) of the type 1A and 1B mRNAs covers all of E5 and the 3′ portions of the E2, E3, and E4 ORFs. The latter ORFs lack ATG triplets beyond splice acceptor site A4. It is presently unclear which protein product is encoded by the type 1A mRNA. Exon-6 which is present in the type 1A mRNAs contains an ATG which is followed by a translational reading frame that is uninterrupted until the splice junction. However, a protein, only 16 amino acids long, would be translated under these conditions. A more likely possibility is that it encodes the E5 protein. If that is the case, the mRNA contains an exceptionally long 5′ untranslated region.

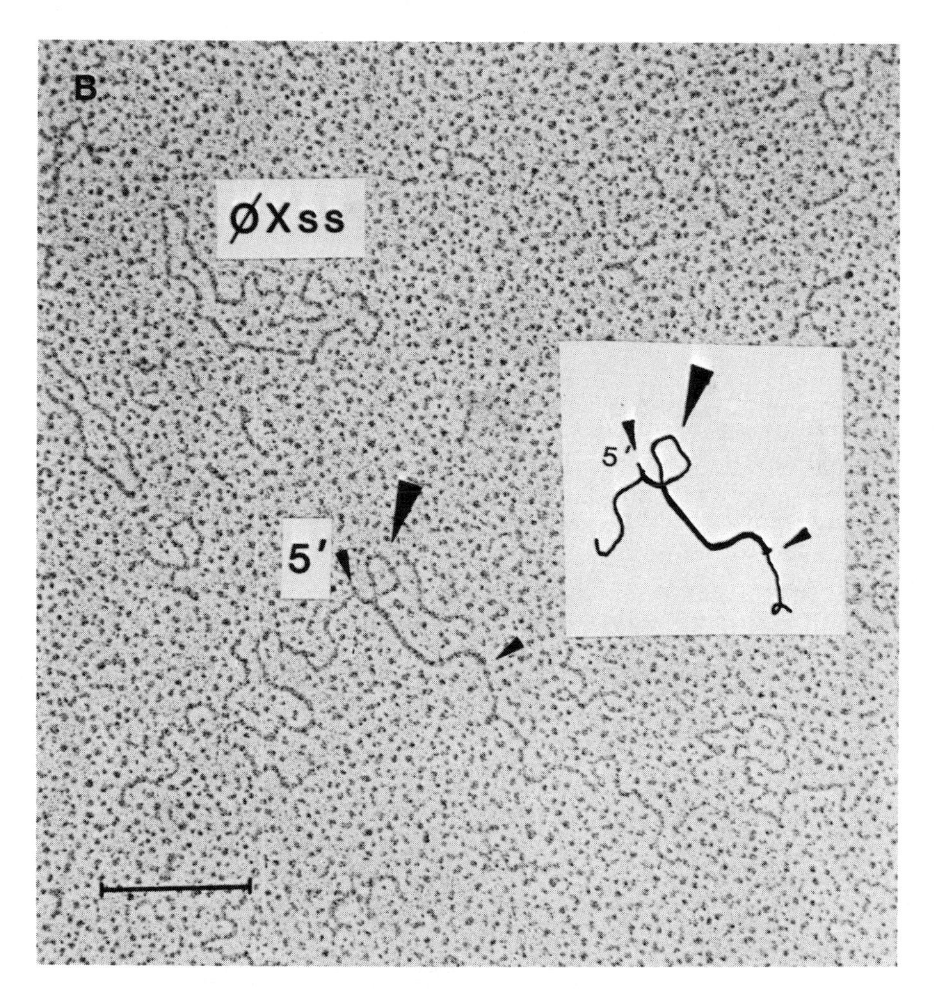

FIGURE 5. (*Continued*)

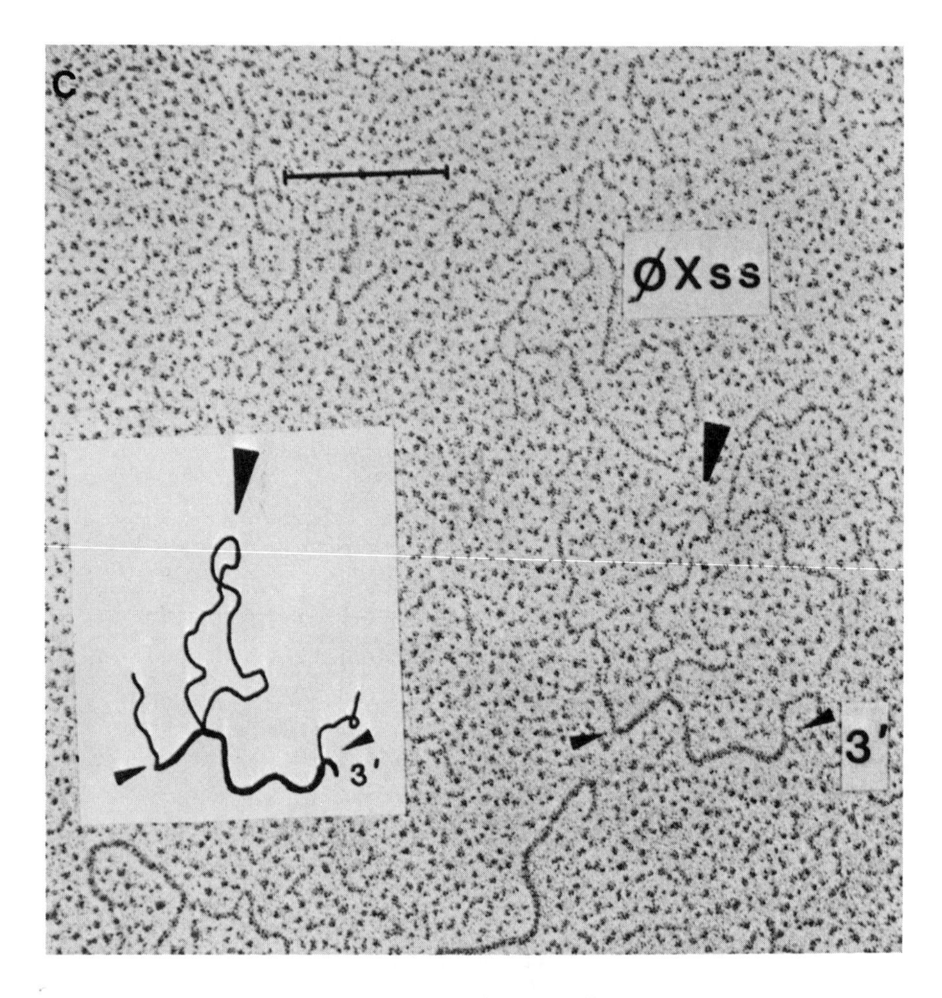

FIGURE 5. (*Continued*)

The type 2 mRNAs which, neglecting the poly(A) tail, have a length of 1.2 kb, resemble the type 1 mRNAs in also containing exon-5, although it is linked to a different 5′-exon which maps much farther upstream (Fig. 4). Electron microscopic heteroduplex analysis has shown that the type 2 mRNAs are heterogeneous with regard to their 5′-exons and at least two subclasses have been identified, designated types 2A and 2B (Stenlund *et al.*, 1985). The structure of the type 2A mRNA has been determined both by S1 nuclease protection technique in combination with heteroduplex mapping (Stenlund *et al.*, 1985) and by sequence analysis of a cDNA clone (Yang *et al.*, 1985). The results show that it is composed of exon-1 and exon-5. The type 2B mRNA has a similar structure although its 5′-exon maps at a different position, namely between coordinates 10 and 15 (Stenlund *et al.*, 1985). The precise structure of the 5′-exon has not been determined although a candidate splice donor site has been mapped at nucleotide 1234 (Stenlund *et al.*, 1985).

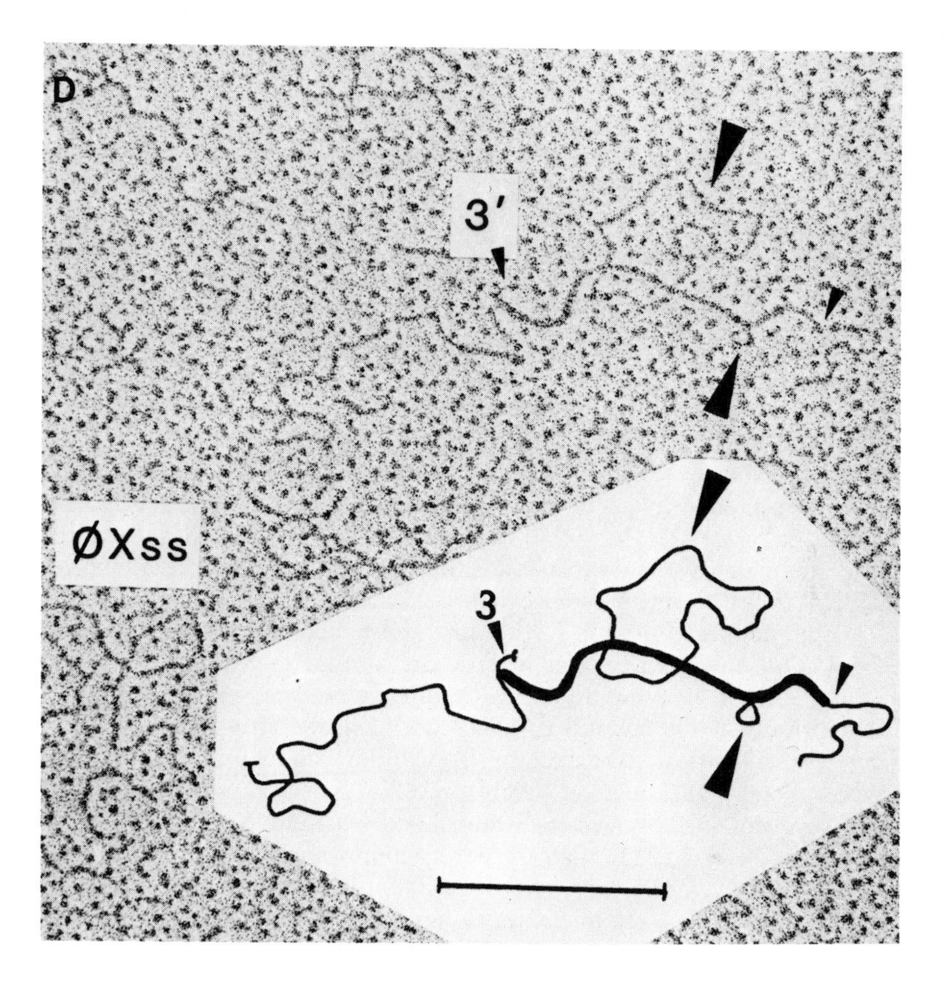

FIGURE 5. (*Continued*)

From its structure it can be predicted that the type 2A mRNAs encode a polypeptide which consists of sequences derived from both the E6 and the E4 ORFs. The coding capacity would be 19K provided that translation initiates at the 5′ proximal ATG in the E6 ORF. The coding capacity of the type 2B mRNA, on the other hand, cannot be predicted since its precise structure is unknown.

The type 3 mRNAs have an estimated length of approximately 1.8 kb, excluding the poly(A) tail. They differ from the other spliced BPV-1 mRNAs in lacking exon-5 (Fig. 4). The latter is replaced by exon-4 which is longer and extends between coordinate 32 and the common polyadenylation site at coordinate 53. This exon appears to be linked to exon-1. The type 3 mRNAs are apparently very rare since exceptionally few molecules of this kind were detected in an electron microscopic study of the mRNA pool, present in BPV-1-transformed cells (Stenlund *et al.*, 1985). An interesting feature of the type 3 mRNA is that it covers the E2 ORF.

The latter contains an ATG close to the 5′-end of exon-4 and the entire E2 ORF might thus be translated from the type 3 mRNAs. Since, however, the splice junction has not yet been mapped at a nucleotide level, another possibility must be considered, namely that translation initiates in exon-1 and then continues into the E2 ORF.

The type 4 mRNA is unique in being composed of three exons, namely exon-1, exon-2, and exon-5 (Fig. 4). Its total length is 1.5 kb, excluding the poly(A) tail, and it can be predicted from the structure that it encodes a polypeptide with a molecular weight of 21K. This polypeptide consists of sequences from both ORF 6 and ORF 7 since the removal of the intron between exon-1 and exon-2 results in an in-phase fusion of these two ORFs. Two subclasses of the type 4 mRNAs appear to exist which differ with regard to the size of the intron that is removed between the middle and the 5′-exon. In a minor population of type 4 mRNAs, 30 additional nucleotides are removed by splicing which reduces the size of exon-2 (designated "exon-2a" in Fig. 3). The E6 and E7 ORFs become fused in phase also in this mRNA class which is predicted to encode a shorter polypeptide (by 10 amino acids) than the other type 4 mRNA subclass.

The type 5 mRNAs are 1.9 kb long and are composed of exon-3 and exon-5. This mRNA thus has a structure similar to that of the type 4 mRNAs, except that the intervening sequence between exon-1 and exon-2 is maintained. The E6 and E7 ORFs will consequently not be fused in the type 5 mRNAs and the expected translation product is the E6 protein product, having a molecular weight of 16K. The structure of the type 5 mRNAs as outlined above was established by using a combination of S1 nuclease and electron microscopic heteroduplex analysis (Stenlund *et al.*, 1985). Yang *et al.* (1985) have identified two mRNAs with a similar structure by sequence studies of cDNA clones. One of the variants differed in having its 5′-end located at nucleotide 7879 instead of nucleotide 89. The observation indicates that promoter elements upstream of coordinate 1.1 sometimes are used to transcribe the E-region. It is, however, unclear whether nucleotide 7879 represents the true cap site for the observed mRNA, since no sequence combinations that are typical for eukaryotic promoters are found near this position. Instead, TATA-boxes are found at coordinate 90 (nucleotide 6859 and 7108) which then might be the promoter. This variant would still be expected to encode the protein product of the E6 ORF (Yang *et al.*, 1985).

Another variant of the type 5 mRNAs differed in having its 5′-end at nucleotide 429 instead of at nucleotide 89. The 5′-end of this mRNA is located in front of the E7 ORF and it may thus encode the protein product of the E7 ORF. It remains, however, to demonstrate that nucleotide 429 represents a bona fide cap site since no 5′-end has been mapped here by the S1 nuclease protection technique. (Stenlund *et al.*, 1985). It is, on the other hand, clear that the CRPV genome contains a promoter in the equivalent position (Danos *et al.*, 1985) and TATA-boxes are found in front of the E7 ORF in the HPV-1a and HPV-6b genomes.

In addition to the spliced cytoplasmic mRNAs, a prominent spliced nuclear RNA species was described by Stenlund *et al.* (1985) (Fig. 4). An intron located between donor site D1 and splice acceptor site A1 is removed from this RNA which has an estimated length of 3.9 kb. It is present in approximately the same quantities as the unspliced RNA precursor which extends from the cap site at nucleotide 89 to the polyadenylation site at coordinate 53. The existence of this spliced nuclear mRNA species suggests that splicing may be retarded at an intermediate stage and, due to its structure, it is likely to be a precursor for the type 4 mRNAs.

In addition to these spliced mRNA species, several colinear RNAs appear to be present in BPV-1-transformed cells. Stenlund *et al.* (1985) have described colinear RNAs which extend from the polyadenylation site at coordinate 53 to either of coordinates 39 (nucleotide 3100), 38 (nucleotide 3000), 32 (nucleotide 2530), or 31 (nucleotide 2430). The RNA species which starts at coordinate 31 may be an unspliced precursor for the type 1 mRNA since it is rather abundant in nuclear RNA and extends from the approximate position of the 5′-end of exon-6 to the polyadenylation site (Stenlund *et al.*, 1985). It is noteworthy that the colinear RNA species which starts at coordinate 39 is heterogeneous and several 5′-ends have been observed between nucleotides 3010 and 3080 which seem to be initiation sites rather than splice acceptor sites. An interesting property of the latter RNAs is that they cover the entire E4 ORF and they might therefore direct the synthesis of the E4 protein. The fact that ATG triplets are located near the 3′-end of the E4 ORF of all papillomavirus genomes so far sequenced suggests that mRNAs indeed should exist, having their capped 5′-end near the start of the E4 ORF.

The exons whose 5′-ends map at coordinates 31 and 32 cover the entire E2 ORF. One of them may correspond to exon-5, present in the type 3 mRNAs.

Yang *et al.* (1985) have also described several colinear RNA species which extend between the polyadenylation site and either of nucleotides 1889, 2360, 2440, and 2534. It is noteworthy that the two latter positions coincide almost exactly with the 5′-ends of the colinear RNAs identified by Stenlund *et al.* (1985). Since the structures of these colinear RNAs were deduced by sequencing cDNA clones, it is impossible to exclude that at least some of them represent truncated cDNA copies of longer mRNA species. Additional work is thus required to establish whether any of the observed colinear RNAs represent functional mRNA species.

B. mRNAs Transcribed in Virus-Producing BPV-1-Induced Fibropapillomas

The mRNA species which are present during a productive BPV-1 infection have been studied by analyzing RNA from bovine warts (Amtmann and Sauer, 1982; Engel *et al.*, 1983). The results show that two sets of RNA are present; one set includes the mRNAs expressed in BPV-1-transformed mouse cells and are presumably synthesized during the "early" phase of the replication cycle. Apart from these mRNAs, several additional RNA species have been described which are unique to the fibropapillomas (Engel *et al.*, 1983). These are transcribed from the same DNA strand as the "early" mRNA species (Amtmann and Sauer, 1982; Heilman *et al.*, 1982) which was already predicted from the nucleotide sequence of the BPV-1 genome (Chen *et al.*, 1982). A characteristic feature of the mRNAs which are produced in productively infected cells is that they have a poly(A) addition site located at coordinate 90 in addition to the one used by "early" mRNAs (Fig. 6). The longest RNA species has its 5'-end at coordinate 90 and covers essentially the entire BPV-1 genome. This RNA is present in a low frequency and is likely to represent an unspliced nuclear RNA precursor.

Baker and Howley (manuscript in preparation) have studied mRNAs, isolated from a BPV-1-induced fibropapilloma by cDNA cloning. Their results identified six mRNA species that appear to be unique for productively infected cells. Five of these mRNAs, which are designated I–V in Fig. 6, are apparently transcribed from a promoter located near position 7250, thus upstream of the promoter that is used for transcription of most "early" mRNAs. Two polyadenylation sites appear to be used in fibropapillomas; one is the polyadenylation site used for the "early" mRNA while the other is located near the 3'-end of the L-region, at position 7175. mRNA species I is a colinear RNA spanning between the "late" promoter and the polyadenylation site in the E-region. mRNA II spans between the

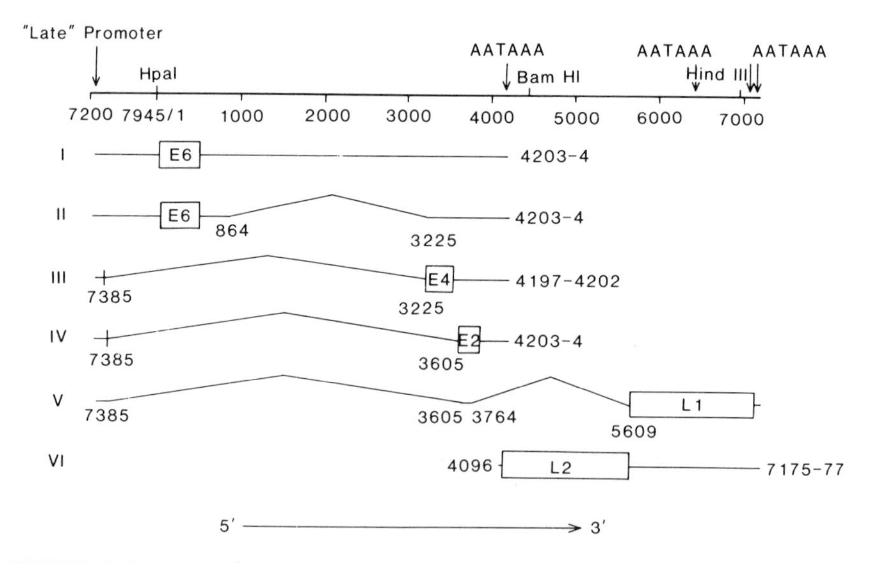

FIGURE 6. Structure of exons present in BPV-1 mRNAs that are unique for productively infected cells (data from Baker and Howley, manuscript in preparation). The exons are indicated as well as the polyadenylation sites. The open translational reading frames which might be translated from the different mRNAs are also shown.

same positions. However, an intron is spliced out between the splice donor site D2 and the splice acceptor A4, as defined in Fig. 3. mRNA species III and IV both contain a short 5′-leader from the "late" promoter region connected to a downstream exon. In species III the splice acceptor A4 appears to be used wheras mRNA IV utilizes a splice acceptor site at position 3605. The latter splice junction is apparently not used in BPV-1-transformed cells. mRNA V consists of three exons (Fig. 6). One is a 5′-leader that is shared with species III and IV. Another extends from a splice acceptor site at 5609, immediately in front of the L1 ORF. The third exon, which is very short, is derived from the 3′-part of the E-region. Finally, RNA species VI seems to be colinear and covers the L2 ORF. Its 5′-end was mapped at position 4096, suggesting that this mRNA, unlike the other "late" mRNAs, is controlled by a promoter within the L-region. Thus papillomaviruses do not seem to display the same type of clear-cut distinction between "early" and "late" genes. Many of the "late" mRNAs apparently contain sequences that are derived from the E-region.

The relationship between the late mRNAs and the ORFs in the L-region is shown in Fig. 6. The RNA species, designated V in Fig. 6, covers the L1 ORF and it is interesting that a typical splice acceptor site is located in front of the first ATG in the L1 ORF (Fig. 2). This mRNA is thus likely to encode the major capsid protein of papillomaviruses (Amtmann and Sauer, 1982; Heilman *et al.*, 1982). A second mRNA species (species VI in Fig. 6) from the L-region covers both the L1 and L2 ORFs. Its 5′-end is located very close to the N-terminal end of the L2 ORF and it is therefore likely to encode the L2 protein.

mRNA species I–IV appear to specify polypeptides encoded by the E-region. Both mRNAs I and II are likely to encode the E6 protein whereas species III and IV seem to encode polypeptides consisting of sequences from the leader and the E4 and E2 ORFs, respectively.

It is clear from the results depicted in Fig. 6 that a specific promoter exists for the L-region mRNAs, located around position 7250 (Baker and Howley, manuscript in preparation). The promoter at coordinate 1 seems to be used for transcription of mRNAs in both BPV-1-transformed cells and productively infected cells. The "late" mRNAs are apparently controlled by a promoter that is located upstream of the E-region. Antitermination would then be required to switch transcription from the "early" to the "late" mode. Two characteristics TATA-motifs are found in the BPV-1 genome near coordinate 90 (nucleotides 6860 and 7109). These do not, however, coincide with the "late" promoter mapped by Baker and Howley (manuscript in preparation). It is thus unclear if promoters are located at these positions.

C. The Transcription Map of the BPV-1 Genome Is Incomplete

Although a very complex pattern has emerged from studies on BPV-1 transcription, it seems likely that the picture still is incomplete and several enigmatic observations need an explanation. An important question concerns the coding capacity of the most abundant mRNA species in transformed cells, the type 1A mRNA. This mRNA lacks ATG triplets in its 3′-exon and translation must therefore initiate in an upstream exon. The leader sequence which is present in type 1A mRNAs contains indeed an ATG triplet. However, if translation is initiated at this ATG, a protein with a predicted length of only 16 amino acids would be synthesized (Yang *et al.*, 1985). An interesting possibility is that it encodes the E5 protein which plays an important role for transformation. Also the structures of the type 1B, type 2B, and type 3 mRNAs need to be elucidated at the molecular level.

Several more mRNAs than shown in Fig. 4 can be predicted to exist from the organization of the BPV-1 genome. No mRNA has yet been detected which has the capacity to encode the E1 protein. The E1 sequences are spliced out from all BPV-1 RNAs except for two nuclear RNAs which, however, are more likely to be RNA precursors than mature mRNAs (Fig. 4). Since the E1 ORF contains sequences which are extremely well conserved between different papillomaviruses, it is likely to be expressed into a protein product and the most probable explanation is that the E1 mRNA has escaped detection due to its very low abundance. Since the E1 ORF of all papillomaviruses so far sequenced contain an ATG triplet near the NH_2-terminus, it is expected that an mRNA species exists in which the NH_2-terminal end of the E1 ORF is located close to the capped 5′-end.

Also the E4 ORFs of all sequenced papillomaviruses contain ATG triplets near their NH_2-terminal ends. It therefore seems likely that an mRNA species should exist which allows translation to initiate at the beginning of the E4 ORF. A class of colinear RNAs that have their 5′-ends located between nucleotides 3010 and 3080 has been identified (Stenlund *et al.*, 1985), which might be unspliced E4 mRNAs.

The E3 ORF provides a problem. No ORF with an equivalent size is present in the corresponding position in any papillomavirus genome other than the BPV-1 genome. It is thus questionable whether it represents a functional exon. Another problem with the E3 ORF in BPV-1 is that it lacks internal ATG triplets and it would have to be connected with a 5′-exon which contains an in-phase ATG in order to be expressed. No splice acceptor site has hitherto been detected beyond nucleotide 3224 and there thus exists no candidate mRNA from which the E3 ORF can be translated. Additional studies are obviously needed to establish whether the E3 ORF is translated or not. This is also true for the E8 ORF.

For transcription of the L-region it remains to be determined whether all mRNAs are controlled by one promoter located at position 7250, or whether promoters also exist within the L-region. It seems possible that the TATA-motif, which is located at nucleotide 4073, immediately preceding the L-region, is part of a functional promoter.

A candidate mRNA for the L1 protein has been detected which consists of a 3′-exon connected to a 5′-leader sequence and a short exon derived from the E-region (Fig. 6). A well-conserved splice acceptor sequence is located immediately in front of the first ATG in the L1 ORF (Fig. 2).

The L2 ORF contains an ATG triplet near its NH_2-terminal end, an mRNA species appears to exist which starts immediately in front of the L2 ORF.

It thus seems likely that future studies will unravel a much more complicated mRNA pattern in cells productively infected or transformed by BPV-1.

D. Transcriptional Organization of the CRPV Genome

CRPV transcription has been studied in several experimental systems including virus-producing tumors in the cottontail rabbit, benign papillomas and carcinomas in the domestic rabbit, and the transplantable VX2 and VX7 carcinomas of the domestic rabbit (Nasseri *et al.*, 1982; Nasseri and Wettstein, 1984a,b; Georges *et al.*, 1984; Danos *et al.*, 1985). The most clear-cut results have been obtained from studies of CRPV transcription in tumors from the domestic rabbit and in the VX2 carcinoma. The latter contains multiple integrated copies of the CRPV genome in a head-to-tail arrangement (McVay *et al.*, 1982; Nasseri *et al.*, 1982; Sugawara *et al.*, 1983; Georges *et al.*, 1984; Nasseri and Wettstein, 1984a,b; Danos *et al.*, 1985). Two major RNA species are present in these cells, having sizes of 1.25 and 2.0 kb (Nasseri *et al.*, 1982; Nasseri and Wettstein, 1984a; Georges *et al.*, 1984; Danos *et al.*, 1985). It is noteworthy that the relative abundance of the two mRNA species differs in benign and malignant tumors (Georges *et al.*, 1984; Nasseri and Wettstein, 1984a,b). In malignant tumors of the domestic rabbit as well as in the VX2 carcinoma, the 1.25-kb mRNA is the most abundant viral mRNA species, in contrast to the benign tumors where the 2.0-kb RNA is more prominent. Both RNA species are spliced and they have a common 3′-exon, located between nucleotides 3714 and 4367 in the CRPV genome (Fig. 7). A common poly(A) addition site for all mRNAs from the E-region is located at the latter position and it is preceded by the polyadenylation signal AATAAA (nucleotide 4348). The 5′ border of the common 3′-exon is somewhat heterogeneous and minor populations of mRNAs appear to exist whose splice acceptor sites map at nucleotides 3745 and 3760 (Nasseri and Wettstein, 1984a; Danos *et al.*, 1985).

The difference between the 1.25- and 2.0-kb mRNAs lies in the structure of their 5′ leaders; both leaders have a common 3′-splice site, located at nucleotide 1371 in the CRPV sequence, whereas their 5′-ends are different. The 5′-end of the 2.0-kb mRNA is located around nucleotide 158, although a minor population of the 2.0-kb mRNAs exists whose 5′-ends map at nucleotide 87 (Danos *et al.*, 1985). The 1.25-kb mRNAs have heterogeneous 5′-ends located between positions 903 and 908. Also in

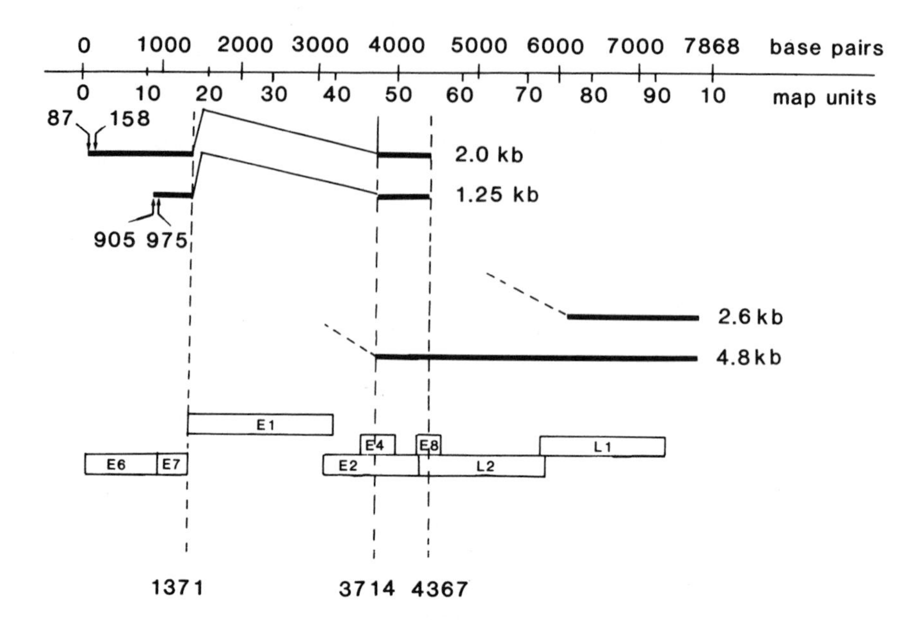

FIGURE 7. Structure of mRNAs expressed in the VX2 carcinoma and CRPV-producing tumors. The exons are indicated by bold lines and the approximate sizes of the mRNAs are indicted at the right-hand side of the figure. The positions of selected splice acceptor and donor sites are also indicated as well as the approximate positions of four different cap sites [Data from Danos *et al.* (1985) and Nasseri and Wettstein (1984a).] The transcripts which map in the L-region are known to contain leader sequences which map in the E-region. The phase location of the leaders are not yet known.

this case a second minor mRNA population exists with 5′-ends between nucleotides 970 and 975 (Danos *et al.*, 1985).

Different promoters appear to be utilized for transcription of the 1.25- and 2.0-kb mRNAs. This conclusion is based on results obtained by the primer extension method (Danos *et al.*, 1985). The sequence TATAT is present at position 59, immediately upstream of one of the cap sites for the 2.0-kb mRNA. A similar sequence, TATAA, occurs at position 130, thus in the appropriate position for the second cap site mapped at nucleotide 158. For the 2.0-kb mRNA the sequence TATAAA has been found at nucleotide 879 at the expected distance from the major cap site, located around nucleotide 905. It is noteworthy that only one functional TATA-sequence has been identified so far in the corresponding region of the BPV-1 genome.

From the structure of the CRPV mRNAs, certain predictions can be made regarding their coding capacities. The 2.0-kb mRNA which starts at nucleotide 87 would cover the entire E6 and E7 ORFs and the 5′-end is located in front of the first ATG of the E6 ORF. It thus seems likely that the 2.0-kb mRNA encodes the E6 protein product. However, the 5′-ends of the majority of the 2.0-kb mRNAs appear to map at nucleotide

158, immediately beyond the first ATG in the E6 ORF. Translation would then have to initiate at the second ATG and the transcription data therefore predict that two different E6 proteins exist, one being 175 and the other being 273 amino acid residues long. The 5′-ends of the 1.25-kb mRNAs are located close to the first ATG in the E7 ORF and this mRNA is thus likely to encode the E7 protein product.

In BPV-1-transformed cells a characteristic mRNA is present in which the E6 and E7 ORFs are fused due to the removal of an intron (the type 4 mRNA). No corresponding mRNA has yet been discovered in CRPV-induced tumors (Nasseri and Wettstein, 1984a; Danos *et al.*, 1985).

Since eukaryotic translation usually initiates at the ATG located closest to the capped 5′-end, it seems unlikely that any sequences are translated from the 3′-exon present in both the 1.25- and 2.0-kb mRNAs. This infers that only ORFs located in the 5′ part of the transforming region of CRPV are translated into protein products in the tumors. A set of minor RNA species has, however, been observed, some of which appear to cover the E2 and E4 ORFs (Danos *et al.*, 1985). It remains to elucidate their molecular structure and also to determine whether these minor RNA species play a role in CRPV transformation.

In the VX7 carcinoma which also contains integrated viral genome copies (Georges *et al.*, 1984; Nasseri and Wettstein, 1984b), a more complex mRNA pattern has been observed (Georges *et al.*, 1984; Nasseri and Wettstein, 1984b). This carcinoma contains in addition prominent 2.1- and 3.1-kb mRNAs as well as several minor mRNAs. Some of those hybridize to fragments from the L-region, indicating that they encode viral capsid polypeptides (Georges *et al.*, 1984). This provides an explanation for the observation that animals bearing the VX7 carcinoma often produce antibodies against virion proteins (Rogers and Rous, 1951; Rogers *et al.*, 1961; Mellors, 1960; Orth *et al.*, 1978). The precise map positions of these mRNAs are, however, not known.

Nasseri and Wettstein (1984) have also studied the mRNAs present in virus-producing papillomas from the cottontail rabbit. In these tumors both "early" and "late" mRNAs are produced, including the 1.25- and 2.0-kb mRNAs which are present in the VX2 transplantable carcinoma and in tumors from the domestic rabbit. Three additional RNA species, having sizes of 0.9, 2.6, and 4.8 kb, were observed which are unique for virus-producing tumors. The precise map position of the 0.9-kb mRNA is unknown. The two larger mRNAs utilize a common poly(A) addition site at coordinate 99 and contain major exons located in the L-region. They are thus likely to encode viral capsid proteins (Fig. 7). The 2.6-kb RNA species is composed of two exons one of which is located between coordinates 79 and 99, thus covering most of the L1 ORF. The second exon is 0.9 kb long and is located in the E-region although its precise position is unknown. Due to its structure the 2.6-kb mRNA is likely to encode the L1 protein although it cannot be excluded that additional E-region sequences are translated from the 5′-exon. The 4.8-kb mRNA is also composed of two exons (Fig. 7). The 3′-exon is 3.7 kb long and covers most of the L-region, as well as sequences from the E-region. The 5′ border of the main exon is located in the E-region and it extends to the poly(A) addition site at coordinate 99 (Fig. 6). The 5′-exon is approximately 1.1 kb long and is located somewhere in the E-region. The 4.8-kb mRNA is heterogeneous and one or two very short introns located at coordinates 79 and/or 61 are spliced out from some species. It is difficult to predict the coding capacity of the 4.8-kb mRNA. One species may encode a protein which includes both L1 and L2 sequences since the intron at coordinate 79 may fuse the L1 and L2 ORFs (Nasseri and Wettstein, 1984a).

E. Common Features between BPV-1 and CRPV Transcription

A comparison between BPV-1 and CRPV gene expression reveal similarities as well as differences. It seems from the nucleotide sequences that a similar transcriptional control region exists in the two genomes, located around coordinate 1. The CRPV genome exhibits a greater complexity with the several 5′-ends mapping in the beginning of the E-region. Primer extension experiments have demonstrated that at least four different cap sites are located in this region, three of which are preceded by TATA-boxes (Danos *et al.*, 1985). Both viruses appear to be expressed differently in transformed/tumor cells, as opposed to productively infected cells. In the transformed cells a subset of mRNAs are synthesized which map in the E-region and which share a common polyadenylation site near the middle of the genome (coordinate 53 in BPV-1 and coordinate 61 in CRPV). The CRPV-induced carcinoma VX7 is an exception since it expresses mRNA sequences also from the L-region (Orth *et al.*, 1978; Georges *et al.*, 1984). In virus-producing cells, additional mRNA species are present which are polyadenylated near the end of the L-region (coordinate 90 in BPV-1 and coordinate 99 in CRPV). Most of the mRNAs which are expressed in transformed cells appear to be spliced in both viruses and a typical feature is that several mRNAs share a common 3′-exon, which maps within the COOH-terminal parts of the E2 and E4 ORFs. This exon is located between coordinates 41 and 53 in BPV-1 and between coordinates 53 and 61 in CRPV. In BPV-1 at least five classes of differentially spliced mRNAs have been recognized in addition to several colinear RNA species. Analyses of RNA from CRPV-induced tumors have so far revealed a less complex pattern and only two major spliced mRNAs, 1.25 and 2.0 kb long, have been characterized in detail. However, recent results (Nasseri and Wettstein, 1984b; Danos *et al.*, 1985) indicate that several additional transcripts are present in CRPV-induced carcinomas, although at a low concentration. The 2.0-kb mRNA of CRPV has an almost identical equivalent in BPV-1-transformed cells, i.e., the mRNA species designated type 5 by Stenlund *et al.* (1985). This mRNA species is composed of a 3′-exon which is linked to a 5′-exon that covers the E6 and E7 ORFs and it is likely to encode the E6 protein product of both viruses. The 1.25-kb mRNA CRPV which is particularly abundant in malignant tumors has a similar structure to that of the 2.0-kb mRNA, the difference being that its 5′-end is located farther downstream. This mRNA seems to be regulated by a separate promoter in CRPV, located around coordinate 12, and it is likely to encode the E7 protein product. Yang *et al.* (1985) have described an mRNA species from BPV-1-transformed cells whose 5′-end is located immediately in front of the E7 ORF and which has a very similar structure as the 1.25-kb mRNA from CRPV. It remains, however, to be demonstrated that the 5′-end of the observed BPV-1 mRNA coincides with the cap site since it has so far only been characterized by sequencing of a cDNA clone. The E7 ORF in BPV-1 contains also an ATG triplet near its NH_2-terminus suggesting that a BPV-1 mRNA species might exist, having a 5′-exon which covers the E7 ORF. On the other hand, an mRNA is known to exist in BPV-1-transformed cells in which the E7 ORF is fused to the E6 ORF. There may thus be no need for separate mRNA, encoding exclusively the E7 protein in BPV-1. It is noteworthy that no intron appears to exist in CRPV which fuses the E6 and E7 ORFs.

A remarkable difference between BPV-1 and CRPV is that the type 1 mRNAs, which are the most abundant mRNA species in BPV-1-transformed cells, appear to have no equivalent in CRPV-induced tumors.

The L-region mRNAs show clear similarities when BPV-1 and CRPV are compared (Figs. 6 and 7). Two major RNA species have been described in cells productively infected with CRPV, being 4.8 and 2.6 kb long. In addition, a 0.9-kb late mRNA may exist in CRPV. The 2.6-kb mRNA in CRPV consists of a 1.7-kb 3′-exon which covers the L1 ORF and it is likely to encode the VP1 protein. This mRNA species is probably equivalent to mRNA species V (Fig. 6) which has been observed in cells productively infected with BPV-1. Both contain a 3′-exon that covers the L1 ORF. In CRPV as in BPV-1, this mRNA species contains a leader sequence, originating from the E-region. The relationship between the 4.8-kb mRNA which is present in cells productively infected with CRPV and the late BPV-1 mRNAs is presently unclear. Different subclasses of the 4.8-kb mRNA seem to exist from which one or two very short introns have been removed one of which appears to fuse the L1 and L2 ORFs. These introns have not been observed in BPV-1.

In summary, it is apparent that the BPV-1 and CRPV genomes are organized and expressed in a similar fashion. There are, however, some clear differences in the structure of the hitherto described mRNAs. The two viruses appear to differ with regard to the number of mRNAs present. It seems likely that some of these discrepancies reflect a lack of information rather than to real differences. The transcriptional maps of the two genomes are likely to become much more complex when additional information accumulates due to the use of more sensitive and accurate analytical methods. It is then likely that the two genomes will turnout to be very similar, also from a transcriptional point of view.

F. Transcriptional Organization of Human Papillomavirus Genomes

So far very little information has been reported concerning the transcription of human papillomavirus genomes. Lehn *et al.* (1984) have studied RNA from condylomata acuminata and Buschke–Löwenstein tumors. They were able to detect viral transcripts in biopsy specimens by Northern blot analysis. One major 1.4-kb species was observed as well as several less prominent 1.7-, 1.85-, 2.7-, 3.2-kb RNAs. Two different 3′-ends were identified, one located somewhere between nucleotides 3917 and 4441 and the other between nucleotides 7232 and 7696. These correspond presumably to the polyadenylation sites for E-region- and L-region-specific mRNAs, respectively. The results infer that the polyadenylation site for the E-region mRNAs is located upstream of the position, predicted by Schwarz *et al.* (1983). Employing different subgenomic fragments as probes it was possible to demonstrate that three "early" transcripts, 1.4, 1.85, and 2.7 kb long, exist in addition to two "late" transcripts, 1.7 and 3.2 kb long. From the analysis it appears that the mRNA patterns in HPV-6- and BPV-1-induced tumors are similar.

Chow and Broker (1984) used a different approach to study HPV-1a transcription. They have cloned the HPV-1a genome in vectors which contain SV40 promoters and the origin for SV40 DNA replication. These constructs were introduced into so-called COS-1 cells which allow the replication of the recombinant molecules. HPV-1a-specific mRNAs were expressed at a low level from the SV40 promoter and electron microscopic heteroduplex analysis was then used to study the viral mRNAs. The results showed that polyadenylation signals were recognized which are located at the end of the E- and L-regions, respectively. Moreover the splice acceptor site that is located in front of the L1 ORF seems to be recognized in this system as well as a splice acceptor site equivalent to A4 in the BPV-1 genome (Fig. 3). Viral mRNAs, expressed in HPV-16 or HPV-18 combining cell lines, have also been studied by various methods (Schwartz *et al.*, 1985). The results show that several spliced mRNAs are expressed consisting of sequences from the E6 ORF connected to cellular sequences.

VII. CONCLUSIONS

The molecular analysis of the organization and expression of papillomavirus genomes has progressed with a remarkable pace. The complete nucleotide sequences are known for several human and animal papillomavirus genomes and approximate transcriptional maps have been established for the BPV-1 and CRPV genomes. It seems from available information that papillomavirus gene expression is more complex than was anticipated. The precise structure of several mRNA species remains to be determined and studies on the control mechanisms which operate on the genome during its productive replication cycle and in papillomavirus-transformed cells are likely to be very rewarding. It is thus expected that the papillomavirus field will continue to advance very rapidly.

ACKNOWLEDGMENTS. The authors are extremely grateful to Jeanette Backman and Ingegerd Schiller for dedicated and patient secretarial work. We would also like to thank Drs. O. Danos, M. Yaniv, F. Wettstein, G. Orth, P. Howley, D. Lowy, D. Groff, W. Lancaster, R. Mitra, G. Sauer, M. Lusky, and M. Botchan for preprints of unpublished work. We are grateful to Göran Magnusson and Chris Welsh for valuable comments on the manuscript. The work performed in the authors' laboratory was supported by grants from the Swedish National Board for Technical Development, the Swedish Cancer Society, the Swedish Medical Research Council, the Swedish Council for Forestry and Agriculture Research, and the Marcus Borgström Foundation.

REFERENCES

Ahola, H., Stenlund, A., Moreno-López, J., and Pettersson, U., 1983, Sequences of bovine papillomavirus type 1 DNA—Functional and evolutionary implications, *Nucleic Acids Res.* **11**:2639–2650.

Ahola, H., Bergman, P., Ström, A.-C., Moreno-Lopez, J., and Pettersson, U., 1986, Organization and expression of the transforming region of the European elk papillomavirus (EEPV), *Gene* **50** (in press).

Amtmann, E., and Sauer, G., 1982, Bovine papilloma virus transcription: Polyadenylated RNA species and assessment of the direction of transcription, *J. Virol.* **43**:59–66.

Banerji, Y., Rusconi, J., and Schaffner, W., 1981, Expression of β-globin gene is enhanced by remote SV40 DNA sequences, *Cell* **27**:299–308.

Boshart, M., Gissmann, L., Ikenberg, H., Kleinheinz, A., Scheurlen, W., and zur Hausen, H., 1984, A new type of papillomavirus DNA: Its presence in genital cancer and in cell lines derived from cervical cancer, *EMBO J.* **3**:1151–1157.

Campo, M. S., Spandidos, D. A., Lang, J., and Wilkie, N. M., 1983, Transcriptional control signals in the genome of bovine papillomavirus type 1, *Nature* **303**:77–80.

Chen, E. Y., Howley, P. M., Levinson, A. D., and Seeburg, P. H., 1982, The primary structure and genetic organization of the bovine papillomavirus type 1 genome, *Nature* **299**:529–534.

Chow, L. T., and Broker, T. R., 1984, Human papilloma virus type 1 RNA transcription and processing in COS-1 cells, in: *Gene Transfer and Cancer* (M. L. Pearson and N. L. Steinberg, eds.) pp. 125–134, Raven Press, New York.

Ciuffo, G., 1907, Innesto positivo con filtrado di verrucae volgare, *G. Ital. Mal. Venereol.* **48**:12–17.

Clad, A., Gissmann, L., Meier, B., and Freese, U. K., 1982, Molecular cloning and partial nucleotide sequence of human papillomavirus type 1a DNA, *Virology* **118**:254–259.

Clertant, P., and Seif, I., 1984, A common function for polyoma virus large-T and papillomavirus E1 proteins? *Nature* **311**:276–279.
Colbère-Garapin, F., Horodniceanu, F., Kourilsky, P., and Garapin, A.-C., 1981, A new dominant hybrid selective marker for higher eukaryotic cells, *J. Mol. Biol.* **150**:1–14.
Cole, S. T., and Streeck, R. E., 1986, Genome organization and nucleotide sequence of human papillomavirus type 33, which is associated with cervical cancer, *J. Virol.* **58**:991–995.
Crawford, L. V., and Crawford, E. M., 1963, A comparative study of polyoma and papilloma viruses, *Virology* **21**:258–263.
Danos, O., and Yaniv, M., 1983, Structure and function of papillomavirus genomes, in: *DNA-virus Oncogenes and Their Action* (G. Klein, ed.), pp. 59–82, Raven Press, New York.
Danos, O., Katinka, M., and Yaniv, M., 1980, Molecular cloning, refined physical map and heterogeneity of methylation sites of papilloma virus type 1a DNA, *Eur. J. Biochem.* **109**:457–461.
Danos, O., Katinka, M., and Yaniv, M., 1982, Human papillomavirus 1a complete DNA sequence: A novel type of genome organization among papovaviridae, *EMBO J.* **1**:231–236.
Danos, O., Engel, L. W., Chen, E. Y., Yaniv, M., and Howley, P. M., 1983, A comparative analysis of the human type 1a and bovine type 1 papillomavirus genomes, *J. Virol.* **46**:557–566.
Danos, O., Georges, E., Orth, G., and Yaniv, M., 1985, Fine structure of the cottontail rabbit papillomavirus mRNAs expressed in the transplantable VX2 carcinoma, *J. Virol.* **53**:735–741.
Dartmann, K., Schwartz, E., Gissmann, L., and zur Hausen, H., 1986, The nucleotide sequence and genome organization of human papilloma virus type 11, *J. Virol.* **151**: 124–130.
de Villiers, E.-M., Gissmann, L., and zur Hausen, H., 1981, Molecular cloning of viral DNA from human genital warts, *J. Virol.* **40**:932–935.
DiMaio, D., Guralski, and Schiller, J. T., 1986, Translatin of open reading frame E5 of bovine papillomavirus is required for its transforming activity, *Proc. Natl. Acad. Sci. USA* **83**:1797–1801.
Dürst, M., Gissmann, L., Ikenberg, H., and zur Hausen, H., 1983, A papillomavirus DNA from a cervical carcinoma and its prevalence in cancer biopsy samples from different geographic regions, *Proc. Natl. Acad. Sci. USA* **80**:3812–3815.
Dvoretzky, I., Shober, R., Chattopadhyay, S. K., and Lowy, D. R., 1980, A quantitative *in vitro* focus forming assay for bovine papilloma virus, *Virology* **103**:369–375.
Engel, L. W., Heilman, C. A., and Howley, P. M., 1983, Transcriptional organization of the bovine papillomavirus type 1, *J. Virol.* **47**:516–528.
Favre M., Breitburd, F., Croissant, O., and Orth, G., 1975, Structural polypeptides of rabbit, bovine and human papillomaviruses, *J. Virol.* **15**:1239–1247.
Ferguson, J., and Davis, R. W., 1975, An electronmicroscopic method for studying and mapping the region of weak sequence homology between SV40 and polyoma DNAs, *J. Mol. Biol.* **94**:135–149.
Finch, J. I., and Klug, A., 1965, The structure of viruses of the papilloma–polyoma type. III Structure of rabbit papilloma virus, *J. Mol. Biol.* **13**:1–12.
Freese, U. K., Schulte, P., and Pfister, H., 1982, Papilloma virus-induced tumors contain a virus-specific transcript, *Virology* **117**:257–261.
Freidman, T., Dolittle, R. F., and Walter, G., 1978, Amino acid sequence homology between polyoma and SV40 tumour antigens deduced from nucleotide sequences, *Nature* **274**:291–293.
Fuchs, P. G., Iftner, T., Weninger, J., and Pfister, H., 1986, Epidermodysplasia verruciformis-associated human papillomavirus 8: Genomic sequence and comparative analysis, *J. Virol.* **58**:626–634.
Georges, E., Croissant, O., Bonneaud, N., and Orth, G., 1984, Physical state and transcription of the cottontail rabbit papillomavirus genome in warts and transplantable VX2 and VX7 carcinomas of domestic rabbits, *J. Virol.* **51**:530–538.

Geraldes, A., 1969, Malignant transformations of hamster cells by cell-free extracts of bovine papillomas (*in vitro*), *Nature* **222:**1283–1284.
Giri, I., Danos, O., and Yaniv, M., 1985, Genomic structure of the cottontail rabbit (Shope) papillomavirus, *Proc. Natl. Acad. Sci. USA* **82:**1580–1584.
Gissmann, L., and zur Hausen, H., 1976, Human papilloma virus DNA: Physical mapping and genetic heterogeneity, *Proc. Natl. Acad. Sci. USA* **73:**1310–1313.
Gissmann, L., Diehl, V., Schultz-Coulon, H.-J., and zur Hausen, H., 1982, Molecular cloning and characterization of human papilloma virus DNA derived from a laryngeal papilloma, *J. Virol.* **44:**393–400.
Groff, D. E., and Lancaster, W. D., 1985, Molecular cloning and nucleotide sequence of deer papillomavirus, *J. Virol.* **56:**85–91.
Hagenbüchle, O., Bovey, R., and Young, R. A., 1980, Tissue-specific expression of mouse α-amylase genes: Nucleotide sequence of isoenzyme mRNAs from pancreas and salivary gland, *Cell* **21:**179–187.
Heilman, C. A., Law, M.-F., Israel, M. A., and Howley, P. M., 1980, Cloning of human papilloma virus genomic DNAs and analysis of homologous polynucleotide sequences, *J. Virol.* **36:**395–407.
Heilman, C. A., Engel, L., Lowy, D. R., and Howley, P. M., 1982, Virus-specific transcription in bovine papillomavirus-transformed mouse cells, *Virology* **119:**22–34.
Jenson, A. B., Rosenthal, J. D., Olson, C., Pass, F., Lancaster, W. D., and Shah, K., 1980, Immunologic relatedness of papillomaviruses from different species, *J. Natl. Cancer Inst.* **64:**495–500.
Jung, A., Sippel, A. E., Grez, M., and Schütz, G., 1980, Exons encode functional and structural units of chicken lysosome, *Proc. Natl. Acad. Sci. USA* **77:**5759–5763.
Kremsdorf, D., Jablonska, S., Favre, M., and Orth, G., 1983, Human papillomaviruses associated with epidermodysplasia verruciformis. II. Molecular cloning and biochemical characterization of human papillomavirus 3a, 8, 10, and 12 genomes, *J. Virol.* **48:**340–351.
Law, M.-F., Lancaster, W. D., and Howley, P. M., 1979, Conserved polynucleotide sequences among the genomes of papilloma viruses, *J. Virol.* **32:**199–207.
Lehn, H., Ernst, T.-M., and Sauer, G., 1984, Transcription of episomal papillomavirus DNA in human condylomata acuminata and Buschke–Löwenstein tumours, *J. Gen. Virol.* **65:**2003–2020.
Lowy, D. R., Dvoretzky, I., Shober, R., Law, M.-F., Engel, L., and Howley, P. M., 1980, *In vitro* tumorigenic transformation by a defined sub-genomic fragment of bovine papilloma virus DNA, *Nature* **287:**72–74.
Lusky, M., and Botchan, M. R., 1984, Characterization of the bovine papilloma virus plasmid maintenance sequences, *Cell* **36:**391–401.
Lusky, M., and Botchan, M. R., 1985, Genetic analysis of the bovine papillomavirus type 1 *trans*-acting replication factors, *J. Virol.* **53:**955–965.
Lusky, M., Berg, L., Weiher, H., and Botchan, M., 1983, Bovine papilloma virus contains an activator of gene expression at the distal end of the early transcription unit. *Mol. Cell. Biol.* **3:**1108–1122.
McVay, P., Fretz, M., Wettstein, F., Stevens, J., and Ito, Y., 1982, Integrated Shope virus DNA is present and transcribed in the transplantable rabbit tumor Vx-7, *J. Gen. Virol.* **60:**271–278.
Matthews, R. E. F., 1982, Classification and nomenclature of viruses, *Intervirology* **17:**1–199.
Meinke, W., and Meinke, G. C., 1981, Isolation and characterization of the major capsid protein of bovine papilloma virus type 1, *J. Gen. Virol.* **52:**15–24.
Mellors, R. C., 1960, Tumor cell location of the antigens of the Shope papillomavirus and the Rous sarcomavirus, *Cancer Res.* **20:**744–746.
Melnick, J. L., Allison, A. C., Butel, J. S., Eckhart, W., Eddy, B. E., Kit, S., Levine, A. J., Miles, J. A. R., Paganbo, J. S., Sachs, L., and Vonka, V., 1974, Papovaviridae, *Intervirology* **3:**106–120.
Moar, M. H., Campo, M. S., Laird, H., and Jarrett, W. H. J., 1981, Persistence of non-intergrated viral DNA in bovine cells transformed *in vitro* by bovine papillomavirus type 2, *Nature* **293:**749–751.

Moreau, P., Hen, R., Wssylyk, B., Everett, R., Gaub, M. P., and Chambon, P., 1981, The SV40 72 base pair repeat has a striking effect on gene expression both in SV40 and other chimeric recombinants, *Nucleic Acids Res.* **9**:6047–6068.

Moreno-López, J., Pettersson, U., Dinter, Z., and Philipson, L., 1981, Characterization of a papilloma virus from the European elk (EEPV), *Virology* **112**:589–595.

Moreno-López, J., Ahola, H., Stenlund, A., Osterhaus, A., and Pettersson, U., 1984, Genome of an avian papillomavirus, *J. Virol.* **51**:872–875.

Nakabayashi, Y., Chattopadhyay, S. K., and Lowy, D. R., 1983, The transforming function of bovine papillomavirus DNA, *Proc. Natl. Acad. Sci. USA* **80**:5832–5836.

Nasseri, M., and Wettstein, F. O., 1984a, Differences exist between viral transcripts in cottontail rabbit papillomavirus induced benign and malignant tumors as well as non-virus producing and virus-producing tumors, *J. Virol.* **51**:706–712.

Nasseri, M., and Wettstein, F. O., 1984b, Cottontail rabbit papillomavirus-specific transcripts in transplantable tumors with integrated DNA, *Virology* **138**:362–367.

Nasseri, M., Wettstein, F. O., and Stevens, J. G., 1982, Two colinear and spliced viral transcripts are present in non-virus-producing benign and malignant neoplasms induced by the Shope (rabbit) papillomavirus, *J. Virol.* **44**:263–268.

Orth, G., Favre, M., and Croissant, O., 1977, Characterization of a new type of human papilloma virus that causes skin warts, *J. Virol.* **24**:108–120.

Orth, G., Breitburd, F., and Favre, M., 1978, Evidence of antigenic determinants shared by the structural polypeptides of (Shope) rabbit papillomavirus and human papillomavirus type 1, *Virology* **91**:243–255.

Pfister, H., Nürnberger, F., Gissmann, L., and zur Hausen, H., 1981, Characterization of a human papillomavirus from epidermodysplasia verruciformis lesions of a patient from Upper Volta, *Int. J. Cancer* **27**:645–650.

Pilacinski, W. P., Glassman, D. L., Krzyzek, R. A., Sadowski, P. L., and Robbins, A. K., 1984, Cloning and expression in *Escherichia coli* of the bovine papillomavirus L1 and L2 open reading frames, *Biotechnology* **1**:356–360.

Proudfoot, N., 1982, The end of the message, *Nature* **298**:516–517.

Rogers S., and Rous, P., 1951, Joint action of a chemical carcinogen and a neoplastic virus to induce cancer in rabbits: Results of exposing epidermal cells to a carcinogenic hydrocarbon at time of infection with the Shope papilloma virus, *J. Exp. Med.* **93**:459–488.

Rogers, S., Kidd, J. G., and Rous, P., 1960, Relationships of the Shope papilloma virus to the cancers it determines in domestic rabbits, *Acta Unio Int. Contra Cancrum* **16**:129–130.

Rösl, F., Waldeck, W., and Sauer, G., 1983, Isolation of episomal bovine papillomavirus chromatin and identification of a DNase I-hypersensitive region, *J. Virol.* **46**:567–574.

Sarver, N., Rabson, M. S., Yang, Y.-C., Byrne, J. C., and Howley, P. M., 1984, Localization and analysis of bovine papillomavirus type 1 transforming functions, *J. Virol.* **S2**:377–388.

Schiller, J. T., Vass, W. C., and Lowy, D. R., 1984, Identification of a second transforming region in bovine papillomavirus DNA, *Proc. Natl. Acad. Sci. USA* **81**:7880–7884.

Schiller, J. T., Vass, W. C., Vousden, K. H., and Lowy, D. R., 1986, The E5 open reading frame of bovine papillomavirus type 1 encodes a transforming gene, *J. Virol.* **57**:1–6.

Schwartz, E., Dürst, M., Demankowski, C., Lattermann, O., Zech, R., Wolfsperger, E., Suhai, S., and zur Hausen, H., 1983, DNA sequence and genome organization of genital human papillomavirus type 6b, *EMBO J.* **2**:2341–2348.

Schwartz, E., Freese, U. K., Gissman, L., Mayer, W., Roggenbuck, B., Stremlau, A., and zur Hausen, H., 1985, Structure and transcription of human papillomavirus sequences in cervical carcinoma cells, *Nature* **314**:111–114.

Seedorf, K., Krammer, G., Durst, M., Suhai, S., and Röwekamp, W. G., 1985, Human papillomavirus type 16 DNA sequence, *J. Virol.* **145**:181–185.

Seif, I., Khoury, G., and Dhar, R., 1979, The genome of human papovavirus BKV, *Cell* **18**:963–977.

Shope, R. E., 1933, Infectious papillomatosis of rabbits; with a note on the histopathology, *J. Exp. Med.* **58**:607–624.

Spalholz, B. A., Yang, Y. C., and Howley, P. M., 1985, Transactivational of a bovine papillomavirus transcriptional regulatory element by the E2 gene product, *Cell* **42**:183–191.

Spandidos, D. A., and Wilkie, N. M., 1983, Host-specificities of papillomavirus, Moloney murine sarcoma virus and simian virus 40 enhancer sequences, *EMBO J.* **2:**1193–1199.

Stenlund, A. J., Moreno-Lopez, J., Ahola, H., and Pettersson, U., 1983, European elk papillomavirus: Characterization of the genome, induction of tumors in animals, and transformation *in vitro, J. Virol.* **48:**370–376.

Stenlund, A., Zabielski, J., Ahola, H., Moreno-Lopez, J., and Pettersson, U., 1985, The messenger RNAs from the transforming region of bovine papilloma virus type 1, *J. Mol. Biol.* **182:**541–554.

Strauss, M. J., Bunting, H., and Melnick, J. L., 1950, Virus-like particles and inclusion bodies in skin papillomas, *J. Invest. Dermatol.* **15:**433–443.

Sugawara, K., Fujinaga, K., Yamashita, T., and Ito, Y., 1983, Integration and methylation of Shope papilloma virus DNA in the transplantable Vx2 and Vx7 rabbit carcinomas, *Virology* **131:**88–89.

Thomas, M., Boiron, M., Tanzer, J., Levy, J. P., and Bernard, J., 1964, *In vitro* transformation of mice cells by bovine papilloma virus, *Nature* **202:**709–710.

Tooze, J., (ed.), 1981, *DNA Tumour Viruses,* Cold Spring Harbor Laboratory, Cold Spring Harbor, N.Y.

Waldeck, W., Rösl, F., and Zentgraf, H., 1984, Origin of replication in episomal bovine papilloma virus type 1 DNA isolated from transformed cells, *EMBO J.* **3:**2173–2178.

Weiher, H., and Botchan, M. R., 1984, An enhancer sequence from bovine papillomavirus DNA consists of two essential regions, *Nucleic Acids Res.* **12:**2901–2916.

Weiher, H., König, M., and Gruss, P., 1983, Multiple point mutations affecting the simian virus 40 enhancer, *Science* **219:**626–631.

Yang, Y. C., Okayama, H., and Howley, P. M., 1985, Bovine papillomavirus contains multiple transforming genes, *Proc. Natl. Acad. Sci. USA* **82:**1030–1034.

CHAPTER 4

The Expression of Papillomaviruses in Epithelial Cells

LORNE B. TAICHMAN AND ROBERT F. LAPORTA

I. INTRODUCTION

Papillomaviruses (PVs) are infectious DNA viruses capable of inducing benign and occasionally malignant growths specifically in keratinizing epithelia. A small number of PVs, such as bovine PV type 1 (BPV-1), are capable of inducing tumors of fibroblasts as well as of epithelial cells. However, only in epithelial cells does papillomavirus undergo vegetative replication. Infection of fibroblasts yields no progeny virus and is therefore a biological dead end.

The focus of this chapter is on the interaction between PVs and keratinocytes, the cells that form a keratinizing epithelium and act as natural hosts for the virus. During transit from the basal layer to the surface, the keratinocyte undergoes a complex series of irreversible changes which are collectively termed keratinization. These cells act in a complex way to provide the metabolic machinery for viral replication. Replication of PVs appears to be closely linked to the keratinization process, and alterations in keratinocyte differentiation can have profound effects on the expression of PV. Investigation of these interactions has been hindered by the lack of a culture model for vegetative viral replication. In spite of

LORNE B. TAICHMAN AND ROBERT F. LAPORTA • Department of Oral Biology and Pathology, School of Dental Medicine, State University of New York, Stony Brook, New York 11794.

this limitation, many important insights have been gained. We begin this chapter with a detailed look at epithelial keratinization followed by a discussion of the interaction between PVs and keratinocytes.

II. EPITHELIUM AND KERATINIZATION

The surface of the body and various internal organs are covered with a stratified squamous epithelium that exhibits different patterns of keratinization. Examples of these patterns are shown in Figs. 1–3. Using the epidermis as an example of fully or orthokeratinizing epithelium (Fig. 1), four distinct strata are observed: basal, spinous, granular, and cornified (Matolsty, 1976). The characteristics of these strata are illustrated in Fig. 4. The basal layer rests on a basement membrane and is the primary site for replicating cells. The spinous layer is characterized by cell enlargement, accumulation of keratin-containing tonofilaments, and numerous intercellular desmosomal junctions which give the surface of the cell its spiny appearance after fixation. In addition, membrane-coating granules appear which are small organelles that have an orderly internal structure consisting of parallel arrays of 20-Å-thick lamellae. Cells of the granular layer are larger in volume and mass, and contain more tonofilaments which now are clustered into bundles. Keratohyalin appears as foci of amorphous material which enmeshes the tonofilaments and is responsible for the cells' characteristic granularity. In the transition from granular to cornified layer, there is a rather abrupt onset of intracellular degradation leading to the loss of nuclei, mitochondria, endoplasmic reticulum, Golgi vesicles, and ribosomes. At this transitional level, membrane-coating granules fuse with the surface of the cell membrane and release membranous lamellae into the intracellular space. In the cornified layer, keratinocytes form flat squames which have a cross-linked cornified envelope applied to the inner surface of the cell membrane. The cytoplasm of these squames is composed entirely of dense bundles of tonofilaments embedded in keratohyalin matrix. The end result of this process is a surface that prevents fluid loss and protects against external trauma and microorganisms.

Historically, keratinization was thought to be a degenerative process initiated and governed by removal of the basal cell from its nutrient source, i.e., when the cell left its position on the basement membrane. Two decades ago, Fukuyama *et al.* (1965) first showed by autoradiography that cells in the spinous and granular layers, as well as the basal layer, incorporated labeled amino acids. It was also observed that some amino acids were preferentially incorporated in the basal and spinous layers while other amino acids were taken up primarily in the granular layers. It was correctly concluded from these observations that keratinocytes in the individual strata were synthesizing different proteins and that keratinization was a process of cell differentiation. Since that time a number

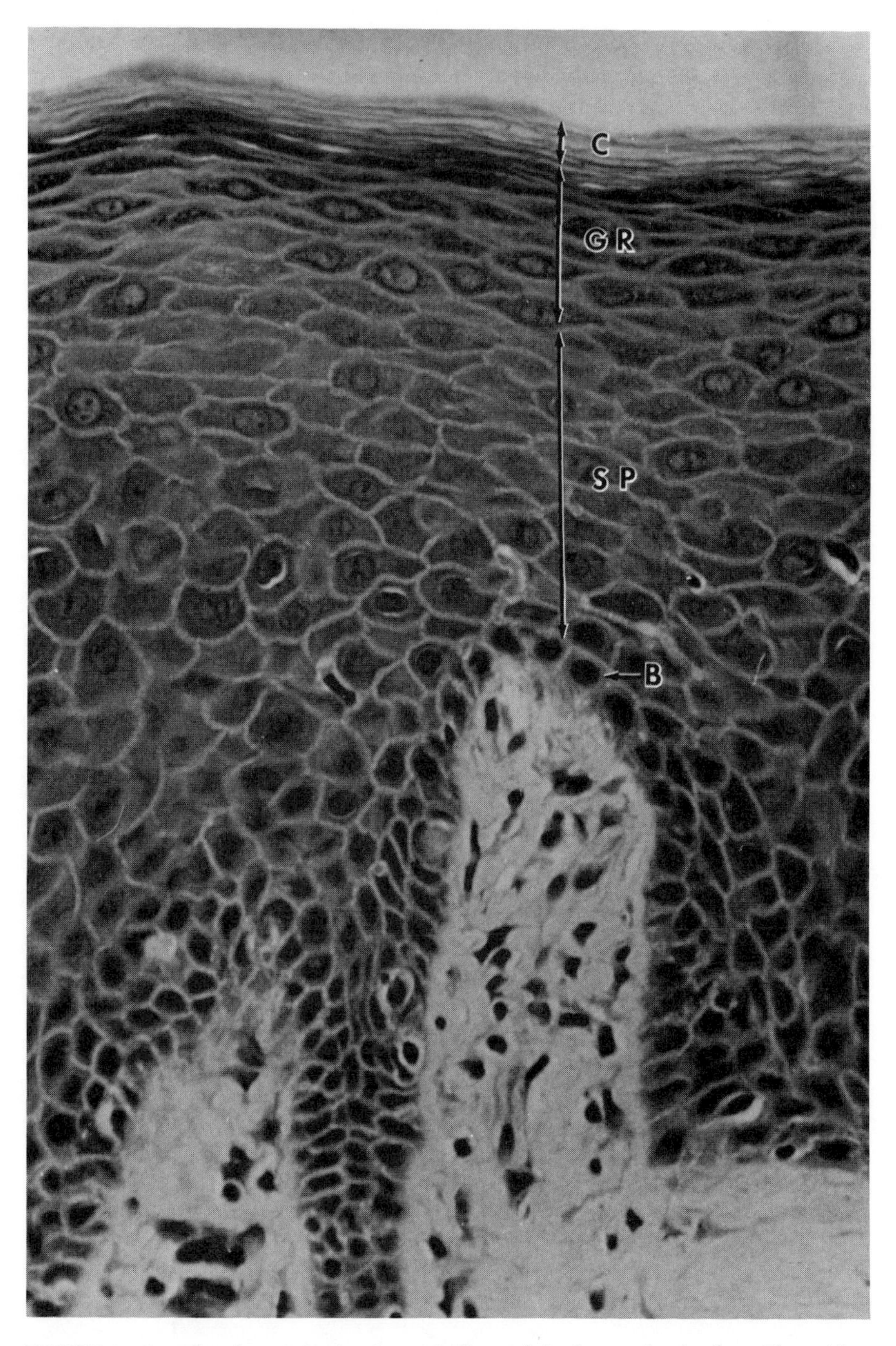

FIGURE 1. Complete keratinization in epithelium of the human hard palate. The epithelium in this section is completely keratinized and exhibits the full spectrum of squamous differentiation including basal (B), spinous (SP), granular (GR), and cornified (C) layers. (Photo courtesy of Dr. Howell O. Archard.)

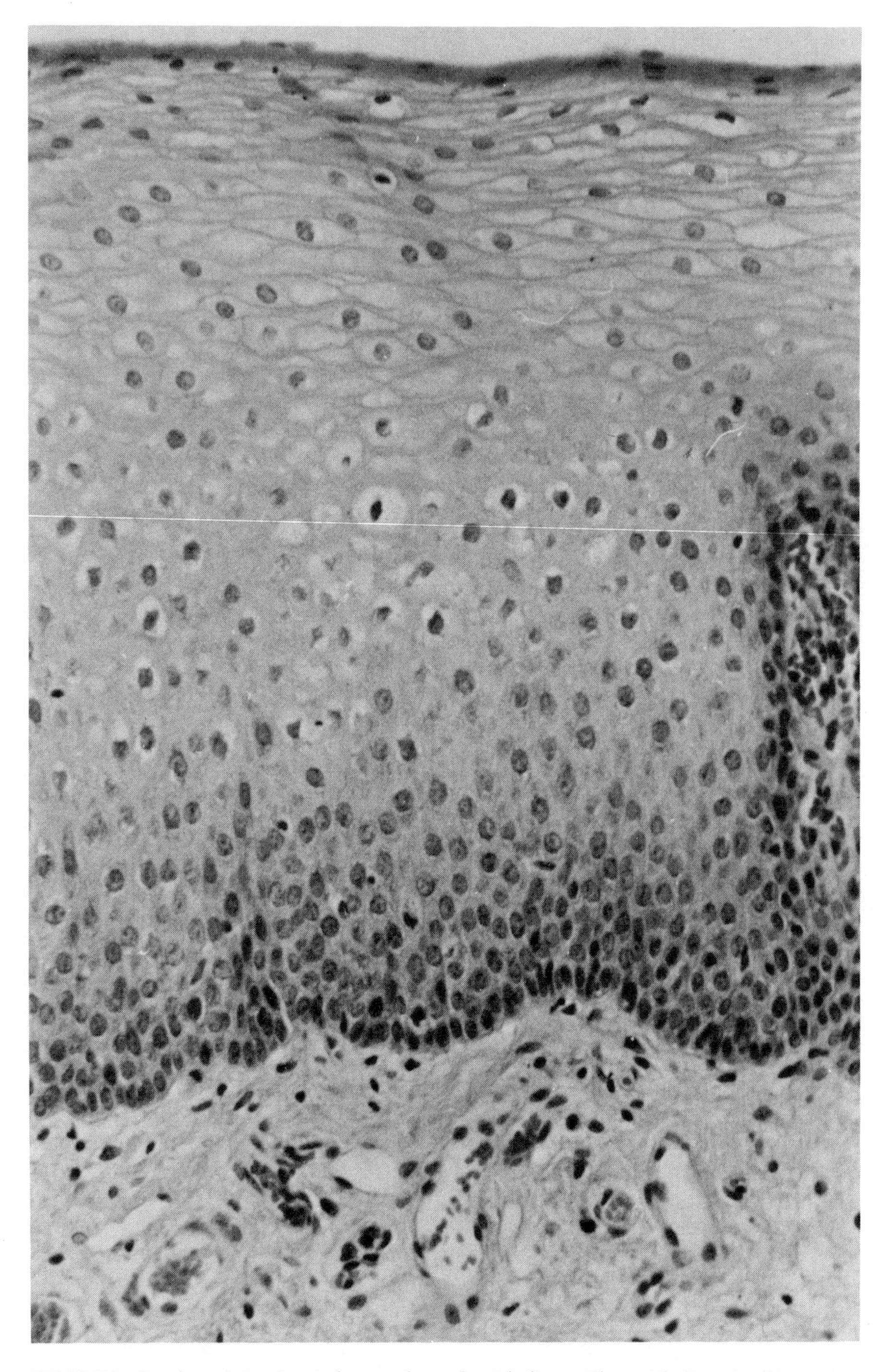

FIGURE 2. Parakeratinization in human buccal epithelium. The epithelium in this section is incompletely keratinized. It does not have a well-developed granular layer and nuclei are present in the cornified layer. (Photo courtesy of Dr. Rosemary Zuna.)

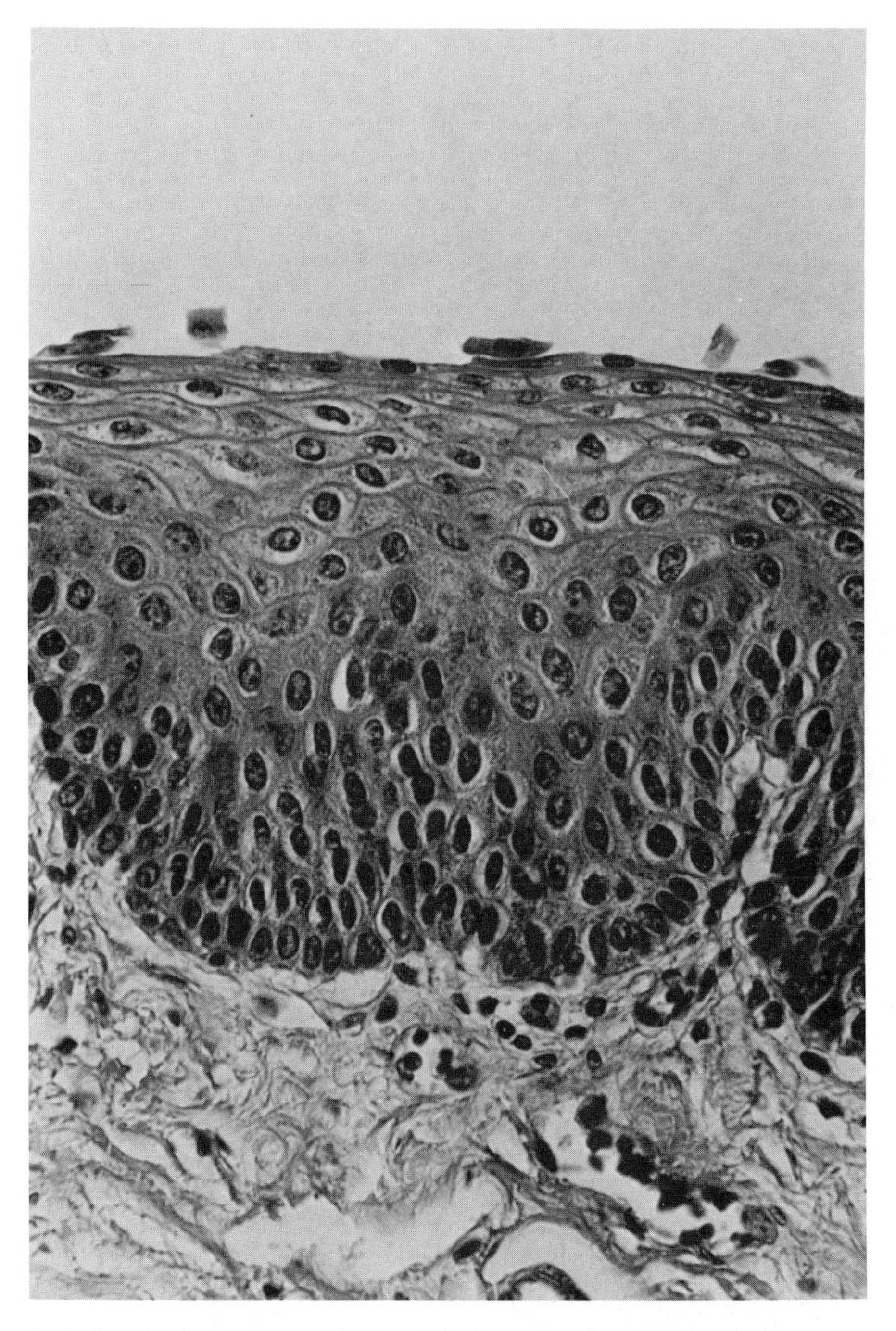

FIGURE 3. Nonkeratinizing epithelium of the human esophagus. The epithelium in this section has no granular or cornified layer. (Photo courtesy of Dr. Howell O. Archard.)

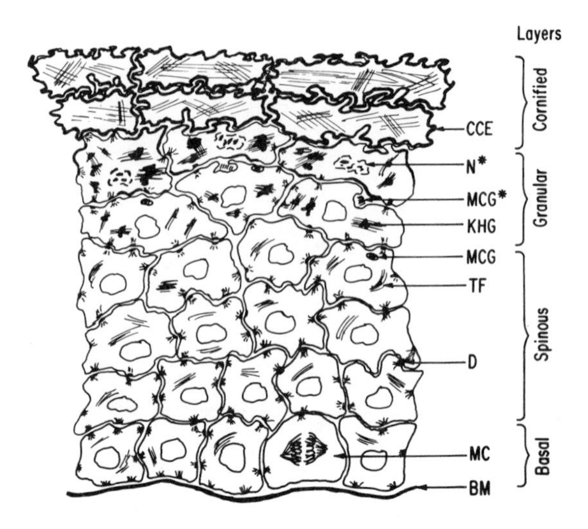

FIGURE 4. Schematic illustration of fully keratinizing epidermis. BM, basement membrane; MC, mitotic cell; D, desmosome; TF, tonofilament; MCG, membrane-coating granule; MCG*, contents of membrane-coating granule being discharged into the intercellular space; KHG, keratohyalin granule; N*, pycnotic nucleus in upper granular layer; CCE, cornified cell envelope.

of studies have shown that keratinizing cells, although no longer capable of replicating, are expressing differentiated traits in a highly regulated and precise manner.

There is no single marker whose expression can be used to define keratinization. Keratinization refers to a series of cellular changes that take place in an ordered sequence. In parakeratinizing epithelia (Fig. 2), keratinization appears incomplete in that pycnotic nuclei are present in the cells of the cornified layer. In "nonkeratinizing" epithelia (Fig. 3), keratinization appears to be arrested at the late spinous–early granular stage. The term "nonkeratinizing" should not be taken to mean that the epithelium is not undergoing differentiation, but rather that the sequence of changes is incomplete as compared to a fully keratinizing surface.

Factors which control the pattern of keratinization are at present poorly understood. It is known that during embryogenesis, inductive influences from the underlying connective tissue play an important role in determining the structure of the epithelium, but the molecular nature of these inductive factors is unclear (for a review see Sengel, 1983). In adult life, continued inductive effects from the connective tissue may be important for some epithelia. For example, the epithelia of the trunk and sole of the guinea pig can respond to inductive effects from grafted subepithelial connective tissue. Epithelia from other sites, such as the tongue, appear to be refractory to redifferentiation (Billingham and Silvers, 1967). Factors such as hormones (e.g., hydrocortisone; Rheinwald and Green, 1975) and vitamins (e.g., vitamin A; Fuchs and Green, 1981) are thought to affect both the rate at which epithelial cells divide and the degree to which they undergo differentiation. Mechanical irritation of palmar and plantar epithelium produces an increase in thickness of the cornified layer (hyperkeratosis); chemical irritation, such as from cigarette smoke, can induce nonkeratinizing epithelium in oral mucosa to undergo complete keratinization (squamous metaplasia).

Three markers of epidermal keratinization have been studied in some detail: keratin proteins, filaggrin, and involucrin. Keratin proteins comprise a family of related water-insoluble proteins (40–70K) which form 10-nm desmosome-associated tonofilaments (Moll *et al.*, 1982). Electrophoretically, these proteins are either acidic or basic, and filament formation involves spontaneous interactions between members from each group (Eichner *et al.*, 1984). Keratin proteins are synthesized in the basal, spinous, and granular layers of the epidermis. In basal keratinocytes, 50 and 58K keratins are produced, and in the spinous and granular layers, additional keratins, 56.5 and 65–67K, are synthesized (Sun *et al.*, 1983). The mRNAs isolated from thin slices at various levels of the epidermis encode for different keratin proteins in an *in vitro* translation system (Fuchs and Green, 1979). Each keratin protein is probably encoded by a unique gene and, therefore, differential synthesis of keratin polypeptides in the various strata is likely to represent differential gene expression at the transcriptional level (Fuchs and Green, 1979, 1980). Filaggrin is a 38K protein which can be isolated from keratohyalin and which has the property of aggregating intermediate filaments into bundles (Dale, 1977). It is synthesized in precursor form (44K) and removal of phosphates leads to its activation (Lonsdale-Eccles *et al.*, 1982). Involucrin is a 92K protein which is one constituent of the cornified envelope. Its synthesis begins in the spinous layer, and in the upper granular layer, it is cross-linked together with other cytoplasmic (Zettergren *et al.*, 1984) and membrane proteins (Simon and Green, 1984) by a calcium-activated transglutaminase to form a chemically resistant cornified envelope (Rice and Green, 1978). The fact that involucrin mRNA is detected only in the upper strata suggests that involucrin synthesis is regulated at the transcriptional level (Watt and Green, 1981). The regulated appearance of these markers has made it evident that keratinization is not a degenerative process, but is a highly ordered sequence of differentiated changes.

The study of keratinization has been given a new dimension with the introduction of techniques for serial and long-term cultivation of keratinocytes. In 1975, Rheinwald and Green demonstrated that mitotically inactivated 3T3 cells could provide substrate support for keratinocyte attachment and early replication while suppressing the growth of human fibroblasts. Keratinocytes derived from newborn human foreskins could thus be maintained for more than 100 cell divisions until senescence (Rheinwald and Green, 1975, 1977; Green, 1978).

The tissue produced *in vitro* is a stratified squamous nonkeratinizing epithelium that resembles morphologically the basal and spinous layers of adult epidermis. Renewal takes place primarily by replication of basal keratinocytes, and terminal differentiation is indicated by withdrawal from the cell cycle, stratification, an increase in cell size and mass, and synthesis of involucrin (Green, 1977; Banks-Schlegel and Green, 1981). The 50 and 58K keratins are present, but the 56.5 and 65–67K keratins are not, indicating the absence of complete keratinization (Nelson and Sun, 1983). Additional keratins (40, 46, 52, and 56K) are seen in cultured keratinocytes that are not found in the intact epidermis (Eichner *et al.*, 1984). Based on the molecular weights of keratins produced, this *in vitro* epithelium has the characteristics of hyperplastic epidermis (Weiss *et al.*, 1984). During differentiation in culture, the rate of protein synthesis increases about twofold and the proportion of keratin protein to total cellular protein remains constant (Taichman and Prokop, 1982). Filaggrin is not synthesized and involucrin is found in differentiated cells shortly after they leave the basal compartment. There is no granular or cornified layer, but in a small number of superficial cells, cornified cell envelopes are seen. The presence of cornified envelopes without a granular or cornified layer indicates that the normal sequence of keratinization is disrupted in this *in vitro* system.

Keratinocytes from a variety of tissues have been grown in culture. Although there are differences in their appearance *in vitro*, these differences are relatively minor (Doran *et al.*, 1980). In the absence of modulating influences from underlying connective tissue and other factors present *in vivo*, keratinocytes express an elemental state of epithelial differentiation. Tissue-specific patterns of keratinization are not irreversibly lost, since injection of cultured keratinocytes into nude (athymic) mice leads to formation of a cyst lined by epithelium that closely resembles the original epithelium in morphology and keratin composition (Doran *et al.*, 1980).

It is not known whether keratinization is initiated in basal cells or occurs upon displacement of the cells from their basal position. When stratification of cultured keratinocytes is blocked by reducing the calcium concentration in the medium, some of the basal cells synthesize involucrin (Watt and Green, 1982). When calcium levels are restored to normal, these involucrin-containing cells are the first ones to be displaced from their basal position. A small population of basal cells in neonatal mouse skin have been shown to contain keratin proteins that are normally found in suprabasalar cells (Schweitzer *et al.*, 1984). These two findings suggest that keratinization can commence while the cells are still in the basal position.

III. BEHAVIORAL CLASSIFICATION OF PVs

PVs are found in a wide variety of animal hosts, and within a number of hosts there is a considerable diversity of PV types (Chapter 1). At the present time, classification of PVs is based on DNA sequence homology and not on any behavioral characteristic of the viral type. The discussion of PV–keratinocyte interactions is complicated by several factors: the fact that PVs infect a wide range of animal hosts, the diversity of viral subtypes in some of these hosts, and the multiplicity of lesions produced by these different subtypes. How is one to analyze the similarities and differences in behavior of various PVs? Since the "aim" of viral infection is to produce progeny, the degree to which a virus is successful at completing its infectious cycle seems to be an appropriate way to catalogue its behavioral interactions. In this chapter we consider all PVs as falling into one of three groups based on the amount of viral progeny present in the benign lesions. The three groups are: *productive,* for lesions in which there are moderate to large amounts of virus present; *weakly productive,* for lesions in which there are minute amounts of virus detected; and *nonproductive,* for lesions in which no detectable virus is produced. Thus, for example, HPV-1, BPV-1, and CRPV, which all induce productive lesions in keratinocytes in their natural hosts, fall within a single group and will be discussed interchangeably in terms of their interactions with keratinocytes.

In Table I the productivity of a number of PVs is summarized. Benign papillomatous lesions of fully keratinizing epithelium produce moderate to large amounts of viral particles. Two exceptions to this are HPV-induced lesions in genital epidermis and BPV-2-induced lesions of the alimentary tract. Benign lesions of certain mucosal surfaces which are incompletely keratinizing (cervix, vagina, urethra, anus, larynx, and oral cavity) and genital epidermis (penis, vulva, and perianal region) are weakly productive. The papillomas induced by CRPV in an unnatural host, the domestic rabbit, are only weakly productive. In instances where benign papillomas progress to malignant carcinomas, the productivity of the malignant lesion is reduced to undetectable levels. BPV-2 lesions of the alimentary tract are nonproductive. The factors which govern the productivity of a lesion will be discussed later in this chapter.

IV. HISTOLOGICAL CHARACTERISTICS OF PV-INDUCED LESIONS

PV-induced lesions in humans appear in three basic types: common warts, flat warts, and genital warts. Common warts (verruca vulgaris) manifest various amounts of papillary hyperplasia with acanthosis, hyperkeratosis, and have a hard roughened surface (Fig. 5). These warts are usually well keratinized and when stained with hematoxylin and eosin, demonstrate basophilic granules in the nuclei of the more superficially located keratinocytes. By electron microscopy, these granules are composed of viral particles arranged in paracrystalline arrays. Mitoses are not limited to the basal layer, but may be observed in the granular layer. This

TABLE I. Productivity of PV-Induced Lesions[a]

PV	Site of infection	Degree of keratinization	Lesions	Comments
1. Productive lesions				
CRPV	Epidermis (cottontail rabbit)	Complete	Papillomas	In domestic rabbits, papillomas are weakly productive
HPV-1	Epidermis (plantar)	Complete	Plantar warts	
2	Epidermis (hand)	Complete	Common warts	
3, 10	Epidermis (arm)	Complete	Juvenile flat warts	
5, 8	Epidermis (arm, trunk)	Complete	Macular lesions in EV	30% progression to malignancy
9, 12, 14, 15, 17, 19, 20–25	Epidermis (arm, trunk)	Complete	Macular lesions in EV	Progression to malignancy variable
BPV-1	Epidermis (teats, penis, and epidermis)	Complete	Fibropapillomas	
2	Epidermis	Complete	Fibropapillomas	
4	Alimentary (esophagus and mouth)	Complete[b]	Papilloma	
5	Epidermis (teat)	Complete	Rice grain fibropapilloma	
6	Epidermis (teat)	Complete	Papilloma	
2. Weakly productive lesions				
HPVX	Mucosa (oral)	Incomplete	Oral focal epithelial hyperplasia	
HPV-6, 11	Epidermis (genital)	Complete	Condyloma acuminata	
	Mucosa (cervix)	Incomplete	Condyloma acuminata	
	(larynx)	Incomplete	Laryngeal warts	
HPV-16, 18	Epidermis (genital)	Complete	Condyloma acuminata	Associated with progression to malignancy
	(cervix)	Incomplete	Condyloma acuminata	Associated with progression to malignancy
3. Nonproductive lesions				
CRPV	Epidermis	Abnormal	Carcinoma	
HPV-5, 8, 14	Epidermis	Abnormal	Carcinoma	In EV
BPV-2	Alimentary (esophagus and mouth)	Complete[b]	Fibropapilloma	

[a] The information in this table represents a compilation of data from the following sources: Howley (1982), Lancaster and Olson (1982), Lutzner *et al.* (1982, 1984), Jarrett *et al.* (1984), Jarrett (1985, and personal communication).
[b] Bovine esophageal epithelium possesses a well-defined cornified layer. The granular layer is unlike the granular layer of the human epidermis.

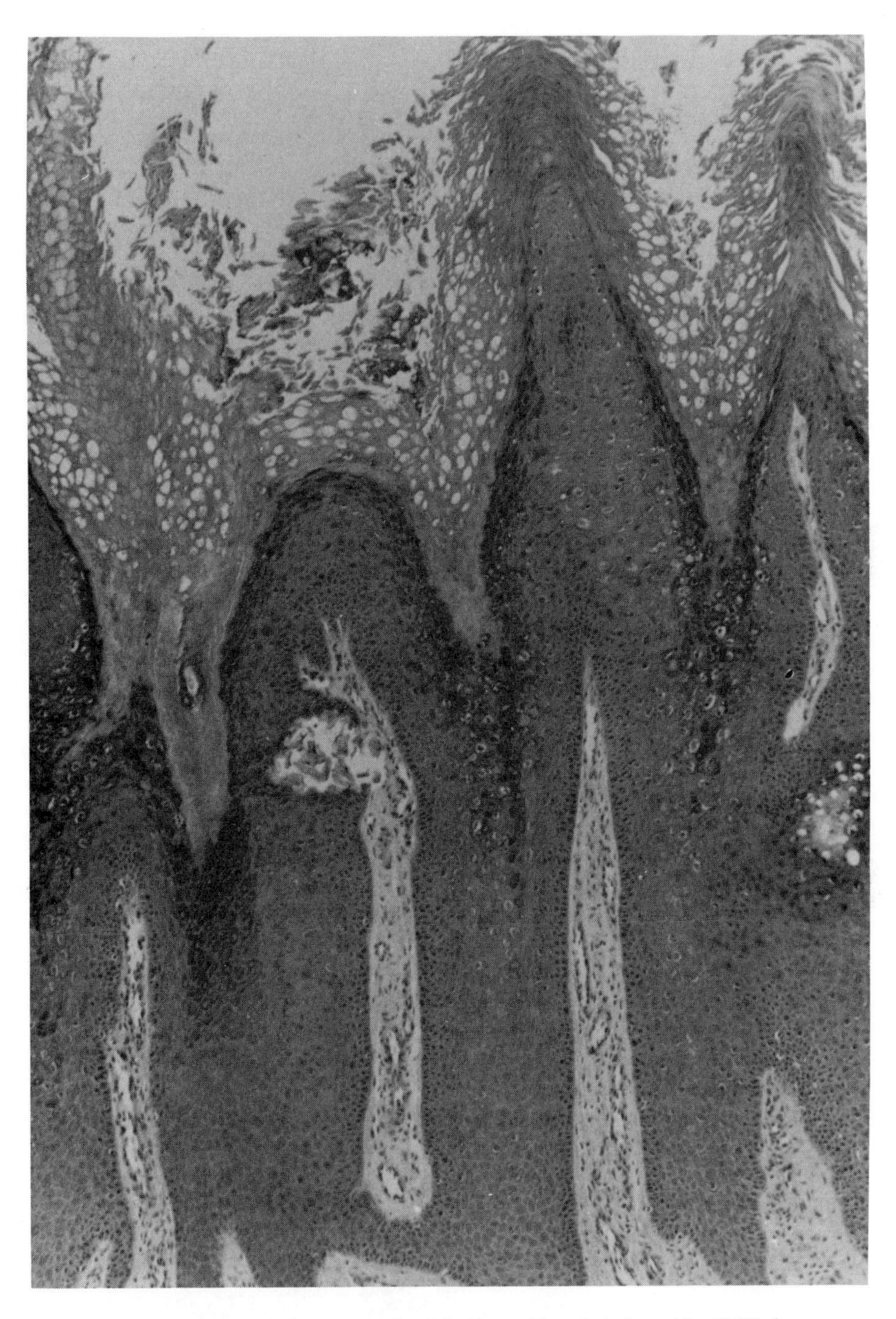

FIGURE 5. Common wart (verruca vulgaris). The epidermis infected by HPV shows papillomatosis, acanthosis (a thickening of the spinous layer), and hyperkeratosis (a thickening of the cornified layer). Viral production is associated with cells that have the perinuclear clearing. (Photo courtesy of Dr. Rosemary Zuna.)

type of wart can occur anywhere on the skin but most often is observed on plantar and palmar surfaces.

Flat warts (verruca plana) (Fig. 6) occur as patches of thickened epithelium without papillary hyperplasia and are seen most often on the arms of children. These lesions may be the color of skin or erythematous, pityriasislike. Spontaneous regression occurs frequently. These warts are uncommon in adults. When they occur with extensive involvement of the body, it is associated with the autosomal recessive disorder, epidermodysplasia verruciformis (Lutzner, 1978). The condition becomes progressively generalized, and in 25% or more of the cases, it progresses to malignancy (see Chapter 8). Verruca plana show hyperkeratosis and acanthosis, but in contrast to common warts there is no papillomatosis. Flat warts contain clear cells which have a perinuclear zone that does not stain. These cells are more evident in flat warts than in common warts (Laurent and Agache, 1975). Viral particles, while not as abundant as in common warts, are present in flat warts.

Genital warts (condylomata acuminata) (Fig. 7) are most frequently found in the cervix, external genitalia, and perianal region. Condylomata occur infrequently in the urethra, bladder, and ureter. Condylomata are characterized by rapid growth and very little keratinization. Mitotic cells

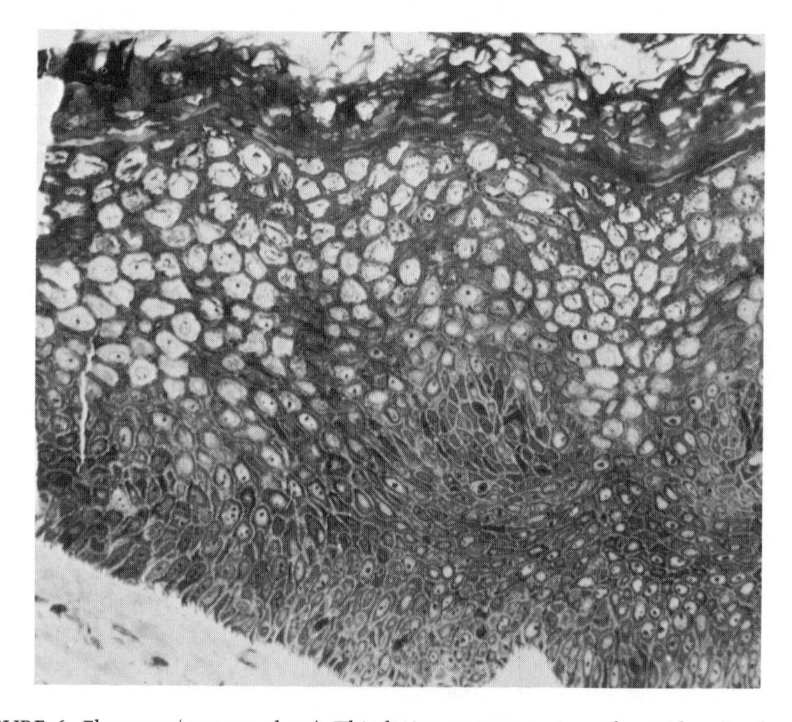

FIGURE 6. Flat wart (verruca plana). This lesion was present on the epidermis of an immunosuppressed renal allograft recipient. There is acanthosis and hyperkeratosis but no papillomatosis. (Photo courtesy of Dr. Marvin Lutzner.)

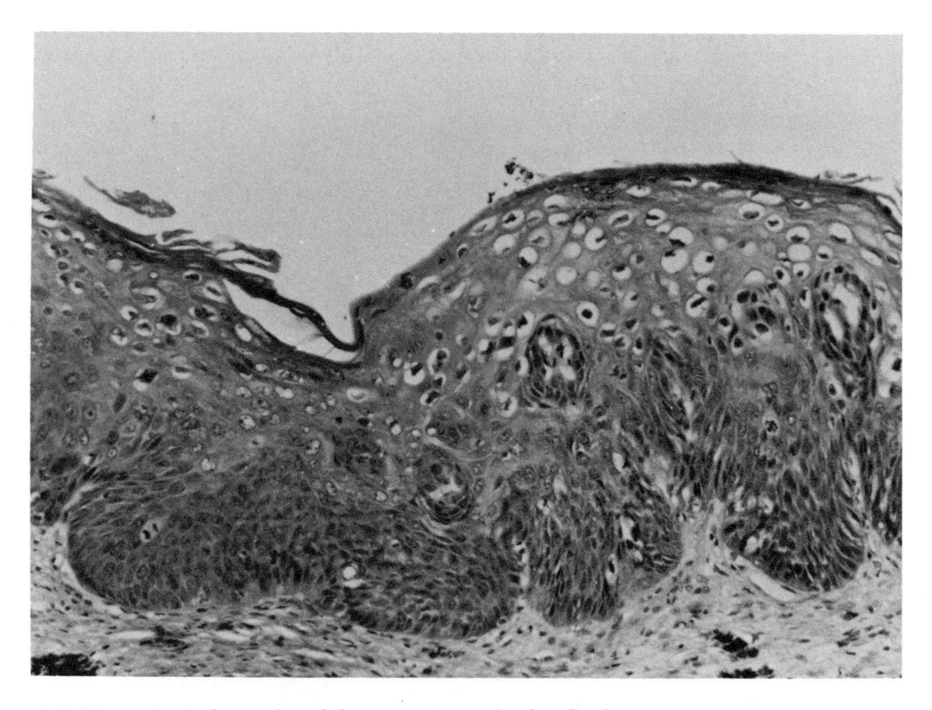

FIGURE 7. Genital wart (condyloma acuminata). This flat lesion was present on a human uterine cervix. There is hyperplasia and prominent koilocytosis. Hyperkeratosis and parakeratosis, although not normal in the cervix, may also occur. Nuclear atypia frequently indicate cervical intraepithelial neoplasia. (Photo courtesy of Dr. Rosemary Zuna.)

are common in basal and parabasal cells, and koilocytes, clear cells of the condyloma, are seen in the upper strata. Viral particles are rare in condylomata, and this along with variation in histology, has caused a great deal of confusion in the classification of these lesions. There is increasing evidence of an association between condylomata and both cervical intraepithelial neoplasia and vulvar intraepithelial neoplasia (see Chapter 9).

V. TRANSFORMATION

Basal keratinocytes are the probable site of viral entry into the epithelium (Fig. 8). Inoculation of viral isolates onto the surface of intact skin is usually unsuccessful and requires some form of abrasion to reach the lower strata (White *et al.*, 1963). Basal cells are nonpermissive for vegetative viral replication and an abortive infection occurs in these cells.

The only detectable effect of PV on basal or germinal cells is to cause transformation. The evidence for this is severalfold (reviewed by Orth *et al.*, 1977): (1) an increase in the mitotic index of basal cells in a papilloma

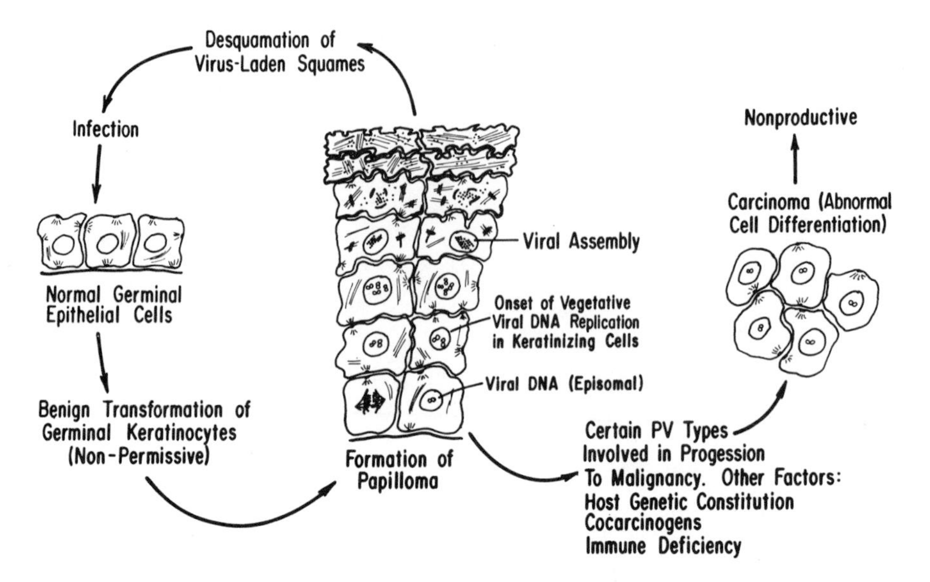

FIGURE 8. Schematic illustration of the life cycle of PV in productive lesions and in lesions that progress to malignancy. (Adapted from Orth *et al.*, 1977.)

(Rashad, 1969); (2) the presence of mitotic cells and cells in S phase in the spinous and granular layers of a papilloma (Shope, 1933; Croissant and Orth, 1967, 1970); (3) the growth of papilloma cells or biopsies of rabbit skin inoculated with CRPV when transplanted to the internal organs of the rabbit or the subcutaneous tissue of mice and hamsters (Rous and Beard, 1935; Greene, 1953); and (4) the enhanced sensitivity of papilloma cells to the carcinogenic effects of DMBA and coal tar (Rous and Friedewald, 1944). In addition, the RNA transcripts present in papillomas induced by CRPV are those that map to the transforming region of the viral genome (Nasseri *et al.*, 1982).

No one has as yet demonstrated PV DNA in basal keratinocytes of papillomas, probably because the number of copies of viral DNA is below the level of detection with *in situ* autoradiography. Keratinocytes cultured from laryngeal warts and therefore derived from the germinative population of papilloma cells, contain about ten copies of HPV-16 DNA per cell (B. Steinberg, personal communication). This is considerably less than that needed for detection by autoradiography. The morphology of basal keratinocytes in papillomas is not different from that in normal tissue (Dunn and Ogilvie, 1968).

There has never been an intranuclear tumor antigen detected in papillomas using the serum from tumor-bearing animals. The reason for this is unclear, though it may be explained by a transforming antigen that is present in undetectable quantities, is only transiently present, or is inaccessible to the antigen-binding cells of the immune system (Orth *et al.*, 1978a). Cells in the circulation may not come into contact with an intranuclear keratinocyte antigen since all the blood supply to the epidermis is confined to the dermis, although Langerhans cells, present in the epidermis, may act as macrophages.

The fact that BPV-1 transforms mouse fibroblastic cell lines in culture has made it possible to define the regions of the PV genome responsible for transformation. Using defined segments of the BPV-1 genome and focus formation with lines of mouse C127 cells, Nakabayashi *et al.* (1983) mapped the transforming function to two discontinuus regions within the 69% *Bam*HI–*Hin*dIII fragment. One region probably represents a control element for gene expression since it can be replaced by a retroviral LTR, i.e., sequences found in retroviruses that control initiation and enhancement of viral transcription. The second region, which cannot be eliminated or substituted, contains an open reading frame and probably encodes a transforming protein(s).

The mechanism of transformation with BPV-1 is unlike other DNA tumor viruses in that BPV-1 does not integrate into the host chromosome but remains as a multicopy plasmid (Lancaster, 1981; Law *et al.*, 1981). Transformation by viral DNA in the plasmid state is not unique to this particular culture system, for PV DNA in papillomas and carcinomas is also found in nonintegrated forms (Lancaster, 1981). Continued presence of plasmid DNA is required for maintenance of transformation. When cells are cured of BPV-1 plasmid DNA by prolonged exposure to interferon, their phenotype reverts to that of a nontransformed cell (Turek *et al.*, 1982). The segments of the BPV-1 genome that maintain the DNA in a plasmid state will be discussed in the next section.

Although mapping of the transformation genes in BPV-1 has been done with nonepithelial cells, it is likely that the same genes are involved in epithelial cell transformation. HPV-1, HPV-5, and CRPV, which do not cause fibromas in their natural hosts, have also been found to induce transformation in cultured mouse C127 cells (Watts *et al.*, 1984). Since the genomic organization of HPV-1 and CRPV is similar to that of BPV-1 (Danos *et al.*, 1984), it is likely that the regions of HPV-1 and CRPV that correspond to the transforming regions in BPV-1 are also responsible for papilloma formation in the natural hosts.

VI. VEGETATIVE VIRAL GROWTH

Vegetative growth of PV can be demonstrated in three ways: (1) by *in situ* autoradiographic detection of PV-specific DNA synthesis; (2) by immunofluorescent detection of capsid antigen indicating activation of the late region of the PV genome; and (3) by electron microscopic detection of intranuclear viral particles indicating complete viral assembly. Using these criteria, vegetative growth of PV has been detected in productive lesions solely in those cells which are keratinized or in the process of keratinization, never in basal cells. Viral capsid antigens have been found in the spinous, granular, and cornified layers (Noyes and Mellors, 1957; Orth *et al.*, 1971). Vegetative viral DNA synthesis has been detected in keratinizing cells (Orth *et al.*, 1971) as early as the first suprabasal layer (Grussendorf and zur Hausen, 1979). The presence of keratohyalin in cells supporting vegetative viral DNA synthesis proves that the keratinocyte is indeed undergoing keratinization and is not simply an arrested cell which has been displaced from its basal position (Croissant *et al.*, 1972). The amount of virus isolated from a single plantar wart can be milligram quantities.

The natural history of productive lesions induced by BPV has been characterized by Jarrett (Jarrett *et al.*, 1984; Jarrett, 1985) in experimental inoculations of cattle. BPVs can be considered in two groups: those which induce fibropapillomas (BPV-1, 2, and 5) and those which produce squamous papillomas (BPV-3, 4, and 6). Fibropapillomas begin as a hyperplasia of fibroblasts in the subepithelial connective tissue. At a later time, the overlying epithelium shows an increased mitotic activity of basal keratinocytes and a gross thickening. In BPV-1 and 2 lesions, the epithelium subsequently penetrates deeply into the connective tissue, whereas in BPV-5 lesions, the epithelial thickening remains plaquelike. The next stage in the pathogenesis of these lesions is a thickening of the granular layer with cellular swelling and distortion. It is at this time and in these granular cells that vegetative viral replication is first seen. In BPV-1 and 2 lesions, cells reaching the cornified layer have nuclei that are completely filled with viral particles. In BPV-4 and 6 lesions, there is an initial thickening of the epithelium due to keratinocyte proliferation. This is followed by selective enlargement of the granular layer whose cells show cytopathic changes and viral particle accumulation. In all of the productive lesions induced by BPV, viral production is seen only in the granular or subcornified layer.

In contrast to productive lesions, some infections of PV are associated with only a minute quantity of progeny virus. In weakly productive lesions of humans, such as anogenital warts (Krzyzek *et al.*, 1980), cervical condylomata (Grussendorf-Conen *et al.*, 1983), and laryngeal warts (Quick *et al.*, 1980), there is insufficient virus in the lesion to form a visible band on cesium chloride density gradients. Viral particles are demonstrable by electron microscopy (Ferenczy *et al.*, 1981; Quick *et al.*, 1980), but in greatly reduced amounts, and what virus is produced is always seen in the nuclei of koilocytic cells (Ferenczy *et al.*, 1981). In weakly productive lesions, all signs of vegetative viral growth are found in the upper strata and never in the lower third of the lesion [e.g., see Oriel and Almeida (1970) for penile, vulval, and anogenital lesions, Braun *et al.* (1982) for laryngeal papillomas, and Morin *et al.* (1981) for cervical lesions]. Whatever viral replication is taking place is confined to cells undergoing terminal differentiation. It is not clear at what level the expression of HPV is blocked in weakly productive lesions. In an analysis of the amount of free DNA and virion DNA in HPV-1, 4, and 6 lesions, the percentage of total viral DNA that was encapsidated was 72.3% for HPV-1 lesions (productive), 13.5% for HPV-4 lesions (moderately productive), and 1.1% for HPV-6 lesions (weakly productive) (Grussendorf-Conen *et al.*, 1983) (productivity of lesions classified by authors). These differences were thought to be the result of either differing amounts of viral protein synthesis, varying efficiencies of encapsidation, or differential stability of virions during isolation.

It is not clear why production of HPV in mucosal surfaces and genital epidermis is reduced. In the absence of a culture model for vegetative growth, it is difficult to analyze the role that viral genes and epithelial factors play in determining the level of viral production. The localization of weakly productive lesions to mucosal surfaces and genital skin would seem to point to epithelial factors. Since vegetative viral growth is confined to differentiating cells, the low level of viral production in mucosal lesions may be related to the incomplete pattern of keratinization exhibited by this epithelium. Genital skin, however, does undergo complete keratinization, so the low level of productivity cannot be related solely to the level of keratinization. Genital skin may be unusual in certain aspects. For example, fibroblasts cultured from genital skin metabolize testosterone at a different rate than do fibroblasts cultured from distant sites (Pinsky *et al.*, 1972; Kaufman *et al.*, 1977).

In past experiments with human volunteers, extracts of warts and condylomata were incubated onto various epithelial surfaces to explore the transmissibility of papillomatous lesions (reviewed by Rowson and Mahy, 1967). In a few instances, extracts from condylomata gave rise to lesions on epidermal surfaces. These experiments might have shed some light on the question of whether the productivity of the lesion was related to viral type or epithelial site, but the viral content of the new lesions was not noted. In addition, heterogeneity of PVs was unknown at the time. Krzyzek *et al.* (1980) have detected HPV-1 and 2 DNA under stringent hybridization conditions in anogenital lesions that were nonproductive for viral particles but contained large amounts of viral DNA. Since these two viral types are normally seen in productive lesions of the epidermis, their presence in weakly productive genital lesions suggests that viral expression in this instance was limited by factors in the anogenital epithelium. The amount of HPV-16 DNA detected in cervical condylomata is much less than the amount recovered from condylomata of the vulval epidermis (L. Gissmann, personal communication). This is another instance where the productivity of a PV type differs in two epithelia that exhibit varying degrees of keratinization.

In cattle BPV-2 induces cutaneous and alimentary fibropapillomas. The cutaneous lesions pass through well-defined stages as described above and are productive of abundant amounts of viral particles. In the alimentary lesions, there is development of the fibroma and downward growth of hyperplastic epithelial cells, but there is no progression to the stage of granular layer thickening and no evidence of viral production at any time

(Jarrett, 1985). The behavior of BPV-2 stands in marked contrast to that of BPV-4 which, in the same epithelium, is productive of viral particles. It would appear that the factors that permit productive replication of BPV-4 in alimentary epithelial cells are ineffective in supporting vegetative replication of BPV-2.

Productive replication of BPV-1 in fibropapillomas is associated with transcription of virus-specific genes. In an analysis of polyadenylated BPV-1-specific RNA molecules, Engel *et al.* (1983) detected transcripts in fibropapillomas that were not present in BPV-1 transformed mouse cells. These transcripts measuring 1.7, 3.7, 3.8, 6.7, and 8.0 kb all hybridized with segments of the genome that code for the major capsid protein and shared a common 3′ terminus at 0.90 map unit. Five transcripts (1.05, 1.15, 1.70, 3.80, and 4.05 kb) common to both fibropapillomas and transformed mouse cells, hybridized to the 69% transforming fragment and shared a 3′ terminus located at 0.53 map unit. Engel *et al.* (1983) suggested that viral transcripts in differentiating keratinocytes of BPV-1-induced papillomas, terminate at a site different from the transcripts made in undifferentiated basal cells. There was insufficient evidence to determine whether these differences arose because of the use of different promoter sequences or different splicing mechanisms.

Replication of PV DNA in basal and keratinizing cells may occur by different mechanisms. Recently, Lusky and Botchan (1984) have defined two regions in the 69% *Bam*HI–*Hin*dIII transforming fragment of BPV-1 that acts to maintain viral DNA in the plasmid state. When ligated to molecules of foreign DNA and transfected into C127 cells harboring resident BPV-1 DNA, these plasmid maintenance regions keep the newly introduced DNA in a nonintegrated state. The suggestion was made that these segments of BPV DNA acted by providing origins of replication. The suggestion was also made that these two putative origins of replication might be under different control, one region active in maintaining the BPV genome in basal keratinocytes and the other active in vegetative replication in keratinizing cells. Steinberg *et al.* (1983) have detected HPV-6 DNA in what appears to be histologically normal epithelium adjacent to HPV-6-containing laryngeal warts. The presence of HPV DNA in normal epithelium confirms the clinical suspicion that PV can exist in a latent state. The mechanism whereby latency is maintained may be through utilization of only one of two putative origins of replication in the PV genome.

It is not known what factors in differentiating keratinocytes are responsible for acquisition of the permissive phenotype. However, studies with other viral systems may help elucidate possible mechanisms. In cultured human keratinocytes, vegetative adenovirus-2 replication does not occur in basal cells, whereas more differentiated suprabasal cells do allow the production of infectious viral particles (LaPorta and Taichman, 1981). In cultured mouse teratocarcinoma cells, expression of SV40 and polyoma virus is dependent on cell differentiation (for review see Maltzman and Levine, 1981). In undifferentiated teratocarcinoma cells, there is no expression of early viral proteins or replication. However, when these cells are induced to differentiate, they become permissive for the production of SV40 early proteins and for lytic replication of polyoma (Teich *et al.*, 1977; Topp *et al.*, 1977). Viral mutants that are able to overcome the host restriction in undifferentiated teratocarcinoma cells contain genetic rearrangements in portions of the genome that may act as regulators of RNA transcription (Fujimura *et al.*, 1981). This suggests that one of the cellular events leading to a permissive phenotype for SV40 and polyoma, and possibly PV, may be a change in the factors that regulate gene transcription.

There is a predilection for HPV types to infect distinct epithelial surfaces. Danos *et al.* (1982, 1983, 1984) have studied the sequences of several PVs and have described putative regulatory regions including polymerase II promoterlike sequences. However, when transfected into 3T6 or HeLa cells which are not associated with productive PV replication, these sequences produced very low to undetectable levels of transcripts (Heilman *et al.*, 1982; Giri *et al.*, 1983). Since the control regions were ineffective in these two cells lines, Danos *et al.* (1984) proposed that PV control elements may require factors present in the differentiating keratinocytes. Danos *et al.* also suggested that the predilection for PVs to infect certain sites may reflect a requirement for cellular factors present in keratinocytes at these sites. PV subtypes may have evolved by adapting to different patterns or keratinocyte differentiation found in specific epithelia.

VII. CYTOPATHIC EFFECTS

The histological features of common and flat warts as well as condylomata have been described in Section IV. Specific cytopathic effects are seen in association with a number of HPV types (Orth *et al.*, 1981; Gross *et al.*, 1982; Laurent *et al.*, 1982). For example, papillomatous lesions of the epidermis induced by either HPV-1, 2, 3, 4, or 7 can be distinguished in three ways: (1) by a number of histological criteria including the degree of papillomatosis, acanthosis, and hyperkeratosis; (2) the amount, size, and distribution of keratohyalin granules; and (3) the pattern of perinuclear cytoplasmic clearing (Orth *et al.*, 1981). Flat warts induced by HPV-3 appear as raised areas of skin, whereas HPV-5-induced flat warts are characterized by flat, erythematous, pityriasislike lesions (Orth *et al.*, 1978b).

Cytopathic effects for some PV-induced lesions begin at different levels in the epidermis depending on the viral type. For example, cytopathic alterations in HPV-1-induced warts are seen early in the spinous layer, whereas in HPV-2 lesions, cytopathic effects are not apparent until the late spinous stage (Gross *et al.*, 1982). One interpretation for this is that the cellular conditions that permit late viral gene expression may not be the same for the two HPV types.

Naturally occurring and artificially induced papillomas of the cottontail rabbit vary in viral content over three log units from virus-poor to virus-rich (Rashad and Evans, 1978). Virus-rich papillomas are characterized by marked fragmentation of the cornified layer and a well-developed granular layer which contains abundant keratohyalin granules. Virus-poor papillomas show a thicker, more homogeneous cornified layer and a poorly defined or nonexistent granular layer. Since these lesions arise in the same sites and from the same virus, the histological differences are likely to have resulted from the different amounts of viral products.

Little is known concerning the effects of PV on specific aspects of keratinocyte metabolism. In HPV-1-induced lesions, there is a disruption of the synthesis of certain keratin proteins and the appearance of large amounts of 16 and 17K proteins in the spinous and granular cells (Croissant *et al.*, 1985). There are subtle changes in the keratin composition of superficial cornified cells in a variety of human papillomas (Staquet *et al.*, 1981). The most common finding is a decrease in the relative amount of 67K keratin. Changes in keratin composition are seen in a variety of epidermal disorders, and these changes seem to correlate with the degree of hyperplasia in the lesion (Weiss *et al.*, 1984). DNA replication in the epidermis takes place in basal cells and to some extent in the immediate suprabasal cells. In rabbit papillomas, incorporation of tritiated thymidine is seen throughout the basal, spinous, and granular layers (Orth *et al.*, 1971), an observation not made in chemically induced papillomas (Rashad, 1969). The incorporation of labeled precursor is not entirely the result of viral DNA synthesis, for mitoses are apparent in CRPV-induced papillomas in cells located in the granular layer containing keratohyalin granules (Croissant and Orth, 1970).

VIII. EXPRESSION OF PV IN CULTURED KERATINOCYTES

Vegetative growth of PV has never been reproducibly documented in culture using either cells grown from warts or cells infected *in vitro* (for review see Taichman *et al.*, 1984). Keratinocytes derived from common hand warts have been grown in culture for up to 70–100 days (Niimura *et al.*, 1975). The cultured cells had no detectable viral antigens or viral particles even though the parent warts did contain viral particles. Primary cultures of domestic rabbit papillomas and human laryngeal papillomas display some mild microscopic abnormalities, but do not support vegetative viral replication (Orth and Croissant, 1968; Steinberg *et al.*, 1982). Keratinocytes cultured from laryngeal papillomas maintain the same copy number of HPV-16 DNA as in the original lesion, but are not productive for viral antigens or viral particles (Steinberg *et al.* 1983). It must be noted that viral particles are rarely seen in the papillomas of the domestic rabbit or human larynx.

HPV-1 particles can infect human foreskin keratinocytes in culture. In the early passes, HPV-1 DNA is detected as monomeric episomes at about 100 copies per cell (LaPorta and Taichman, 1982). There is no evidence for viral DNA integration into the cellular genome. There is also no indication of vegetative viral replication, late gene expression, or cell transformation. HPV-1 DNA replication was demonstrated by the incorporation of labeled thymidine. However, with repeated passages, the amount of viral DNA diminishes and is eventually lost (Reilly *et al.*, 1985). Thus, HPV-1 DNA is not stably maintained in foreskin keratinocytes. In addition, the amount of HPV-1 DNA per cell does not increase during keratinocyte differentiation in culture. Since cell DNA replication ceases upon terminal differentiation, the lack of increase in HPV-1 DNA during differentiation must mean that viral DNA does not replicate in terminally differentiated cells. Thus, one of the blocks to vegetative growth of PV in cultured keratinocytes is the inability of differentiating cells to support viral DNA replication. There is also an inability to maintain HPV-1 DNA as a stable episome over many passages.

Since keratinocytes in culture do not differentiate fully, attempts have been made to induce late viral gene expression by enhancing the degree of keratinization. Vitamin A is an inhibitor of keratinization *in vitro*. Its removal from the medium leads to synthesis of keratin proteins normally seen in the upper strata and formation of a well-defined stratum corneum (Fuchs and Green, 1981). When HPV-1-infected keratinocytes are grown in medium lacking vitamin A, there is no evidence for production of HPV viral proteins or particles (LaPorta, 1982).

Cultured keratinocytes re-form an epithelium similar to the original tissue when injected into the subepithelial tissue of nude mice (Doran *et al.*, 1980). Keratinocytes infected with HPV-1 do not contain demonstrable viral particles or viral antigens when induced to form cysts in nude mice (LaPorta, 1982). The injected cells may not have had sufficient time to manifest lytic viral growth since *in vivo* warts take 2–3 months to develop following inoculation (Rowson and Mahy, 1967) and the cysts in the nude mice begin to regress after 4–6 weeks.

Another approach to induce productive viral replication in keratinocytes has been to grow HPV-infected cells in the presence of the tumor promoter TPA (12-*O*-tetradecanoylphorbol 13-acetate). Tumor promoters are known to enhance the amount of BPV-1 DNA synthesized in nontransformed infected mouse fibroblasts (Amtmann and Sauer, 1982). Infected keratinocytes grown in the presence of 10^{-7} M TPA showed a two- to fourfold increase in the amount of viral DNA relative to cellular DNA, but no viral particles were seen in these cells (LaPorta, 1982).

Vegetative growth of PV in papillomas occurs in cells which are differentiating but nevertheless are derived from transformed cells. PVs are atypical in this regard. DNA tumor viruses, such as SV40 or polyoma virus, do not undergo vegetative replication in the cells they transform.

Depending on the origin of the cells, there is either transformation or viral replication. Burnett and Gallimore (1983) transfected cloned HPV-1 DNA into cultured fetal trunk keratinocytes along with SV40 origin-minus DNA. One clone was isolated which was transformed, presumably by the SV40, and which contained integrated HPV-1 DNA and small amounts of virus-specific RNA. There was, however, no evidence of viral antigens, vegetative viral DNA synthesis, or viral particles. Integration of the HPV-1 genome into the host chromosome probably precluded its undergoing lytic replication.

Cells cultured directly from papillomas are usually not immortalized (Orth and Croissant, 1968; Niimura *et al.*, 1975; Steinberg *et al.*, 1982). Although not remarked upon in the literature, investigators have generally found that keratinocytes present in papillomas are difficult to culture and senesce before any useful experiments can be done. Papilloma cells cultured from human laryngeal warts contain HPV-16 DNA, but these cells cannot routinely be grown beyond four passes (Steinberg *et al.*, 1982). There has been a report, however, of a permanent line of keratinocytes established from a human laryngeal papilloma. These cells are unable to grow in soft agar or produce tumor nodules in nude mice (Leventon-Kriss *et al.*, 1983). The recent discovery that a new HPV type, HPV-18, is present in the human cervical carcinoma line designated HeLa may signify that immortalization may have originally been related to the HPV-18 genome (Boshart *et al.*, 1984).

The current thinking in the study of transformation is that there is a spectrum of phenotypes for the transformed cell. The methods available for selecting transformed cells are ones which select for the most malignant phenotype. Typically, transformed fibroblasts are immortalized, can grow in soft agar, have reduced media requirements, have altered density-dependent growth characteristics, can form tumors in nude mice, and have reduced levels of fibronectin and plasminogen activator. However, not all transformed cells share this spectrum of traits (Steinberg *et al.*, 1978).

The phenotypic traits of transformed keratinocytes have not been studied to the same extent as those of transformed fibroblasts. One of the most striking differences is the alteration in density-dependent growth. Normal keratinocytes are not contact inhibited and typically form a multilayered tissue. Transformed keratinocytes no longer maintain this characteristic but become density dependent in their growth and form a monolayer (Rheinwald and Beckett, 1981). Keratinocytes transformed *in vitro* by SV40 or murine sarcoma viruses are immortalized, have reduced requirements for serum and growth factors, and are unable to grow in soft agar (Steinberg and Defendi, 1979; Weissman and Aaronson, 1983). Keratinocytes cultured from naturally occurring squamous carcinomas are immortalized, with a few exceptions cannot grow in soft agar, do not form tumors in nude mice, and do not form a multilayered tissue *in vitro* (Rheinwald and Beckett, 1980, 1981).

The difficulty in isolating transformed keratinocytes, either from papilloma cells cultured *in vitro* (Orth and Croissant, 1968; Niimura *et al.*, 1975; Steinberg *et al.*, 1982) or from normal keratinocytes infected with HPV-1 (LaPorta and Taichman, 1982), is puzzling. In papillomas, there is a benign transformation and certain aspects of the normal differentiation program occur in the papilloma cells. Keratinocyte lines established from a variety of malignant squamous cell carcinomas exhibit blocks to the expression of a number of *in vitro* markers of keratinization, namely stratification, involucrin synthesis (Rheinwald and Beckett, 1981), and withdrawal from the cell cycle (Albers and Taichman, unpublished findings). The malignant cells continue to synthesize keratin proteins, but some of the keratins that are specific to the original tissue are not made (Nelson and Sun, 1983). Immortalization, which is so vital to the study of transformed cells *in vitro*, may not be a property of PV-transformed keratinocytes. The methods currently available for selecting and maintaining malignantly transformed keratinocytes may be unsuitable for cultivating the type of transformed cell induced by PV *in vivo*.

IX. EXPRESSION OF PV IN MALIGNANT CELLS

The progression from a benign papilloma to a malignant carcinoma is accompanied by cessation of productive PV replication (Syverton, 1952). This phenomenon has been examined with CRPV where progression to malignancy occurs in 25% of tumor-bearing cottontail rabbits and 75% of tumor-bearing domestic rabbits (Kreider and Bartlet, 1981). Malignancy usually begins at the periphery of a papilloma as an ulcerated region with rolled margins. In the cottontail rabbit, there are about 1000 copies of CRPV DNA per cell in the papilloma, whereas in the carcinoma, the copy number drops to about 80 per cell (Watts *et al.*, 1983). In the domestic rabbit, there are 200–300 copies of CRPV DNA per cell in both benign and malignant lesions. Benign papillomas in the domestic rabbit are only weakly productive for CRPV replication. In the carcinomas of either rabbit there are no rearrangements or deletions of the CRPV genome. All of the viral DNA is present as monomeric episomes with the exception of a minor amount that is present as multimeric episomes (Wettstein and Stevens, 1982; Watts, 1983). In certain instances, viral DNA in the carcinoma is found in the integrated state (Wettstein and Stevens, 1982). Two major transcripts (1.3 and 2.0K) are present in both the benign papilloma and the malignant carcinoma, although the relative amounts of each are different in the two tumors (Nasseri *et al.*, 1982).

In the CRPV–rabbit system, host genetic factors determine if the virus will cause a benign proliferation of the skin or, alternatively, will progress to a cancerous state. In both the cottontail and the domestic rabbit, the papillomas are produced on fully keratinizing epidermis, although in the latter animal, benign lesions are weakly productive of PV particles. The nature of the host factors regulating PV replication is unknown. When the benign papillomas progress to carcinomas, the pattern of keratinization is profoundly altered. Thus, transformation associated with productive viral replication correlates with continued keratinocyte differentiation, whereas nonproductive carcinomas show loss of normal cellular differentiation.

Epidermodysplasia verruciformis (EV) is a rare human disease in which there are progressive, generalized, flat warts and pityriasis versicolorlike lesions associated with a number of HPV types (Lutzner *et al.*, 1984; Orth, this volume). In these lesions, viral particles are readily seen, indicating productive HPV replication (Ruiter and Van Mullem, 1966; Aaronson and Lutzner, 1967; Jablonska *et al.*, 1968). In a large number of these patients, the benign lesions progress to carcinoma *in situ* or to squamous cell carcinoma. Among the HPV types that are recovered from the benign lesions, HPV-5, 8, and 14 are found with regularity in the squamous cell carcinomas, suggesting that these three HPVs have a high oncogenic potential (Lutzner *et al.*, 1984). Metastasized squamous cell carcinomas of patients with EV have been found to contain HPV-5 in an unintegrated form (Ostrow *et al.*, 1982). In the malignant lesions, viral particles and antigens can no longer be detected although viral DNA persists (Aaronson and Lutzner, 1967; Jablonska *et al.*, 1972, 1976).

The inverse relationship between productivity of a PV-induced lesion and the degree of neoplasia is graphically illustrated in the case of the human cervix where benign lesions and low-grade cervical intraepithelial neoplasias coexist in the same specimen (Reid, 1983). In flat condylomata or subclinical infections of the cervix, 36% of the lesions are positive for viral antigens whereas in cervical intraepithelial neoplasia, only 9% of the lesions contain viral antigens (Reid *et al.*, 1984). There is also statistical and histopathological evidence suggesting that condylomata, which may exist as a subclinical PV infection, and cervical intraepithelial neoplasia represent the extremes of a single disease spectrum (Crum *et al.*, 1983; Reid *et al.*, 1984).

The cessation of productive PV replication that accompanies malignant conversion is consistent with the hypothesis that productive viral replication requires a certain level or pattern of keratinization. Histopathic changes seen in malignant lesions indicate that the process of keratinization is deranged. In fact, the degree to which the normal pattern of keratinization is lost correlates well with the aggressiveness of the neoplastic lesion. Whatever the factors are that confer a permissive phenotype upon a differentiating keratinocyte in a productive lesion, these factors are absent or rendered ineffective when the cell becomes malignant.

It is not clear how the malignant phenotype actually alters PV replication. Experiments currently under way in the authors' laboratory are aimed at exploring the interrelationship between PV replication and malignancy. When two lines of squamous cell carcinoma (SCC12 and SCC25) were infected with HPV-1 particles, multiple copies of episomal viral DNA were detected in the nucleus and were observed to incorporate tritiated thymidine (Reilly *et al.*, 1985). However, after 12–15 cell doublings, the amount of viral DNA diminished to undetectable levels.

X. CONCLUSIONS

In this chapter we have attempted to correlate the expression of PVs with the differentiative changes that take place in keratinizing epithelia. It is evident that vegetative replication of all PVs, whether in productive or weakly productive lesions, takes place in keratinizing cells. Although formal proof is lacking, it seems reasonable to conclude that vegetative replication of PV requires factors that are found only in terminally differentiating keratinocytes. It is also evident that some association exists between the pattern of epithelial keratinization and the productivity of the PV-induced lesion. Lesions that produce moderate to large amounts of virus are found only in epithelia which undergo complete keratinization. PV infections of mucosal surfaces where keratinization is incomplete are weakly productive. Malignant conversion of benign papillomas is accompanied by the cessation of all vegetative PV replication. However, complete keratinization is not always associated with productive PV infections. In humans, genital skin is unusual in that while fully keratinizing, it does not support productive viral replication beyond minute amounts. In domestic rabbits, CRPV-induced epidermal lesions are weakly productive and BPV-2 infections of fully keratinizing alimentary epithelium in cattle are nonproductive. Other factors such as PV type and host genetic constitution clearly play a role, although at this stage in our understanding of PV biology, it is difficult to envisage how these factors might operate.

From an evolutionary point of view, the strategy used by PVs to produce infectious progeny is ingenious. By infecting, but not undergoing lytic replication in basal keratinocytes, the germinative pool of cells in the epithelium is spared from the cytopathic effects of late viral gene expression. Virus production takes place only in terminally differentiated cells that are no longer in the replicative pool and that are destined to be shed into the environment. In this way spread of mature viral particles is achieved and continued growth of host keratinocytes is ensured.

There appears to be a specific association of certain HPV types with distinct areas of skin. One possible reason for this specificity may be that different PV types have evolved in response to the various types of keratinization found in different body areas. Danos *et al.* (1984) made this suggestion following an analysis of the similarities in sequence homology among HPV-1, HPV-6, BPV-1, and CRPV. It was noted that HPV-1 was closer to CRPV than to HPV-6, and that BPV-1 showed the least homology with the other three viruses. Danos *et al.* pointed out that HPV-1 and CRPV infect epidermal keratinocytes in their respective hosts, whereas HPV-6 infects genital epithelium and BPV-1 shows unique biological features such as the induction of fibropapillomas. The fact that the homology seemed to relate more to the type of epithelium infected than to the host led Danos *et al.* to suggest that the main pressure determining PV evolution and divergence is keratinocyte differentiation.

Many of the questions concerning the control of PV expression in keratinocytes await an *in vitro* model for vegetative viral growth. It is precisely the complexity of the interaction between PV and the keratinocyte that has rendered a culture model so elusive. An understanding of the expression of PV will surely lead to new insights into the biology of keratinization.

REFERENCES

Aaronson, C. M., and Lutzner, M. A., 1967, Epidermodysplasia verruciformis and epidermoid carcinoma: Electron microscopic observations, *J. Am. Med. Assoc.* **201:**775–777.

Amtmann, E., and Sauer, G., 1982, Activation of non-expressed bovine papilloma virus genomes by tumor promoters, *Nature* **296:**675–677.

Banks-Schlegel, S., and Green, H., 1981, Involucrin synthesis and tissue assembly by keratinocytes in natural and cultured human epithelia, *J. Cell Biol.* **90:**732–737.

Billingham, R. E., and Silvers, W. K., 1967, Studies on the conservation of epidermal specificities of the skin and certain mucosa in adult mammals, *J. Exp. Med.* **125:**429–446.

Boshart, M., Gissmann, L., Ikenberg, H., Kleinheinz, A., Scheurlen, W., and zur Hausen, H., 1984, A new type of papillomavirus DNA, its presence in genital cancer biopsies and in cells derived from genital cancer, *EMBO J.* **3:**1151–1157.

Braun, L., Kashima, H., Eggleston, J., and Shah, K., 1982, Demonstration of papillomavirus antigens in paraffin sections of laryngeal papillomas, *Laryngoscope* **92:**640–643.

Burnett, S. T., and Gallimore, P. H., 1983, Establishment of a human keratinocyte cell line carrying complete human papilloma type-1 genomes: Lack of vegetative viral synthesis upon keratinization, *Virology* **64:**1509–1520.

Croissant, O., and Orth, G., 1967, Mise en évidence par autoradiographie d'une synthèse d'ADN dans les cellules en voie de kératinisation des papillomes provoqués par le virus de Shope chez le lapin domestique, *C.R. Acad. Sci.* **265:**1341–1346.

Croissant, O., and Orth, G., 1970, Contrôl de l'activité mitotique en voie de kératinisation de papillomes en croissance provoqués par le virus de Shope chez le lapin domestique, *C.R. Acad. Sci.* **270:**2609–2612.

Croissant, O., Dauguet, C., Jeanteur, P., and Orth, G., 1972, Application de la technique d'hybridization moleculaire *in situ* à la mise en evidence au microscope electronique, de la replication vegetative de l'ADN viral dans les papillomes provoqués par le virus de Shope chez le lapin cottontail, *C.R. Acad. Sci.* **274:**614–617.

Croissant, O., Breitburd, F., and Orth, G., 1985, Specificity of cytopathic effect of cutaneous human papillomaviruses, in: *Clinics in Dermatology* (S. Jablonska and G. Orth eds.), Vol. 3, pp. 43–55, J. B. Lippincott, Philadelphia.

Crum, C. P., Egawa, K., Barron, B., Fenoglio, C. M., Levine, R. U., and Richart, R. M., 1983, Human papillomavirus infection (condyloma) of the cervix and cervical intraepithelial neoplasia: A histopathologic and statistical analysis, *Gynecol. Oncol.* **15:**88–94.

Dale, B. A., 1977, Purification and characterization of a basic protein from the stratum corneum of mammalian epidermis, *Biochim. Biophys. Acta* **491:**193–204.

Danos, O., Katinka, M., and Yaniv, M., 1982, Human papillomavirus la complete DNA sequence: A novel type of genome organization among papovaviridae, *EMBO J.* **1:**231–236.

Danos, O., Engel, L. W., Chen, E. Y., and Yaniv, M., 1983, A comparative analysis of the human type 1a and bovine type 1 papillomavirus genomes, *J. Virol.* **46:**557–566.

Danos, O., Giri, I., Thierry, F., and Yaniv, M., 1984, Papillomavirus genomes: Sequences and consequences, *J. Invest. Dermatol.* **83:**7s–11s.

Doran, T. I., Vidrich, A., and Sun, T.-T., 1980, Intrinsic and extrinsic regulation of skin, corneal, and esophageal epithelial cells, *Cell* **22:**17–25.

Dunn, A. E. G., and Ogilvie, M. M., 1968, Intranuclear virus particles in human genital wart tissue: Observations on the ultrastructure of the epidermal layer, *J. Ultrastruct. Res.* **22:**282–295.

Eichner, R., Bonitz, P., and Sun, T.-T., 1984, Classification of epidermal keratins according to their immunoreactivity, isoelectric point, and mode of expression, *J. Cell Biol.* **98:**1388–1396.

Engel, L. W., Heilman, C. A., and Howley, P. M., 1983, Transcriptional organization of bovine papilloma type 1, *J. Virol.* **47:**516–528.

Ferenczy, A., Braum, L., and Shah, K. V., 1981, Human papillomavirus (HPV) in condylomatous lesions of the cervix, *Am. J. Surg. Pathol.* **5:**661–670.

Fuchs, E., and Green, H., 1979, Multiple keratins of cultured human keratinocytes are translated from different mRNA molecules, *Cell* **17:**573–582.

Fuchs, E., and Green, H., 1980, Changes in keratin gene expression during terminal differentiation of the keratinocyte, *Cell* **19:**1033–1044.

Fuchs, E., and Green, H., 1981, Regulation of terminal differentiation of cultured human keratinocytes by vitamin A, *Cell* **25:**617–625.

Fujimura, F. K., Deininger, P. L., Friedmann, T., and Linney, E., 1981, Mutation near the polyoma DNA replication origin permits productive infection of F9 embryonal carcinoma cells, *Cell* **23:**809–814.

Fukuyama, K., Nakamura, T., and Bernstein, I. A., 1965, Differentially localized incorporation of amino acids in relation to epidermal keratinization in the newborn rat, *Anat. Rec.* **152:**525–536.

Giri, I., Jouanneau, J., and Yaniv, M., 1983, Comparative studies of the expression of linked *Escherichia coli gpt* gene and BPV1 DNA in transfected cells, *Virology* **127:**385–396.

Green, H., 1977, Terminal differentiation of cultured human epidermal cells, *Cell* **11:**405–416.

Green, H., 1978, Cyclic AMP in relation to proliferation of the epidermal cell: A new view, *Cell* **15:**801–811.

Greene, H. S. N., 1953, The induction of Shope papilloma in transplants of embryonic rabbit skin, *Cancer Res.* **13:**58–63.

Gross, G., Pfister, H., Hagedorn, M., and Gissmann, L., 1982, Correlation between human papillomavirus type and histology of warts, *J. Invest. Dermatol.* **78:**160–164.

Grussendorf, E. I., and zur Hausen, H., 1979, Localization of viral DNA replication in sections of human warts by nucleic acid hybridization with complementary RNA of human papilloma virus type 1, *Arch. Dermatol. Res.* **264:**55–63.

Grussendorf-Conen, E. I., Gissmann, L., and Holters, J., 1983, Correlation between content of viral DNA and evidence of mature virus particles in HPV-1, HPV-4, and HPV-6 induced virus acanthomata, *J. Invest. Dermatol.* **81:**511–513.

Heilman, C. A., Engle, L., Lowy, D. R., and Howley, P. M., 1982, Virus specific transcription in bovine papilloma virus transformed mouse cells, *Virology* **119:**22–34.

Howley, P. M., 1982, The human papillomaviruses, *Arch. Pathol. Lab. Med.* **106:**429–432.

Jablonska, S., Biczysko, W., Jakubowicz, K., and Dabrowski, J., 1968, On the viral etiology of epidermodysplasia verruciformis Lewandowsky–Lutz: Electron microscopic studies, *Dermatologica,* **137:**113–125.

Jablonska, S., Dabrowski, J., and Jakubowicz, K., 1972, Epidermodysplasia verruciformis as a model in studies on the role of papovaviruses in oncogenesis, *Cancer Res.* **32:**583–589.

Jablonska, S., Maciejewski, W., Dabrowski, J., and Langer, A., 1976, Epidermodysplasia verruciformis, in: *Biomedical Aspects of Human Wart Virus Infection* (M. Prunieras, ed.), pp. 113–121, Foundation Merieux, France.

Jarrett, W. F. H., 1985, The natural history of bovine papillomavirus infections, in: *Advances in Viral Oncology* (G. Klein, ed.), Vol. 5, pp. 83–101, Academic Press, New York.

Jarrett, W. F. H., Campo, M. S., O'Neil, B. W., Laird, H. M., and Coggins, L. W., 1984, A novel bovine papillomavirus (BPV-6) causing true epithelial papillomas of the mammary gland skin: A member of a proposed new BPV group, *Virology* **136:**255–264.

Kaufman, M., Straisfeld, C., and Pinsky, L., 1977, Expression of androgen-responsive properties in human skin fibroblast strains of genital and nongenital origin, *Somat. Cell Genet.* **3:**17–25.

Kreider, J. W., and Bartlet, G. L., 1981, The Shope papilloma–carcinoma complex of rabbits: A model system of neoplastic progression and spontaneous regression, *Adv. Cancer Res.* **35:**81–108.

Krzyzek, R. A., Watts, S. L., Anderson, D. L., Faras, A. J., and Pass, F., 1980, Anogenital warts contain several distinct species of human papillomavirus, *J. Virol.* **36:**236–244.

Lancaster, W. D., 1981, Apparent lack of integration of bovine papilloma virus DNA in virus induced equine and bovine tumor cells and virus transformed mouse cells, *Virology* **108:**251–255.

Lancaster, W. D., and Olson, C., 1982, Animal papillomaviruses, *Microbiol. Rev.* **46:**191–207.

LaPorta, R. F., 1982, Effect of cell differentiation on expression of viral agents in cultured human keratinocytes, Doctoral thesis, State University of New York, Stony Brook.

LaPorta, R. F., and Taichman, L. B., 1981, Adenovirus type-2 infection of human keratinocytes: Viral expression dependent upon the state of cellular maturation, *Virology* **110:**137–146.

LaPorta, R. F., and Taichman, L. B., 1982, Human papilloma viral DNA replicates as a stable episome in cultured human epidermal keratinocytes, *Proc. Natl. Acad. Sci. USA* **79:**3393–3397.

Laurent, R., and Agache, P., 1975, Ultrastructure of clear cells in human viral warts, *J. Cutan. Pathol.* **2:**140–148.

Laurent, R., Kienzler, J. L., Croissant, O., and Orth, G., 1982, Two anatomoclinical types of warts with plantar localization: Specific cytopathological effects of papillomavirus, *Arch. Dermatol. Res.* **274:**101–111.

Law, M.-F., Lowy, D. R., Dvoretzky, I., and Howley, P. M., 1981, Mouse cells transformed by bovine papillomavirus contain only extrachromosomal viral DNA sequences, *Proc. Natl. Acad. Sci. USA* **78:**2727–2731.

Leventon-Kriss, S., Ben-Shoshan, J., Barzilay, Z., Shahar, A., and Leventon, G., 1983, Human epithelial cell line established from a child with juvenile laryngeal papillomatosis, *Isr. J. Med. Sci.* **19:**505–514.

Lonsdale-Eccles, J. D., Teller, D. C., and Dale, B. A., 1982, Characterization of a phosphorylated form of the intermediate filament-aggregating protein filaggrin, *Biochemistry* **21:**5940–5948.

Lusky, M., and Botchan, M. R., 1984, Characterization of the bovine papilloma virus and plasmid maintenance sequences, *Cell* **39:**391–401.

Lutzner, M. A., 1978, Epidermodysplasia verruciformis. An autosomal recessive disease characterized by viral warts and skin cancer. A model for viral oncogenesis, *Bull. Cancer* **65:**169–182.

Lutzner, M., Kuffer, P., Blanchet-Bardon, C., and Croissant, O., 1982, Different papillomaviruses as the causes of oral warts, *Arch. Dermatol.* **118:**393–399.

Lutzner, M. A., Blanchet-Bardon, C., and Orth, G., 1984, Clinical observations, virologic studies, and treatment trials in patients with epidermodysplasia verruciformis, a disease induced by specific human papillomaviruses, *J. Invest. Dermatol.* **83:**18s–25s.

Maltzman, W., and Levine, A. J., 1981, Viruses as probes for development and differentiation, in: *Advances in Virus Research,* (Lauffer, M. A., Bang, F. B., Maramorosch, K., and Smith, K. M., eds.), Vol. 26, pp. 65–116, Academic Press, New York.

Matolsty, A. G., 1976, Keratinization, *J. Invest. Dermatol.* **67**:20–25.

Moll, R., Franke, W. W., Schiller, D. L., Geiger, B., and Krepler, R., 1982, The catalog of human cytokeratins: Patterns of expression in normal epithelia, tumors and cultured cells, *Cell* **31**:11–24.

Morin, C., Braum, L., Casas-Cordero, M., Shah, K. V., Roy, M., and Meisel, A., 1981, Confirmation of the papillomavirus etiology of condylomatous cervix lesions by the peroxidase–antiperoxidase technique, *J. Natl. Cancer Inst.* **66**:831–835.

Nakabayashi, Y., Chattopadhyay, S. K., and Lowy, D. R., 1983, The transforming function of bovine papillomavirus DNA, *Proc. Natl. Acad. Sci. USA* **80**:5832–5836.

Nasseri, M., Wettstein, F. O., and Stevens, J. G., 1982, Two colinear and spliced viral transcripts are present in non-virus-producing benign and malignant neoplasms induced by Shope (rabbit) papilloma virus, *J. Virol.* **44**:263–268.

Nelson, W. G., and Sun, T.-T., 1983, The 50- and 58-kdalton keratin classes as molecular markers for stratified squamous epithelia: Cell culture studies, *J. Cell Biol.* **97**:244–251.

Niimura, M., Pass, F., Wooley, R., and Soutor, C. A., 1975, Primary tissue culture of human wart derived epidermal cells (keratinocytes), *J. Natl. Cancer Inst.* **54**:563–569.

Noyes, W. F., and Mellors, R. C., 1957, Fluorescent antibody detection of the antigens of the Shope papillomavirus in papillomas of the wild and domestic rabbit, *J. Exp. Med.* **106**:556–562.

Oriel, J. D., and Almeida, J. D., 1970, Demonstration of papilloma virus particles in human genital warts, *Br. J. Vener. Dis.* **46**:37–42.

Orth, G., and Croissant, O., 1968, Caractères des cultures de première explantation de cellules de papillomes provoqués par le virus de Shope chez le lapin domestique, *C.R. Acad. Sci.* **266**:1084–1087.

Orth, G., Jeanteur, P., and Croissant, O., 1971, Evidence for and localization of vegetative viral DNA replication by autoradiographic detection of RNA–DNA hybrids in sections of tumors induced by Shope papilloma virus, *Proc. Natl. Acad. Sci. USA* **68**:1876–1880.

Orth, G., Breitburd, F., Favre, M., and Croissant, O., 1977, Papillomaviruses: Possible role in human cancer, in: *Origins of Human Cancer* (H. H. Hiatt, J. D. Watson, and J. A. Winston, eds.), pp. 1043–1068, Cold Spring Harbor Laboratory, Cold Spring Harbor, N.Y.

Orth, G., Jablonska, S., Breitburd, F., Favre, M., and Croissant, O., 1978a, The human papillomaviruses, *Bull. Cancer* **65**:151–164.

Orth, G., Jablonska, S., Favre, M., Croissant, O., Jarzabek-Chorzelska, M., and Rzesa, G., 1978b, Characterization of two new types of human papillomaviruses in lesions of epidermodysplasia verruciformis, *Proc. Natl. Acad. Sci. USA* **75**:1537–1541.

Orth, G., Jablonska, S., Favre, M., Croissant, O., Obalek, S., Jarzabek-Chorzelska, M., and Jibard, N., 1981, Identification of papillomaviruses in butcher's warts, *J. Invest. Dermatol.* **76**:97–102.

Ostrow, R. S., Bender, M., Niimura, M., Seki, T., Kawashima, M., Pass, F., and Faras, A. J., 1982, Human papillomavirus DNA in cutaneous primary and metastasized squamous cell carcinomas from patients with epidermodysplasia verruciformis, *Proc. Natl. Acad. Sci. USA* **79**:1634–1638.

Pinsky, L., Finkelberg, R., Straisfeld, C., Zilahi, B., Kaukman, M., and Hall, G., 1972, Testosterone metabolism by serially subcultured fibroblasts from genital and nongenital skin from individual human donors, *Biochem. Biophys. Res. Commun.* **46**:364–369.

Quick, C. A., Krzyzek, R. A., Watts, S. L., and Faras, A. J., 1980, Relationship between condylomata and laryngeal papillomata, *Ann. Otol.* **89**:467–471.

Rashad, A. L., 1966, DNA synthesis in viral and in chemically-induced papillomas and carcinomas as revealed by T-H^3 radioautography, *Proc. Am. Assoc. Cancer Res.* **7**:57–63.

Rashad, A. L., 1969, Radioautographic evidence of DNA synthesis in well differentiated cells of human skin papillomas, *J. Invest. Dermatol.* **53**:356–362.

Rashad, A. L., and Evans, C. A., 1978, Histological features of virus rich and virus poor Shope papillomas of cottontail rabbits, *Cancer Res.* **37**:1855–1860.

Reid, R., 1983, Genital warts and cervical cancer. II. Is human papillomavirus infection the trigger to cervical carcinogenesis? *Gynecol. Oncol.* **15:**239–252.

Reid, R., Crum, C. P., Herschman, B. R., Fu, Y. S., Braum, L., Shad, K. V., Agronow, S. J., and Stanhope, R. C., 1984, Genital warts and cervical cancer. III. Subclinical papillomaviral infection and cervical neoplasia are linked by a spectrum of continuous morphologic and biologic change, *Cancer* **53:**943–953.

Reilly, S. S., Albers, K. M., and Taichman, L. B., 1985, Replication of HPV1 in malignant keratinocytes, in: *UCLA Symposium on Molecular and Cellular Biology,* Vol. 32: *Papillomavirus: Molecular and Clinical Aspects* (P. M. Howley and T. R. Broker, eds.), pp. 427–436, Alan R. Liss, New York.

Rheinwald, J. G., and Beckett, M. A., 1980, Defective terminal differentiation in culture as a consistent and selectable character of malignant human keratinocytes, *Cell* **22:**629–632.

Rheinwald, J. G., and Beckett, M. A., 1981, Tumorigenic keratinocyte lines requiring anchorage and fibroblast support cultured from human squamous cell carcinomas, *Cancer Res.* **41:**1657–1663.

Rheinwald, J. G., and Green, H., 1975, Serial cultivation of strains of human epidermal keratinocytes: The formation of keratinizing colonies from single cells, *Cell* **6:**331–343.

Rheinwald, J. G., and Green, H., 1977, Epidermal growth factor and the multiplication of human epidermal keratinocytes, *Nature* **265:**421–424.

Rice, R. H., and Green, H., 1978, Relation with protein synthesis and transglutaminase activity to formation of the crosslinked envelope during terminal differentiation of the cultured human epidermal keratinocyte, *J. Cell Biol.* **76:**705–711.

Rous, R., and Beard, J. W., 1935, The progression to carcinoma of virus induced rabbit papilloma (Shope), *J. Exp. Med.* **62:**523–548.

Rous, R., and Friedewald, W. F., 1944, The effect of chemical carcinogens on virus induced rabbit carcinomas, *J. Exp. Med.* **79:**511–537.

Rowson, K. E. K., and Mahy, B. W. J., 1967, Human papova (wart) virus, *Bacteriol. Rev.* **31:**110–131.

Ruiter, M., and Van Mullem, J., 1966, Demonstration by electron microscopy of an intranuclear virus in epidermodysplasia verruciformis, *J. Invest. Dermatol.* **47:**247–252.

Schweitzer, J., Kinjo, M., Furstenburger, G., and Winter, H., 1984, Sequential expression of mRNA-encoded keratin sets in neonatal mouse epidermis: Basal cells with properties of terminally differentiating cells, *Cell* **37:**159–170.

Sengel, P., 1983, Epidermal–dermal interactions during formation of skin and cutaneous appendages, in: *Biochemistry and Physiology of the Skin* (L. A. Goldsmith, ed.), pp. 102–131, Oxford University Press, London.

Shope, R. E., 1933, Infectious papillomatosis of rabbits; with a note on the histopathology, *J. Exp. Med.* **58:**607–624.

Simon, M., and Green, H., 1984, Participation of membrane-associated proteins in the formation of the cross-linked envelope of the keratinocyte, *Cell* **36:**827–834.

Staquet, M. J., Viac, J., and Thivolet, J., 1981, Keratin polypeptide modification induced by human papilloma viruses (HPV), *Arch. Dermatol. Res.* **271:**83–90.

Steinberg, B. M., Pollack, R., Topp, W. C., and Botchan, M. R., 1978, Isolation and characterization of T antigen negative revertants from a line of transformed rat cells containing one copy of the SV40 genome, *Cell* **13:**19–32.

Steinberg, B. M., Abramson, A. L., and Meade, R. P., 1982, Culture of human laryngeal papilloma cells *in vitro, Otolaryngol. Head Neck Surg.* **90:**728–735.

Steinberg, B. M., Topp, W. C., Schneider, P. S., and Abramson, A. L., 1983, Laryngeal papillomavirus infection during clinical remission, *N. Engl. J. Med.* **308:**1261–1264.

Steinberg, M. L., and Defendi, V., 1979, Altered pattern of growth and differentiation in human keratinocytes infected by simian virus 40, *Proc. Natl. Acad. Sci. USA* **76:**801–805.

Sun, T.-T., Eichner, R., Nelson, W. G., Tseng, S. C. G., Weiss, R. A., Jarvinen, M., and Woodcock-Mitchell, J., 1983, Keratin classes: Molecular markers for different types of epithelial differentiation, *J. Invest. Dermatol.* **81:**109s–115s.

Syverton, J. T., 1952, The pathogenesis of the rabbit papilloma-to-carcinoma sequence, *Ann. N.Y. Acad. Sci.* **54:**1126–1140.

Taichman, L. B., and Prokop, C., 1982, Synthesis of keratin protein during maturation of cultured human keratinocytes, *J. Invest. Dermatol.* **78:**464–467.

Taichman, L. B., Breitburd, F., Croissant, O., and Orth, G., 1984, The search for a culture system for papillomavirus, *J. Invest. Dermatol.* **83:**2s–6s.

Teich, N. M., Weiss, R. A., Martin, G. R., and Lowry, D. R., 1977, Virus infection of murine teratocarcinoma lines, *Cell* **12:**973–982.

Topp, W., Hall, J. D., Rifkin, D., Levine, A. J., and Pollack, R., 1977, The characterization of SV40 transformed cell lines derived from mouse teratocarcinomas: Growth properties and differentiated characteristics, *J. Cell. Physiol.* **93:**269–276.

Turek, L. P., Byrne, J. C., Lowy, D. R., Dvoretzky, I., Friedman, R. M., and Howley, P. M., 1982, Interferon induces morphologic reversion with elimination of extrachromosomal viral genomes in bovine papillomavirus transformed mouse cells, *Proc. Natl. Acad. Sci. USA* **79:**7914–7918.

Watt, F. M., and Green, H., 1981, Involucrin synthesis is correlated with cell size in human epidermal cultures, *J. Cell Biol.* **90:**738–742.

Watt, F. M., and Green, H., 1982, Stratification and terminal differentiation of cultured epidermal cells, *Nature* **295:**434–436.

Watts, S. L., Ostrow, R. S., Phelps, W. C., Prince, J. T., and Faras, A. J., 1983, Free cottontail rabbit papillomavirus DNA persists in warts and carcinomas of infected rabbits and in cells in culture transformed with virus or viral DNA, *Virology* **125:**127–138.

Watts, S. L., Phelps, W. C., Ostrow, R. S., Zachow, K. R., and Faras, A. J., 1984, Cellular transformation by human papillomavirus DNA *in vitro, Science* **225:**634–636.

Weiss, R. A., Eichner, R., and Sun, T.-T., 1984, Monoclonal antibody analysis of keratin expression in epidermal diseases: A 48 and 56-kdalton keratin as molecular markers for hyperproliferative keratinocytes, *J. Cell Biol.* **98:**1397–1402.

Weissman, B. E., and Aaronson, S. A., 1983, BALB and Kirsten murine sarcoma viruses alter growth and differentiation of EGF-dependent BALB/c mouse epidermal keratinocyte lines, *Cell* **32:**599–606.

Wettstein, F. O., and Stevens, J. G., 1982, Variable-sized free episomes of Shope papilloma virus DNA are present in all non-virus-producing neoplasms and integrated episomes are detected in some, *Proc. Natl. Acad. Sci. USA* **79:**790–794.

White, D. O., Huebner, R. J., Rowe, W. P., and Traub, R., 1963, Studies on the virus of rabbit papilloma. 1. Methods of assay, *Aust. J. Exp. Biol.* **41:**41–50.

Zettergren, J. G., Peterson, L. L., and Wuepper, K. D., 1984, Keratolinin: The soluble substrate of epidermal transglutaminase from human and bovine tissue, *Proc. Natl. Acad. Sci. USA* **81:**238–242.

CHAPTER 5

Papillomavirus Transformation

PETER M. HOWLEY AND RICHARD SCHLEGEL

I. INTRODUCTION

The papillomaviruses have been studied extensively since the early 1930s when Shope initiated his studies on CRPV (Shope, 1933). Detailed studies on the biology and molecular genetics of this group of viruses have been severely hampered, however, due in large part to the lack of a tissue culture system for the *in vitro* propagation of these viruses. Cellular transformation of rodent cells by some papillomaviruses has provided cell culture systems for investigators to study the viral functions involved in cellular proliferation, cellular transformation, and plasmid maintenance. Although a few studies have now demonstrated transformation with certain of the HPV types, the analysis of papillomavirus-transforming functions has largely been performed with a subgroup of papillomaviruses which are able to induce fibroblastic tumors in hamsters. These viruses are listed in Table I. Of these, BPV-1 is the best studied. BPV-1 can induce fibroblastic tumors in a variety of heterologous hosts other than hamsters, including horses (Olson *et al.*, 1969; Lancaster *et al.*, 1977), mice (Boiron *et al.*, 1964), and pika (Puget *et al.*, 1975). One characteristic of the viruses listed in Table I is that each induces fibropapillomas consisting of a proliferative fibroblastic component as well as a proliferative squamous epithelial component. Thus, the ability to induce fibroblastic tumors in heterologous hosts would appear to correlate with the ability of an in-

PETER M. HOWLEY AND RICHARD SCHLEGEL • Laboratory of Tumor Virus Biology, National Cancer Institute, Bethesda, Maryland 20892.

TABLE I. Papillomaviruses Tumorigenic in Hamsters

	Reference
Bovine papillomavirus (types 1 and 2)	Friedman *et al.* (1963)
Deer papillomavirus	Koller and Olson (1972)
Ovine papillomavirus	Gibbs *et al.* (1975)
European elk papillomavirus	Stenlund *et al.* (1983)

dividual papillomavirus to induce fibroblastic proliferation as a part of the normal benign lesion induced in its natural host.

Most of the viruses listed in Table I have demonstrated the ability to readily transform susceptible rodent cells in tissue culture. This subgroup of viruses has been the principal group studied in the analysis of papillomavirus transformation functions. A list of papillomaviruses which have been reported to induce transformation of mammalian cells *in vitro* is given in Table II. Although BPV-1 has been the prototype of the papillomaviruses used for studying transformation, BPV-2, EEPV, and DPV readily transform certain rodent tissue culture cell lines *in vitro*. Additionally, BPV-4, which is associated with esophageal papillomatosis in cattle, has been reported to transform NIH 3T3 cells in culture (Campo and Spandidos, 1983). CRPV and its cloned DNA have been reported to transform C127 cells and NIH 3T3 cells in culture (Watts *et al.*, 1983),

TABLE II. *In Vitro* Transformation by a Papillomavirus

Virus	Cell	Transforming agent	Selection	Reference
BPV[a]	Bovine conjunctiva cells	Virus	Morphology	Black *et al.* (1963)
	Mouse embryo	Virus	Morphology	Boiron *et al.* (1964), Thomas *et al.* (1964)
	Hamster embryo	Virus	Morphology	Geraldes (1969)
BPV-1	NIH 3T3 and C127	Virus	Focus	Dvoretzky *et al.* (1980)
	NIH 3T3 and C127	Cloned DNA	Focus	Lowy *et al.* (1980)
BPV-2	NIH 3T3 and C127	Virus	Focus	Dvoretzky *et al.* (1980)
	Bovine conjunctiva and palate cells	Virus	Passage	Moar *et al.* (1981)
	NIH 3T3 and C127	Cloned DNA	Focus	Lowy *et al.* (1980), Campo and Spandidos (1983)
BPV-4	NIH 3T3	Cloned DNA	Focus	Campo and Spandidos (1983)
EEPV	C127	Virus	Focus	Stenlund *et al.* (1983)
DPV	NIH 3T3	Virus	Focus	Groff *et al.* (1983)
CRPV	C127 and NIH 3T3	Virus, cloned DNA	Focus	Watts *et al.* (1983)
HPV-1	C127	Cloned DNA	Focus	Watts *et al.* (1984)
HPV-5	C127	Cloned DNA	Focus	Watts *et al.* (1984)
HPV-16	NIH 3T3	Cloned DNA	Focus	Yasumoto *et al.* (1986)

[a] BPV from a cutaneous fibropapilloma but not typed.

and several HPVs have now been reported to transform cultured rodent cells *in vitro* (Table II). As discussed in this chapter, it has been possible to genetically define the genes in the BPV which are associated with *in vitro* transformation of rodent cells. The gene products for each of the two defined BPV-1 transforming genes have been identified. It is not known what the functions of these genes and their encoded proteins are in benign fibropapillomas associated with BPV-1. It is also not known what role the analogous genes in the viruses that are associated with carcinogenic progression may have in the progression from a benign papilloma to a carcinoma.

The study of papillomavirus-induced cell transformation could potentially provide insights into functions required for full papillomavirus replication. Virus induced alterations of cell functions which are involved in cellular transformation are probably also involved in the viral replication cycle. As in the case of the well-studied papovavirus, polyoma virus, the cell transformation and viral maturation functions seem to be tightly coupled (Benjamin, 1970). The middle T-antigen function of polyoma which is crucial for transformation and tumorigenicity, is also essential for posttranslational modification of the capsid proteins required for efficient virus assembly. Cell transformation in the case of the polyomaviruses may represent a consequence of gene functions necessary to facilitate viral growth.

This chapter will focus on the various aspects of papillomavirus-mediated transformation. Although other papillomaviruses have been shown capable of transforming rodent cells in culture, the major emphasis of this chapter will be on BPV-1 which has served as the prototype for studying papillomavirus transforming functions. In the summary and discussion, we will attempt to relate the papillomavirus transforming functions to their possible role in carcinogenesis and tumor progression.

II. TRANSFORMATION BIOLOGY

In vitro morphological transformation was first described for BPV in the early 1960s (Black *et al.*, 1963; Thomas *et al.*, 1964; Boiron *et al.*, 1964) (see Table II). Two independent studies demonstrated that a focus assay could be used with mouse cells to study BPV-1 transformation (Meischke, 1979; Dvoretzky *et al.*, 1980). Subsequently, cellular transformation by BPV-1 has been studied extensively using mouse C127 cells, NIH 3T3 cells, rat 3T3 cells, CREF cells, and hamster embryo fibroblasts (Dvoretzkey *et al.*, 1980; Morgan and Meinke, 1980; Binetruy *et al.*, 1982; Grisoni *et al.*, 1984; Babiss and Fisher, 1986).

Many studies involving BPV-1 transformation have utilized the mouse C127 cell line. This is a cell line that was originally derived from a mammary tumor from an R-III mouse. The cells are of epithelial origin and have a very flat phenotype. Early passage C127 cells are nontumorigenic in nude mice and are not anchorage independent. Transformation by BPV-1 or by the cloned BPV-1 DNA leads to an altered phenotype which is readily detected (Fig.1). The cells become more refractile, spindle-shaped, and lose their contact inhibition. BPV-1-transformed C127 cells have a fully transformed phenotype as assayed by anchorage independence and tumorigenicity in nude mice (Dvoretzky *et al.*, 1980). The C127 cell line has been particularly useful in defining and mapping the two independent transforming functions now recognized for BPV-1 which are discussed later in this chapter. BPV-1-transformed hamster embryo cells have also been shown to be anchorage independent (Morgan and Meinke, 1980).

Analysis of BPV-1-transformed Fischer rat fibroblast cells has indi-

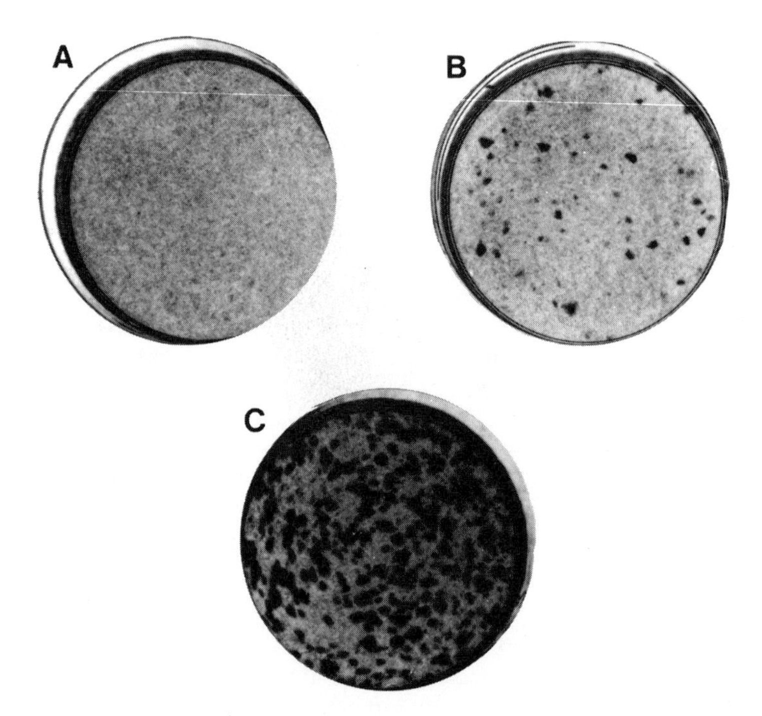

FIGURE 1. BPV-1-transformed foci of C127 cells. Each of the plates was transfected with the indicated DNA using calcium coprecipitation followed by a 15% (vol/vol) glycerol shock as described by Sarver *et al.* (1982). The 60-mm plates were stained 21 days after transfection with 1% methylene blue after fixation with 70% isopropyl alcohol. Salmon sperm DNA was used as a carrier. (A) A plate transfected with salmon sperm DNA was the negative control and has no foci. (B) This plate was transfected with 1 μg of the plasmid pdBPV-1(142-6) consisting of the full *Bam*HI linearized BPV-1 genome cloned into the *Bam*HI site of pML2d (Sarver *et al.*, 1982). (C) This plate was transfected with 1 μg of the purified recircularized BPV-1 DNA insert from p142-6. As noted in the literature (Sarver *et al.*, 1982, 1985), BPV-1 DNA freed from the prokaryotic pML2d sequences and recircularized transforms mouse C127 cells five to ten times more efficiently than the DNA linked to pML2d DNA.

cated that such transformation may involve a stepwise process (Grisoni *et al.*, 1984; Babiss and Fisher, 1986). Grisoni *et al.* found that BPV-1-transformed FR 3T3 cells selected either by focus or by colony formation grew poorly in suspension and remained highly serum dependent, indicating that they were not fully transformed. In addition, these cells did not produce detectable levels of plasminogen activator, and maintained an organized cytoskeleton, characteristic of a normal fibroblast. These cells were able to induce tumors in syngeneic animals, and cells derived from these tumors had a more advanced transformed phenotype. Cells selected by their tumorigenicity were able to grow efficiently in agar suspension and they produced high levels of plasminogen activator. The cells still maintained a high serum dependence for growth in culture and maintained an extended cytoskeletal structure. Babiss and Fisher also found a progression in the transformed phenotype of BPV-1-transformed CREF cells that were tumor derived (Babiss and Fisher, 1986). Detailed studies of the molecular events involved in transformation progression using Fischer rat embryo cells have not been done. It is unclear whether or not progression involves alterations in the resident viral DNA and the expression of the viral genome, or changes in the host DNA.

One characteristic of BPV-1-transformed mouse and rat cells is that integration of the viral DNA is not required for either initiation or maintenance of the transformed state. In cloned lines of BPV-1-transformed mouse C127 cells, Law *et al.* (1981) demonstrated that all of the viral DNA sequences present within these cell lines were present as extrachromosomal circular molecules. BPV-1 genomes were present as multiple copies (10 to 200 copies per diploid genome). No evidence was found in that study for host–viral DNA junction fragments and it was concluded that integration of the viral DNA was not a prerequisite for the induction of transformation in these mouse cells. The viral DNA is also largely, if not exclusively, extrachromosomal in mouse and rat cells transformed by cloned BPV-1 DNA (Law *et al.*, 1981; Lancaster, 1981; Binetruy *et al.*, 1982; Sarver *et al.*, 1982). The viral functions involved in BPV-1 plasmid replication and the apparent faithful partitioning of these plasmids to daughter cells will be discussed later in this chapter. It seems likely, however, that the viral functions involved in the stable plasmid replication and partitioning are manifestations of the nonproductive latent infection of cells by the papillomaviruses. In the case of BPV-1, these latent DNA replication functions would appear to be characteristic of the fibroblasts in the dermal portion of the fibropapillomas. In addition, these functions may also be involved in the stable DNA replication and partitioning that may be required for the virus to remain persistently and latently infected within the basal cells of the squamous epithelial component of a wart. Integration of the viral DNA in these transformed cells, however, is not precluded. Several studies examining recombinant molecules containing BPV-1 have demonstrated the integration of BPV-1 sequences (DiMaio *et al.*, 1984; Sarver *et al.*, 1985). Allshire and Bostock

(1986) have reported evidence of integrated BPV-1 DNA molecules in mouse C127 cells transformed with either purified BPV-1 DNA or recombinant molecules containing the BPV-1 genome.

Regardless of the actual molecular and cellular events occurring during BPV-1 transformation, it has been shown that the maintenance of this induced state is dependent on the continued presence and presumed expression of the BPV-1 genome (Turek *et al.*, 1982). When BPV-1-transformed C127 cells were treated with mouse L-cell interferon, there was a decrease in the average number of viral genomes per cell to a level of one-third to one-eighth that of the control cells after 60 cell divisions. Flat revertants were observed in the treated cultures and not in the untreated cultures. These flat revertants had lost all characteristics of the transformed cells in that they grew to a decreased cell density characteristic of nontransformed C127 cells and had lost anchorage independence. These cells were shown to be "cured" of their BPV-1 DNA. Since the cells could be retransformed by BPV-1, they did not arise from cellular mutations making them refractory to BPV-1 gene functions.

Studies with recombinant plasmids containing the full BPV-1 genome and a second independent selective marker have indicated that the presence and the expression of the BPV-1 genome *per se* is not sufficient for the expression of the transformed phenotype (Law *et al.*,1983; Matthias *et al.*, 1983; Meneguzzi *et al.*, 1984). These studies utilized a series of different BPV-1 vectors in which the neomycin resistance gene cloned from Tn5 was expressed from a mammalian cell transcriptional promoter. Cells were selected for expression of the bacterial neomycin resistance gene utilizing the drug G418 which is toxic to mammalian cells but is detoxified by the neomycin resistance gene. In these studies, G418-resistant clones were isolated and demonstrated to contain extrachromosomal recombinant molecules containing BPV-1 DNA. With varying degrees of efficiency, these G418-resistant clones then progressed to develop the gross morphological characteristics of transformation. The delayed appearance of the transformed phenotype in these experiments indicates that some intracellular event or series of events subsequent to the establishment of transcriptionally active BPV-1 hybrid plasmids is required for the manifestation of the transformed phenotype. This intracellular event could be the accumulation of a virus-specific product and could indicate a requirement for a threshold level of BPV-1 product for the transformed phenotype. Indeed, in support of this notion is the finding that a threshold level of BPV-1 transcription also appears to be required for the manifestation of the transformed phenotype. A class of mutants mapping to the E6 and E7 open reading frames (ORFs) of the virus (to be discussed below) results in a lower copy number of the BPV-1 plasmids in the transformed cells and lower levels of transcription (Lusky and Botchan, 1985; Berg *et al.*, 1986). These cells have a flat nontransformed appearance.

III. TRANSCRIPTION

The analysis of transcription in BPV-1-transformed cells has been somewhat hampered by the very low levels of RNA present within the transformed cells. A map of the BPV-1 genome is shown in Fig. 2. This map is based on the genetic information derived from the analysis of the 7945 base pairs of BPV-1 genome (Chen *et al.*, 1982; Ahola *et al.*, 1983). Sequence analysis revealed that all of the ORFs greater than approximately 380 bases in size are located on one strand (Chen *et al.*, 1982). This led to the prediction that all of the transcripts for the papilloma-

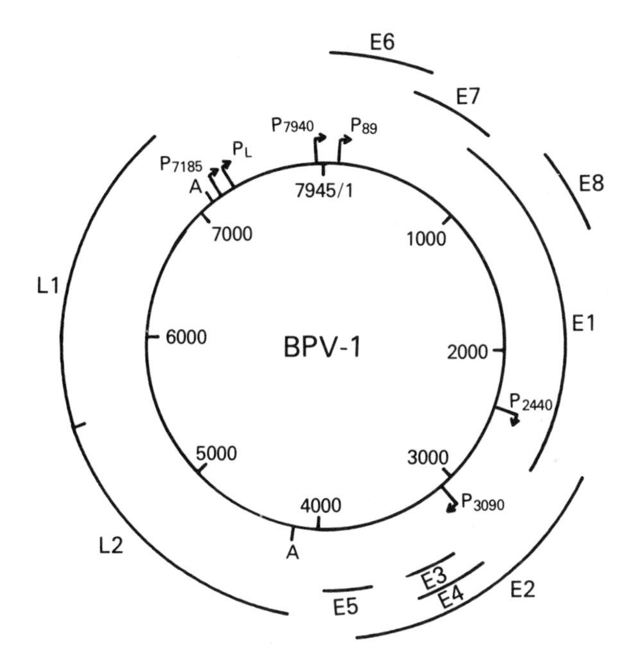

FIGURE 2. Genomic organization of BPV-1 DNA. The entire 7945-bp circular genome has been sequenced (Chen *et al.*, 1982), and the positions of the base pairs in units of 1000 are noted on the inside of the genome. The 69% transforming region sufficient for transformation and plasmid maintenance extends clockwise from a unique *Hind*III site at base 6959 to a unique *Bam*HI site at base 4451 (Lowy *et al.*, 1980). The ORFs contained in this region are designated E1 through E8 and are positioned outside the genomic map in the appropriate translation frames. The L1 and L2 ORFs are expressed only in productively infected bovine fibropapillomas (Engel *et al.*, 1983). The small arrows and "P" (for "promoter") correspond to the major 5′-ends of mRNA classes found in BPV-1-transformed cells (Ahola *et al.*, 1983; Yang *et al.*, 1985a; Stenlund *et al.*, 1985; Pettersson *et al.*, 1986; Baker and Howley, manuscript submitted). Classes with RNAs mapping to approximately bases 7185, 7945, 89, 2440, and 3090 have been found in warts and in transformed cells. A wart-specific promoter (PL) found active only in warts directs the synthesis of RNAs with heterogeneity of 5′-ends mapping around base 7250. "A" designates the sites of polyadenylation used in transformed cells at base 4203 and in warts for the L1 and L2 region RNAs at base 7175.

viruses would be derived from one strand. Analysis of transcription in transformed cells as well as in productively infected fibropapillomas has shown this prediction to be correct (Heilman *et al.*, 1982; Amtmann and Sauer, 1982; Engel *et al.*, 1983).

The initial studies with the cloned BPV-1 DNA established that the entire viral genome was not required to effect stable transformation and plasmid maintenance; a specific subgenomic fragment extending from the unique *Hin*dIII site to the unique *Bam*HI site (the 69% transforming fragment) was sufficient (Lowy *et al.*, 1980; Howley *et al.*, 1980). The ORFs that map in this 69% fragment are labeled the E ORFs (Fig. 2.). Two ORFs located in the remaining 31% segment, not required for transformation or for autonomous plasmid maintenance, are labeled L1 and L2 (Fig. 2). Heilman *et al.* (1982) established that all of the transcripts expressed in BPV-1-transformed mouse cells were derived from the 69% transforming fragment, were transcribed from the same strand, and shared a common 3′-end at approximately base 4200 ("A" in Fig. 2). There is a polyadenylation site (AATAAA) located at base 4180 in the genome (Chen *et al.*, 1982) and sequence analysis of cDNA clones has shown polyadenylation occurs at base 4203 (Yang *et al.*, 1985a). Northern blot analysis of the RNAs from transformed cells revealed multiple species of viral RNAs ranging in size from approximately 1000 bases to 4000 bases. Subsequent studies have shown that the patterns of transcription are complex and that there are multiple species of RNAs generated by differential splicing. In addition, several different viral promoters function in transformed cells. The most detailed analyses of the viral transcripts in transformed cells come from an electron microscopic mapping of the RNAs (Stenlund *et al.*, 1985) and from direct cDNA cloning (Yang *et al.*, 1985a). These studies indicated that, in addition to the existence of differential splicing, different classes of viral RNAs could be grouped based on their 5′-ends. One class of RNAs has 5′-ends which map to approximately base 89 (Ahola *et al.*, 1983; Yang *et al.*, 1985a), just downstream from a TATA box located at base 58 in the BPV-1 genome. In addition, another class of RNAs was identified with 5′-ends mapping around base 2443 (Yang *et al.*, 1985a; Stenlund *et al.*, 1985). A class of RNAs with 5′-ends mapping to base 3090 has also been noted (Pettersson *et al.*, 1986). Subsequent studies have revealed additional promoters in the long control region (LCR) located upstream of the E6 ORF. Additional 5′-ends of mRNAs in transformed cells have been mapped by primer extension studies to base 7185 and to a heterogeneous cluster around base 7940 (Baker and Howley, manuscript submitted). These putative promoters are all indicated in Fig. 2.

The virus-specific polyadenylated RNA species found in transformed mouse cells appear to be a subset of the RNAs found in productively infected bovine fibropapillomas (Engel *et al.*, 1983). In addition to the species of RNAs found in transformed cells, there are RNA species which are derived from the 31% fragment which is transcriptionally inactive in transformed cells. The additional RNA species found in productively infected warts are presumably only transcribed in the terminally differentiated cells of the warts and these RNA species have a common 3'-terminus at base 7175 where a polyadenylation site (AATAAA) is localized. A wart-specific promoter which is active only in productively infected cells has been mapped to base 7250 (Baker and Howley, manuscript submitted). A more detailed analysis of the RNA structures is provided in Chapter 3 on papillomavirus transcription.

The genomic organization of BPV-1 is quite similar to other papillomaviruses which have been sequenced. The sequences of some of these other papillomaviruses are compared with BPV-1 in the Appendix of this volume. Each of these other papillomaviruses which has been sequenced also contains ORFs on only one strand. There are characteristic homologies in the nucleotide and deduced amino acid sequences shared by the papillomaviruses, particularly within the four largest ORFs (E1, E2, L1, and L2). At the 5'-end of the transforming region of BPV-1 is a region approximately 1000 bases in size which was originally referred to as the noncoding region but now is referred to as the LCR. This region has been shown to contain the origin of replication for BPV-1 minichromosomes in transformed cells (Waldeck *et al.*, 1984) and contains the major DNase I hypersensitive site for the genome (Rösl *et al.*, 1983, 1986). As mentioned above, this region contains several different promoters which are active in transformed cells, and an additional promoter which is active only in productivity infected warts (Baker and Howley, manuscript submitted). Analysis of cDNAs of wart-specific RNAs indicates that a short coding exon is contained in this region; therefore, this region formally cannot be considered as noncoding (Baker and Howley, manuscript submitted). Evidence exists that this region may also contain coding exons for RNAs expressed in transformed cells (M. Botchan, personal communication).

The BPV-genome also contains enhancer elements which may be involved in the control of viral gene expression in transformed cells. Enhancers are transcriptional regulatory elements which are able to increase the transcriptional activity of promoters in mammalian cells. An element with enhancer activity was first described in the BPV-1 genome by Lusky *et al.* (1983). When linked to the herpesvirus thymidine kinase (TK) gene, this enhancer increased the efficiency of stable transformation of a Rat-2 cell line from a TK^- to a TK^+ phenotype. In addition, this element could substitute for the SV40 72-bp repeat enhancer in a sensitive indirect assay for SV40 large T expression (Lusky *et al.*, 1983). This element, referred to as the distal enhancer, is located between bases 4390 and 4451, approximately 200 bases 3' to the polyadenylation site used in transformed cells (Lusky *et al.*, 1983; Campo *et al.*, 1983; Weiher and Botchan, 1984). Although in the initial studies of this element it was argued that the expression of the early genes of BPV-1 was dependent on this enhancer, further analysis of this element has shown that it is not *cis* essential for transformation or for plasmid maintenance in mouse

C127 cells (Howley *et al.*, 1985). Although this element does have enhancer activity as illustrated by the studies referred to above, and by its ability to activate the rat preproinsulin gene in mouse C127 cells (Sarver *et al.*, 1985), it is not clear what role, if any, it may have in BPV-1 gene expression. It is possible that it may have a physiologic role in BPV-1 gene expression which is bypassed in the DNA transfection experiments utilized to assess its function. It is certainly possible that in any natural viral infection, the expression of the BPV-1 early region may be ordered and regulated, and that the distal enhancer as a constitutive enhancer could be required for the expression of an immediate early class of genes. Studies of the temporal expression of the various BPV-1 promoters in mammalian cells following viral infection have not been carried out.

An additional enhancer element has been described in the LCR of the BPV-1 genome. This element has the property of being markedly activated by the E2 gene product (Spalholz *et al.*, 1985). The role of the E2 gene product presumably through its effect on the transactivation of this conditional enhancer will be discussed later in this chapter.

IV. TRANSFORMING FUNCTIONS

A. Genetics

In the initial studies carried out with the cloned BPV-1 genome, it was shown that a specific 69% subgenomic fragment encompassed by the unique *Hind*III and *Bam*HI sites was sufficient for inducing transformation of C127 cells or NIH 3T3 cells, albeit at a somewhat lowered efficiency than the full-length BPV-1 genome (Lowy *et al.*, 1980; Howley *et al.*, 1980). Mutagenesis studies of the full-length viral genome and experiments with subgenomic fragments expressed from surrogate promoters have mapped regions of the viral genome involved in cellular transformation (Nakabayashi *et al.*, 1983; Sarver *et al.*, 1984; Lusky and Botchan, 1984; Schiller *et al.*, 1984, 1986; Yang *et al.*, 1985b; DiMaio *et al.*, 1986a,b; Groff and Lancaster, 1986). In addition to regions which have essential transcriptional control sequences, several regions have been identified which apparently encode proteins involved in transformation. Two independent coding segments are required for the full transformed phenotype (Sarver *et al.*, 1984). One of these regions maps to the 3′ ORFs which includes the E2, E3, E4, and E5 ORFs (Nakabayashi *et al.*, 1983; Sarver *et al.*, 1984; Schiller *et al.*, 1986). The expression of the 3′ ORFs, however, is insufficient for the full transformed phenotype. Expression of the 3′ ORFs from BPV-1 promoters located in the LCR or from a surrogate promoter will lead to efficient focus formation. These transformed cells, however, have only a partially transformed phenotype; they are not anchorage independent and are not highly tumorigenic in nude mice (Sarver *et al.*, 1984). Sarver *et al.* (1984) demonstrated that the integrity of the E6/E7 region of the BPV-1 genome is also required for the full transformed phenotype. Thus, mutagenesis studies revealed that two independent coding domains of the BPV-1 genome are required for the full transformed phenotype.

It was not clear from these initial studies whether these analyses were defining two independent transforming genes or two different exons of the same transforming gene. This ambiguity was resolved from the analysis of the virus-specific transcripts in BPV-1-transformed cells. Yang *et al.* (1985a) constructed a cDNA library using mRNA from BPV-1-transformed C127 cells employing a vector system which permitted expression of the various cDNAs directly in mammalian cells (Okayama and Berg, 1983). This vector system can generate full-length cDNA clones which are positioned behind the SV40 early promoter and positioned upstream from the SV40 late region polyadenylation site, allowing functional analysis of the individual cDNA clones directly in mammalian cells. In this study it was possible to determine the structure of viral RNAs present in BPV-1-transformed cells and to assess their function directly. Sequence analysis of the cDNA clones revealed that the RNAs are generated by differential splicing as described in Chapter 3. Six independent species of cDNAs could be discerned based on the splicing patterns and on the basis of the 5′-ends of the clones (Yang *et al.*, 1985a). Two different species of cDNAs were each able to independently transform mouse C127 cells (Fig. 3). One of the clones contained the E6 ORF intact. Those cDNAs which contain the E6 ORF spliced to the E7 ORF or those which contained only the E7 ORF intact were unable to transform mouse C127 cells. A second species of cDNAs was also able to transform mouse C127 cells. These cDNAs contained all the 3′ ORFs intact and were unspliced. Apparently, these are transcribed from the promoter directing the synthesis of RNAs with 5′-ends at base 2443. Interestingly, the two viral functions encoded by the E6 ORF and the 3′ ORFs acted synergistically in transforming mouse C127 cells. Cotransformation experiments using a combination of the E6 cDNA clone and a 3′ ORF clone led to a higher number of foci than transfection with either of the transforming cDNAs alone. The effect was more than additive, indicating a cooperative effect between the transformation functions encoded by each of these two classes of cDNA clones. In addition, the foci induced in the cotransformation experiment were larger and appeared faster than those induced by either the E6 cDNA class or the 3′ ORF cDNA class of clones alone (Yang *et al.*, 1985a). The expression of the E6 ORF from the strong surrogate promoter, the Harvey sarcoma virus LTR, has been shown to be sufficient for morphological transformation of C127 cells alone (Schiller *et al.*, 1984).

A critical domain for transformation maps to the E5 ORF downstream from the methionine codon at base 3879. Disruption of this downstream portion of the E5 ORF by introduction of an in-frame termination codon (Rabson *et al.*, 1986) or by frameshift mutants (Schiller *et al.*, 1986; Groff and Lancaster, 1986; DiMaio *et al.*, 1986b) has a marked inhibitory effect

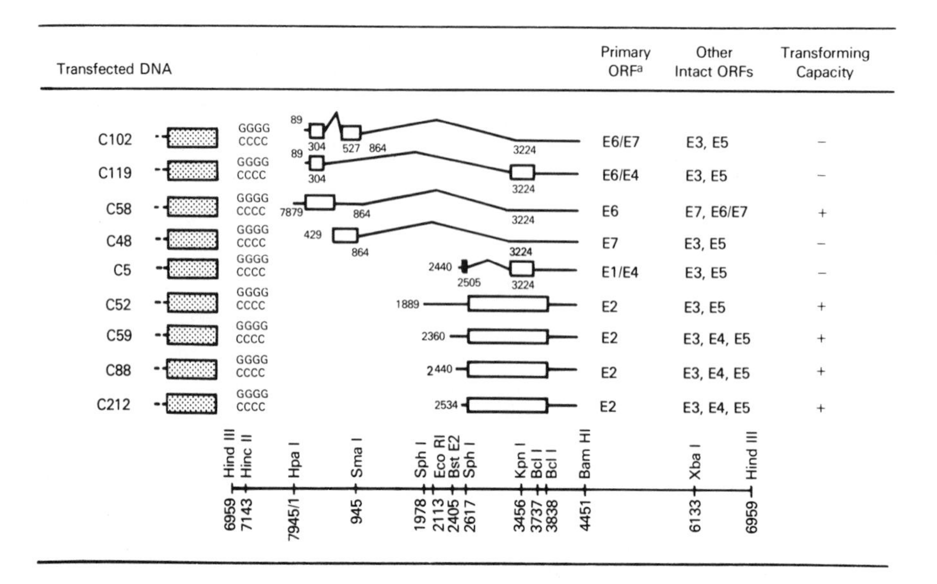

FIGURE 3. Structures of various BPV-1 cDNAs and their transforming capacities. Potential coding exons following an ATG are indicated by the open boxes (except for C5 in which E1/E4 contains no ATG and could represent a truncated clone). Other intact portions of the cDNA clones are represented by the horizontal solid lines. Intervening sequences are indicated by the slanted lines. The numbers outside of the coding regions represent the 5′-ends of the cDNAs and the splice acceptor and donor sites. Each of the cDNAs ends at base 4203 (See Fig. 2). The primary ORF, defined as the first major ORF containing an ATG (except for E1/E4 for C5), is designated immediately to the right. Other intact ORFs contained in the cDNAs are also listed. The stippled box to the left represents the SV40 early promoter and splice sequences. The individual transforming capacity of each of these cDNAs on C127 cells is noted on the right (Yang *et al.*, 1985a).

on transformation efficiency. Mutations that affect the E5 ORF upstream of the AUG methionine codon at base 3879 appear to have little effect on transformation efficiency (Sarver *et al.*, 1984).

The E2 gene also appears to be essential for efficient transformation in the full wild-type BPV-1 genome (Sarver *et al.*, 1984; Lusky and Botchan, 1985; DiMaio *et al.*, 1986a). Since the E2 gene has been shown to encode a product which can function to transactivate transcriptional regulatory sequences in the LCR (Spalholz *et al.*, 1985), the E2 gene product could have an indirect role in the induction of transformation. The requirement for the E2 product in transformation can be circumvented by the use of the surrogate promoter such as the SV40 early promoter or the Harvey sarcoma virus LTR (Yang *et al.*, 1985b; Schiller *et al.*, 1986).

Genetic studies and complementation analyses of mutants in the 3′ ORFs indicate that an additional gene is likely to be involved in BPV-1 transformation. The 3′ ORF cDNA C59 containing the 3′ ORFs intact is capable of complementing some, but not all, of the E2 mutants in *trans*

(Rabson *et al.*, 1986). C127 cells expressing a C59 cDNA containing the intact E2 ORF but mutated in the E5 ORF at the *Bst*XI site are not transformed but do express the E2 gene product (Rabson *et al.*, 1986). These cells can be efficiently transformed by E2 mutants mapping in the 5′ portion of the E2 ORF but not by mutants mapping in the 3′ portion of the E2 ORF (Rabson *et al.*, 1986). The 3′ ORF mutants which cannot be complemented in this cell line are mutated at sites downstream from the splice acceptor at base 3225 and map to the region of the genome where the E2 and E4 ORFs overlap (Fig. 2). It was not clear from this study whether this additional gene mapped to the E2 or the E4 ORF although mutagenesis studies in progress reveal that it is the 3′ portion of the E2 ORF and not the E4 ORF.

B. E6 Gene

The BPV-1 E6 gene is one of two transforming genes which has been identified (Schiller *et al.*, 1984; Yang *et al.*, 1985a). It is sufficient for the transformation of C127 cells by itself when expressed from the Harvey sarcoma virus LTR (Schiller *et al.*, 1984). In addition, a cDNA containing the E6 ORF intact expressed from the SV40 early promoter is sufficient for transforming mouse C127 cells (Yang *et al.*, 1985a). Furthermore, the E6 gene acts synergistically with the 3′ ORF transforming functions in inducing transformation of mouse C127 cells (Yang *et al.*, 1985a). With the full BPV-1 genome, the integrity of the E6 ORF appears to be essential for the full transformed phenotype. In the complete BPV-1 genome, deletions eliminating the E6 ORF decrease the efficiency of transformation somewhat, and have a marked effect on the transformed phenotype (Sarver *et al.*, 1984). The E6 ORF is required for anchorage independence and for full tumorigenicity in nude mice (Sarver *et al.*, 1984).

The E6 protein has recently been identified in BPV-1-transformed cells (Androphy *et al.*, 1985). A fusion polypeptide containing the amino-terminal end of the phage CII protein linked to the E6 protein was expressed in *E. coli* and used to generate antibodies to the E6 gene product. Antibodies to this fusion protein were capable of specifically immunoprecipitating a 15.5K BPV-1 E6 protein from cells transformed by the E6 gene. The E6 protein has been identified by biochemical fractionation in both the nuclear and membrane fractions of these transformed cells (Androphy *et al.*, 1985).

The E6 ORF is conserved in the genomes of each of the papillomaviruses that has been sequenced. The predicted amino acid composition of the E6 gene products of each of the papillomaviruses analyzed contains the motif Cys-X-X-Cys repeated four times and the spacings of these elements are conserved (see the Appendix). Based on the presence of this repeated motif and the high lysine and arginine content of the BPV-1 E6 protein, Androphy *et al.* (1985) suggested that the E6 protein may be a nucleic acid-binding protein. Direct evidence of E6 and DNA binding, however, has not been demonstrated.

The role of the E6 gene in benign papillomas has not been defined. It should be pointed out, however, that the papillomavirus E6 genes are generally maintained intact and expressed as RNAs in papillomavirus-associated carcinomas and cell lines. Thus, E6 could potentially be an important function in the malignant progression of a benign wart to a malignant carcinoma. It is selectively retained and is transcriptionally active in carcinomas induced by the Shope papillomavirus (Nasseri and Wettstein, 1984; Danos *et al.*, 1985). In human cervical carcinomas, the E6 region of the associated HPV genomes are also generally expressed (Schwarz *et al.*, 1985).

C. E5 Gene

The E5 gene is capable of independently transforming mouse C127 cells and NIH 3T3 cells. Mutations in the E5 ORF downstream from the methionine codon at base 3879 have a dramatic effect, knocking out BPV-1 transformation of mouse C127 cells or NIH 3T3 cells (Groff and Lancaster, 1986; Rabson *et al.*, 1986; DiMaio *et al.*, 1986b). Behind either the Harvey sarcoma virus LTR or the SV40 early region promoter, the E5 ORF downstream from the first methionine codon is sufficient for inducing efficient transformation (Yang *et al.*, 1985b; Schiller *et al.*, 1986). The E5 transforming protein of BPV-1 has recently been identified (Schlegel *et al.*, 1986). Antibodies were generated against a synthetic peptide corresponding to the COOH-terminus of the E5 ORF. This antiserum was useful in identifying a 7K protein in BPV-1-transformed cells which contained the E5 ORF intact (Fig. 4). C127 cells transformed by the E6 ORF did not contain this protein and mutations in the E5 ORF downstream from the first AUG eliminated the ability of mouse cells to synthesize this protein. The amino acid composition of the 44-amino-acid E5 gene product and its predicted structure are shown in Fig. 5. The protein is strikingly hydrophobic (Fig. 6) and cell fractionation studies have localized the polypeptide predominantly to cellular membranes.

The E5 ORF is strongly conserved among those papillomaviruses (BPV-1, BPV-2, and DPV) which induce both epidermal and dermal cell proliferation. The putative E5 proteins encoded by those papillomaviruses which produce purely epidermal cell proliferations (e.g., HPV-1 and CRPV) show little homology to the E5 protein of BPV-1, suggesting that the latter E5 protein may have a role in stimulating dermal fibroblastic proliferation. The ability of the product of the BPV-1 E5 ORF to readily transform NIH 3T3 cells (of presumed fibroblastic origin) rather than C127 cells (of presumed epithelial origin) is consistent with this hypothesis (Schiller *et al.*, 1986; R. Schlegel, unpublished observations). The phenotype of C127

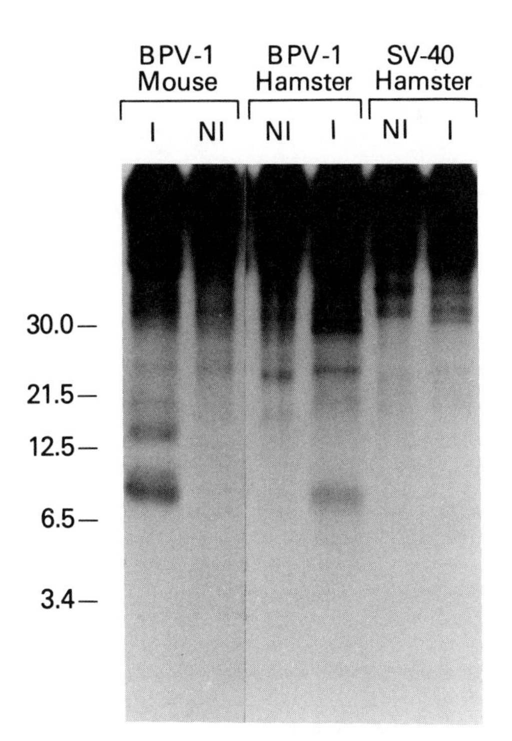

FIGURE 4. Immunoprecipitation of the E5 7K protein using antipeptide antiserum. Mouse C127 and LSH hamster embryo cells transformed by BPV-1 were double-labeled with [^{35}S]methionine and cysteine and used for immunoprecipitation with normal rabbit or immune rabbit antipeptide antiserum. LSH hamster cells transformed by SV40 were used as controls. Only the mouse and hamster cells transformed by BPV-1 contain the 7K protein. NI, nonimmune rabbit serum; I, immune antipeptide antiserum.

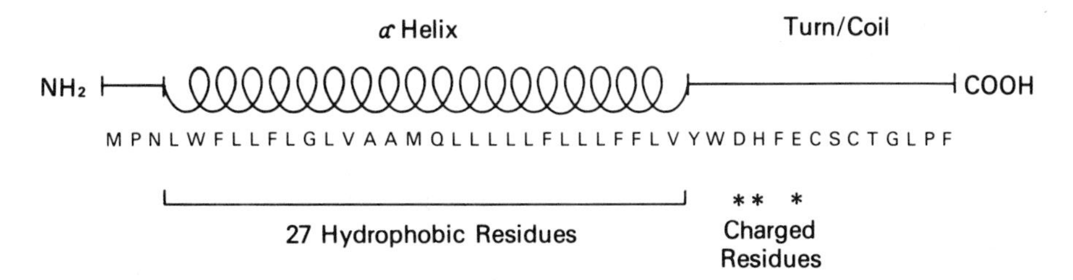

FIGURE 5. Predicted sequence and structure of the BPV-1 E5 protein. From the first methionine in the E5 ORF (nucleotide 3879), the E5 protein polypeptide is predicted to consist of 44 amino acids. The amino-half of the protein contains 27 strongly hydrophobic amino acids (extending from residue 4 to 30) which strongly favor an α-helical conformation. This leucine-rich region is of sufficient length and hydrophobicity (see Fig. 6) to function potentially as a transmembrane domain. It is also possible, however, that this domain does not span the membrane bilayer but rather forms a loop and permits both amino and carboxy-terminal ends of the molecule to be on the same side of the membrane. The carboxy-terminal potion of E5 (starting at tyrosine at residue 31) consists of helix-destabilizing and charged residues which favor a turn or coil conformation. This nonhelical region contains two cysteines which may permit E5 dimer formation (R. Schlegel, unpublished results). A synthetic peptide corresponding to the last 20 amino acids of E5 was used to generate E5-specific antiserum in rabbits.

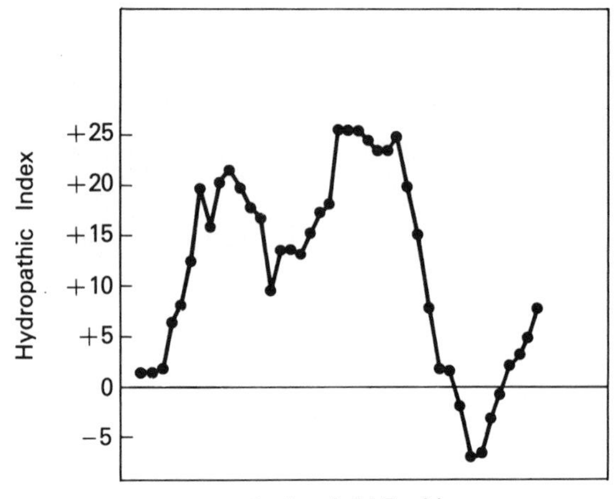

FIGURE 6. Hydropathic index of the predicted E5 protein. The hydropathic index of the E5 protein was plotted using the hydropathy scale for amino acids developed by Kyte and Doolittle (1982). Positive and negative hydropathic indices represent hydrophobic and hydrophilic regions. Domains of proteins which are membrane-associated often exhibit an index of +20 or greater. The first two-thirds of E5 is strongly hydrophobic with hydropathic indices ranging as high as +25. The carboxy-terminus of E5 is slightly hydrophilic (representing the presence of aspartic acid, histidine, and glutamic acid).

cells transformed by the E5 and E6 transforming genes of BPV-1 are different. Morphological characteristics of these independently transformed cells are depicted in Fig. 7.

D. E2 Gene

In a full-length BPV-1 background, mutations in the E2 ORF have a marked effect on the efficiency of cellular transformation (Sarver *et al.*, 1984; Lusky and Botchan, 1985; DiMaio *et al.*, 1986a). The effect of these mutations on transformation, however, may be indirect. The E2 gene has recently been shown to encode a function capable of transactivating an inducible enhancer element located in the viral LCR (Spalholz *et al.*, 1985). When directed from a surrogate promoter such as the SV40 early promoter or the Harvey sarcoma virus LTR, mutations in the E2 ORF have minimal effect, if any, on transformation efficiency (Yang *et al.*, 1985b; Schiller *et al.*, 1986). Thus, it seems likely that the effect of E2 on transformation is indirect and could be due to the transcriptional activation of the inducible enhancer element(s) in the LCR which is in turn required for transcriptional activation of other papillomavirus genes which may be directly involved in transformation. Recent studies by Depak Bastia have demonstrated that the E2 gene product of BPV-1 is a DNA-binding protein and binds directly to sequences in the viral LCR

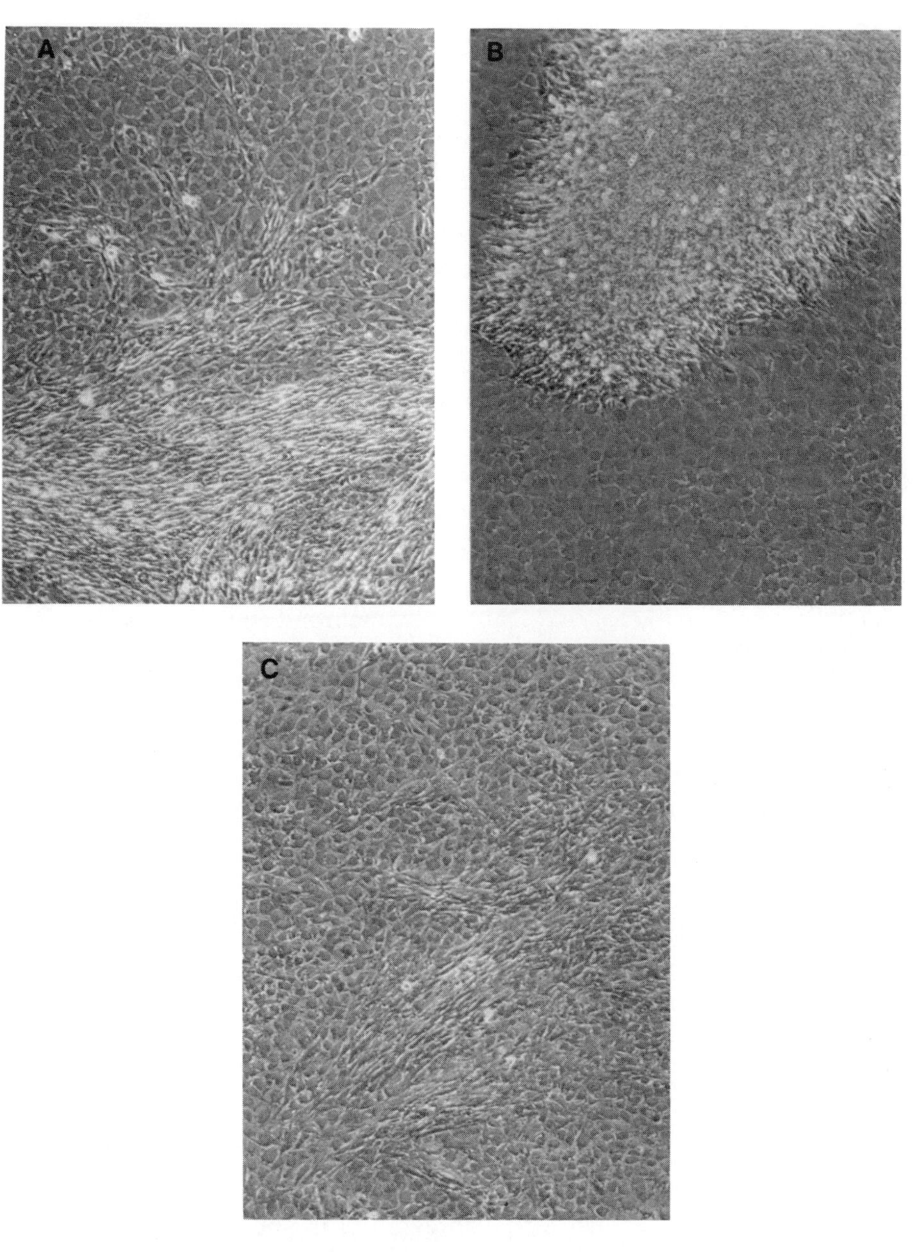

FIGURE 7. Morphology of transformed foci of mouse C127 cells induced by the entire BPV-1 genome or by the E6 and E5 transforming genes of BPV-1. C127 mouse cells were transfected with (A) the intact BPV-1 genome (142-6); (B) the E6 transforming gene of BPV-1 expressed from the Moloney sarcoma virus LTR; and (C) the E5 transforming gene expressed from the Harvey murine sarcoma virus LTR. C127 cells transformed by intact BPV-1 DNA exhibit dense focus formation with spindle-shaped cells which appear to intercalate into the nontransformed monolayer. In contrast, cells transformed by the E6 gene display only dense focus formation without the spindle-cell invasion of the monolayer; cells transformed by the E5 gene show only spindle-cell formation without dense focus formation.

(Moska and Bastia, manuscript submitted). Whether the specific DNA-binding capacity of the E2 gene product is directly involved in transcriptional activation has not been established.

The transactivation function of the E2 gene is not peculiar to BPV-1. Recently, it has also been shown that the E2 gene product of HPV-16 is involved in the transactivation of an inducible element in the HPV-16 LCR (Phelps and Howley, manuscript submitted). In addition, the papillomavirus E2 functions can work on heterologous LCR elements. For instance, the BPV-1 E2 gene product can transactivate the HPV-16 LCR element and, conversely, the HPV-16 E2 gene product can transactivate the BPV-1 LCR element (W. Phelps and P. Howley, manuscript submitted).

V. PLASMID MAINTENANCE FUNCTIONS

The benign epithelial and fibroepithelial tumors induced by papillomaviruses contain the viral DNA as extrachromosomal plasmids. The viral DNA exists as a stable low-copy-number plasmid in the dermal fibroblasts of a bovine fibropapilloma (Lancaster and Olson, 1980). It is generally believed that the viral DNA is also present as a low-copy-number plasmid in the basal cells of the epidermis in warts. Vegetative viral DNA replication does not occur in these basal cells or in the keratinocytes of the lower levels of the epidermis (Howley, 1983). In tumors induced by BPV-1, the DNA is also predominantly, if not exclusively, extrachromosomal. This has been shown in fibroblastic tumors induced by BPV-1 in Syrian hamsters and in equine sarcoid induced in horses by BPV-1 (Lancaster and Olson, 1980).

The viral genome also exists as a stable multicopy plasmid in transformed mouse C127 cells or NIH 3T3 cells (Law *et al.*, 1981; Lancaster, 1981). In addition, it has been shown to exist as a stable extrachromosomal plasmid in transformed Fischer fibroblasts (Binetruy *et al.*, 1982; Babiss and Fisher, 1986). Law *et al.* (1981) demonstrated that integration of the viral genome was not required for the induction of BPV-1 transformation of C127 cells. No evidence of viral–host junctions indicative of integration of the viral genome was found in clonal transformed cell lines of C127 cells. Further evidence that integration is not required for transformation is provided by Turek *et al.* (1982) who showed that treatment of BPV-1-transformed mouse C127 cells led to a decrease in the copy number of the extrachromosomal BPV-1 plasmids. Furthermore, long-term treatment of cultures with interferon induced phenotypic revertants which no longer contained BPV-1 DNA (Turek *et al.*, 1982).

Plasmid replication of the BPV-1 genome in rodent cells appears, therefore, to be a good model for the latent and persistent infection of cells by papillomaviruses. The DNA is maintained as a stable multiple-copy plasmid. Isolation and characterization of subclones of BPV-1-transformed cells, in general, have revealed subclones which have a copy number approximately that of the parent cell lines (Howley, unpublished observations). Thus, the BPV-1 DNA plasmid copy number appears to be controlled within transformed cells once stable transformation is established and the partitioning of the viral DNAs to the daughter cells appears to be faithfully maintained.

The establishment of BPV-1 plasmids in rodent cells is complex. BPV-1 transformation of mouse cells follows single-hit kinetics (Dvoretzky *et al.*, 1980) and yet the DNA replicates as a multicopy nuclear plasmid (Law *et al.*, 1981). Clearly, there must be an initial replication of BPV-1 DNA genomes from one or a few copies to a level which will be stably maintained in the transformed cells. The pathway by which the viral DNA establishes and maintains copy number control in this latent state has not been elucidated. Botchan has shown than BPV-1 plasmids replicate once per cell cycle after they have been established as a stable multicopy plasmid (M. Botchan, personal communication). Immediately following infection or transfection of mouse cells, however, BPV-1 plasmid replication appears not to be tied to the cell cycle (Botchan *et al.*, 1986). Both *cis*- and *trans*-acting functions are involved in the stable replication of BPV-1 plasmids in mouse cells.

A. *Cis* Elements

Electron microscopy of BPV-1 replication intermediates in transformed cells has revealed the center of replication bubbles at a discrete site within the BPV-1 LCR (Waldeck *et al.*, 1984). This site maps closely, if not identically, with a plasmid maintenance sequence (PMS1) mapped by Lusky and Botchan (1984). A PMS is defined as a minimal sequence required for the maintenance of BPV-1 plasmids in cells providing BPV-1 replication functions in *trans.* In this 1984 study, Lusky and Botchan mapped two regions which they refer to as PMS1 and PMS2 which they determined as sufficient for the maintenance of exogenous plasmids in BPV-1-transformed cells. PMS1 mapped to the Cla C fragment in the LCR and PMS2 mapped within the E1 ORF. Subsequently, Botchan and others have shown that these PMS sequences are, however, by themselves, not sufficient for maintaining plasmids in BPV-1-transformed cells but that they also require a functional enhancer element. The BPV-1 distal enhancer located adjacent to the *Bam*HI site is sufficient for providing this function and recently Lusky and Botchan have mapped an additional sequence in the BPV-1 31% nontransforming region which can also provide this function (Lusky and Botchan, 1986). They refer to this sequence as a replication enhancer.

B. *Trans* Functions

Genetic analyses have identified two complementation groups within the viral genome which encode factors involved in BPV-1 plasmid replication and maintenance. Mutations in the 3′ portion of the E1 ORF abolish plasmid replication, leading to integration of the viral DNA (Sarver *et al.*, 1984; Lusky and Botchan, 1985). In addition, mutations in the E6 and E6/E7 genes result in DNAs which are able to maintain as plasmids, but only in low copy number. The viral plasmids cannot maintain a copy number of higher than 5 per cell (Lusky and Botchan, 1985; Berg *et al.*, 1986). In the full genome background, mutations in the E2 gene also lead to the integration of the viral DNA into the host chromosome (Sarver *et al.*, 1984; Rabson *et al.*, 1986; DiMaio *et al.*, 1986a; Groff and Lancaster, 1986). Furthermore, Rabson *et al.* (1986) have shown that the E2 function, when provided in *trans*, allows plasmid replication. It seems possible that the role of E2 in plasmid replication as in transformation may be indirect rather than direct, mediated through the transcriptional activation of gene functions required directly for plasmid maintenance. The argument that the function may be indirect comes from the identification of the E2 function in transactivating a transcriptional element in the LCR (Spalholz *et al.*, 1985) and the recognition that under certain experimental conditions, some E2 mutants are able to be maintained extrachromosomally as stable plasmids (Lusky and Botchan, 1985).

VI. SUMMARY AND CONCLUSIONS

The lack of an *in vitro* system for the productive propagation of papillomaviruses in tissue culture has severely impeded studies on this important group of tumor viruses. Indeed, detailed studies on the molecular biology of these viruses could not begin until the molecular cloning of the viral genomes of individual papillomaviruses in the late 1970s. The natural host for the productive replication of the papillomaviruses is the squamous epithelial cell, and the natural tumors that this group of viruses is associated with are squamous cell carcinomas. Cellular transformation by the papillomaviruses was recognized first for the BPVs in the early 1960s. In those early studies, attempts to define tumor antigens using the antisera of tumor-bearing experimental animals were unfruitful; consequently, this transformation system was effectively forgotten until the late 1970s when the cloned genomes of the papillomaviruses became available. Since the cellular host for these viruses is the squamous epithelial cell, purists may question the relevance of the transformation of rodent fibroblasts or of other immortalized cells by these viruses. However, cellular transformation has provided the only biological system for investigators to begin to dissect the viral functions associated with the induction of cellular proliferation and with the latent infection of mammalian cells.

In this chapter we have reviewed and discussed the data indicating that certain papillomaviruses encode gene products that are capable of morphologically transforming mouse cells in tissue culture. Two genes appear to be directly involved in cellular transformation. These are the E6 gene which is highly conserved among all papillomaviruses with respect to the conservation of the motif Cys-X-X-Cys and its spacing in the predicted viral protein, and the E5 gene which may be specifically associated with papillomaviruses which can induce fibroplasia as well as proliferation of epithelial cells. The model system we have focused on in this chapter, and which we have been studying in our laboratory, is that of BPV-1. Sequence analysis of the other papillomaviruses indicates that the overall genomic organization of each of the papillomaviruses is quite similar to that of BPV-1, validating the use of this virus as the prototype virus for defining the functions of the papillomaviruses. Each of the papillomaviruses contains a control region sitting upstream of a series of ORFs which encode the gene products that share a predicted limited amino acid homology with BPV-1 putative gene products, strongly suggesting that the genes defined in BPV-1 have their analogues in these other papillomaviruses.

Papillomaviruses normally induce benign lesions in their natural hosts. What then is the function of each of the genes involved in malignant cellular transformation as measured on NIH 3T3 cells or on mouse C127 cells? It is likely that each of these products is involved in a specific manner with the cellular proliferation characteristic of a benign wart. Given the high conservation of the E6 ORF in all papillomaviruses, it seems reasonable to suggest that this protein may be involved in the normal proliferation of squamous epithelial cells. The high conservation of the E5 ORF in those viruses which induce fibropapillomas and which induce fibroblastic tumors in hamsters (Table I) suggests that the E5 gene product may have a direct role in inducing dermal fibroblastic proliferation. In BPV-1 the integrity of the E2 gene is required for efficient transformation. It seems likely that the role of E2 in transformation may be indirect, and that the E2 gene product may be required for the transcriptional transactivation of an inducible element located in the LCR, which is in turn necessary for the activation of genes directly involved in cellular transformation, namely E5 and/or E6.

The evidence associating papillomaviruses with naturally occurring cancers in their natural hosts is overwhelming. In animal systems the association with CRPV and cutaneous carcinomas in experimentally infected domestic or cottontail rabbits provides a useful model for studying the association of papillomaviruses with benign lesions which have the potential for malignant progression. In cattle, BPV-4 is associated with esophageal papillomatosis. Those cattle infected by BPV-4 which feed on bracken fern containing a radiomimetic substance have a high risk of developing esophageal and foregut squamous cell carcinomas (Jarrett *et al.*, 1978). In humans, HPVs are associated with a variety of squamous cell carcinomas including cutaneous carcinomas in patients with epidermodysplasia verruciformis, cervical and anogenital carcinomas, and some laryngeal and oral carcinomas. These human cancers and their association with HPVs are discussed in detail in the other chapters in this volume. In general, only certain of the over 41 HPVs now recognized are associated with benign lesions which have a prospensity for malignant progression. It is clear that malignant progression of these benign lesions is a relatively rare event and that the expression of the papillomavirus gene functions, *per se,* cannot be solely responsible for the malignant progression. It is tempting to extrapolate from the transformation functions mapped and studied in the transforming papillomaviruses to carcinomatous progression associated with papillomaviruses in an attempt to define those viral functions involved. Each of the papillomavirus/carcinoma associations discussed above, however, has unique features. For instance, in the carcinomas associated with BPV-4 found in cattle, the viral genome cannot be found. Consequently, there is no evidence for the presence of the viral genome or continued viral transcription in the associated carcinomas. This contrasts with the CRPV-associated cutaneous carcinomas in rabbits and the various HPV-associated carcinomas in which specific papillomavirus DNAs can readily be identified in the carcinomas. In each of these cases, the viral DNAs are usually transcriptionally active. In the carcinomas in which the papillomaviruses are transcriptionally active, it is usually the E6 region of the genome which is expressed, suggesting that the continued expression of this gene which is analogous to one of the BPV-1 transforming genes may be required for the maintenance of the carcinomatous state.

The CRPV-associated cutaneous carcinomas in rabbits are similar to BPV-1-transformed cells in that the viral DNA can exist as extrachromosomal molecules. This is similar to the HPV-associated cutaneous carcinomas in humans with epidermodysplasia verruciformis. This contrasts with the HPV-16 and HPV-18 sequences associated with human cervical carcinoma and human anogenital carcinomas. In these tumors, the DNA is generally integrated although some extrachromosomal molecules may also be detected. In cell lines established from human cervical carcinomas, the HPV-16 and HPV-18 DNAs are totally integrated (Schwarz *et al.*, 1985; Yee *et al.*, 1985). Integration is generally associated with the disruption of the E2 or E1 ORFs. In the case of the E2 ORF, this likely results in the knocking out of a transcriptional transactivating function. Evidence is accumulating in a number of laboratories that the E1 ORF may encode a gene product that is also involved in the regulation of gene expression by the papillomaviruses. Hence, integration of the viral DNA to disrupt the E1 or E2 ORF could result in mutations which lead to the unregulated expression of the viral genome due to the absence of the normal transcriptional regulatory factors in *trans.*

In these cases in which integration of the papillomavirus genomes has occurred into the host chromosomes, one must consider the formal possibility that the integration has introduced viral enhancer elements in the vicinity of cellular genes whose activation or improperly controlled expression leads to cellular transformation. To date, however, there is no evidence that this potential mechanism of enhancer insertion is involved in papillomavirus-associated carcinomatous progression.

ACKNOWLEDGMENT. We are grateful to Ms. Nan Freas for her assistance in the preparation of the manuscript.

REFERENCES

Ahola, H., Stenlund, A., Moreno-López, J., and Pettersson, U., 1983, Sequences of bovine papillomavirus type 1 DNA—Functional and evolutionary implications, *Nucleic Acids Res.* **11:**2639–2650.

Allshire, R. C., and Bostock, C. J., 1986, Structure of bovine papillomavirus type 1 DNA in a transformed mouse cell line, *J. Mol. Biol.* **188:**1–13.

Amtmann, E., and Sauer, G., 1982, Bovine papillomavirus transcription: Polyadenylated RNA species and assessment of the direction of transcription, *J. Virol.* **43:**59–66.

Androphy, E., Schiller, J. T., and Lowy, D., 1985, Identification of the protein encoded by the E6 transforming gene of bovine papillomavirus, *Science* **230:**442–445.

Babiss, L. E., and Fisher, P. B., 1986, Characterization of Fischer rat embryo (CREF) cells transformed by bovine papillomavirus type 1, *Virology* **154:**180–194.

Benjamin, T. L., 1970, Host range mutants of polyoma virus, *Proc. Natl. Acad. Sci. USA* **67:**394–399.

Berg, L. J., Singh, K., and Botchan, M., 1986, Complementation of a bovine papillomavirus low-copy-number mutant: Evidence for a temporal requirement of the complementing gene, *Mol. Cell. Biol.* **6:**859–869.

Binetruy, B., Meneguzzi, G., Breathnach, R., and Cuzin, F., 1982, Recombinant DNA molecules comprising bovine papilloma virus type 1 DNA linked to plasmid DNA are maintained in a plasmidial state both in rodent fibroblasts and in bacterial cells, *EMBO J.* **1:**621–628.

Black, P. H., Hartley, J. W., Rowe, W. P., and Huebner, R. J., 1963, Transformation of bovine tissue culture cells by bovine papilloma virus, *Nature* **199:**1016–1018.

Boiron, M., Levy, J. P., Thomas, M., Friedman, J. C., and Bernard, J., 1964, Some properties of bovine papillomavirus, *Nature* **201:**423–424.

Botchan, M., Berg, L., Reynolds, J., and Lusky, M., 1986, The bovine papillomavirus replicon, *Ciba Symp.* **120:**53–64.

Campo, M. S., and Spandidos, D. A., 1983, Molecularly cloned bovine papillomavirus DNA transforms mouse fibroblasts *in vitro, J. Gen. Virol.* **64:**549–557.

Campo, M. S., Spandidos, D. A., Lang, J., and Wilkie, N. M., 1983, Transcriptional control signals in the genome of bovine papillomavirus type 1, *Nature* **303:**77–80.

Chen, E. Y., Howley, P. M., Levinson, A. D., and Seeburg, P. H., 1982, The primary structure and genetic organization of the bovine papillomavirus (BPV) type 1, *Nature* **299:**529–534.

Danos, O., Georges, E., Orth, G., and Yaniv, M., 1985, Fine structure of the cottontail rabbit papillomavirus mRNAs expressed in the transplantable VX2 carcinoma, *J. Virol.* **53:**735–741.

DiMaio, D., Corbin, V., Sibley, E., and Maniatis, T., 1984, High-level expression of a cloned HLA heavy chain gene introduced into mouse cells on a bovine papillomavirus vector, *Mol. Cell. Biol.* **4:**340–350.

DiMaio, D., Metherall, J., Neary, K., and Guralski, D., 1986a, Nonsense mutation in open reading frame E2 of bovine papillomavirus DNA, *J. Virol.* **57**:475–480.

DiMaio, D., Guralski, D., and Schiller, J. T., 1986b, Translation of open reading frame E5 of bovine papillomavirus is required for its transforming activity, *Proc. Natl. Acad. Sci. USA* **83**:1797–1801.

Dvoretzky, I., Shober, R., Chattopadhyay, S., and Lowy, D. R., 1980, A quantitative *in vitro* focus forming assay for bovine papilloma virus, *Virology* **103**:369–375.

Engel, L. W., Heilman, C. A., and Howley, P. M., 1983, Transcriptional organization of the bovine papillomavirus type 1, *J. Virol.* **47**:516–528.

Friedman, J. C., Levy, J. P., Lasneret, J., Thomas, M., Boiron, M., and Bernard, J., 1963, Induction de fibromes sous-cutanés chez le hamster doré par inoculation d'extraits à cellulaires de papillomes bovins, *C.R. Acad. Sci.* **257**:2328–2331.

Geraldes, A., 1969, Malignant transformation of hamster cells by cell-free extracts of bovine papilloma (*in vitro*). *Nature* **222**:1283–1285.

Gibbs, E. P. J., Smale, C. J., and Lawman, M. J. P., 1975, Warts in sheep, *J. Comp. Pathol.* **85**:327–334.

Grisoni, M., Meneguzzi, G., DeLapeyriere, O., Binetruy, B., Rassoulzadegan, M., and Cuzin, F., 1984, The transformed phenotype in culture and tumorigenicity of Fischer rat fibroblast cells (FR3T3) transformed with bovine papillomavirus type 1, *Virology* **135**:406–416.

Groff, D. E., and Lancaster, W. D., 1986, Genetic analysis of the 3' early region transformation and replication functions of bovine papillomavirus type 1, *Virology* **150**:221–230.

Groff, D. E., Sundberg, J. P., and Lancaster, W. D., 1983, Extrachromosomal deer fibromavirus DNA in deer fibromas and virus transformed mouse cells, *Virology* **131**:546–550.

Heilman, C. A., Engel, L., Lowy, D. R., and Howley, P. M., 1982, Virus-specific transcription in bovine papillomavirus transformed mouse cells, *Virology* **119**:22–34.

Howley, P. M., 1983, The molecular biology of papillomavirus transformation, *Am. J. Pathol.* **113**:414–421.

Howley, P. M., Law, M.-F., Heilman, C. A., Engel, L. W., Alonso, M. C., Lancaster, W. D., Israel, M. A., and Lowy, D. R., 1980, Molecular characterization of papillomavirus genomes, in: *Viruses in Naturally Occurring Cancers* (M. Essex, G. Todaro, and H. zur Hausen, eds.), pp. 233–247, Cold Spring Harbor Laboratory, Cold Spring Harbor, N.Y.

Howley, P. M., Schenborn, E. T., Lund, E., Byrne, J. C., and Dahlberg, J. E., 1985, The bovine papillomavirus distal "enhancer" is not *cis*-essential for transformation or for plasmid maintenance, *Mol. Cell. Biol.* **5**:3310–3315.

Jarrett, W. F. H., McNeil, P. E., Grimshaw, W. T. R., Selman, I. E., and McIntyre, W. I. M., 1978, High incidence area of cattle cancer with a possible interaction between an environmental carcinogen and papilloma virus, *Nature* **274**:215–217.

Koller, L. D., and Olson, C., 1972, Attempted transmission of warts from man, cattle, and horses, and of deer fibroma, to selected hosts, *J. Invest. Dermatol.* **58**:366–368.

Kyte, J., and Doolittle, R., 1982, A simple method for displaying the hydrophobic character of a protein, *J. Mol. Biol.* **157**:105–132.

Lancaster, W. D., 1981, Apparent lack of integration of bovine papillomavirus DNA in virus-induced equine and bovine tumor cells and virus-transformed mouse cells, *Virology* **108**:251–255.

Lancaster, W. D., and Olson, C., 1980, State of bovine papillomavirus DNA in connective-tissue tumors, *Cold Spring Harbor Conf. Cell Prolif.* **7**:223–232.

Lancaster, W. D., Olson, C., and Meinke, W., 1977, Bovine papilloma virus: Presence of virus-specific DNA sequences in naturally occurring equine tumors, *Proc. Natl. Acad. Sci. USA* **74**:524–528.

Law, M.-F., Lowy, D. R., Dvoretzky, I., and Howley, P. M., 1981, Mouse cells transformed by bovine papillomavirus contain only extrachromosomal viral DNA sequences, *Proc. Natl. Acad. Sci. USA* **78**:2727–2731.

Law, M.-F., Byrne, J. C., and Howley, P. M., 1983, A stable bovine papillomavirus hybrid plasmid that expresses a dominant selective trait, *Mol. Cell. Biol.* **3**:2110–2115.

Lowy, D. R., Dvoretzky, I., Shober, R., Law, M.-F., Engel, L., and Howley, P. M., 1980, *In vitro* tumorigenic transformation by a defined subgenomic fragment of bovine papillomavirus DNA, *Nature* **287**:72–74.

Lusky, M., and Botchan, M., 1984, Characterization of the bovine papilloma virus plasmid maintenance sequences, *Cell* **36**:391–401.

Lusky, M., and Botchan, M. R., 1985, Genetic analysis of the bovine papillomavirus type 1 *trans*-acting replication factors, *J. Virol.* **53**:955–965.

Lusky, M., and Botchan, M. R., 1986, Transient replication of BPV-1 plasmids: Cis and trans requirements, *Proc. Natl. Acad. Sci. USA* **83**:3609–3613.

Lusky, M., Berg, L., Weiher, H., and Botchan, M., 1983, The bovine papillomavirus contains an activator of gene expression at the distal end of the transcriptional unit, *Mol. Cell. Biol.* **3**:1108–1122.

Matthias, D. D., Bernard, H. U., Scott, A., Brody, G., Hashimoto-Gotoh, T., and Schutz, G., 1983, A bovine papillomavirus vector with a dominant resistance marker replicates extrachromosomally in mouse cells and *E. coli* cells, *EMBO J.* **2**:1487–1492.

Meischke, H. R. C., 1979, *In vitro* transformation by bovine papilloma virus, *J. Gen. Virol.* **43**:473–487.

Meneguzzi, G., Binetruy, B., Grisoni, M., and Cuzin, F., 1984, Plasmidial maintenance in rodent fibroblasts of a BPV-1–pBR322 shuttle vector without immediately apparent oncogenic transformation of the recipient cells, *EMBO J.* **3**:365–371.

Moar, M. H., Campo, M. S., Laird, H., and Jarrett, W. F. H., 1981, Persistence of non-integrated viral DNA in bovine cells transformed *in vitro* by bovine papillomavirus type 2, *Nature* **293**:749–751.

Morgan, D. M., and Meinke, W., 1980, Isolation of clones of hamster embryo cells transformed by the bovine papilloma virus, *Curr. Microbiol.* **3**:247–251.

Nakabayashi, Y., Chattopadhyay, S. K., and Lowy, D. R., 1983, The transforming function of bovine papillomavirus DNA, *Proc. Natl. Acad. Sci. USA* **80**:5832–5836.

Nasseri, M., and Wettstein, F. O., 1984., Differences exist between viral transcripts in cottontail rabbit papillomavirus-induced benign and malignant tumors as well as non-virus-producing and virus-producing tumors, *J. Virol.* **51**:706–712.

Okayama, H., and Berg, P., 1983, A cDNA cloning vector that permits expression of cDNA inserts in mammalian cells, *Mol. Cell. Biol.* **3**:280–289.

Olson, C., Gordon, D. C., Robl, M. G., and Lee, K. P., 1969, Oncogenicity of bovine papillomavirus, *Arch. Environ. Health* **19**:827–837.

Pettersson, U., Ahola, H., Stenlund, A., Bergman, P., Ustav, M., and Moreno-López, J., 1986, Organization and expression of the genome of bovine papillomavirus type 1, *Ciba Symp.* **120**:23–34.

Puget, A., Favre, M., and Orth, G., 1975, Induction de tumeurs fibroblastiques cutanées où sous-custanées chez l'Ochotone afghan (Ochotono rufescens rufescens) par inoculation du virus du papillome bovin, *C.R. Acad. Sci.* **280**:2813–2816.

Rabson, M. S., Yee, C., Yang, Y.-C., and Howley, P. M., 1986, Analysis of the bovine papillomavirus type 1 3′ early region transformation and plasmid maintenance function, *J. Virol.* **60**:626–634.

Rösl, R., Waldeck, W., and Sauer, G., 1983, Isolation of episomal bovine papillomavirus chromatin and identification of a DNase I-hypersensitive region, *J. Virol.* **46**:567–574.

Rösl, R., Waldeck, W., Zentgraf, H., and Sauer, G., 1986, Properties of intracellular bovine papillomavirus chromatin, *J. Virol.* **58**:500–507.

Sarver, N., Byrne, J. C., and Howley, P. M., 1982, Transformation and replication in mouse cells of a bovine papillomavirus/pML2 plasmid vector that can be rescued in bacteria, *Proc. Natl. Acad. Sci. USA* **79**:7147–7151.

Sarver, N., Rabson, M. S., Yang, Y.-C., Byrne, J. C., and Howley, P. M., 1984, Localization and analysis of bovine papillomavirus type 1 transforming functions, *J. Virol.* **52**:377–388.

Sarver, N., Muschel, R., Byrne, J. C., Khoury, G., and Howley, P. M., 1985, Enhancer dependent expression of the rat preproinsulin genes in BPV-1 vectors, *Mol. Cell. Biol.***5**:3507–3516.

Schiller, J., Vass, W. C., and Lowy, D. R., 1984, Identification of a second transforming region of bovine papillomavirus, *Proc. Natl. Acad. Sci. USA* **81:**7880–7884.

Schiller, J., Vousden, K., Vass, W. C., and Lowy, D. R., 1986, The E5 open reading frame of bovine papillomavirus type 1 encodes a transforming gene, *J. Virol.* **57:**1–6.

Schlegel, R., Wade-Glass, M., Rabson, M. S., and Yang, Y.-C., 1986, The E5 transforming gene of bovine papillomavirus encodes a small hydrophobic polypeptide, *Science* **233:**464–466.

Schwarz, E., Freese, U. K., Gissmann, L., Mayer, W., Roggenback, B., Stremlau, A., and zur Hausen, H., 1985, Structure and transcription of human papillomavirus sequences in cervical carcinoma cells, *Nature* **314:**111–114.

Shope, R. E., 1933, Infectious papillomatosis of rabbits, with a note on the histopathology, *J. Exp. Med.* **58:**607–624.

Spalholz, B. A., Yang, Y.-C., and Howley, P. M., 1985, Transactivation of a bovine papillomavirus transcriptional regulatory element by the E2 gene product, *Cell* **42:**183–191.

Stenlund, A., Moreno-López, J., Ahola, H., and Pettersson, U., 1983, European elk papillomavirus: Characterization of the genome, induction of tumors in animals, and transformation *in vitro*, *J. Virol.* **48:**370–376.

Stenlund, A., Zabielski, J., Ahola, H., Moreno-López, J., and Pettersson, U., 1985, The messenger RNAs from the transforming region of bovine papillomavirus type 1, *J. Mol. Biol.* **182:**541–544.

Thomas, M., Boiron, M., Tanzer, J., Levy, J. P., and Bernard, J., 1964, *In vitro* transformation of mice cells by bovine papilloma virus, *Nature* **202:**709–710.

Turek, L. P., Byrne, J. C., Lowy, D. R., Dvoretzky, I., Friedman, R. M., and Howley, P. M., 1982, Interferon induces morphologic reversion with elimination of extrachromosomal viral genomes in bovine papillomavirus transformed mouse cells, *Proc. Natl. Acad. Sci. USA* **79:**7914–7918.

Waldeck, W., Rösl, F., and Zentgraf, H., 1984, Origin of replication in episomal bovine papilloma virus type 1 DNA isolated from transformed cells, *EMBO J.* **3:**2173–2178.

Watts, S. L., Ostrow, R. S., Phelps, W. C., Prince, J. T., and Faras, A. J., 1983, Free cottontail rabbit papillomavirus DNA persists in warts and carcinomas of infected rabbits and in cells in culture transformed with virus or viral DNA, *Virology* **125:**127–138.

Watts, S. L., Phelps, W. C., Ostrow, R. S., Zachow, K. R., and Faras, A. J., 1984, Cellular transformation of human papillomavirus DNA *in vitro*, *Science* **225:**634–636.

Weiher, H., and Botchan, M. R., 1984, An enhancer sequence from bovine papillomavirus DNA consists of two essential regions, *Nucleic Acids Res.* **12:**2901–2916.

Yang, Y.-C., Okayama, H., and Howley, P. M., 1985a, Bovine papillomavirus contains multiple transforming genes, *Proc. Natl. Acad. Sci. USA* **82:**1030–1033.

Yang, Y.-C., Spalholz, B. A., Rabson, M. S., and Howley, P. M., 1985b, Dissociation of transforming and transactivation functions for bovine papillomavirus type 1, *Nature* **318:**575–577.

Yasumoto, S., Burkhardt, A. L., Doniger, J., and DiPaolo, J. A., 1986, Human papillomavirus type 16 DNA-induced malignant transformation of NIH 3T3 cells, *J. Virol.* **57:**572–577.

Yee, C., Krishnan-Hewlett, I., Baker, C. C., Schlegel, R., and Howley, P. M., 1985, Presence and expression of human papillomavirus sequences in human cervical carcinoma cell lines, *Am. J. Pathol.* **119:**361–366.

CHAPTER 6

PAPILLOMAVIRUSES AND CARCINOGENIC PROGRESSION I

Cottontail Rabbit (Shope) Papillomavirus

FELIX O. WETTSTEIN

I. INTRODUCTION

The etiological agent of rabbit papillomatosis, a disease brought to the attention of Shope by a hunter, was identified as a virus over 50 years ago (Shope and Hurst, 1933). As later shown, the Shope rabbit papilloma virus—now usually named cottontail rabbit papillomavirus (CRPV)—represented the first DNA virus known to cause tumors. Five biological phenomena brought considerable attention to CRPV and its biology: (1) the ability to induce tumors (papillomas); (2) the disappearance of detectable virus from many tumors, a phenomenon called "masking"; (3) the spontaneous regression of virus-induced tumors in some animals; (4) the progression of the papillomas to malignant, invasive, metastasizing carcinomas; and (5) the synergism between virus and chemical carcinogens. Finally, an early, though unsuccessful, experiment in gene therapy with CRPV involving hyperarginemic patients was undertaken (Terheggen *et al.*, 1975). The rationale for this experiment was based on the assumption that increased levels of arginase in virus-induced rabbit papillomas was due to the expression of a virus-coded enzyme (Rogers, 1959; Rogers and Moore, 1963). Subsequent measurements of the arginase activity in 9,10-dimethyl-1,2-benzanthracene (DMBA)-induced papillomas showed similarly increased arginase levels (Orth *et al.*, 1967; Satoh *et al.*, 1967) and the enzyme was indistinguishable from highly purified rabbit liver arginase (Orth *et al.*, 1971b; Vielle-Breitburd and Orth, 1972).

FELIX O. WETTSTEIN • Department of Microbiology and Immunology, UCLA School of Medicine, Los Angeles, California 90024.

Progress in the understanding of these phenomena was slow since the virus cannot be grown in tissue culture and, until recently, no inbred rabbit strains were available. Much progress has been made with the application of techniques of modern molecular biology but still our understanding of the molecular events of tumor induction is fragmentary and little is known about the progression to malignancy.

II. BIOLOGICAL PROPERTIES OF THE SYSTEM

A. Virus Multiplication and Tumor Induction

1. Animal Specificity

The rabbit papillomatosis is endemic in the states of the Mississippi valley (Kreider and Bartlett, 1981) but cottontail rabbits in the east (New York; Syverton, 1952) as well as in the west (California; Stevens and Wettstein, 1979) are equally susceptible and produce similar amounts of virus as those trapped in Kansas. The yield of virus isolated from different cottontail rabbits is quite variable and generally higher from naturally infected animals than from laboratory-infected ones (Kidd, 1938, 1939).

CRPV has a narrow host range with respect to virus multiplication. Good virus yields are only obtained from papillomas of cottontail rabbits. Domestic rabbit papillomas may yield some virus depending on the strain of CRPV utilized (Shope, 1935); for some strains no evidence of virus replication can be found (Stevens and Wettstein, 1979). Infectious virus has also been obtained from papillomas induced in jackrabbits, snowshoe hares (Beard and Rous, 1935), and the host range has even been extended to rats though under special experimental conditions (Kreider and Bartlett, 1981).

In contrast to virus multiplication, tumor induction in domestic rabbits is as efficient as in cottontail rabbits. Since cottontail rabbits are not readily available and more delicate to keep in captivity, domestic rabbits have become the preferred species for most animal studies with CRPV, particularly those concerned with tumor induction. Further, papillomas on domestic rabbits are more likely to progress to malignant invasive and finally metastasizing tumors than papillomas on cottontail rabbits (Syverton, 1952) (Fig. 1).

2. Tissue Specificity

Tumor induction by CRPV is uniquely specific for the hairy epithelium (Kidd and Parsons, 1936). Early observations employing standard methods of infection indicated that papillomas arose from areas between hair follicles and sebaceous glands (Rous and Beard, 1934). A modified

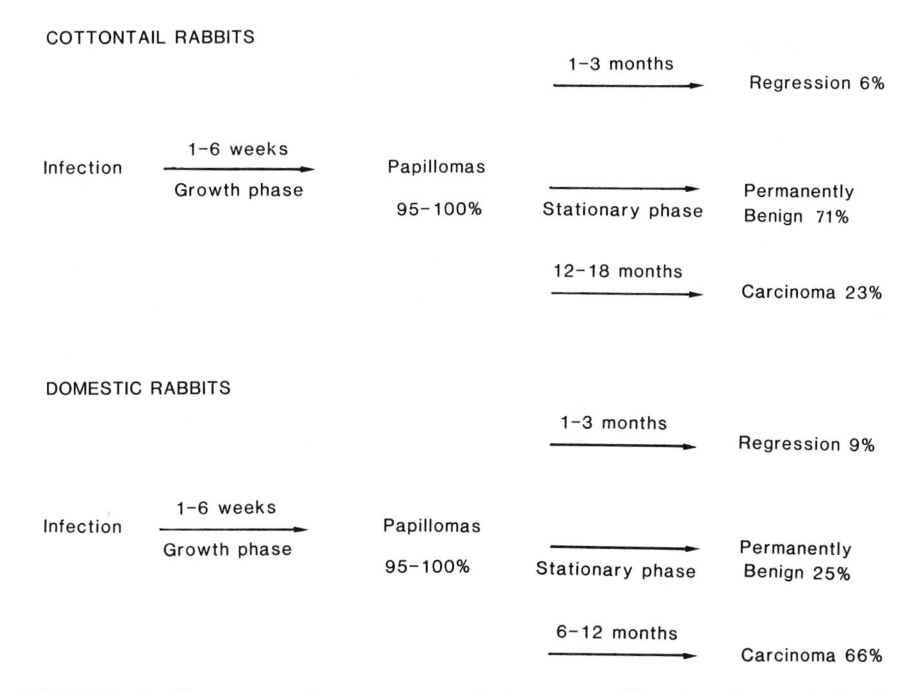

FIGURE 1. Papilloma-to-carcinoma sequence in experimentally infected cottontail and domestic rabbits. The schematic presentation is derived from data of Syverton *et al.* (1950a,b) and Syverton (1952).

procedure of infection permitted observations unobscured by local necrosis caused by scarification, and these suggested that the target cells were cells migrating from the hair follicles repopulating connective tissue areas denuded by scarification (Friedewald, 1944). The modified infection procedure together with turpentine-induced hyperplasia also resulted in a greatly increased efficacy of infection from an estimated 56×10^6 particles per infectious unit to as low as 2000 particles per unit (Friedewald, 1944).

3. Cell Specificity and Properties of Infected Cells

Early observations relating to altered properties of CRPV-infected cells came from experiments with tumors and tumor explants. It was noted that [^{3}H]thymidine incorporation into the basal and granulocyte layer was increased (Croissant and Orth, 1967, 1970). The altered cellular growth properties of virus-induced papillomas could also be shown by their ability to grow when implanted as homografts into different internal organs where normal skin did not grow (Rous and Beard, 1934; Coman, 1946).

Attempts to establish cell lines from normal rabbit skin (Orth and Croissant, 1967) and from CRPV-induced tumors proved difficult and usually resulted in short-term cultures. Only minor morphological and ultrastructural differences were found between primary cells derived from normal skin or from papillomas (Orth and Croissant, 1968). Infection of epithelial cell cultures with CRPV did not result in obvious morphological transformation but the cells developed into papillomas when introduced into the cheek pouch of hamsters (Kreider *et al.*, 1967). Further, it was noted that epithelial cell outgrowth from papilloma explants into culture exhibited more vigorous growth than that from normal skin and outgrowing cells from normal skin grew better when infected with CRPV (Coman, 1946). However, viral infection did not result in establishment of permanent cell lines. Finally, primary skin cultures infected with CRPV or derived from papillomas behaved differently from uninfected cultures in their response to superinfection by other viruses (Chardonnet *et al.*, 1978), although evidence of overgrowth with fibroblasts in these experiments was noticeable.

A permanent cell line has been derived from a cottontail rabbit papilloma by Kniazeft, Nelson-Rees, and Owens in 1964 and is available from the American Tissue Culture Collection. These cells do not induce tumors in nude mice (Nasseri and Wettstein, unpublished findings). A cell line derived from a transplantable carcinoma VX2* was maintained for several years and the cells were tumorigenic in nude mice (McVay *et al.*, 1982). More recently, cell lines have been established from NIH 3T3 and C127 cells morphologically transformed by transfection with CRPV DNA or infection with CRPV and these lines were also tumorigenic in nude mice (Watts *et al.*, 1983).

B. Tumor Progression

Early experiments suggested that the papilloma-to-carcinoma progression was characteristic of the events in domestic rabbits. In one experiment, seven of ten domestic rabbits infected with CRPV and observed over 200 days developed carcinomas and the malignant growth appeared in each of them at multiple sites. In addition, in five of the animals metastatic tumors had developed in regional lymph nodes and lung. None of ten cottontail rabbits infected in parallel developed any cancer (Rous and Beard, 1935b). Subsequently, more extensive experiments involving large numbers of experimentally infected cottontail and domestic rabbits revealed a quantitative rather than qualitative difference in progression to carcinomas between cottontail and domestic rabbits (Syverton *et al.*, 1950a,b; Syverton, 1952) (Fig. 1). A general finding of these experiments

* In McVay *et al.* (1982), this tumor was referred to as VX7. Subsequent detailed analysis indicated that it was VX2 (Nasseri and Wettstein, 1984b).

was also that no infectious virus could be obtained from carcinomas developing on cottontail rabbits and this was not the result of a general immunological reaction directed against the virus since virus usually could be recovered from papillomas of animals which simultaneously had carcinomas (Syverton, 1952). Two differences were noted between naturally and experimentally infected cottontail rabbits. The first, and not surprising, was that in naturally infected animals the first carcinomas appeared within 9 rather than 12 months; the percentage of animals developing carcinomas, however, was rather similar in both groups. The second difference was that tumor regression of natural papillomas continued over much longer periods of time and the percentage was ultimately higher. Similar rates of progression with domestic rabbits were also reported by Kreider (1980).

C. Interaction with Chemical Carcinogens

The long delay after viral infection in the appearance of malignant tumors made it difficult to establish a role of the virus in the malignant development. The observed correlation between increasing size of the virus inoculum and increasing chance to develop malignant tumors (Rous and Kidd, 1936) and the clearly established continued expression of viral genetic information indicated by the induction of antibodies to viral antigens in at least one malignant tumor, the VX7 transplantable carcinoma (Rogers *et al.*, 1950, 1960; Orth *et al.*, 1978), pointed toward a continued viral contribution in the malignant development. A link between progression to malignancy and virus was also suggested by synergism with carcinogens, which when applied under conditions where they normally would not induce malignant tumors could greatly accelerate the progression to malignancy of virus-induced papillomas.

First, it was observed that the compound scarlet R did stimulate the growth of virus-induced papillomas (Rous and Beard, 1934). In addition, hyperplasia induced by this compound as well as by turpentine could promote virus papilloma formation in the hyperplastic skin even when virus was given intravenously (Rous and Beard, 1934; Friedewald, 1944). On rabbit skin made hyperplastic by painting with tar, a carcinogen, tumor growth could again be induced by intravenous injection of virus. However, the growth arising from hyperplastic skin developed into highly anaplastic cancers in periods as short as 3 weeks (Rous and Kidd, 1936; Kidd and Rous, 1937). When no virus was introduced, no malignant growth developed, and if tar papillomas were present they usually regressed when treatment was stopped (Roud and Beard, 1935a). The synergism of virus and carcinogen in this progression was also illustrated by the fact that tarred skin hashed and implanted as a homograft did not result in any growth unless it was exposed to virus before implantation, in which case tumor growth developed (Kidd and Rous, 1937).

When 20-methylcholanthrene or 9,10-dimethyl-1,2-benzanthracene is applied in a single dose to the skin following viral infection, the growth of papillomas is rather inhibited. But in the carcinogen-treated sites, carcinomatous growth developed rather suddenly after 4 to 5 months (Rogers and Rous, 1951). The rate of progression to malignant tumors was also increased when a carcinogen was applied to growing virus-induced papillomas. With 20-methylcholanthrene as the carcinogen, usually several independent cancers developed on a single papillomatous growth within about 2 months under conditions where application of the carcinogen to normal skin would result in carcinomas only after 9 months to 2 years without the intermediate stage of papilloma (Rous and Friedewald, 1941).

Two phenomena appear to be operational. First, some compounds by inducing hyperplasia appear to increase the number of cells susceptible to viral infection or increase the access of virus to susceptible cells and some also promote the growth of papillomas. Second, with carcinogens the time from infection to the appearance of malignant tumors is shortened. The outcome is dramatic with tar, a compound which induces hyperplasia and is also a carcinogen. The accelerated progression to malignancy by carcinogens does not represent an overall acceleration of the natural progression. This is indicated by the effect of 20-methylcholanthrene; this compound rather inhibits papilloma development (Rogers and Rous, 1951). Finally, the fact that pretreatment of skin with a carcinogen leads to an accelerated progression indicates that viral genetic information is required at least for the initiation of the carcinogenic development. Further, it also suggests that the carcinogen probably does not act upon the viral DNA and this would be in agreement with observations in the natural progression where DNA isolated from malignant tumors did not result in an altered mode of tumor induction when compared to DNA from benign tumors (Ito, 1962).

D. Regression

Regression appears to be particularly striking in cottontail rabbits infected naturally. In a survey involving about 200 naturally infected animals, those observed up to 6 months showed a 20% regression rate and among those observed over 6 months the regression rate was 36%. In contrast, the regression among experimentally infected cottontail rabbits was only 6% as indicated in Fig. 1 (Syverton, 1952).

Genetic differences may play some role in the regression pattern. The naturally infected cottontail rabbits from Kansas represented a *Sylvilagus floridanus* subspecies (*S. floridanus alacar* Bangs) different from that of the experimentally infected animals from northwestern New York (*S. floridanus mearnsi* Allen). In addition, other factors such as the size of the inoculum and, more important, the area infected do influence tumor development (Evans *et al.*, 1962a). Further, animals from endemic areas

(Kansas) may have some residual immunity from previous infection which could influence papilloma development.

Tumor regression in general is systemic (Kidd, 1938), probably the result of immunological mechanisms (Evans *et al.*, 1962a). Tumor loss due to mechanical action particularly in cottontail rabbits which have a more delicate skin than domestic rabbits can obscure this general phenomenon (Kreider and Bartlett, 1981). More important, however, in animals with different-sized tumors, small ones may regress while large ones may be affected transiently or not at all. The size of the abraded and inoculated area determines the outcome rather than the amount of virus inoculated (Evans *et al.*, 1962a).

The immunity involved in regression is not directed against the virus. Rabbits immune to challenge with virus are not resistant to challenges with autologous skin infected in organ culture, and, in addition, progressor and regressor rabbits have similar antiviral titers (Kreider, 1963). The finding that regression of tumors induced with DNA is high in regressor rabbits and low in progressors (Evans and Ito, 1966) further supports the notion that the virus itself is not the target. In agreement with this are also results of experiments with vaccines prepared from domestic rabbit papillomas which contain little, if any, virus and vaccines prepared from cottontail papillomas; both vaccines increased the frequency of regression (Evans *et al.*, 1962b).

The nature of the immune response has not been elucidated. Passive transfer of sera from animals with regressing tumors did not increase the regression rate in recipient animals nor did serum from tumor-bearing animals result in enhancement of tumor progression (Evans *et al.*, 1962a). In contrast, *in vitro* cytotoxicity experiments suggested the presence of blocking antibodies in progressor but not regressor sera (Hellstrom *et al.*, 1969); thus, the question about the existence of enhancing or blocking antibodies remains open. Some role of leukocytes in regression is suggested by the fourfold higher leukocyte infiltration in regressing papillomas as compared to progressing ones (Kreider, 1980). More conclusive results concerning a cellular immune response will await the use of inbred rabbits now available.

III. MOLECULAR ASPECTS OF THE SYSTEM

A. The Virus and Viral Genome

The first highly purified preparations of virus were obtained by differential centrifugation and banding in CsC1 (Breedis *et al.*, 1962). The complete virions band at a density of 1.34 g/ml. The particles have an icosahedral symmetry and a diameter of 55 nm. The viral capsid is more likely composed of 72 capsomeres (Finch and Klug, 1965), although estimates of 42 (Williams *et al.*, 1960) or 92 (Mattern, 1962) have been

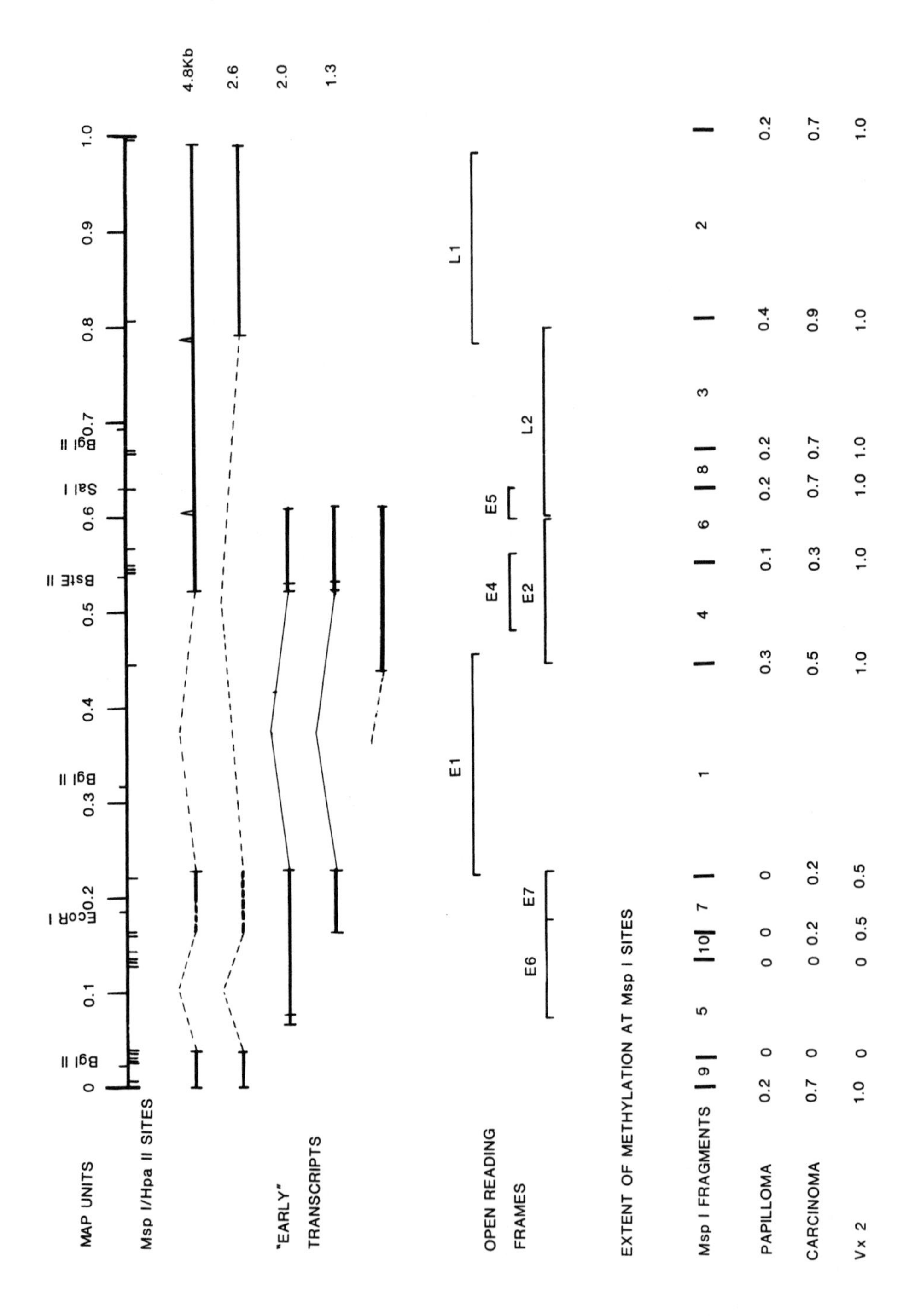
MAP UNITS
Msp I/Hpa II SITES
"EARLY" TRANSCRIPTS
OPEN READING FRAMES
EXTENT OF METHYLATION AT Msp I SITES
Msp I FRAGMENTS
PAPILLOMA
CARCINOMA
Vx 2
0 0.1 0.2 0.3 0.4 0.5 0.6 0.7 0.8 0.9 1.0
Bgl II
EcoR I
Bgl II
BstE II
Sal I
Bgl II
4.8Kb
2.6
2.0
1.3
E6 E7 E1 E4 E2 E5 L2 L1
9 5 10 7 1 4 6 8 3 2
0.2 0 0 0 0 0.3 0.1 0.2 0.2 0.4 0.2
0.7 0 0 0.2 0.2 0.5 0.3 0.7 0.7 0.9 0.7
1.0 0 0 0.5 0.5 1.0 1.0 1.0 1.0 1.0 1.0

FIGURE 2. Transcription map and sites of methylation in CRPV-induced rabbit tumors. (Top) Restriction map of CRPV DNA [note that nucleotide 1 of the sequence determined by Giri *et al.* (1985) is located at map position 0.05]. (Upper middle) Major transcripts of cottontail rabbit papillomas (4.8, 2.6, 2.0, 1.3, and 0.9 kb). Only the 2.0- and 1.3-kb RNA designated "early" transcripts are present in non-virus-producing papillomas and carcinomas of domestic rabbits (Nasseri and Wettstein, 1984a) and they are identical to the 2.0- and 1.25-kb RNA mapped by Georges *et al.* (1984) and Danos *et al.* (1985). The 4.8-, 2.6-, and 0.9-kb RNA unique to virus-producing cottontail rabbit papillomas have not been mapped completely. The 3′-proximal exons of the 4.8- and 2.6-kb RNA have been mapped (Nasseri and Wettstein, 1984a). The leader exon of the two RNA species separated by sedimentation in glycerol gradients has been identified by S1 mapping with probes labeled 3′ and 5′ at the *Bgl*II site (map position 0.02) and the 5′-end of the common leader represents the cap site as determined by primer extension (Wettstein and Nasseri, unpublished findings). Similarly, the ends of the 2.0 (two ends)- and 1.3-kb RNA represent the cap sites. Heavy lines designate exons and light brackets represent introns. Dashed lines represent segments not definitively mapped. (Lower middle) Open reading frames of CRPV DNA (Giri *et al.*, 1985). (Bottom) Extent of methylation at *Msp*I sites. The map position of the largest ten *Msp*I fragments is indicated and the extent of methylation at the fragment boundaries is given below for a domestic rabbit papilloma, a carcinoma, and the transplantable VX2 carcinoma (Wettstein and Stevens, 1983). Several fragment boundaries represent multiple *Msp*I sites (see map at top). For these boundaries the average extent of methylation at individual sites could be higher than indicated.

reported. The virus contains a double-stranded supercoiled DNA with a molecular weight of about 5×10^6 (Watson and Littlefield, 1960; Crawford and Crawford, 1963; Kleinschmidt *et al.*, 1965). The sequence of a cloned genome has been determined and consists of 7868 bp (Giri *et al.*, 1985). The molecular weight of most independent isolates agrees very well with the size calculated from the sequence and only minor variations exist with respect to restriction sites. Exceptions are a virus isolate with a molecular weight of 6.1×10^6 (Murphy *et al.*, 1981) and the major viral plasmid also of 6.1×10^6 (9.4 kb) present in the cottontail rabbit papilloma epidermis-derived cell line SflEp available from the American Tissue Culture Collection (Nasseri and Wettstein, unpublished findings). In both cases the restriction maps generated are not detailed enough to determine whether the increased size is due to a partial duplication of the genome or whether the two DNA species are identical.

The viral proteins have not been characterized definitively. A major protein of 54,000 (Favre *et al.*, 1975) or 60,000 daltons (Spira *et al.*, 1974) has been identified; the different values obtained probably resulted from differences in the methods employed. The lower estimate is more in agreement with the size of the open reading frames L1 or L2 (See Fig. 2), each of which has a coding capacity of about 500 amino acids. L1 is more likely coding for this protein since the most abundant mRNA unique to virus-producing tumors maps to this region of the genome (Nasseri and Wettstein, 1984a) and in analogy to findings with BPV. In BPV the assignment of L1 as the gene for the major structural protein was made based on the experimentally determined amino acid composition of the major structural protein (Meinke and Meinke, 1981) and the amino acid composition determined from the DNA sequence (Chen *et al.*, 1982; Ahola *et al.*, 1983). The virus also contains a minor larger protein, proteins migrating with cellular histones, and several minor species migrating between the major protein and the histones (Favre *et al.*, 1975). Peptide mapping of two minor proteins revealed that they were related and a fraction of their common peptide spots were also present in the peptide map of the major viral proteins. This could indicate that minor proteins are at least in part coded by sequences coding for the major capsid protein (L1) and in part from other sequences, possibly L2 (Feitelson *et al.*, 1985).

B. State of the Viral DNA

One of the puzzling aspects in the progression of papillomas to carcinomas was the loss of viral infectivity, a phenomenon called virus "masking" (Kidd and Rous, 1940). A similar situation exists in domestic rabbit tumors except here, infectious virus can usually not even be detected in papillomas (Shope and Hurst, 1933; Syverton, 1952). Direct evidence for the continued presence of viral nucleic acid in these tumors was provided by the ability of nucleic acid extracts to induce tumors (Ito, 1960). DNA extracts from domestic rabbit tumors were less efficient in tumor induction than those of cottontail rabbits (Ito and Evans, 1961) and among domestic rabbit tumors, carcinoma DNA was less efficient than papilloma DNA (Ito, 1962). The progression of DNA-induced papillomas to carcinomas was essentially the same as that of virus-induced papillomas and there was no apparent difference with respect to progression between DNA isolated from papillomas or carcinomas. Further, DNA isolated from virus-negative domestic rabbit papillomas and carcinomas gave rise to virus-producing papillomas in cottontail rabbits (Ito, 1962). These findings indicate that DNA propagated in virus-negative tumors retains full biological activity, and further suggest that progression of papillomas to carcinomas is not the consequence of a genetic change in the viral DNA.

Quantitative information about the viral DNA content of different tumors was obtained by measuring the reassociation kinetics of viral DNA in DNA extracts of tumors or by estimating band intensities in Southern blots relative to standards. The values for domestic rabbit papillomas range from about ten to a few hundred gene copies per diploid cell DNA equivalent (Stevens and Wettstein, 1979; Sugawara *et al.*, 1983; Watts *et al.*, 1983; Georges *et al.*, 1984) and values are similar for domestic rabbit carcinomas (Stevens and Wettstein, 1979; Watts *et al.*, 1983). The values for virus-producing cottontail rabbit papillomas fall between 1000 and 10,000 (Stevens and Wettstein, 1979; Sugawara *et al.*, 1983; Watts *et al.*, 1983) and were much lower for a cottontail rabbit carcinoma [80 viral gene copies per cell (Watts *et al.*, 1983)]. The relative distribution of viral DNA among various cell layers is quite different between virus-producing and non-virus-producing tumors. In cottontail rabbit papillomas the viral DNA can readily be detected by cytohybridization but only in keratinizing cells (Orth *et al.*, 1971a, 1977; Stevens and Wettstein, 1979) although it has been suggested that the basal cells of the epidermis may harbor the viral genome (Orth *et al.*, 1977). The presence of viral DNA in basal cells is supported by findings with domestic rabbit tumors, where all tumor cells contained equivalent numbers of viral genomes (Stevens and Wettstein, 1979).

1. Multimeric Plasmid Formation and DNA Integration

The state of the viral DNA in domestic rabbit tumors was investigated by Southern blotting. A first analysis revealed a complex multiband pattern in carcinomas with only a fraction of the viral DNA exhibiting the mobility of supercoiled (form I) or nicked (form II) DNA (Wettstein and Stevens, 1980). More detailed analysis of domestic rabbit papillomas and carcinomas showed that in papillomas the majority of viral DNA was present as nicked or supercoiled DNA. In carcinomas usually a majority of the DNA existed in the form of multimeric large circles which could be converted into multimeric linear forms by S1 nuclease digestion (Wettstein and Stevens, 1982). It is possible that the extent of multimeric plasmid formation is somewhat dependent on virus and rabbit strains employed (Watts *et al.*, 1983). A head-to-tail arrangement of the DNA in the multimeric plasmids is indicated by their conversion into unit-length linear DNA by single-cut enzymes. In a minority of tumors there was also a fraction of viral DNA, at most 25%, migrating with cellular DNA. Integration of this DNA was suggested most convincingly by two-dimensional agarose gel electrophoresis under conditions where nicked and supercoiled circular DNA of varying size is separated from linear DNA. Integration of this DNA was further indicated by the presence of integration bands after digestion with single-cut restriction endonucleases (Wettstein and Stevens, 1982).

Quite unexpected results were obtained when the DNA isolated from the transplantable carcinomas VX2 and VX7 was analyzed. VX2 was established by transfer of a metastatic tumor from a virus-infected domestic rabbit into an adult rabbit. The tumor never yielded infectious virus but for a limited number of passages it elicited neutralizing antibodies to the virus (Kid and Rous, 1940; Rous *et al.*, 1952; Rogers *et al.*, 1960). VX7 was established by transfer of a metastatic cancer of a domestic rabbit into a newborn domestic rabbit (Rogers *et al.*, 1950, 1960). Viral infectivity could be detected at least in the first transplant generation (Rogers *et al.*, 1950) and the tumor has continuously induced neutralizing antibodies to the virus so far up to the 165th transplant generation (Orth *et al.*, 1978) and viral antigen in nuclei was detectable by fluorescent antibodies in a low number of cells and with a low intensity in the 47th transplant generation (Mellors, 1960). VX2* with 10–40 viral gene copies per cell was found to contain only integrated DNA arranged in head-to-tail tandem repeats (McVay *et al.*, 1982; Sugawara *et al.*, 1983; Georges *et al.*, 1984). The number of integration bands detected (up to 6) suggests that the viral DNA is integrated in as many as three different sites (Nasseri and Wettstein, 1984b). VX7 contains 100–400 viral gene copies per cell also integrated in head-to-tail tandem repeats (Georges *et al.*, 1984).

The significance or cause for the accumulation of multimeric plasmids in carcinomas is unknown. The existence of integrated head-to-tail tandem repeats in the VX2 and VX7 tumors could indicate that these result from integration of oligomeric plasmids although other mechanisms could lead to their formation.

2. DNA Methylation

Preliminary evidence of a difference in viral DNA methylation between papillomas and carcinomas was obtained from digestion of tumor DNA with the endonuclease *Sal*I. Digestion with this single-cut enzyme converted papilloma DNA completely into unit-sized linear DNA while carcinoma DNA digests still contained large amounts of linear dimers as

* In the references of McVay *et al.* (1982) and Sugawara *et al.* (1983), this tumor was referred to as VX7. Subsequent detailed analysis suggested that the tumors analyzed rather represent the VX2 carcinoma (Nasseri and Wettstein, 1984b).

well as trimers and tetramers (Wettstein and Stevens, 1982). No comparable phenomenon was observed with the single-cut enzyme *Eco*RI. Since the recognition sequence of *Sal*I, but not *Eco*RI, contains a CG sequence—the site of methylation in eukaryotic cells (Ehrlich and Wang, 1981)—and *Sal*I digestion is inhibited by methylation at the C residue, incomplete digestion with *Sal*I did suggest that carcinoma DNA was partially methylated. More detailed information was obtained by digestion with the isoschizomeric endonucleases *Msp*I and *Hpa*II both recognizing the sequence CCGG. The former is not inhibited by methylation at the C residue next to G while the latter is inhibited. Such analysis revealed that cottontail rabbit papilloma DNA was essentially not methylated; domestic rabbit papilloma DNA was methylated to a variable but intermediate level; carcinoma DNA was more highly methylated and VX2* DNA was very highly methylated (Sugawara *et al.*, 1983; Wettstein and Stevens, 1983). Employing *Msp*I fragments as subgenomic probes in hybridization of Southern blots of *Hpa*II exhaustively digested DNA, the degree of methylation at different *Msp*I sites could be estimated (Fig. 2, bottom) (Wettstein and Stevens, 1983). There are limitations to the quantitation shown in Fig. 2 since some of the fragment boundaries represent multiple *Msp*I sites. For these positions the degree of methylation could be underestimated. The data clearly show the increasing methylation from the domestic rabbit papilloma to the virus-induced carcinoma and ultimately to the transplantable carcinoma VX2. Further, the viral genome is not methylated uniformly. Sites in the area of the 5′-proximal exons of the 1.3- and 2.0-kb RNA are unmethylated in papillomas. In the carcinoma and in VX2, sites close to the 5′-end of the 1.3-kb RNA become increasingly methylated but an area around the 5′-end of the 2.0-kb RNA appears to remain unmethylated. Interestingly, when tissue culture lines were derived from VX2, the pattern of methylation in the cell lines maintained for 2 years remained unchanged from that in the parent tumors (Wettstein and Stevens, 1983).

C. Viral Transcripts

The first evidence for continued DNA transcription in non-virus-producing domestic rabbit papillomas and carcinomas came from single-phase liquid hybridizations of labeled CRPV DNA with total tumor RNA. The transcripts were of low abundance: The R_0t values ranged from 200 to 1000 and they were at most homologous to 10–12% of the labeled whole genomic probe (Wettstein and Stevens, 1981) which is equivalent to transcription of 1.6 to 1.9 kb, values in good agreement with more detailed subsequent studies. However, no systematic differences between papillomas and carcinomas were found and similar values were obtained for the transplantable carcinoma VX2* (McVay *et al.*, 1982). Analysis by

* See footnote on p. 178.

Northern blotting of RNA from domestic rabbit tumors revealed two major spliced transcripts of about 1.3 and 2.0 kb and the 2.0-kb RNA was relatively more prominent compared to the 1.3-kb RNA in virus-induced papillomas than in carcinomas (Nasseri *et al.*, 1982; Nasseri and Wettstein, 1984a). The relative difference in the abundance of the 1.25-KB (1.3 kb) and 2.0-kb RNA is particularly striking in the VX2 tumor which has very little 2.0-kb RNA (Georges *et al.*, 1984). The transcripts were characterized in more detail by R-loop mapping (Georges *et al.*, 1984) and by S1 and *Exo*VII nuclease mapping (Nasseri and Wettstein, 1984a; Danos *et al.*, 1985) and their map location is shown in Fig. 2.

With S1 mapping, two slightly different 2.0-kb transcripts could be identified; a minor one extending about 70 nucleotides farther upstream than the major one, can code for a full-length E6 protein, while the major one could only initiate translation of E6 at a second ATG codon which would give rise to a protein 97 amino acids shorter (Nasseri and Wettstein, 1984a; Danos *et al.*, 1985). The comparison of a domestic rabbit carcinoma, a cottontail rabbit and a domestic rabbit papilloma revealed about a fourfold more prominent, larger 2.0-kb RNA in the papillomas than in the carcinoma (Nasseri and Wettstein, 1984a); however, subsequent analyses of three additional domestic rabbit papillomas and three carcinomas indicated that there was no systematic difference between benign and malignant tumors (Wettstein and Nasseri, unpublished findings).

Minor transcripts not yet completely mapped have been detected in VX2 which could transcribe the E2–E4 region (Danos *et al.*, 1985). The 3′-proximal exon of a transcript(s) from the E2–E4 region has been mapped with S1 nuclease in virus-induced domestic rabbit tumors. The exon extends from 50 nucleotides upstream of the beginning of E2 to the polyadenylation site common to the 1.3 (1.25)- and 2.0-kb RNA (See Fig. 2). The transcripts containing this exon whose leader has not yet been mapped, could code for E2 and less likely for E4. Quantitation of the amount of the RNA containing this exon present in papillomas and carcinomas did not reveal tumor-specific differences (Wettstein and Nasseri, unpublished findings).

The analyses of viral transcripts in papillomas and carcinomas thus far have not revealed any qualitative differences. Differences were, however, noted in the relative amounts of the major 1.3 (1.25)- and 2.0-kb RNA, the former being relatively more abundant in carcinomas (Nasseri *et al.*, 1982; Georges *et al.*, 1984; Nasseri and Wettstein, 1984a; Danos *et al.*, 1985).

D. Conclusion

The molecular analysis of domestic rabbit papillomas and carcinomas has revealed consistent differences with respect to the state of DNA and its methylation and a quantitative difference in the relative abundance of the two major transcripts. The differences observed, however, do not lead to an obvious explanation of tumor progression. Preliminary data indicate that the abundance of a minor transcript(s) which could code for the E2–E4 region involved in transformation of NIH 3T3 and C127 cells by BPV-1 (Nakabayashi *et al.*, 1983; Sarver *et al.*, 1984) is similar in both tumor types. Second, the 2.0-kb transcripts which could code for the E6 ORF implicated in transformation of C127 cells independent of E2–E4 (Schiller *et al.*, 1984) are rather decreased in carcinomas, certainly relative to the second major transcript of 1.3 (1.25) kb. Further, the relative amount of the minor 2.0-kb RNA which could code for a full-length E6 protein compared to the major transcript coding for a smaller E6 protein is similar in benign and malignant tumors. Finally, the 1.3 (1.25)-kb RNA is increased in its abundance relative to the 2.0-kb RNA in carcinomas compared to papillomas. The sequencing (Giri *et al.*, 1985) and S1 mapping data (Nasseri and Wettstein, 1984a; Danos *et al.*, 1985) suggest that this transcript could only code for an E7 protein. Interestingly, the 1.3 (1.25)-kb RNA contains sequences in the 3′-proximal exon which could code for the carboxy-terminal segment of E2 and the amino acid sequence of this segment was shown to have some homology to the *c-mos* oncogene (Danos and Yaniv, 1984; Giri *et al.*, 1985). Although the 5′-proximal exon of the 1.3 (1.25)-kb RNA extends past the termination codon of E7, partial suppression of the single termination codon could give rise to an E7–E2 fusion protein provided that the splicing of the 1.3-kb RNA aligns the proper reading frames; this has not yet been determined. Although suppressor tRNAs are present in mammalian cells (Hatfield *et al.*, 1982), the proposal is speculative and would suggest increased suppression in malignant tumors. Verification of such a mechanism would depend on the isolation of the protein.

A major deficiency in our knowledge remains the lack of information about the "early" proteins expressed in these tumors. Apparently animals bearing tumors do not produce antibodies or not in sufficient quantity to permit "early" protein detection (Stevens and Wettstein, 1979). There have been reports claiming the detection of virus-specific antigens in non-virus-producing cells (Noyes and Mellors, 1957; Mellors, 1960; Osato and Ito, 1968). The interpretation of such results is difficult. First, it has been shown that very sensitive methods (Watts *et al.*, 1983) permitted the detection of virus particles at least in some tumors normally considered non-virus-producing. Second, the non-virus-producing VX7 tumor still elicits antibodies to the major viral structural protein (Orth *et al.*, 1978) and the VX2 tumors contain a 4.8-kb transcript as a minor species which appears to be identical to the major 4.8-kb transcript unique to virus-producing cottontail rabbit papillomas (Nasseri and Wettstein, 1984b).

The parameters most clearly changing with tumor progression are multimeric plasmid formation and DNA methylation. The formation of multimeric plasmids with the concomitant reduction or almost complete loss of monomeric plasmid, at least in some carcinomas, could indicate that over extended periods—months to years—plasmids with multiple origins of replication are retained preferentially: ultimately, transplantable carcinomas, propagated over decades, may have persisted only because the viral DNA became integrated. As indicated, multimeric plasmids could represent the precursors to integrated head-to-tail tandem repeats. The data on the state of viral DNA do not suggest that DNA integration is a prerequisite for the progression to malignancy since no evidence of integration was found in several carcinomas. It appears more likely that as tumors progress, multimeric plasmids are preferentially formed or maintained and conditions for integration may be more favorable.

The significance of increasing DNA methylation, considered to play some role in gene inactivation (Razin and Riggs, 1980), is more difficult to assess. Potentially, a differential regulation of "early" transcripts by methylation is possible since at least three different cap sites (promoter regions) are utilized by the transcripts mapped (Nasseri and Wettstein, 1984a; Danos *et al.*, 1985). In the progression to carcinomas, the area of the genome around the 5′-end of the 1.3 (1.25)-kb RNA becomes partially methylated while the genomic segment around the 5′-end of the 2.0-kb RNA remains unmethylated. Transcription changes rather in the opposite direction with the 1.3-kb RNA becoming relatively more abundant. This could suggest that DNA methylation is not casually related to this change. However, there are limitations to the method employed (*Msp*I/*Hpa*II digestion) since the degree of methylation can only be determined for a small fraction of all potential methylation sites [CG sequences (Ehrlich and Wang, 1981)]. Methylation could be related to what appear to be increasingly unfavorable conditions for monomeric plasmid maintenance. At present, no transcripts for E1, the ORF implicated in plasmid maintenance in BPV-1-transformed cells (Nakabayashi *et al.*, 1983; Lusky and Botchan, 1984), have been identified. Possibly, DNA methylation could further reduce transcription of this ORF. Further, additional minor transcripts not yet mapped may be important in tumor development. Experiments with DNA deletions have indicated that sequences extending several hundred base pairs upstream from the "early" region and located within the "late" region are essential for tumor formation (Nasseri and Wettstein, unpublished findings). The sequences have not been mapped and their function(s) as well as a possible effect of methylation on them is unknown. Finally, although DNA methylation increases in a time scale comparable to tumor progression, the causative changes for progression may be entirely confined to changes in the cellular genome as suggested by the synergism between virus and carcinogens.

ACKNOWLEDGMENTS. The author is very grateful to Jeannie Martinez for the expert secretarial work and to Margerita Tayag for the art work, and thanks to Drs. I. Giri, O. Danos, M. Yaniv, G. Orth, and U. Pettersson for preprints.

REFERENCES

Ahola, H., Stenlund, A., Moreno-López, J., and Pettersson, U., 1983, Sequences of bovine papillomavirus type 1 DNA—Functional and evolutionary implications, *Nucleic Acids Res.* **11**:2639–2650.

Beard, J. W., and Rous, P., 1935, Effectiveness of the Shope papilloma virus in various American rabbits, *Proc. Soc. Exp. Biol. Med.* **33**:191–193.

Breedis, C., Berwick, L., and Anderson, T., 1962, Fractionation of Shope papilloma virus in cesium chloride density gradients, *Virology* **17**:84–94.

Chardonnet, Y., Greenland, T., Lyon, M., and Sohier, R., 1978, Interactions of the Shope papilloma virus with some other DNA viruses, *Bull. Cancer* **65**:195–206.

Chen, E. Y., Howley, P. M., Levinson, A. D., and Seeburg, P. H., 1982, The primary structure and genetic organization of the bovine papillomavirus type 1 genome, *Nature* **299**:529–534.

Coman, D. R., 1946, Induction of neoplasia *in vitro* with a virus: Experiments with rabbit skin grown in tissue culture and treated with Shope papilloma virus, *Cancer Res.* **6**:602–607.

Crawford, L. V., and Crawford, E. M., 1963, A comparative study of polyoma and papilloma viruses, *Virology* **21**:258–263.

Croissant, O., and Orth, G., 1967, Autoradiographic evidence of DNA synthesis in cells differentiating toward keratinization in domestic rabbit papillomas induced by the Shope virus, *C.R. Acad. Sci.* **265**:1341–1343.

Croissant, O., and Orth, G., 1970, Control of the mitotic activity in cells differentiating toward keratinization in domestic rabbit papillomas induced by the Shope virus, *C.R. Acad. Sci.* **270**:2609–2612.

Danos, O., and Yaniv, M., 1984, A homologous domain between the C-mos gene product and a papilloma virus polypeptide with putative role in cellular transformation, in: *Cancer Cells*, Vol. 2 (G. F. Vande Woulde, A. J. Levine, W. C. Topp, and J. D. Watson, eds.), pp. 291–294, Cold Spring Harbor Laboratory, Cold Spring Harbor, N.Y.

Danos, O., Georges, E., Orth, G., and Yaniv, M., 1985, Fine structure of the cottontail rabbit papillomavirus mRNAs expressed in the transplantable VX2 carcinoma, *J. Virol.* **53**:735–741.

Ehrlich, M., and Wang, R. Y.-H., 1981, 5-Methylcytosine in eukaryotic DNA, *Science* **212**:1350–1357.

Evans, C. A., and Ito, Y., 1966, Antitumor immunity in the Shope papilloma–carcinoma complex of rabbits. III. Response to reinfection with viral nucleic acid, *J. Natl. Cancer Inst.* **36**:1161–1166.

Evans, C. A., Weiser, R. S., and Ito, Y., 1962a, Antiviral and antitumor immunologic mechanisms operative in the Shope papilloma–carcinoma system, *Cold Spring Harbor Symp. Quant. Biol.* **27**:453–462.

Evans, C. A., Gorman, L. R., Ito, Y., and Weiser, R. S., 1962b, Antitumor immunity in the Shope papilloma–carcinoma complex of rabbits. II. Suppression of a transplanted carcinoma, VX7, by homologous papilloma vaccine, *J. Natl. Cancer Inst.* **29**:287–292.

Favre, M., Breitburd, F., Croissant, O., and Orth, G., 1975, Structural polypeptides of rabbit, bovine and human papilloma viruses, *J. Virol.* **15**:1239–1247.

Feitelson, M. A., Wettstein, F. O., and Stevens, J. G., 1985, Analysis and tryptic peptide characterization of the structural proteins of the Shope (rabbit) papillomavirus, in: *Papillomaviruses, Molecular and Clinical Aspects* (P. M. Howley and T. R. Broker, eds.), pp. 327–340, Alan R. Liss, New York.

Finch, J. T., and Klug, A., 1965, The structure of the viruses of the papilloma–polyoma type. III. Structure of the rabbit papilloma virus, *J. Mol. Biol.* **13**:1–12.

Friedewald, W. F., 1944, Certain conditions determining enhanced infection with the rabbit papilloma virus, *J. Exp. Med.* **80**:65–82.

Georges, E., Croissant, O., Bonneaud, W., and Orth, G., 1984, Physical state and transcripts of the cottontail rabbit papillomavirus genome in warts and transplantable VX2 and VX7 carcinoma of domestic rabbits, *J. Virol.* **51**:530–538.

Giri, I., Danos, O., and Yaniv, M., 1985, Structure of the genome of the cottontail rabbit (Shope) papillomavirus, *Proc. Natl. Acad. Sci. USA* **82:**1580–1584.

Hatfield, D., Diamond, A., and Dudock, B., 1982, Opal suppressor serine t-RNAs from bovine liver form phosphoseryl-t-RNA, *Proc. Natl. Acad. Sci. USA* **79:**6215–6219.

Hellstrom, I., Evans, C. A., and Hellstrom, K. E., 1969, Cellular immunity and its serum-mediated inhibition in Shope-virus-induced rabbit papillomas, *Int. J. Cancer* **4:**601–607.

Ito, Y., 1960, A tumor-producing factor extracted by phenol from papillomatous tissue (Shope) of cottontail rabbits, *Virology* **12:**596–601.

Ito, Y., 1962, Relation of components of papilloma virus to papilloma and carcinoma cells, *Cold Spring Harbor Symp. Quant. Biol.* **27:**387–394.

Ito, Y., and Evans, C. A., 1962, Induction of tumors in domestic rabbits with nucleic acid preparations from partially purified Shope papilloma virus and from extracts of the papillomas of domestic and cottontail rabbits, *J. Exp. Med.* **114:**485–500.

Kidd, J. G., 1938, The course of virus induced rabbit papillomas as determined by virus, cells and host, *J. Exp. Med.* **64:**63–96.

Kidd, J. G., 1939, The masking effect of extravasated antibody on the rabbit papilloma virus (Shope), *J. Exp. Med.* **70:**583–604.

Kidd, J. G., and Parsons, R. J., 1936, Tissue affinity of Shope papilloma virus, *Proc. Soc. Exp. Biol. Med.* **35:**438–440.

Kidd, J. G., and Rous, P., 1937, Effect of the papilloma virus (Shope) upon the tar warts of rabbits, *Proc. Soc. Exp. Biol. Med.* **37:**518–520.

Kidd, J. G., and Rous, P., 1940, A transplantable rabbit carcinoma originating in a virus-induced papilloma and containing the virus in masked or altered form, *J. Exp. Med.* **71:**813–838.

Kleinschmidt, A. K., Kass, S. J., Williams, R. C., and Knight, C. A., 1965, Cyclic DNA of Shope papilloma virus, *J. Mol. Biol.* **13:**749–756.

Kreider, J. W., 1963, Studies on the mechanism responsible for the spontaneous regression of the Shope rabbit papilloma, *Cancer Res.* **23:**1593–1599.

Kreider, J. W., 1980, Neoplastic progression of the Shope rabbit papilloma, *Cold Spring Harbor Conf. Cell Prolif.* **7:**283–299.

Kreider, J. W., and Bartlett, G. L., 1981, The Shope papilloma–carcinoma complex of rabbits: A model system of neoplastic progression and spontaneous regression, *Adv. Cancer Res.* **35:**81–110.

Kreider, J. W., Breedis, C., and Curran, J. S., 1967, Interaction of Shope papilloma virus and rabbit skin cells *in vitro*. I. Immunofluorescent localization of virus inocula, *J. Natl. Cancer Inst.* **38:**921–931.

Lusky, M., and Botchan, M. R., 1984, Characterization of the bovine papillomavirus plasmid maintenance sequences, *Cell* **36:**391–401.

McVay, P., Fretz, M., Wettstein, F. O., Stevens, J. G., and Ito, Y., 1982, Integrated Shope virus DNA is present and transcribed in the transplantable rabbit tumor VX7, *J. Gen. Virol.* **60:**271–278.

Mattern, C. F. T., 1962, Polyoma and papilloma viruses: Do they have 42 or 92 subunits? *Science* **137:**612–613.

Meinke, W., and Meinke, G. C., 1981, Isolation and characterization of the major capsid protein of bovine papillomavirus type 1, *J. Gen. Virol.* **52:**15–24.

Mellors, R. C., 1960, Tumor cell localization of the antigens of the Shope papilloma virus and the Rous sarcoma virus, *Cancer Res.* **20:**744–746.

Murphy, M. F., Potter, H. L., Abraham, J. M., Morgan, D. M., and Meinke, W. J., 1981, Analysis of a restriction endonuclease map for a rabbit papillomavirus DNA, *Curr. Microbiol.* **5:**349–352.

Nakabayashi, Y., Chattopadhyay, S. K., and Lowy, D. R., 1983, The transforming function of bovine papillomavirus DNA, *Proc. Natl. Acad. Sci. USA* **80:**5832–5836.

Nasseri, M., and Wettstein, F. O., 1984a, Differences exist between viral transcripts in cottontail rabbit papillomavirus-induced benign and malignant tumors as well as non-virus-producing and virus-producing tumors, *J. Virol.* **51:**706–712.

Nasseri, M., and Wettstein, F. O., 1984b, Cottontail rabbit papillomavirus-specific transcripts in transplantable tumors with integrated DNA, *Virology* **138:**362–367.

Nasseri, M., Wettstein, F. O., and Stevens, J. G., 1982, Two colinear and spliced viral transcripts are present in non-virus-producing benign and malignant neoplasms induced by the Shope (rabbit) papilloma virus, *J. Virol.* **44:**263–268.

Noyes, W. F., and Mellors, R. C., 1957, Fluorescent antibody detection of the antigens of the Shope papilloma virus in papillomas of the wild and domestic rabbit, *J. Exp. Med.* **106:**555–562.

Orth, G., and Croissant, O., 1967, Morphology and ultrastructure of tissue culture cells derived from rabbit epidermis, *C.R. Acad. Sci.* **265:**2149–2152.

Orth, G., and Croissant, O., 1968, Characteristics of primary tissue culture derived from Shope virus induced domestic rabbit papillomas, *C.R. Acad. Sci.* **266:**1084–1087.

Orth, G., Vielle, F., and Changeux, J.-P., 1967, On the arginase of the Shope papillomas, *Virology* **31:**729–732.

Orth, G., Jeanteur, P., and Croissant, O., 1971a, Evidence for and localization of vegetative viral DNA replication by autoradiographic detection of RNA–DNA hybrids in sections of tumors induced by Shope papilloma virus, *Proc. Natl. Acad. Sci. USA* **68:**1876–1880.

Orth, G., Jibard, N., and Vielle-Breitburd, F., 1971b, Purification and properties of L-arginase in papillomas induced by the rabbit papilloma virus, *C.R. Acad. Sci.* **273:** 1171–1174.

Orth, G. R., Breitburd, F., Favre, M., and Croissant, O., 1977, Papilloma viruses: Possible role in human cancer, in: *Origins of Human Cancer* (J. Watson, ed.), pp. 1043–1068, Cold Spring Harbor Laboratory, Cold Spring Harbor, N.Y.

Orth, G., Breitburd, F., and Favre, M., 1978, Evidence for antigenic determinants shared by the structural polypeptides of (Shope) rabbit papillomavirus and human papillomavirus type 1, *Virology* **91:**243–255.

Osato, T., and Ito, Y., 1968, Immunofluorescence studies of Shope papilloma virus in cottontail rabbit kidney tissue cultures, *Proc. Soc. Exp. Biol. Med.* **128:**1025–1029.

Razin, A., and Riggs, A. D., 1980, DNA methylation and gene function, *Science* **210:**604–610.

Rogers, S., 1959, Induction of arginase in rabbit epithelium by the Shope rabbit papilloma virus, *Nature* **183:**1815–1816.

Rogers, S., and Moore, M., 1963, Studies on the mechanism of action of the Shope rabbit papilloma virus. I. Concerning the nature of the induction of arginase in the infected cells, *J. Exp. Med.* **117:**521–542.

Rogers, S., and Rous, P., 1951, Joint action of chemical carcinogen and a neoplastic virus to induce cancer in rabbits, *J. Exp. Med.* **93:**459–488.

Rogers, S., Kidd, J., and Rous, P., 1950, An etiological study of the cancers arising from the virus induced papillomas of domestic rabbits, *Cancer Res.* **10:**237.

Rogers, S., Kidd, J. G., and Rous, P., 1960, Relationships of the Shope papilloma virus to the cancers it determines in domestic rabbits, *Acta Unio Int. Contra Cancrum* **16:**129–130.

Rous, P., and Beard, J. W., 1934, A virus-induced mammalian growth with the characters of a tumor (the Shope rabbit papilloma). I. The growth on implantation within a favorable host, *J. Exp. Med.* **60:**701–722.

Rous, P., and Beard, J. W., 1935a, A comparison of the tar tumors of rabbits and the virus-induced tumor, *Proc. Soc. Exp. Biol. Med.* **3:**358–360.

Rous, P., and Beard, J. W., 1935b, The progression to carcinoma of virus-induced rabbit papillomas (Shope), *J. Exp. Med.* **62:**523–548.

Rous, P., and Friedewald, W. F., 1941, Carcinogenesis on rabbit papillomas due to virus, *Science* **94:**495–496.

Rous, P., and Kidd, J. G., 1936, The carcinogenic effect of a virus upon tarred skin, *Science* **83:**468–469.

Rous, P., Kidd, J. G., and Smith, W. E., 1952, Experiments on the cause of the rabbit carcinomas derived from virus-induced papillomas, *J. Exp. Med.* **96:**159–174.

Sarver, N., Rabson, M. S., Yang, Y.-C., Byrne, J. C., and Howley, P. M., 1984, Localization and analysis of bovine papillomavirus type 1 transforming function, *J. Virol.* **52:**377–388.

Satoh, P. S., Yoshida, T. O., and Ito, Y., 1967, Studies on the arginase activity of Shope papilloma: Possible presence of isozymes, *Virology* **33:**354–356.

Schiller, J. T., Vass, W. C., and Lowy, D. R., 1984, Identification of a second transforming region in bovine papilloma-virus DNA, *Proc. Natl. Acad. Sci. USA* **81:**7880–7884.

Shope, R. E., 1935, Serial transmission of virus of infectious papillomatosis in domestic rabbits, *Proc. Soc. Exp. Biol. Med.* **32:**830–832.

Shope, R. E., and Hurst, E. W., 1933, Infectious papillomatosis of rabbits, *J. Exp. Med.* **58:**607–623.

Spira, G., Estes, M. K., Dreesman, G. R., Butel, J. S., and Rawls, W. E., 1974, Papovavirus structural polypeptides: Comparison of human and rabbit papilloma viruses with simian virus 40, *Intervirology* **3:**220–231.

Stevens, J. G., and Wettstein, F. O., 1979, Multiple copies of Shope virus DNA are present in cells of benign and malignant non-virus-producing neoplasms, *J. Virol.* **30:**891–898.

Sugawara, K., Fujinaga, K., Yamashita, T., and Ito, Y., 1983, Integration and methylation of Shope papilloma virus DNA in the transplantable VX2 and VX7 rabbit carcinomas, *Virology* **13:**88–99.

Syverton, J. T., 1952, The pathogenesis of the rabbit papilloma-to-carcinoma sequence, *Ann. N.Y. Acad. Sci.* **54:**1126–1140.

Syverton, J. T., Dascomb, H. E., Koomenjr, J., Wells, E. B., and Berry, G. P., 1950a, The virus induced rabbit papilloma-to-carcinoma sequence. I. The growth pattern in natural and experimental infections, *Cancer Res.* **10:**379–384.

Syverton, J. T., Dascomb, H. E., Wells, E. B., Koomenjr, J., and Berry, G. P., 1950b, The virus-induced rabbit papilloma-to-carcinoma sequence. II. Carcinomas in the natural host, the cottontail rabbit, *Cancer Res.* **10:**440–444.

Terheggen, H. G., Lowenthal, A., Lavinha, F., Colombo, J. B., and Rogers, S., 1975, Unsuccessful trial of gene replacement in arginase deficiency, *Z. Kinderheilkd.* **119:**1–3.

Vielle-Breitburd, F., and Orth, G., 1972, Rabbit liver L-arginase, *J. Biol. Chem.* **247:**1227–1235.

Watson, J. D., and Littlefield, J. W., 1960, Some properties of DNA from Shope papilloma virus, *J. Mol. Biol.* **2:**161–165.

Watts, S. L., Ostrow, R. S., Phelps, W. C., Prince, J. T., and Faras, A. J., 1983, Free cottontail rabbit papillomavirus DNA persists in warts and carcinomas of infected rabbits and in cells in culture transformed with virus or viral DNA, *Virology* **125:**127–138.

Wettstein, F. O., and Stevens, J. G., 1980, Distribution and state of viral nucleic acid in tumors induced by Shope papilloma virus, in: *Viruses in Naturally Occurring Cancers* (M. Essex and H. zur Hausen, eds.), pp. 301–307, Cold Spring Harbor Laboratory, Cold Spring Harbor, N.Y.

Wettstein, F. O., and Stevens, J. G., 1981, Transcription of the viral genome in papillomas and carcinomas induced by the Shope virus, *Virology* **109:**448–451.

Wettstein, F. O., and Stevens, J. G., 1982, Variable-sized free episomes of Shope papilloma virus DNA are present in all non-virus-producing neoplasms and integrated episomes are detected in some, *Proc. Natl. Acad. Sci. USA* **79:**790–794.

Wettstein, F. O., and Stevens, J. G., 1983, Shope papilloma virus DNA is extensively methylated in non-virus-producing neoplasms, *Virology* **126:**493–504.

Williams, R. C., Kass, S. J., and Knight, C. A., 1960, Structure of Shope papilloma virus particles, *Virology* **12:**48–58.

CHAPTER 7

PAPILLOMAVIRUSES AND CARCINOGENIC PROGRESSION II

The *Mastomys natalensis* Papillomavirus

EBERHARD AMTMANN AND KLAUS WAYSS

I. THE ANIMAL SYSTEM

Praomys (*Mastomys*) *natalensis* (A. Smith, 1834) is commonly called the multimammate mouse. The animal (Fig. 1), which is intermediate in size between mouse and rat, is one of the most common rodents in Africa south of the Sahara (Davis and Oettlé, 1958). In 1939, breeding of wild animals was started in South Africa (Davis and Oettlé, 1958) at the Medical Ecology Center, Johannesburg, in order to use them in plague research. In the following years, several times wild *Mastomys* were crossed with the laboratory colonies. In 1966, the Institute for Parasitology, Giessen, West Germany, started inbreeding chamois-colored mutants with red eyes. The ancestors of the partly inbred colony at the German Cancer Research Center, Heidelberg, were derived from the colony in Giessen in 1969.

The interest of cancer researchers was stimulated by Oettlé's detection that 30 to 50% of *Mastomys* that were older than 1 year suffered from stomach cancer (Oettlé, 1957). This tumor type is extremely rare in other laboratory rodents. As the incidence of stomach cancer was rather variable in different colonies (in the Heidelberg colony, no case of spontaneous stomach cancer has occurred to date), a genetic or exogenous factor such as viruses or carcinogens in the food of the animals was postulated (Simmers *et al.*, 1968). The factor(s) responsible for this high

EBERHARD AMTMANN • Institute for Virus Research, German Cancer Research Center, 6900 Heidelberg, Federal Republic of Germany. KLAUS WAYSS • Institute for Pathology, German Cancer Research Center, 6900 Heidelberg, Federal Republic of Germany.

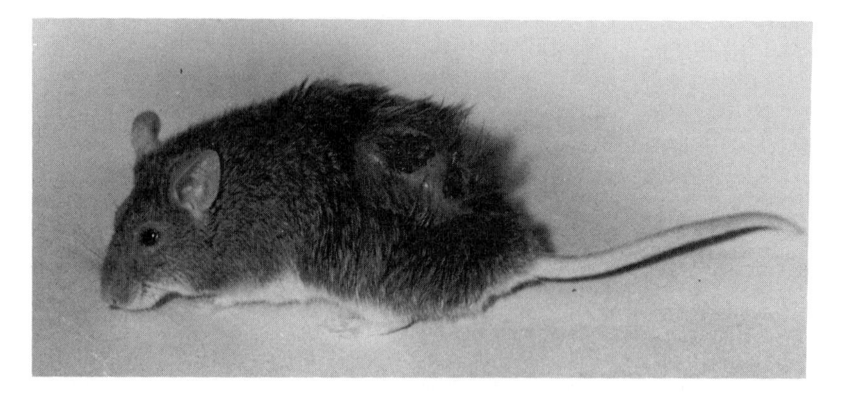

FIGURE 1. *Mastomys natalensis.* A 14-month-old female bearing a spontaneous keratoacanthoma.

rate of stomach cancer has not been identified, although it is likely that a papillomavirus is enveloped in the etiology of these tumors. We have detected extrachromosomal MnPV DNA in various tissues from such animals, whereas colonies that did not display spontaneous tumor formation at an unusually high rate, were free from MnPV (Amtmann *et al.*, 1987).

II. EPITHELIAL SKIN TUMORS IN *MASTOMYS*

Spontaneously occurring epithelial skin tumors in *Mastomys* were described first by Burtscher *et al.* in 1973. The authors classified these tumors as keratoacanthomas. As tumors of the same histology could be induced by cell-free filtrates of homogenated tumor material, a viral etiology of the keratoacanthomas in connection with a special (genetically determined?) susceptibility of the yellow mutant strain GRA Giessen was proposed.

From 20 tumor-bearing *Mastomys,* 54 skin tumors were analyzed in detail by Rudolph and Thiel (1976). Most of these tumors were localized on the head (45 cases). Only 1 was found on either the neck or the chest. Two were detected on the back, 3 on the legs, and 2 on the tail. These tumors were histologically classified as keratoacanthomas (12), papillomas (8), epithelial proliferations (5), and kertatinizing squamous carcinomas. The carcinomas infiltrated the surrounding tissues, namely the muscles, the bones, and the brain. Metastasis or tumor growth in inner organs was, however, never observed. A section through a typical keratoacanthoma is shown in Fig. 2.

Almost one-half of the tumors failed to give a uniform picture by histological examination. They were composed of keratoacanthoma, pap-

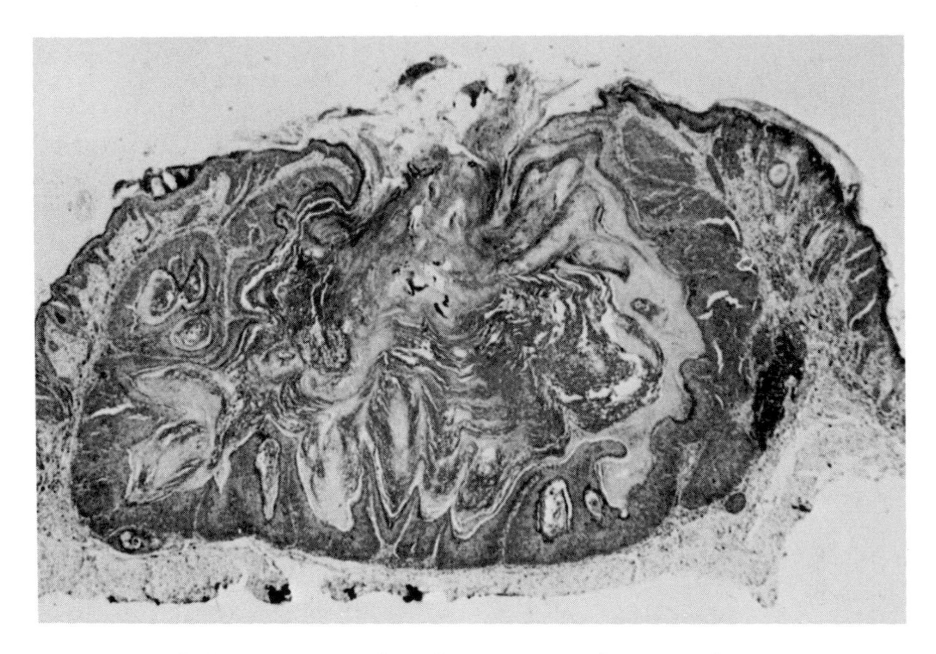

FIGURE 2. Section through a spontaneous keratoacanthoma.

illoma, and epithelial proliferation. In two cases, parts of the tumors were identified as carcinomas.

The rate of spontaneous transition from benign tumors to carcinomas seems to be correlated with the genetic disposition of the host, similarly as is the case in the CRPV system. While 11% of the tumors in the Giessen colony were of a malignant type, in the Heidelberg colony (which was split from the Giessen strain, before it was inbred) in contrast, no spontaneous malignant conversion has ever been observed.

III. IDENTIFICATION OF MnPV AS THE CAUSATIVE AGENT OF SKIN TUMORS

Reinacher *et al.* (1978) searched for virus particles in 26 spontaneous tumors (13 keratoacanthomas, 3 squamous cell carcinomas, and 10 papillomas). In all cases, regardless of the tumor type, they found virus particles resembling papovaviruses in identical sites. In the nuclei of keratinized cells of the stratum granulosum and stratum corneum, large numbers of virus particles were present. The virus was called *Mastomys natalensis* papillomavirus (MnPV). From benign as well as from malignant tumors, virus particles could be isolated by CsCl gradient centrifugation. When the purified virions were used for the infection of the scarified skin of young *Mastomys*, 36% of the infected animals (a total of 38) developed tumors (Müller and Gissmann, 1978). The virally induced tumors were histologically indistinguishable from spontaneous tumors. Again, virus particles with the same morphology as the inoculum could be isolatd (Müller and Gissmann, 1978). When the authors treated the virions prior to infection with a hyperimmune serum which was raised in rabbits against purified MnPV, no tumors developed in 30 animals during an observation period of 1 year.

IV. CHARACTERIZATION OF THE VIRUS

The morphology of MnPV is quite similar to that of other papillomaviruses. The virion has an average diameter of 52 nm and is composed of capsomers that appear to be hollow cylinders with a diameter of 7 nm. By polyacrylamide gel eletrophoresis, 12 different virion proteins were identified. The major capsid protein which represents about 75% of the total protein was estimated to have a molecular weight of 56,000 (Müller and Gissmann, 1978). When centrifuged to equilibrium in CsCl gradients, the density of MnPV was found to be 1.34 g/ml. The virus particles contain circular superhelical DNA of 4.9×10^6 molecular weight. The G + C content of the viral DNA is 50%, which is unusually high for the papillomaviruses (Müller and Gissmann, 1978). A restriction endonuclease cleavage map of the viral DNA is shown in Fig. 3. When electrophoret-

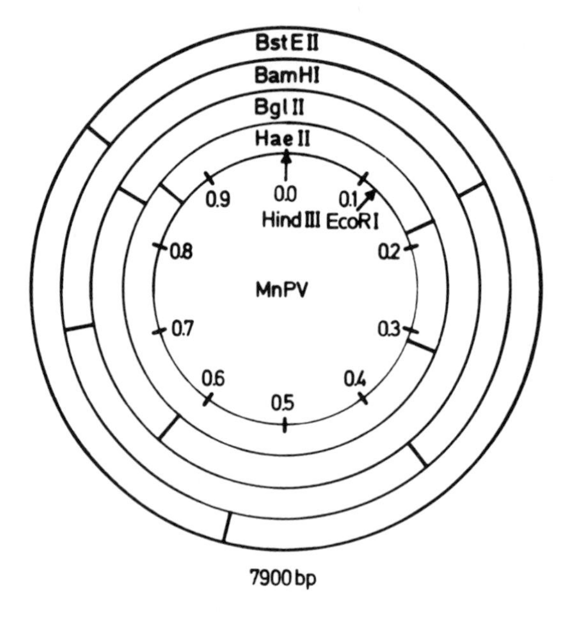

FIGURE 3. Restriction endonuclease cleavage map of MnPV DNA.

ically separated MnPV DNA restriction fragments were hybridized to 32p-labeled BPV-1 DNA under relaxed conditions (20% formamide, 42°C), two regions displaying homology to BPV-1 could be revealed. A rather strong signal was obtained with fragments spanning 0.72 and 0.83 unit on the MnPV map while a region with weak homology was located between 0.19 and 0.4 map unit (Amtmann, unpublished findings).

V. PERSISTENCE OF MnPV GENOMES IN NORMAL TISSUES

In the *Mastomys* colony at the German Cancer Research Center, Heidelberg, the tumor incidence was independent of contacts between tumor-free and tumor-bearing animals. After weaning from the mother at the age of 4 weeks, all animals were kept in separate cages. As we suspected that infection had occurred early in the life of *Mastomys* before weaning, long latency periods (most tumors appear in animals older than 1 year) of MnPV seemed to be the rule.

In order to search for potential latent MnPV genomes, we extracted the total DNA from the skin of 6-month-old animals. The DNA was electrophoresed on agarose gels, transferred to nitrocellulose filters, and hybridized to ^{32}P-labeled cloned MnPV DNA. In all skin-DNA samples from five males and five females, extrachromosomal MnPV DNA could be found. In addition, in four cases viral DNA was present also in tissues other than skin. The data for one of these cases, a pregnant female, are shown in Fig. 4. Superhelical, relaxed circular, and linear MnPV DNA molecules are visible in skin (lane a), muscle (lane b), liver (lane c), colon (lane d) and an embryo. Frequently, prestomach, esophagus, larynx, and bronchial tissue were also positive for MnPV DNA. The lymphocytes from five animals were negative (Amtmann *et al.*, 1984).

There was no evidence for integration of viral DNA into the host genome. Digestion with single-cut or multi-cut restriction endonucleases produced exclusively the authentic cleavage patterns of MnPV virion DNA (see Fig. 4). If multiple integration sites of viral DNA existed, the Cellular DNA should give a hybridization signal in the uncleaved samples. From experiments with SV40-transformed mouse 3T3 cells, we conclude that an integrated DNA fragment corresponding to 10% of the MnPV genome should be detectable in an uncleaved DNA sample (Amtmann, unpublished findings). The activities of the restriction endonucleases *Hpa*II and *Hha*I are sensitive to DNA methylation. As the viral genomes which were present in normal skin and in the embryo were cleaved by these enzymes to DNA fragments of the same size as the DNA which was isolated from virions, it is unlikely that DNA hypermethylation is involved in the regulation of viral functions in normal tissues.

By *in situ* hybridizations with nick-translated MnPV DNA, no cells could be revealed that contain exceptionally high numbers of MnPV genomes. In contrast, the grains in the autoradiogram were homogeneously

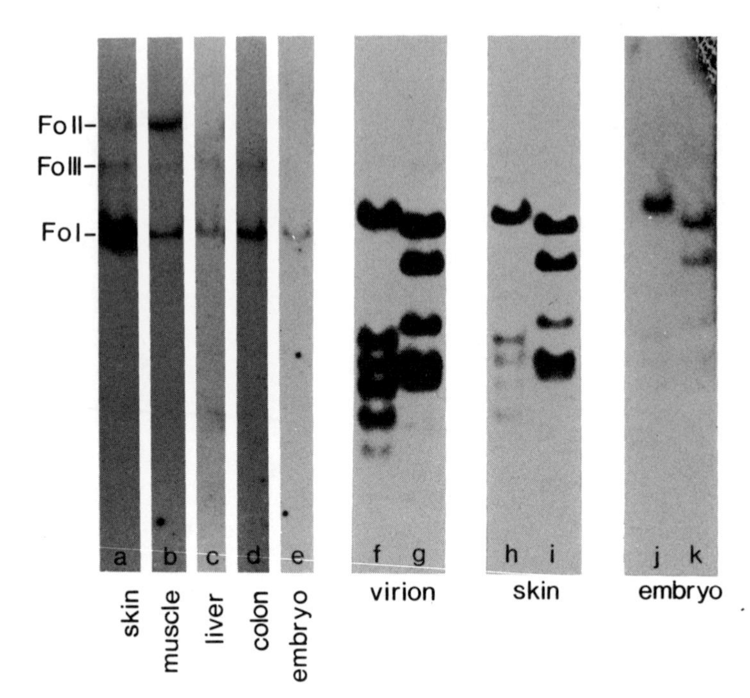

FIGURE 4. Detection of MnPV-specific DNA sequences in normal tissues of *Mastomys natalensis*. The DNA preparations in lanes a–e were uncleaved; those in lanes f, h, and j were cleaved with *Hpa*II, and those in lanes g, i, and k were cleaved with *Hha*I. The DNA samples were isolated from: lane a, skin; lane b, muscle; lane c, liver; lane d, colon; lane e, embryo; lanes f and g, mouse 3T3 cells mixed with 200 pg MnPV virion DNA. (From Amtmann *et al.*, 1984, with permission.)

distributed over the whole tissue section. In the keratinized layers of tumors, on the other hand, highly MnPV DNA-positive cells that also reacted with antisera against MnPV were present (Reinacher *et al.*, 1978; Amtmann, unpublished findings). These data indicate that the MnPV genomes detected by Southern blotting in normal skin persist, indeed, in the majority of the cells rather than in a few virus-producing cells that behave like normal, untransformed cells in the tissue.

VI. ACCUMULATION OF VIRAL DNA IN NORMAL SKIN DURING AGING

In our *Mastomys* colony, spontaneous skin tumors appeared at a rate that was strictly dependent on the age of the animals. As a rule, animals younger than 50 weeks were devoid of tumors. However, with increasing age the number of tumor bearers increased rapidly. For example, of 10 animals aged 16 months, 8 bore tumors. The DNA that was isolated from

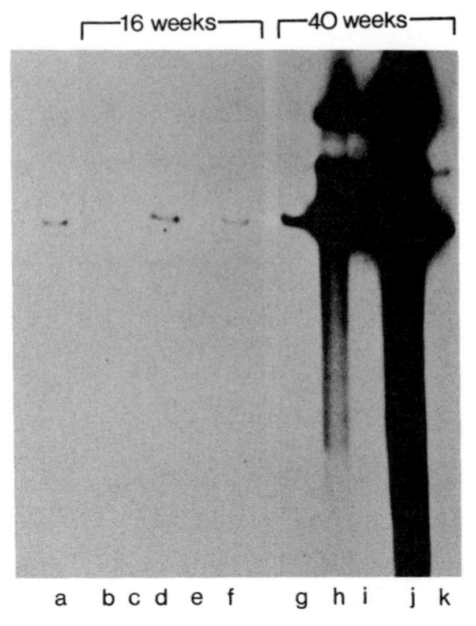

FIGURE 5. Comparison of the viral DNA content in the skin of 16- and 40-week-old *Mastomys*. Southern blot hybridization with ^{32}P-labeled MnPV DNA. a, 10 μg of 3T3 DNA mixed with 25 pg of MnPV DNA; b–f, 10 μg of DNA isolated from 16-week-old animals; g–k, 10 μg of DNA isolated from the skin of 40-week-old animals; all DNA samples were uncleaved.

16- and 40-week-old animals was analyzed in Southern blot experiments. It was obvious that the number of viral genomes per skin cell had drastically increased within the period of 24 weeks (Fig. 5). In order to follow the kinetics of MnPV DNA accumulation, from 50 newborn animals, groups of 10 (5 males and 5 females) were sacrificed at intervals of 16 weeks. The DNA that was isolated from the skin of these animals was analyzed in Southern blots and the number of viral genomes per cell was determined by densitometer scanning of the x-ray films. In only 9 of 14 embryos and 6 of 10 skin samples from 16-week-old animals could MnPV DNA sequences be revealed (at a level of detection ≥ 0.1 genome equivalent per cell). During aging the viral DNA content increased logarithmicaly from 0.2 genome equivalent per cell at 16 weeks to 3000 at 80 weeks (see Table I) (Amtmann *et al.*, 1984).

TABLE I. Accumulation of Viral DNA in Skin Cells during Aging

Age	Mean number of genome equivalents per cell	Range	Tested animals/ animals bearing tumors
Embryo	0.1	<0.1–2	14/0
16	0.2	<0.1–10	10/0
32	10	0.5–500	10/0
48	100	1–300	10/0
64	800	20–2000	10/5
80	3000	500–5000	5/2

VII. MnPV GENOME EXPRESSION IN TUMORS

The RNA isolated from skin samples that were proven to be normal by histological examination was compared in Northern blots with the RNA isolated from tumors and preneoplastic lesions. While MnPV-specific RNA sequences were revealed in all of ten tumor specimens (Fig. 6, lane f) and in two preneoplastic lesions (lane d) although each skin cell contained more than 1000 molecules of MnPV DNA (lane a). These data cannot exclude a temporary or a low-level expression of viral genes in the normal skin, which might be responsible for the accumulation of viral DNA during aging. In several skin samples, which contained only 5- to 10-fold fewer MnPV genomes per cell than did the tumors, no viral RNA sequences were detectable. As the level of mRNA detection was at least 100-fold lower than the amount of transcripts that were present in the

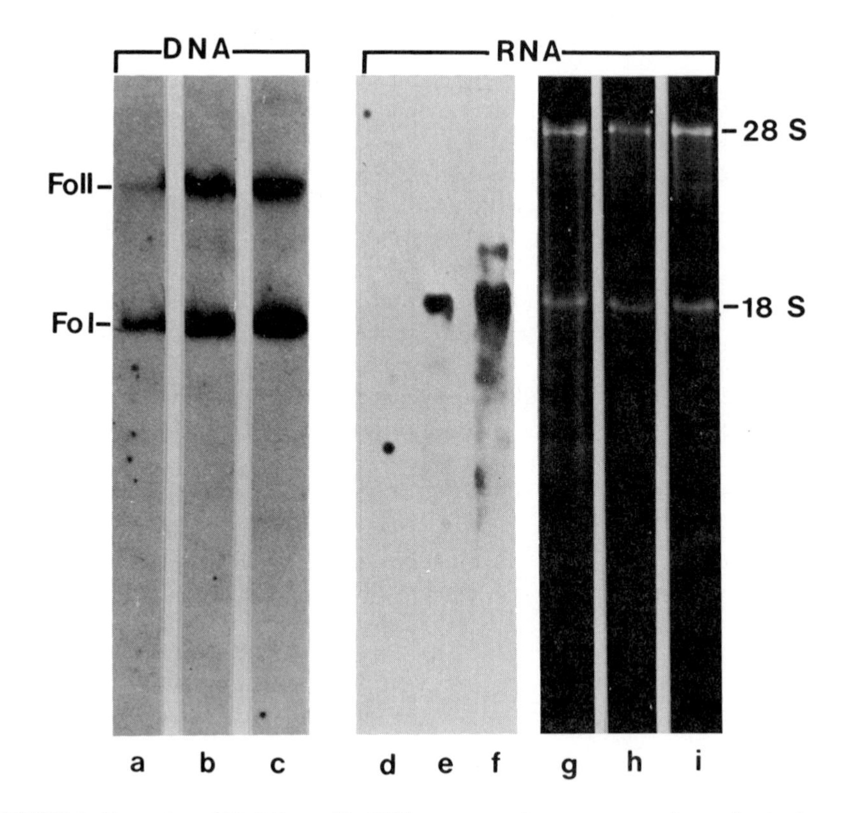

FIGURE 6. Detection of MnPV-specific RNA sequences in tumorous and acanthotic tissues. Lane a, DNA, and lanes d and g, RNA isolated from normal skin of a tumor-free animal; lane b, DNA, and lanes e and h, RNA isolated from skin of a tumor-free animal (by histological examination, mutiple areas of acanthosis and hyperplasia were detected in an adjacent sample); lane c, DNA, and lanes f and i, RNA isolated from a keratoacanthoma. (From Amtmann *et al.*, 1984, with permission.)

tumors, the viral genomes in normal skin were transcribed (if at all) at a much lower frequency, as most of the viral DNA that can be isolated from tumors is found in virus particles which precludes its transcriptional activity.

VIII. INDUCTION OF MnPV BY TUMOR PROMOTERS

Unexpressed papillomavirus genomes could be activated by the tumor promoter 12-*O*-tetradecanoylphorbol-13-acetate (TPA) *in vitro* (Amtmann and Sauer, 1982). This fact prompted us to study the effect of TPA on MnPV genomes *in vivo.* Young *Mastomys* (6 weeks old) were treated topically with TPA twice a week. The first tumor appeared after 14 weeks of TPA treatment at the site of promoter application. After 44 weeks the first animal of the acetone-treated group ($N = 28$) developed a tumor. By then 19 (66%) of the TPA animals ($N = 29$) were already bearing tumors (Fig. 7). All tumors were identified as keratoacanthomas histologically identical with spontaneous tumors (Amtmann *et al.*, 1984).

The viral DNA content of the skin of animals that had been treated

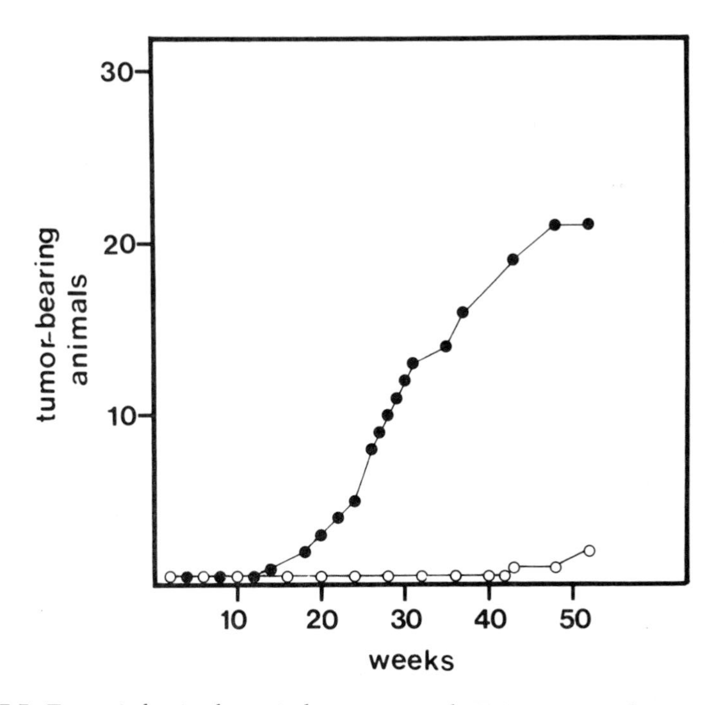

FIGURE 7. Tumor induction by topical treatment with TPA. TPA was administered to the dorsal skin of 6-week-old *Mastomys* twice weekly (29 animals; ●). A control group (28 animals; ○) was treated with 0.2 ml acetone. The animals were examined for tumors weekly. All tumors arising were identified histologically as keratoacanthomas. (From Amtmann *et al.*, 1984, with permission.)

for 8 weeks with TPA was stimulated by 100-fold in comparison to that of acetone-treated animals. The skin cells of six control animals (17 weeks old) revealed, as expected from the aging experiment, about 0.2 MnPV genome per cell, while the viral DNA had replicated in the TPA-treated skin to a mean value of 50 copies per cell. In all cases the viral DNA content was 30- to 250-fold higher in the treated back skin than in the untreated skin from the chest of the same animal (Amtmann *et al.*, 1984). After 16 weeks of treatment, the viral DNA content in TPA-induced tumors was even 5- to 10-fold higher than in the TPA-treated skin of the same animal (Fig. 8).

In the skin of mice, chemical carcinogenesis can be divided into three steps (Fürstenberger *et al.*, 1981). The first step, termed initiation, is mediated by a single dose of a carcinogen. The following steps are the first-stage promotion, for which a single dose of a full tumor promoter, like TPA, is sufficient, and, as final step, the chronic second-stage promotion. A potent second-stage promoter is 12-*O*-retinoylphorbol-13-ace-

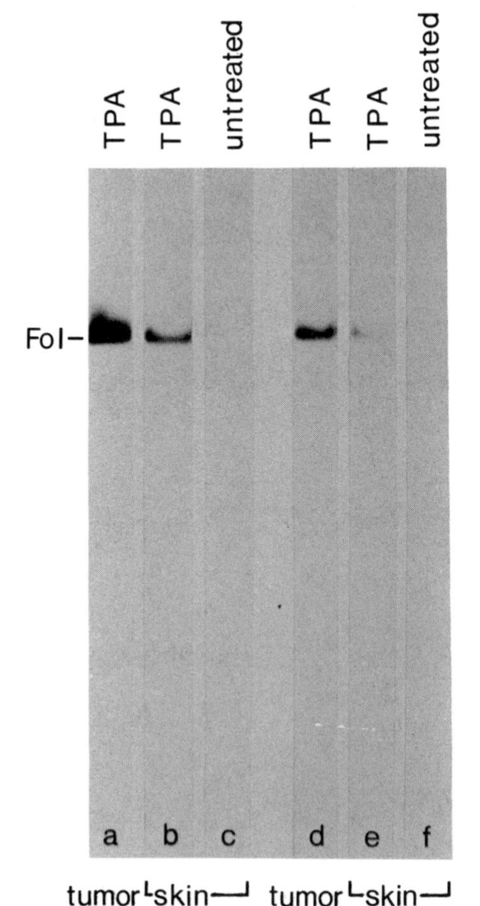

FIGURE 8. Comparison of TPA-treated and untreated skin. After 16 weeks of treatment with TPA, the DNA was isolated from tumors, treated normal skin, and untreated skin of the same animal. After gel electrophoresis the DNA was hybridized to ^{32}p-labeled MnPV DNA. All samples were uncleaved.

tate (RPA). Similarly as *in vitro* (Amtmann and Sauer, 1982), RPA was an *in vivo* inductor of MnPV as potent as TPA (data not shown). When *Mastomys* were treated with RPA, tumors appeared at the same rate and the viral DNA content was stimulated to the same extent as when the animals were treated with TPA. Neither TPA, RPA, nor the carcinogen 9,10-dimethyl-1,2-benz(*a*)anthracene (DMBA) induced the conversion of benign keratoacanthomas to carcinomas after chronic application of these compounds. DMBA had no effect on the viral DNA content in *Mastomys* skin even when applied chronically. There was also no synergism between DMBA and spontaneous tumor formation: Chronic treatment with DMBA, initiated on 6-week-old animals, led finally to a variety of skin tumor types, in which neither MnPV-specific DNA- nor RNA-sequences could be revealed, while in 30-week-old animals exclusively virus-producing benign keratoacanthomas arose. The rate of appearance and the number of keratoacanthomas did not differ significantly from control groups (Amtmann *et al.*, 1987). We conclude from these results that the activation of MnPV *in vivo* is mediated by a cellular mechanism that is correlated to second-stage tumor promotion. Transition from benign keratoacanthomas to malignant tumors is not induced by tumor promoters or DMBA.

IX. INDUCTION OF ANTIBODIES AGAINST MnPV IN *MASTOMYS*

In the case of MnPV, virions were observed in keratoacanthomas as well as in epidermal hyperplasia (Reinacher *et al.*, 1978). As virus is already produced in the earliest clinical manifestation of MnPV infection, it was important to know whether the onset of virus synthesis coincided with lesion formation. As after TPA treatment tumors arose much earlier in the life of *Mastomys,* we also studied whether the application of TPA could lead to an earlier onset of virus production. The detection of antibodies which are directed against MnPV virions can be regarded as an indicator of virus production in the animal from which this serum was isolated. Animals were statistically distributed to groups of ten. At the age of 10 weeks (at this time, chronic TPA paint treatment was initiated in five groups), the first two groups of animals were sacrificed and the titer of antibodies directed against MnPV was determined. In no case, and the same was true at the age of 12 weeks, were antibodies detected (see Table II). At the age of 14 weeks, three animals of the TPA group were positive. At the age of 18 weeks when the first animals of the control group became positive, all TPA-treated *Mastomys* carried antibodies against MnPV. We conclude from these findings that virus production (perhaps with a very low efficiency) begins at least half a year before clinically detectable lesions (such as hyperplasia) are present. Antibodies directed against MnPV can be detected several weeks earlier in TPA-treated *Mas-*

TABLE II. Titers of Antibodies[a] against MnPV in Tumor-Free Animals of Different Age

Age	Mean serum titer[b] (control/TPA treated[c])	No. of positive animals (control/TPA treated[c])
10	0/0	0/0
12	0/0	0/0
14	0/0.4	0/3
18	0.5/6.2	2/10
26	1.8/18.5	5/10
30	2.7/22	9/10
42	3.4/n.d.	10/n.d.
50	3.8/n.d.	10/n.d.

[a] Calculated from the highest dilution which was positive in indirect immunofluorescence using FITC-labeled goat antibodies to mouse IgG. Serum was collected from sacrificed animals and serial 1:2 dilutions (in 0.9% NaCl) were tested for staining of nuclei in the peripheral cells of keratoacanthomas.
[b] Ten animals in each group.
[c] Six-week-old animals were treated twice a week with 20 nmole TPA, for 16 weeks.

tomys and the final antibody titers are more than sixfold higher. It can be concluded that TPA induces the expression of some or all MnPV-specific genes which lead to the replication of MnPV DNA and the synthesis of MnPV virions.

REFERENCES

Amtmann, E., and Sauer, G., 1982, Activation of non-expressed bovine papilloma virus genomes by tumour promoters, *Nature* **296:**675–677.

Amtmann, E., Volm, M., and Wayss, K., 1984, Tumour induction in the rodent *Mastomys natalensis* by activation of endogenous papilloma virus genomes, *Nature* **308:**291–292.

Amtmann, E., Randeria, J. D. and Wayss, K., 1987, The interaction of papilloma viruses with carcinogens and tumor promoters, in: *Cancer Cells,* Vol. 5, Cold Spring Harbor Laboratory, Cold Spring Harbor, N.Y.

Burtscher, H., Grünberg, W., and Meingassner, G., 1973, Infektiöse Keratoakanthome der Epidermis bei Praomys (*Mastomys*) *natalensis, Naturwissenschaften* **60:**209–210.

Davis, D. H. S., and Oettlé, A. G., 1958, The multimammate mouse *Rattus* (*Mastomys*) *natalensis* Smith: A laboratory adapted African wild rodent, *Proc. Zool. Soc. London* **131:**293–299.

Fürstenbrger, G., Berry, D. L., Sorg, B., and Marks, F., 1981, Skin tumor promotion by phorbol esters is a two-stage process, *Proc. Natl. Acad. Sci. USA* **78:**7722–7726.

Müller, H., and Gissmann, L., 1978, *Mastomys natalensis* papilloma virus (MnPV), the causative agent of epithelial proliferations: Characterization of the virus particle, *J. Gen. Virol.* **41:**315–323.

Oettlé, A. G., 1957, Spontaneous carcinoma of the glandular stomach in *Rattus* (*Mastomys*) *natalensis, Br. J. Cancer* **11:**415–433.

Reinacher, M., Müller, H., Thiel, W., and Rudolph, R. L., 1978, Localization of papillomavirus and virus-specific antigens in the skin of tumor-bearing *Mastomys natalensis* (GRA Giessen), *Med. Microbiol. Immunol.* **165:**93–99.

Rudolph, R. L., and Thiel, W., 1976, Pathologische Anatomie und Histologie von spontanen, epithelialen Hauttumoren bei *Mastomys natalensis, Zentralbl. Veterinaermed. Reihe B* **23:**143–150.

Simmers, M. H., Ibsen, K. H., and Berk, J. E., 1968, Concerning the incidence of "spontaneous" stomach cancer in Praomys (*Mastomys*) *natalensis, Cancer Res.* **28:**1573–1576.

CHAPTER 8

Epidermodysplasia Verruciformis

GÉRARD ORTH

I. INTRODUCTION

Epidermodysplasia verruciformis (EV) is a rare lifelong skin disease, which begins during infancy or childhood. It is induced by numerous specific types of human papillomaviruses (HPVs), sometimes including the HPVs associated with flat warts in the general population. EV is characterized by refractory, disseminated skin lesions resembling flat warts, or presenting as macules of various colors. Cutaneous carcinomas *in situ* or invasive carcinomas usually of Bowen's type, appear in a high proportion of the patients, generally at an early age. HPV DNA sequences, usually HPV type 5, are regularly detected in EV carcinomas. EV is a multifactorial disease which involves genetic, immunological, and extrinsic factors, in addition to specific HPVs. This has been suggested by the parental consanguinity and the sibling involvement observed in some cases, the impaired cell-mediated immunity reported in most patients, and the usual localization of skin cancers in light-exposed areas. Reviews on EV have been published by Maschkilleisson (1931), Sullivan and Ellis (1939), Midana (1949), Jablonska and Milewski (1957), Bilancia (1961), Jablonska *et al.* (1966, 1972), Oehlschlaegel *et al.* (1966), Relias *et al.* (1967), Tsoitis *et al.* (1973), Rueda and Rodriguez (1976), Kaufmann *et al.* (1978), Lutzner (1978), Orth *et al.* (1980), Jablonska and Orth (1985), and Orth (1986).

After EV was first described as a congenital epidermal anomaly by

GÉRARD ORTH • Unité des Papillomavirus, Unité INSERM 190, Institut Pasteur, 75724 Paris Cedex 15, France.

Lewandowsky and Lutz in 1922, the nosological entity of EV became a matter of controversy for the next four decades. Some authors considered EV as a congenital defect of epidermal differentiation (genodermatosis), resulting in a vacuolar degeneration of the epidermis and involving a predisposition to the development of skin cancers (Maschkilleisson, 1931; Waisman and Montgomery, 1942; Midana, 1949; Lazzaro *et al.*, 1966; Oehlschlaegel *et al.*, 1966; Relias *et al.*, 1967). Others considered EV as a particular form of generalized verrucosis (Hoffmann, 1926; Kogoj, 1926; Sullivan and Ellis, 1939; Jablonska and Milewski, 1957).

The role of papillomavirus in the genesis of benign EV lesions was ascertained by successful auto- and heteroinoculation experiments (Lutz, 1946; Jablonska and Milewski, 1957; Jablonska and Formas, 1959; Jablonska *et al.*, 1966, 1972), and by the regular observation of viral particles with a morphology similar to those observed in warts (Ruiter and Van Mullem, 1966; Aaronson and Lutzner, 1967; Baker, 1968; Califano and Caputo, 1968; Jablonska *et al.*, 1968, 1972; Schellander and Fritsch, 1970; Grupper *et al.*, 1971; Rajagopalan *et al.*, 1972; Rueda and Rodriguez, 1972, 1976; Duverne *et al.*, 1973; Yabe and Sadakane, 1975; Hammar *et al.*, 1976; Prystowsky *et al.*, 1976; Prawer *et al.*, 1977; Kaufmann *et al.*, 1978; Laurent *et al.*, 1978). EV was considered as a model for the study of viral oncogenesis in humans (Jablonska *et al.*, 1966, 1972; Ruiter, 1969; Grupper *et al.*, 1971). The role of the wart virus in the production of EV cancers remained uncertain, however, for two main reasons: the malignant transformation of skin warts in the general population was exceptional and viral particles were only irregularly detected in early malignant lesions and never observed in invasive EV cancers (Aaronson and Lutzner, 1967; Jablonska *et al.*, 1970; Ruiter and Van Mullem, 1970a; Grupper *et al.*, 1971; Delescluse *et al.*, 1972; Yabe and Sadakane, 1975; Rueda and Rodriguez, 1976; Yabe *et al.*, 1978; Orth *et al.*, 1980; Lutzner *et al.*, 1984). Actinic radiations were supposed to be a major factor, if not the only one, in the development of EV carcinomas (Schellander and Fritsch, 1970; Ruiter, 1973).

The demonstration of the plurality of HPVs associated with skin warts (Gissmann *et al.*, 1977; Orth *et al.*, 1977a,b, 1978a,b,c) and that of the existence of distinct specific HPV types associated with EV (Orth *et al.*, 1978c) led to reconsideration of the role of the virus in the pathogenesis of the disease. It soon became apparent that the morphology of EV lesions and the risk of their malignant conversion depended on the HPV type present in the lesions (Orth *et al.*, 1978c, 1979; Jablonska *et al.*, 1979b; Kienzler *et al.*, 1979), and that at least one of the HPV types specifically associated with EV, HPV-5, actually had an oncogenic potential (Orth *et al.*, 1980). Since then, numerous EV-specific HPVs (at least 15 types) have been characterized (Kremsdorf, 1980; Pfister *et al.*, 1981; Kremsdorf *et al.*, 1982a, 1983, 1984; Green *et al.*, 1982; Ostrow *et al.*, 1982; Yutsudo *et al.*, 1982; Gassenmaier *et al.*, 1984; Kawashima *et al.*,

1986b) and the oncogenic potential of HPV-5 has been increasingly supported by its association with the majority of EV carcinomas (Orth *et al.*, 1980; Ostrow *et al.*, 1982; Pfister *et al.*, 1983; Orth, 1986; Van Voorst Vader *et al.*, 1986).

In this chapter we shall review the clinical aspects of EV, the biochemical and biological properties of the multiple HPV types specifically associated with the disease, the factors controlling the expression of the pathogenicity and the oncogenicity of these viruses and, finally, the viral mechanisms involved in the malignant transformation of EV lesions.

II. CLINICAL ASPECTS OF EV

A. Clinical Course of the Disease

EV usually begins during infancy or early childhood (Sullivan and Ellis, 1939; Midana, 1949; Bilancia, 1961; Lazzaro *et al.*, 1966; Jablonska *et al.*, 1972, 1979a; Lutzner, 1978; Jablonska and Orth, 1985). The first lesions generally arise on the dorsa of the hands and/or the forehead, then spread progressively to the limbs, neck, and trunk. The lesions may remain unchanged for many years, or undergo modifications of their aspect and color. Some may disappear (Marill and Grould, 1967; Prystowsky *et al.*, 1976), but a total regression of the lesions has never been observed in the form of EV associated with EV-specific HPVs. Such a regression has been reported in two patients suffering from an EV variety associated with HPV-3 (Jablonska and Orth, 1985). EV lesions are refractory to conventional wart treatments, and only irregular or transient results have been obtained with α-interferon (Blanchet-Bardon *et al.*, 1981; Androphy *et al.*, 1984; Blanchet-Bardon and Lutzner, 1985), retinoids (Guilhou *et al.*, 1980; Claudy *et al.*, 1982; Blanchet-Bardon and Lutzner, 1985), or transfer factor (Vasily *et al.*, 1984).

Multiple cutaneous cancers, most frequently localized in sun-exposed areas, develop in a certain proportion of the patients. The cancers usually arise during the second, third, or fourth decade of life. In a survey of 147 cases (Lutzner, 1978), about one-third of the patients suffering from EV developed cancers, and the average interval between the onset of the benign lesions and the cancer development was 24 years. The cancer incidence, as deduced from literature data, has probably been underestimated. Some of the EV cases reported concern young patients, without cancer but with rare follow-up examinations, and some cases of more or less generalized forms of disseminated flat warts and/or common warts may have been misdiagnosed as EV, thus decreasing the actual cancer incidence (Rueda and Rodriguez, 1976; Kaufmann *et al.*, 1978). Higher incidences have been reported in some series of patients (Rueda and Rodriguez, 1976; Lutzner *et al.*, 1984). Nevertheless, it appears that cancers are rare in black-skin EV patients (two cases so far reported) (Bloc *et al.*, 1971; Jacyk and Subbuswamy, 1979; Lutzner *et al.*, 1984).

B. Morphology of EV Lesions

The typical form of EV, associated with EV-specific HPVs, is characterized by the coexistence of lesions resembling flat warts and presenting as macules (Orth *et al.*, 1978c, 1979; Jablonska *et al.*, 1979a,b; Jablonska and Orth, 1985). The lesions are usually widely disseminated, if not almost generalized. Flat wartlike lesions are small in size (diameter 2–6 mm), sometimes confluent, with the color of the skin, or pink, red, brown, white. Macular lesions, more or less scaly, are red, reddish-brown, brown, or white (Fig. 1), and may coalesce into large patches with polycyclic borders. The brownish or achromic macules resemble lesions of pityriasis versicolor, a fungal infection of the skin. The flat wartlike lesions are usually observed on the dorsa of the hands, the forearms, face, and feet, while the reddish macules and the pityriasis versicolorlike lesions are most often distributed on the forehead, face, neck, arms, shoulders, and trunk (Jablonska and Orth, 1985). The palms and scalp are but

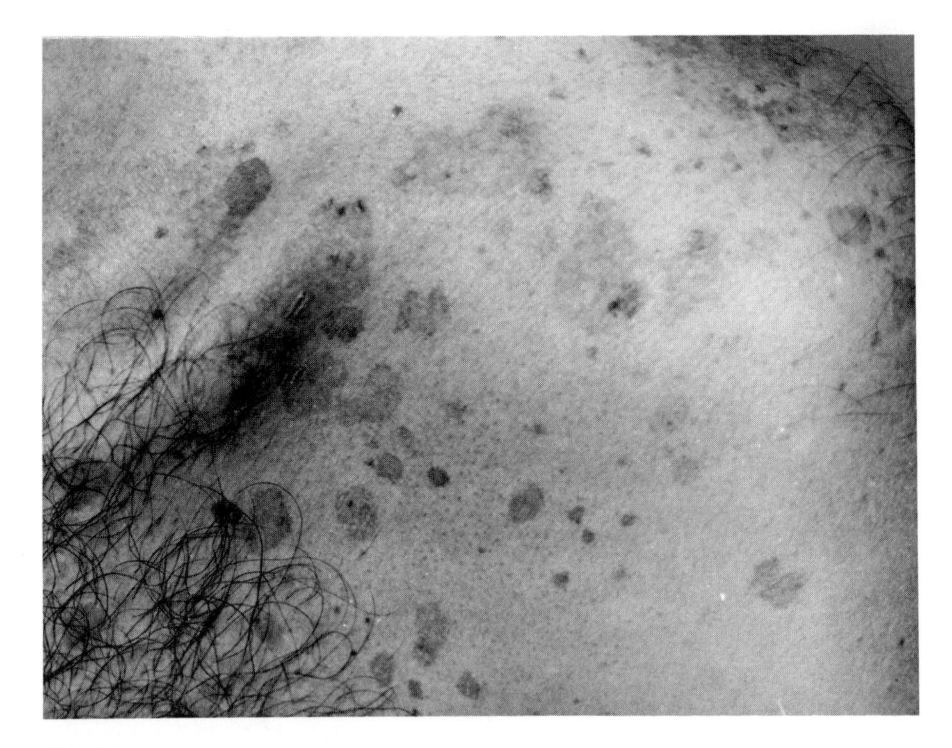

FIGURE 1. Cutaneous macular lesions of epidermodysplasia verruciformis. Courtesy of Professor S. Jablonska.

rarely involved, and the mucous membranes are free of lesions as a rule. Some patients may show an even greater polymorphism of skin lesions, some resembling lichen planus papules (Midana, 1949), or appearing as psoriasiform confluent plaques, especially on the knees (Duverne *et al.*, 1973; Kienzler *et al.*, 1979). Seborrheic keratoses have been observed (Schellander and Fritsch, 1970; Rueda and Rodriguez, 1976), especially in black-skin patients (Jacyk and Subbuswamy, 1979), but their relationships with HPVs remain unclear. Typical common warts are only occasionally found in EV patients, and condylomata acuminata have only exceptionally been observed (Midana, 1949; Schellander and Fritsch, 1970; Lutzner, 1978; Lutzner *et al.*, 1984).

Lesions converting to malignancy appear as light or dark brown, keratotic, sometimes papillomatous patches, or as erythematous, crusty plaques, resembling actinic keratoses or lesions of Bowen's disease. Such lesions may be converted into fungating or invasive, ulcerated tumors, which are squamous cell carcinomas. The premalignant and malignant lesions are usually multiple, and are most often localized on the face. More than half of the cancers develop on the forehead. Some may also be found on the dorsa of the hands and in the presternal region (Rueda and Rodriguez, 1976; Lutzner, 1978). Cancers are seldom observed in unexposed areas. An example is a black African patient who showed multiple carcinomas *in situ* on the scrotum, and invasive cancers on the buttocks and on one of the palms (Lutzner *et al.*, 1984). Metastases of EV carcinomas have only been occasionally described (Depaoli, 1971; Ostrow *et al.*, 1982; Lutzner *et al.*, 1984) but follow-up studies of many patients are not available. EV carcinomas may extend progressively to the whole forehead and scalp (Glinski *et al.*, 1976; Kaminski *et al.*, 1985), or invade the skull and brain (Oehlschlaegel *et al.*, 1966; Lutzner *et al.*, 1984). A worsening effect of x-rays has been stressed (Lutzner *et al.*, 1984; Jablonska and Orth, 1985).

A certain EV form associated with HPV-3 and HPV-10, types also found in flat warts in the general population, is characterized by the presence of disseminated flat warts and, occasionally, confluent, elevated brownish plaques, observed mainly on the extremities and the face (Orth *et al.*, 1978c, 1979; Jablonska *et al.*, 1979a,b; Jablonska and Orth, 1985). The malignant transformation of these lesions seems exceptional, but HPV-10 genomes have been found in a vulvar Bowen's carcinoma of an EV patient (Green *et al.*, 1982). Some authors believe that this form, which is not associated with EV-specific HPVs, corresponds to disseminated flat warts and should not be regarded as EV (Rueda and Rodriguez, 1976; Kaufmann *et al.*, 1978). However, such a case has been described in a family with five other typical EV cases (Fig. 2) (Orth *et al.*, 1980), and HPV-3 has been found, with a relatively high frequency in patients infected with EV-specific HPVs (Orth, 1986). This may indicate that the EV phenotype also includes a greater susceptibility to HPV-3 infection.

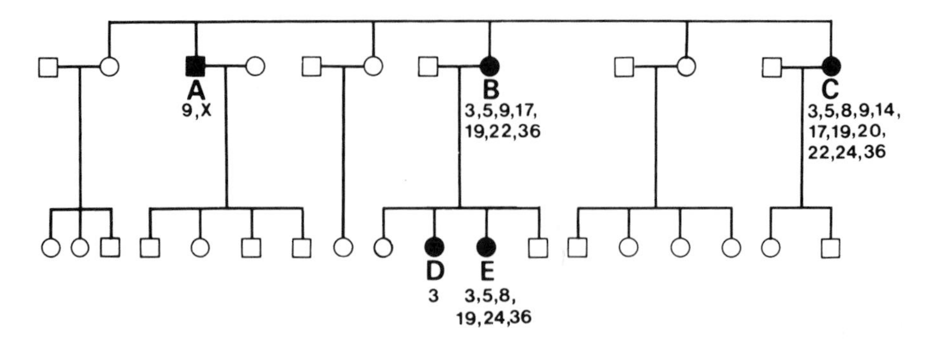

FIGURE 2. Familial epidermodysplasia verruciformis. Family tree from Glinski *et al.* (1976), Jablonska *et al.* (1979a), Orth *et al.* (1980), and Wojtulewicz-Kurkus *et al.* (1985). Numbers correspond to the types of HPV identified in benign lesions (G. Orth and S. Jablonska, unpublished results).

III. HISTOLOGY AND ULTRASTRUCTURE OF EV LESIONS

A. Benign Lesions

The different morphological types of benign lesions induced by EV-specific HPVs have a common characteristic histological pattern (Fig. 3) (Rueda and Rodriguez, 1972, 1976, 1983; Kaufmann *et al.*, 1978; Orth *et al.*, 1980; Jablonska and Orth, 1985). Large "dysplastic" cells with pale-stained cytoplasm are present in the spinous and granular layers of the epidermis, either in nests or filling the major part of the epidermis in a bandlike arrangement. The clear cells of the granular layers contain prominent round or ovoid keratohyalin granules scattered in the cytoplasm. The nuclei often show an intranuclear vacuole and marginated chromatin. The cytopathic effect begins in the suprabasal layers, and the size of the cells increases progressively as they migrate toward the superficial layers. Moderate acanthosis is observed in the flat wartlike lesions. The cornified layers, which cover the dysplastic cells, show parakeratosis and hyperkeratosis.

Electron microscope studies have shown that the pale-stained, characteristic EV cells have lost their cytoplasmic organelles and contain only rare keratin filament bundles. The desmosome–filament complexes, which join the cells, are preserved. The keratohyalin granules are not associated with tonofilaments. The nuclei of some EV cells contain a clear space, sometimes large, with a finely fibrillar structure. The chromatin and nucleoli are marginated. The assembly of viral particles occurs at the periphery of these clear nuclear spaces in cells containing keratohyalin granules. Viral particles often assemble in a crystalline array, and may entirely fill the nucleus of superficial cells (Califano and Caputo, 1968; Grupper *et al.*, 1971; Rueda and Rodriguez, 1972, 1976; Pruniéras, 1975; Kaufmann *et al.*, 1978; Laurent *et al.*, 1978; Croissant *et al.*, 1985). The

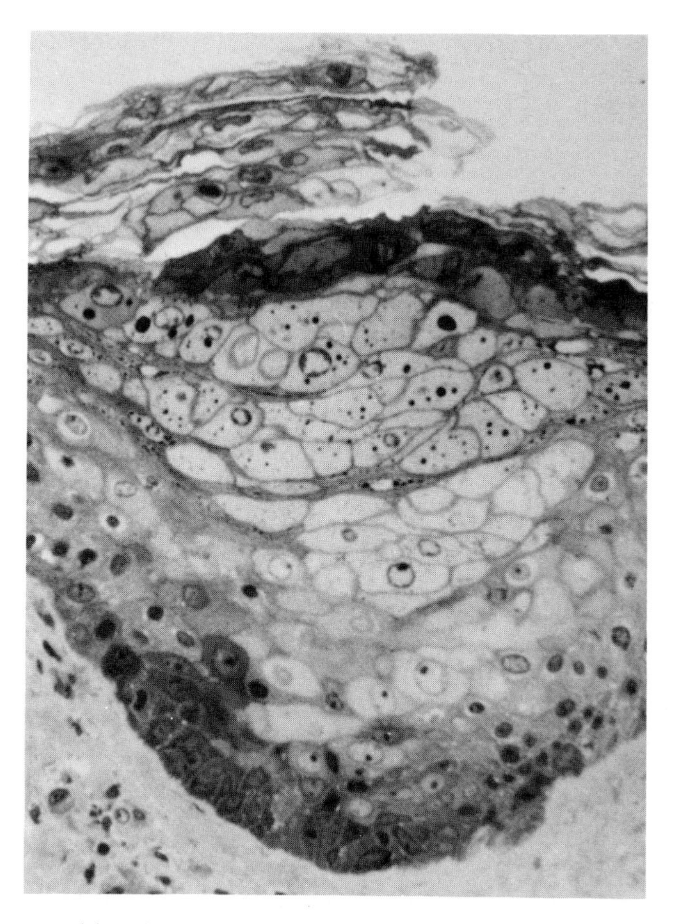

FIGURE 3. Typical histological pattern of a benign lesion of epidermodysplasia verruciformis. Plastic-embedded 1-μm-thick section. Methylene blue–basic fuchsin; × 400.

intranuclear clear spaces observed in EV cells seem specifically associated with the replication of EV HPVs and have been assumed to contain the precursor components of the viral particles (Rueda and Rodriguez, 1972, 1976, 1983).

In situ molecular hybridization experiments, using tritiated EV-specific HPV RNA or DNA probes, have shown that the first cellular alterations of the deeper layers correspond to the onset of the vegetative viral DNA replication (Orth *et al.*, 1980; Croissant *et al.*, 1985). Structural viral proteins are detected, by immunofluorescence, only in the nuclei of the dysplastic cells of the more superficial granular layers (Orth *et al.*, 1978c). Incorporation of tritiated thymidine in EV lesion specimens has shown that DNA synthesis is detectable, by autoradiography, in the dysplastic spinous and granular cells (Langner *et al.*, 1968; Delescluse *et al.*, 1970, 1972). This suggests that vegetative viral DNA replication proceeds during the whole abnormal terminal differentiation process of EV cells, and/or that cellular DNA replication, repressed as a rule in normal keratinizing epidermal cells, is maintained in EV clear cells. Unexpectedly and in contrast to differentiating normal keratinocytes, no incorporation of tritiated uridine has been detected in dysplastic cells (Delescluse *et al.*, 1973).

The histological and ultrastructural features, characteristic of EV lesions, reveal a specific cytopathic effect connected with the replication of EV-specific HPVs and resulting, in particular, in a disturbed expression of keratins, a major differentiation product of keratinocytes. This cytopathic effect is distinct from that observed in HPV-3- and HPV-10-induced flat warts, the latter being characterized by a progressive perinuclear clarification process, pushing cytoplasmic organelles, keratohyalin granules, and tonofilaments toward the periphery of the cell (Rueda and Rodriguez, 1972, 1976; Kaufmann *et al.*, 1978; Laurent *et al.*, 1978; Croissant *et al.*, 1985).

B. Malignant Lesions

EV malignant lesions appear as carcinomas *in situ*, squamous cell carcinomas, and, occasionally, basal cell carcinomas. The intraepidermal carcinomas have the histological features described for Bowen's disease lesions (Lever and Schaumburg-Lever, 1975), with a disorganized structure of the epidermis, pleomorphic cells, enlarged and hyperchromatic nuclei, abnormal mitotic figures, multinucleated cells, and cells showing individual keratinization (dyskeratotic cells). The epidermis is thickened and parakeratotic. Clear cells may be observed in the superficial layers of the epidermis, suggesting that carcinomas derive from benign EV lesions (Jablonska *et al.*, 1966; Oehlschlaegel *et al.*, 1966; Ruiter and Van Mullem, 1966, 1970b; Ruiter, 1969; Delescluse *et al.*, 1972; Rueda and Rodriguez, 1972, 1976, 1983; Jablonska and Orth, 1985).

Invasive squamous cell carcinomas often retain the features of Bowen's carcinomas (Fig. 4). Large, lightly stained cells with irregular and large nuclei resembling EV cells are often observed. Individual cell keratinization and formation of parakeratotic horn pearls are usual features (Jablonska *et al.*, 1966; Oehlschlaegel *et al.*, 1966; Ruiter and Van Mullem, 1970b; Ruiter, 1973; Rueda and Rodriguez, 1976, 1983; Jablonska and Orth, 1985).

Some authors have observed viral particles, usually in small amounts, in rare cells of the superficial layers of intraepidermal carcinomas (Ruiter and Van Mullem, 1970a,b; Yabe and Sadakane, 1975; Rueda and Rodriguez, 1976; Yabe *et al.*, 1978), but such particles have not been detected by other authors in similar tumors (Aaronson and Lutzner, 1967; Jablonska *et al.*, 1968, 1970; Delescluse *et al.*, 1972). Viral particles have never been found in invasive cancers (Ruiter and Van Mullem, 1970a; Deles-

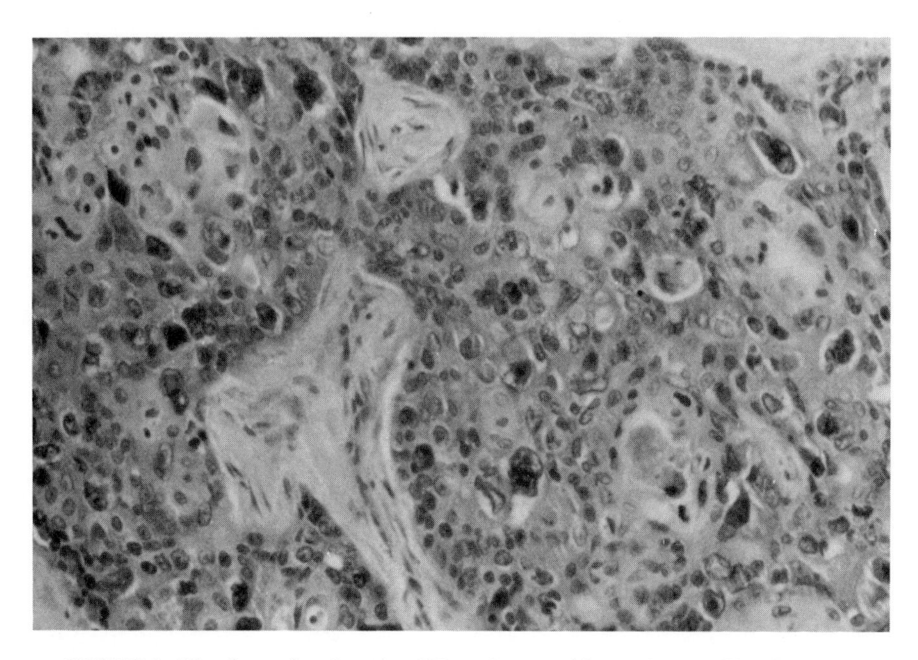

FIGURE 4. Histology of an invasive EV carcinoma of Bowen's type. H & E; × 175.

cluse *et al.*, 1972; Yabe and Sadakane, 1975; Rueda and Rodriguez, 1976; Orth *et al.*, 1980; Lutzner *et al.*, 1984).

IV. GENETIC FACTORS IN EV

EV shows no racial predisposition, and is observed with similar frequency in both sexes. No correlation between EV and HLA-A and -B antigens has been found (Wojtulewicz-Kurkus *et al.*, 1985), but the possible association of EV with products of HLA-D loci, involved in immune responses, has not been investigated.

The importance of genetic factors in EV is suggested by the high frequency of parental consanguinity in EV patients and the frequently familial occurrence of the disease (Midana, 1949; Bilancia, 1961; Lazzaro *et al.*, 1966; Lutzner, 1978). Eleven percent of the 147 EV cases reviewed by Lutzner (1978) had consanguineous parents (usually first cousins). The frequency of consanguineous marriages is usually lower than 1% in Europe and the United States, but a higher frequency is observed in some small communities and in religious, geographic, and ethnic isolates (Vogel and Motulsky, 1982). Forty percent of the 42 EV cases first reported were from Japan (Sullivan and Ellis, 1939), a country where the occurrence of consanguinity is high (about 8%) (Vogel and Motulsky, 1982). Homozygotes of a rare disease are more frequent among children of consanguineous marriages than in the general population, and the higher rate of consanguinity among parents of EV patients thus suggests that the disease may be recessively inherited.

The existence of one (or more) rare recessive EV gene(s) is also supported by the frequently familial occurrence of the disease. Ten percent of the EV families whose histories have been reviewed by Lutzner (1978) had more than one sibling affected, and in families where information was available, 31% (40 of 129) of the siblings suffered from EV. The spouses of heavily infected patients never become affected. The frequency of siblings with EV (about 25%) and the sex ratio (about 1) also support that EV is an autosomal recessive disorder. A sex limitation to women was suggested by Rajagopalan *et al.* (1972) who described a family of 11 siblings with 4 healthy boys and 6 of 7 girls with EV, but usually both sexes are equally affected in EV families (Lutzner, 1978). The occurrence of EV in two successive generations has only been exceptionally reported (Lutzner, 1978). One such family has been followed for about 30 years (Jablonska and Milewski, 1957; Jablonska *et al.*, 1966, 1968, 1979a, 1982b; Glinski *et al.*, 1976; Orth *et al.*, 1978c, 1979, 1980; Kremsdorf, 1980; Kremsdorf *et al.*, 1982a, 1983, 1984; Wojtulewicz-Kurkus *et al.*, 1985). The members of this family (Fig. 2) are characterized by a susceptibility to a great number of EV-specific HPVs and to HPV-3. A recessive mode of transmission would presuppose that the spouse of patient B was a heterozygote.

Another mode of EV transmission has recently been suggested by the study of a family in which only the males were affected (Androphy *et al.*, 1984, 1985; Vasily *et al.*, 1984). The proband was born of nonconsanguineous parents. None of his seven siblings and ten children were affected. Four of the eight sons born from his daughters had EV associated with HPV-3 and HPV-8, the latter being an EV-specific HPV. This observation strongly suggests an X-linked, recessive inheritance of the disease (Androphy *et al.*, 1985).

Another argument to be taken into account when discussing the role of genetic factors in EV is the association of the disease with mental retardation and other congenital abnormalities observed in some patients (Midana, 1949; Bilancia, 1961; Lazzaro *et al.*, 1966; Lutzner, 1978). Eleven (7.5%) of the 147 patients whose history was reviewed by Lutzner (1978) were mentally retarded, most of them showing severe retardation and some of them requiring institutionalization. Neurological disorders were observed only in a minority of these patients. Other congenital abnormalities affecting the skin, teeth, or bones were observed in only one or two patients (Lutzner, 1978). One EV patient, with severe mental retardation, showed dystrophic changes of the nails as well as skeletal abnormalities (Ruiter and Van Mullem, 1966). A cousin of this patient suffered from skeletal abnormalities, while another one was mentally retarded. None of the latter suffered from EV. In another EV family, the first cousin of a mentally retarded EV patient had two children suffering from mental retardation and a normal one who, himself, had a mentally retarded child and a child suffering from EV (L. A. Rueda, personal communication). The association of mental retardation with EV in some patients and the segregation of these diseases in some of the patients' relatives may reflect a linkage between an EV gene and one of the recessively transmitted genes involved in mental retardation, rather than the pleiotropic expression of a subgroup of EV genes as proposed by Lutzner (1978) by analogy with xeroderma pigmentosum.

No chromosomal abnormalities have been observed in the few EV patients studied (Lutzner, 1978). The probable existence of two recessive modes of EV transmission, autosomal and X-linked, suggests that the EV phenotype, that is an abnormal susceptibility to specific HPV types, depends on the abnormal alleles of at least two genes localized on different chromosomes. The nature of the gene defects remains unknown.

V. IMMUNOLOGICAL FACTORS IN EV

EV is characterized by the lifelong persistence of widespread skin lesions induced by specific HPV types and by the development of HPV-associated skin carcinomas, usually in sun-exposed sites. EV patients are not abnormally prone to bacterial, fungal, or viral infections, including infections with HPV types which induce skin and genital warts in the general population, with the exception of HPV-3 and HPV-3-related types associated with flat warts (Jablonska and Orth, 1985; Orth, 1986). The incidence of cancers other than skin carcinomas does not seem to be abnormally high. Only single cases of rhabdomyosarcoma (Okamoto *et al.*, 1970), malignant reticulosis (Sayag *et al.*, 1966), lymphoma (Lutzner, 1978), and hepatocarcinoma (Van Voorst Vader *et al.*, 1986) have been reported in EV patients. EV may, thus, involve a specific impairment of immune functions allowing the recognition of, and the response to, EV HPV-encoded antigens.

The immune responses to HPV infection are still little understood, but the available evidence points to a prominent contribution of cell-mediated immune responses to wart cell antigens, rather than that of humoral immunity, in the control of the growth and regression of warts. Patients with genetic or acquired defects of cell-mediated immunity (CMI) show an increased incidence of warts and skin cancers (Koranda *et al.*, 1974; Spencer and Andersen, 1979; Penn, 1980; Jablonska *et al.*, 1982a; Boyle *et al.*, 1984; Lutzner, 1985), some presenting an EV-like syndrome (Koranda *et al.*, 1974; Spencer and Andersen, 1979; Lutzner *et al.*, 1980, 1983). The systemic regression of flat warts has been shown to involve the invasion of the epidermis by mononuclear cells, chiefly cytotoxic T lymphocytes and macrophages (Jablonska *et al.*, 1982a; Tagami *et al.*, 1985a,b).

The skin immune protection against intracellular pathogens and neoplasms is most likely achieved by a specialized set of skin-associated lymphoid tissues which interact with the systemic immune apparatus. A major role has been assigned to antigen-presenting epidermal cells and T lymphocytes which display an affinity for skin and draining lymph nodes (Streilen, 1983, 1985). Studies of rodent models have shown that ultraviolet radiations in the sunburn spectrum have a profound *in vivo* effect on immune responses. This includes the induction of unresponsiveness to contact allergens, like dinitrochlorobenzene (DNCB), and the systemic enhancement of susceptibility to UV-induced tumors (Fisher and Kripke, 1982; Streilen, 1983, 1985; Kripke and Morrison, 1985). It has been proposed that this involves the generation of specific suppressor T cells by UV-resistant antigen-presenting epidermal cells, distinct from the UV-sensitive antigen-presenting Langerhans cells involved in the contact hypersensitivity reactions (Streilen, 1985; Kripke and Morrison, 1985).

So far, the immunological studies have mostly been aimed at an evaluation of the general immune status of EV patients, but little is known about their specific local and systemic immune responses to EV HPV-associated antigens.

A. Humoral Immunity

With the exception of a case associated with congenital IgM deficiency (Gavanou *et al.*, 1976) and of another case showing common variable immunodeficiency (Goldes *et al.*, 1984), EV patients have been found to have normal immunoglobulin levels (Lutzner, 1978). The prevalence of antibodies against pathogens, in particular measles and rubella viruses, has been found normal (Jablonska *et al.*, 1979b). The presence of anti-HPV antibodies, detected by using viral particles purified from pooled warts as antigens, has been reported to be slightly less frequent in EV patients than in patients presenting other wart types (Pyrhönen *et al.*, 1980). Anti-HPV-1 antibodies were found in 6 of 13 EV patients (Orth *et al.*, 1980), an incidence similar to that observed in patients with skin warts (Kienzler *et al.*, 1983). In a series of 7 patients with typical EV form, 3 of the 4 patients infected with both EV HPVs and HPV-3 had anti-HPV-3 antibodies while 3 patients infected only with EV-specific HPVs had none, as indicated by an immunodiffusion test using purified HPV-3 particles. In the latter series, anti-EV-specific HPV antibodies were detected in 2 of 3 cases, using virions purified from pooled EV lesions as antigens. Out of 5 EV patients infected only with HPV-3 or related types, only 1 had detectable anti-HPV-3 antibodies [data from Orth *et al.* (1980) corrected for the types of infecting HPVs]. The low antibody titers or the absence of detectable antibodies in some patients could favor reiterative infections and progressive dissemination of the lesions.

B. Cell-Mediated Immunity

Most EV patients studied were reported to have an impaired CMI (Jablonska and Orth, 1985). Among these, some patients with inherited or acquired CMI defects may have represented EV phenocopies. EV has been reported in a patient with dyskeratosis congenita (Bundino *et al.*, 1978), in a patient with common variable immunodeficiency (Goldes *et al.*, 1984), in two patients with lepromatous leprosy (Jacyk and Lechner, 1984), and in renal allograft recipients under immunosuppressive therapy (Koranda *et al.*, 1974; Lutzner *et al.*, 1980, 1983). An impairment of the CMI has generally been revealed in other patients by a reduced lymphocyte responsiveness to mitogens and by abnormal responses to skin tests (Glinski *et al.*, 1976, 1981; Prawer *et al.*, 1977; Jablonska *et al.*, 1979a, 1982a,b; Guilhou *et al.*, 1980; Lutzner *et al.*, 1984). The most constant feature has been an anergy to sensitization to DNCB. It should be noticed, however, that these *in vivo* and *in vitro* tests, including DNCB sensitization, did not disclose any abnormality in a certain number of patients (Prystowsky *et al.*, 1976; Kienzler *et al.*, 1979; Claudy *et al.*, 1982; Androphy *et al.*, 1984).

No correlation has been observed between the CMI defect of EV patients and the type of infecting HPV, or the development of skin cancers (Jablonska *et al.*, 1979a, 1982a). In a series of seven patients infected with EV HPVs, who had or had had carcinomas, and six patients with an HPV-3-associated form of EV without carcinoma, most of them had a reduced percentage of E rosette-forming lymphocytes and a reduced lymphocyte responsiveness to phytohemagglutinin. All patients were found anergic to DNCB sensitization, using various challenging doses of DNCB (12.5 to 100 μg), with the exception of one patient in each group who responded only to the highest doses (Glinski *et al.*, 1981).

So far, no studies have been carried out on the CMI response of EV patients to EV HPV-specific viral or cellular antigens, with the exception of an investigation on the immune response to HPV-3 purified virions in nine EV patients using a leukocyte migration inhibition factor test (Haftek *et al.*, 1985). Four of five patients infected only with HPV-3 or with HPV-10, an HPV-3-related type, showed a response, but none of four patients infected with EV HPVs and developing skin cancers showed a response to HPV-3 antigens, even though three of them were also infected with HPV-3 (Haftek *et al.*, 1985).

Natural killer (NK) cells represent an immunologically nonspecific cell-mediated defense against tumor cells and virus-infected cells (Herberman, 1983). In a study of six EV patients, four patients who had had multiple skin carcinomas were reported to have an increased NK cell activity (Kaminski *et al.*, 1985). In a series of seven patients, NK cell activity was found normal and no significant changes were observed during a 4-week treatment with α-interferon (Androphy *et al.*, 1984).

C. Regression of EV

Regression has never been observed in a typical EV form. However, some lesions have been reported to improve or disappear in a few cases, in particular in a patient with preserved CMI (Prystowsky *et al.*, 1976). In a case of EV associated with an HPV-5-related type in a setting of common variable immunodeficiency, the benign lesions improved only transiently after a successful T-lymphocyte graft (Goldes *et al.*, 1984).

Complete regression has been reported in two patients with HPV-3-induced EV after disease durations of 17 and 20 years (Jablonska *et al.*, 1982b; Haftek *et al.*, 1985). In the first patient, the lesions began to regress during treatment with aromatic retinoids and simultaneous repeated scrapings of the lesions (Haftek *et al.*, 1985). In the other patient, a familial case (patient D, Fig. 2), regression occurred after two deliveries (Jablonska *et al.*, 1982b). In contrast to the systemic regression of HPV-3-induced flat warts, in the latter two cases regression occurred progressively and without inflammation. In both patients, nonspecific CMI, as evaluated by *in vitro* tests, became normal and the patient's response to DNCB challenge became positive (Haftek *et al.*, 1985; Jablonska and Orth, 1985). This raises the question as to whether the CMI defects observed in EV patients are primary or only secondary to massive HPV infection.

For obvious reasons, little information is available on the immunological status of EV patients at the onset of their disease, if one excludes the patients with genetic or acquired immune defects discussed above. However, impaired immune responses at the time of infection were likely for some patients, as suggested by their case reports. An immunological tolerance resulting from neonatal infection (Nash, 1985) could be involved when the onset of EV was allegedly reported to have occurred shortly after birth (Lewandowsky and Lutz, 1922; Prystowsky *et al.*, 1976; Lutzner, 1978). An impairment of immune responses should also be considered for EV patients in whom the disease was reported to have begun after an overexposure to the sun (Relias *et al.*, 1967; Kienzler *et al.*, 1979), or after an infectious disease, like measles (Midana, 1949), or lepromatous leprosy (Jacyk and Lechner, 1984).

VI. HPV TYPES ASSOCIATED WITH EV

A. Multiplicity and Specificity

After the demonstration of the presence, in EV lesions, of intranuclear viral particles with a morphology similar to that of wart virus, EV was regarded as a genodermatosis in which an anomaly of the skin resulted in a decreased resistance to a virus presumably identical or related to that of common warts (Ruiter and Van Mullem, 1966). However, the specific histological and ultrastructural features of EV lesions led to the hypothesis that EV could be associated with a papillomavirus distinct from that found in skin warts, especially flat warts (Rueda and Rodriguez, 1972, 1976). The discovery of the plurality of HPV types infecting the skin (Gissmann *et al.*, 1977; Orth *et al.*, 1977a,b, 1978a,b) and the observation that deep plantar warts and hand common warts could constitute distinct diseases caused by different HPV types (HPV-1 and HPV-2, respectively) (Orth *et al.*, 1977a,b, 1978b) rendered the hypothesis likely. This possibility was first substantiated by serological studies. The virions isolated from the lesions of two EV patients and from pools of warts showed no or only weak antigenic cross-reactions by immune electron microscopy and complement fixation tests (Pass *et al.*, 1977). In addition, no antigenic relationship was detected by immunofluorescence, using anti-HPV-1 or anti-HPV-2 antisera and sections of EV lesions (Orth *et al.*, 1977b).

The biochemical and antigenic characterization of the HPVs present in the lesions of 11 patients showed unequivocally that EV was associated with at least two HPV types, HPV-3 and HPV-5 (the latter having initially been called HPV-4), distinct from HPV-1 and HPV-2 previously characterized in skin warts (Orth *et al.*, 1978c). No detectable DNA sequence homology was found among HPV types 1, 2, 3 and 5 by RNA–DNA filter hybridization experiments using RNAs transcribed *in vitro* from HPV-1 or HPV-2 DNAs, or DNAs purified from virions obtained from pooled lesions of individual EV patients. No antigenic relationship was shown between HPV types 1, 2, 3, and 5 by the immunofluorescence and immunodiffusion techniques, using antisera prepared against purified viral particles. A genetic heterogeneity of HPV-3 and HPV-5 was disclosed by comparison of HPVs present in different EV patients by means of reciprocal serological studies (Pass *et al.*, 1977; Orth *et al.*, 1978c), cross-hybridization experiments (Orth *et al.*, 1978c, 1979), and restriction enzyme analysis (Orth *et al.*, 1980). In addition, it seemed clear that EV patients were infected by more than one HPV type (Orth *et al.*, 1978c, 1980).

Because of the multiplicity of HPVs infecting a single EV patient and because of the limited amount of material available, the characterization of EV HPVs only became possible after molecular cloning of the viral genomes (Kremsdorf, 1980; Pfister *et al.*, 1981; Kremsdorf *et al.*, 1982a,b, 1983, 1984; Green *et al.*, 1982; Ostrow *et al.*, 1982, 1983a; Brackmann *et al.*, 1983; Tsumori *et al.*, 1983; Gassenmaier *et al.*, 1984). This led to the identification of 16 HPV types (Table I) on the basis of cross-hybridization levels of less than 50% among the cloned HPV DNAs under the most stringent hybridization conditions (Coggin and zur Hausen, 1979). Two types, HPV-3 and HPV-10, have also been found in flat warts in the general population (Jablonska *et al.*, 1985). Fourteen types, HPV-5, 8, 9, 12, 14, 15, 17, 19–25, seem to be specifically associated with EV since they have not been found in skin warts, and only exceptionally in premalignant and malignant skin lesions in the general population (Lutzner *et al.*, 1983; Jablonska *et al.*, 1985; Pfister *et al.*, 1985; Kawashima *et al.*,

TABLE I. HPV Types Cloned from Lesions of EV Patients

HPV type[a]	EV HPV subgroup[b]	Patients origin	References[c]
3	A	Poland	5, 7, 8
		USA	9
5	B	Poland	1, 4
		Japan	6
		Turkey	11
8	B	Poland	1, 5, 8
		Upper Volta	2
		Turkey	11
9	C	Poland	1, 4
10	A	Poland	4
		USA	3
12	B	Poland	5, 8
14	B	Poland	5, 12
		Japan	10
		France	12
15	C	Poland	5, 12
17	C	Columbia	12
		Italy	12
19	B	Holland	12
		Turkey	11
20	B	Columbia	12
		Turkey	11
21	B	Poland	12
22	B	Italy	12
23	B	Poland	12
24	D	Holland	12
25	B	Turkey	11

[a] HPV types 5, 8, 9, 12, and 10 were described first as types 5b, 5c, 5a, 5d (Orth *et al.*, 1980), and 3f (Orth *et al.*, 1981), respectively.
[b] EV HPV types showing from 5% to 40% DNA sequence homology with HPV types 3, 5, 9, and 24 taken as prototypes have been classified into subgroups A, B, C, and D, respectively (Kremsdorf *et al.*, 1983, 1984).
[c] 1, Kremsdorf (1980); 2, Pfister *et al.* (1981); 3, Green *et al.* (1982); 4, Kremsdorf *et al.* (1982a); 5, Kremsdorf *et al.* (1982b); 6, Ostrow *et al.* (1982); 7, Brackmann *et al.* (1983); 8, Kremsdorf *et al.* (1983); 9, Ostrow *et al.* (1983a); 10, Tsumori *et al.* (1983); 11, Gassenmaier *et al.* (1984); 12, Kremsdorf *et al.* (1984).

1986a,b; Scheurlen *et al.*, 1986). An HPV-5-related type, HPV-36, first obtained and cloned from an actinic keratosis of a non-EV patient, was later found only in EV lesions (Kawashima *et al.*, 1986b). Two HPV-9-related types, HPV-37 and HPV-38, have recently been cloned from a keratoacanthoma and a melanoma (Scheurlen *et al.*, 1986), but their role in EV has not been determined.

B. Biochemical Properties

1. Structural Polypeptides

Little is known about the structural polypeptides of the EV-associated HPVs. In an unpublished study (M. Favre, S. Jablonska, and G. Orth), the virions purified from pooled lesions of single EV patients displayed a major polypeptide with an apparent molecular weight of 55K (HPV-3) or 52K (EV HPVs), a minor component with a molecular weight of about 70K (HPV-3) or 76K (EV HPVs), and four histonelike low-molecular-weight polypeptides. Such a composition is similar to that reported for other HPVs (Orth and Favre, 1985).

2. Characterization of EV HPV Genomes

Our present knowledge on the EV-associated HPVs is based on the biochemical properties of their genomes. The HPV DNAs were usually cloned in *E. coli* after insertion into plasmid pBR322. The viral DNAs were usually obtained from purified viral particles (Pfister *et al.*, 1981; Green *et al.*, 1982; Ostrow *et al.*, 1982; Tsumori *et al.*, 1983), or selectively extracted from lesions and purified by equilibrium centrifugation in CsCl gradients and/or sedimentation in sucrose gradients in the presence of ethidium bromide (Kremsdorf, 1980; Kremsdorf *et al.*, 1982a, 1983, 1984; Brackmann *et al.*, 1983). Alternatively, some EV HPV DNAs were cloned from total DNA extracted from benign or malignant lesions after insertion into phage lambda DNA (Ostrow *et al.*, 1983a; Gassenmaier *et al.*, 1984).

The sizes of the EV HPV genomes fall into two classes of molecules differing by about 500 nucleotide pairs, as deduced from the electrophoretic migration of cloned and uncloned DNAs in agarose gels and from length measurements under the electron microscope (Orth *et al.*, 1980; Kremsdorf *et al.*, 1982a, 1983, 1984). The genomes of HPV types 3, 5, 8, 12, 14, 19–21 have a greater size (about 7700 nucleotide pairs) than those of HPV types 9, 15, 17, 22, 23 (about 7200 nucleotide pairs) (Kremsdorf *et al.*, 1984). The G + C contents of the cloned DNAs of HPV types 3, 5, 8–10, 12 have values of 45.5, 40.9, 43.6, 41.6, 44.2, and 41.8%, respectively, as deduced from their buoyant densities (Kremsdorf *et al.*, 1983).

The genomes of the EV-associated HPVs show variable extents of homology, from none to about 40%, as determined by hybridization at saturation in liquid phase followed by nuclease S1 digestion (Kremsdorf *et al.*, 1982a, 1983, 1984; Brackmann *et al.*, 1983; Gassenmaier *et al.*, 1984). Four HPV groups have been defined on the basis of less than 5% cross-hybridization between the genomes of the HPV types belonging to different groups (Kremsdorf *et al.*, 1983, 1984). The first group includes HPV-3 and HPV-10, also found in flat warts in the general population (Jablonska *et al.*, 1985) and showing a cross-hybridization level of 35% (Kremsdorf *et al.*, 1982a, 1983; Green *et al.*, 1982). The second group includes HPV types 5, 8, 12, 14, 19–23, 25, and 36, with cross-hybridization levels ranging from 5 to 40% (Kremsdorf *et al.*, 1983, 1984; Gassenmaier *et al.*, 1984; Kawashima *et al.*, 1986b). The third group includes HPV types 9, 15, 17, with cross-hybridization levels varying from 6 to 22% (Kremsdorf *et al.*, 1984). HPV-37, which shows a 20% cross-hybridization with HPV types 15 and 17, and HPV-38, which shows a weak cross-hybridization under stringent conditions with HPV types 9, 15, 17,

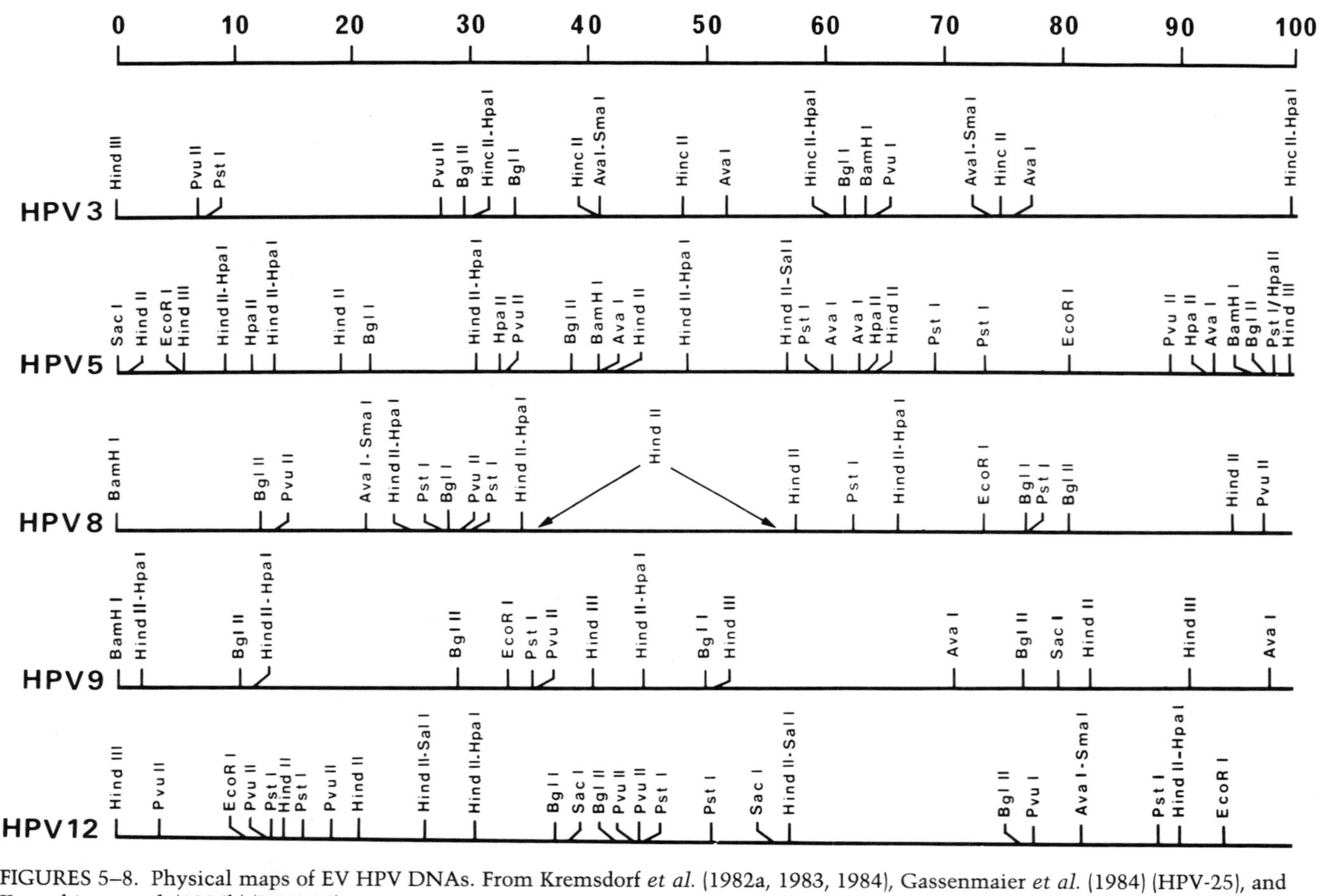

FIGURES 5–8. Physical maps of EV HPV DNAs. From Kremsdorf *et al.* (1982a, 1983, 1984), Gassenmaier *et al.* (1984) (HPV-25), and Kawashima *et al.* (1986b) (HPV-36).

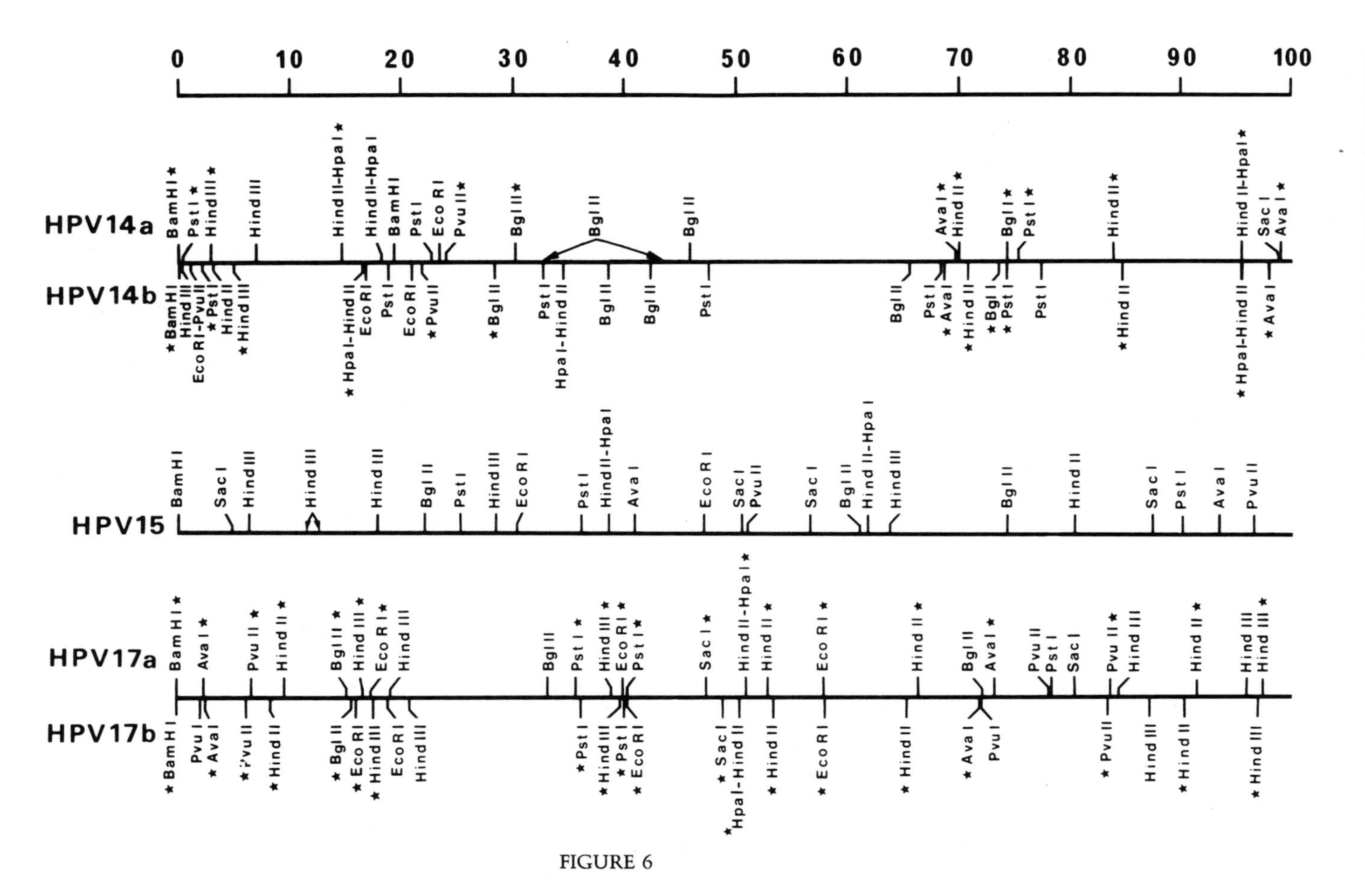
0
10
20
30
40
50
60
70
80
90
100
HPV14a
HPV14b
HPV15
HPV17a
HPV17b

FIGURE 6

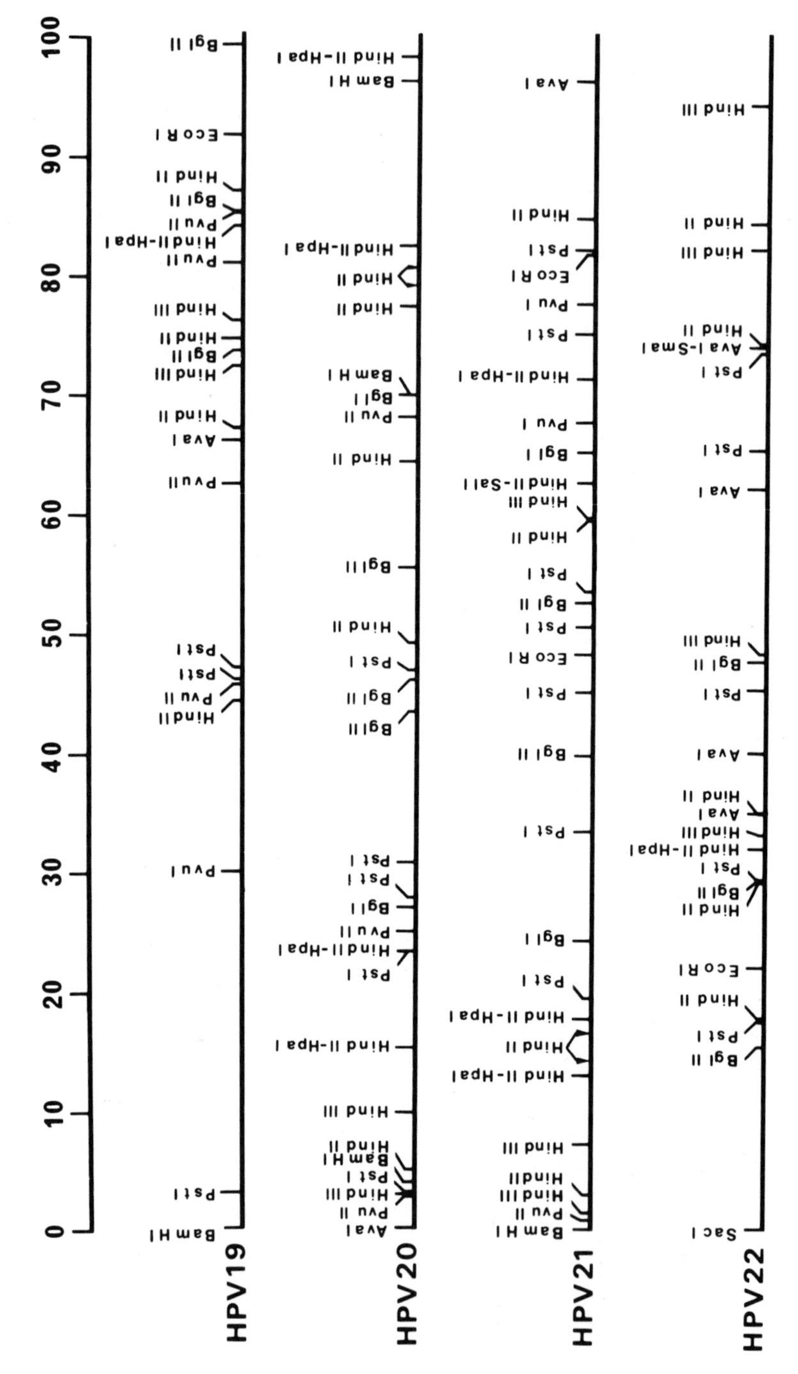

FIGURE 7

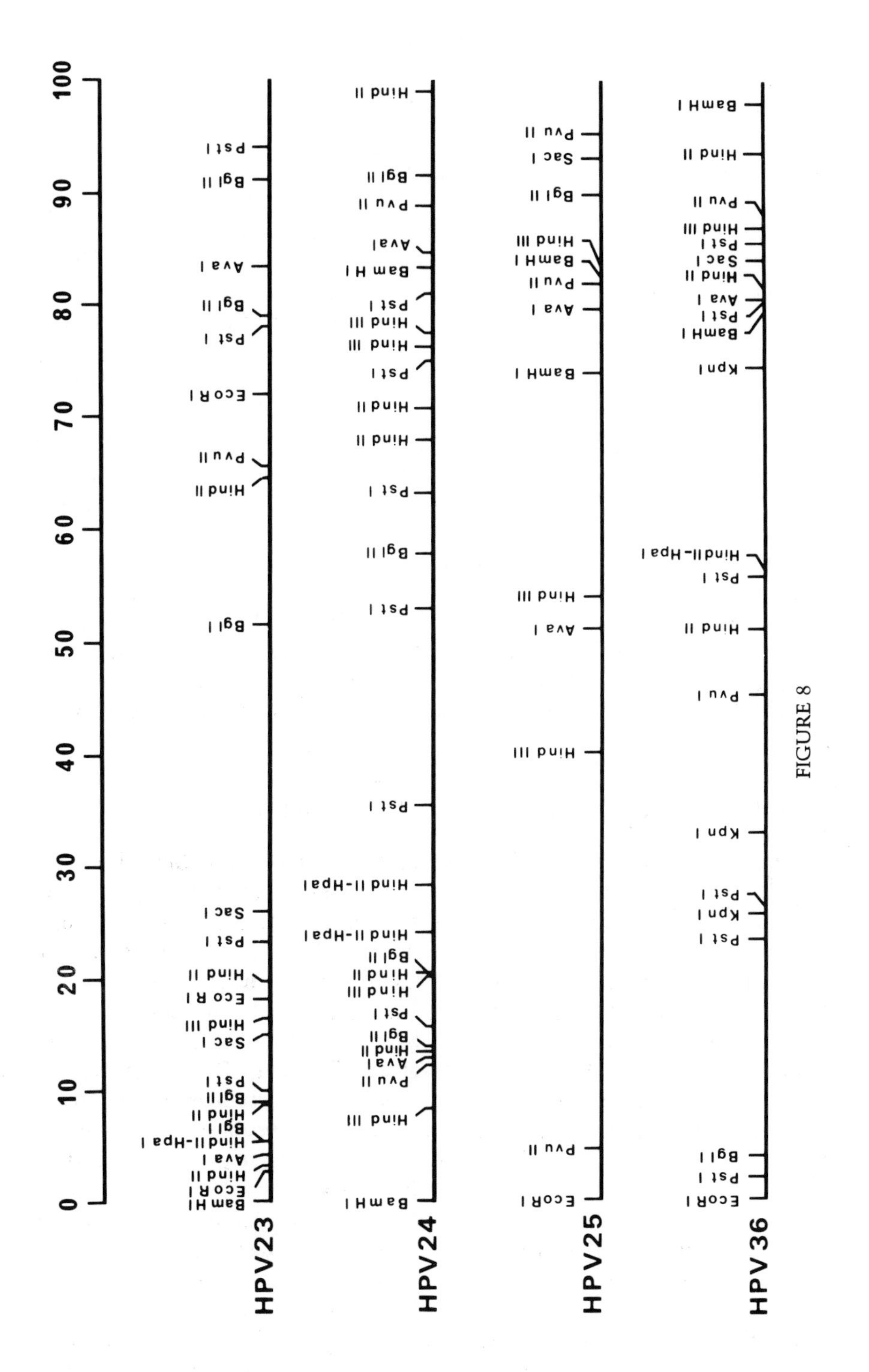
0
10
20
30
40
50
60
70
80
90
100
HPV23
BamHI
EcoRI
HindII
AvaI
HindII-HpaI
BglI
HindII
BglII
PstI
SacI
HindIII
EcoRI
HindII
PstI
SacI
BglI
HindII
PvuII
EcoRI
PstI
BglII
AvaI
BglII
PstI
HPV24
BamHI
HindIII
PvuII
AvaI
HindII
BglII
PstI
HindIII
HindII
BglII
HindII-HpaI
HindII-HpaI
PstI
PstI
BglII
PstI
HindII
HindII
PstI
HindIII
HindIII
PstI
BamHI
AvaI
PvuII
BglII
HindII
HPV25
EcoRI
PvuII
HindIII
AvaI
HindIII
BamHI
AvaI
PvuII
BamHI
HindIII
BglII
SacI
PvuII
HPV36
EcoRI
PstI
BglI
PstI
KpnI
PstI
KpnI
PvuI
HindII
PstI
HindII-HpaI
KpnI
BamHI
PstI
AvaI
HindII
SacI
PstI
HindIII
PvuII
HindII
BamHI

FIGURE 8

and 37 (Scheurlen *et al.*, 1986), should belong to this group, although their roles in EV are not known. The fourth group consists, so far, of a single type, HPV-24 (Kremsdorf *et al.*, 1984).

No obvious similarities have been detected among the restriction endonuclease cleavage maps of the EV-associated HPVs (Figs. 5 to 8) (Kremsdorf, 1980; Pfister *et al.*, 1981, 1983; Kremsdorf *et al.*, 1982a,b, 1983, 1984; Ostrow *et al.*, 1982, 1983a; Yutsudo *et al.*, 1982, 1985; Tsumori *et al.*, 1983; Gassenmaier *et al.*, 1984; Kawashima *et al.*, 1986b). Evidence for some conserved sites has been found in HPV types showing the highest levels of cross-hybridization. Five of the twenty-one sites localized in the HPV-3 DNA seem to be also present in HPV-10 DNA (Kremsdorf *et al.*, 1983). Similarly, 7 of the 28 sites localized in HPV-20 DNA seem to be present in HPV-21 DNA (Fig. 7) (M. Favre and G. Orth, unpublished observation). Subtypes or variants have been recognized on the basis of substantial or minor differences in the cleavage maps of different isolates of HPV-3 (Kremsdorf *et al.*, 1983; Ostrow *et al.*, 1983a), HPV-5 (Kremsdorf *et al.*, 1982a,b; Ostrow *et al.*, 1982; Pfister *et al.*, 1983), HPV-8 (Kremsdorf, 1980; Pfister *et al.*, 1981; Kremsdorf *et al.*, 1983; Gassenmaier *et al.*, 1984), HPV-14 (Kremsdorf *et al.*, 1982b, 1984; Yutsudo *et al.*, 1982; Tsumori *et al.*, 1983), HPV-17 (Kremsdorf *et al.*, 1984; Yutsudo *et al.*, 1985), HPV-19 and HPV-20 (Gassenmaier *et al.*, 1984; Kremsdorf *et al.*, 1984). The polymorphism of the restriction endonuclease cleavage sites of distinct EV HPV isolates is illustrated for HPV-5 (Fig. 9), HPV-14 and HPV-17 (Fig. 6).

3. Evolutionary Relationships of EV HPV Genomes

Analysis of the relationships between EV HPVs and other HPV types, by liquid phase or blot hybridization experiments performed under stringent conditions (around T_m $-20°C$), has shown no or little cross-hybridization between EV-specific HPVs and the HPV types associated with lesions of the skin and the mucous membranes in the general population (Kremsdorf, 1980; Orth *et al.*, 1980; Pfister *et al.*, 1981, 1983; Green *et al.*, 1982; Ostrow *et al.*, 1982, 1983b; Kremsdorf *et al.*, 1982a, 1983, 1984; Brackmann *et al.*, 1983; Gassenmaier *et al.*, 1984). Blot hybridization experiments, performed under conditions of lower stringency (around T_m $-40°C$) which allow the detection of conserved sequences among papillomaviruses (Heilman *et al.*, 1980), have revealed regions of partial homology among EV HPVs and other HPV types. A much greater extent of cross-hybridization has been observed among EV-specific HPVs than between EV HPVs and the other HPV types, including HPV-3 and HPV-10 (Fig. 10) (M. Favre and G. Orth, unpublished observations).

The close evolutionary relationship between EV-specific HPVs has also been shown by the electron microscopic analysis of heteroduplex DNA molecules formed under various conditions of stringency (O. Croissant, N. Bonneaud, G. Pehau-Arnaudet and G. Orth, unpublished results;

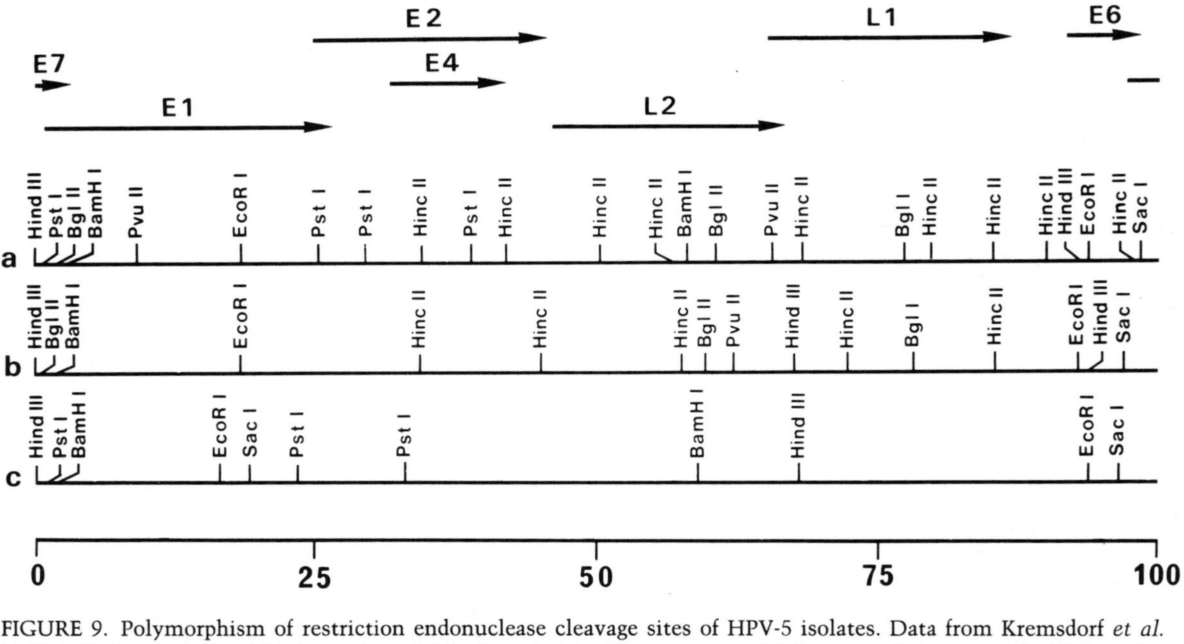

FIGURE 9. Polymorphism of restriction endonuclease cleavage sites of HPV-5 isolates. Data from Kremsdorf *et al.* (1982a), D. Kremsdorf and M. Favre (unpublished results) (*a*), Ostrow *et al.* (1982) (*b*), and Pfister *et al.* (1983) (*c*). The alignment of the maps with the map of the open reading frames of the HPV-8 genome (Fuchs *et al.*, 1986) has been deduced from the electron microscopic analysis of heteroduplex molecules formed between HPV-5 and HPV-8 DNAs (O. Croissant *et al.*, unpublished results).

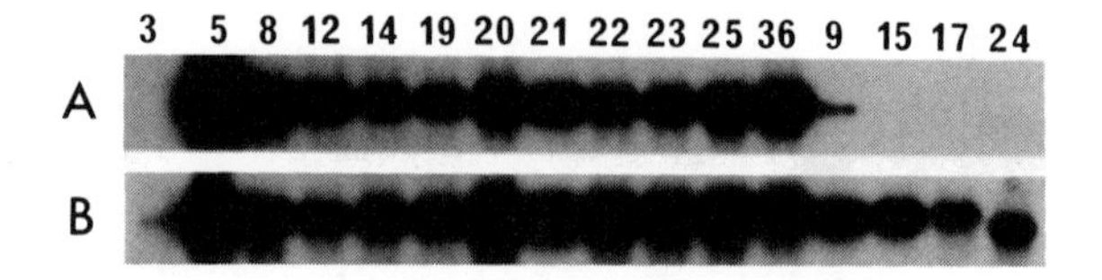

FIGURE 10. DNA sequence homology among EV HPV DNAs, as detected by blot hybridization. Blots of cloned HPV DNAs (5 ng) were hybridized with ^{32}P-labeled HPV-5 DNA in stringent conditions (T_m −20°C) (A) or nonstringent conditions (T_m −40°C) (B). Numbers indicate HPV types. Significant cross-hybridization in stringent conditions is only observed with the types belonging to the same group as HPV 5 (Table I). In nonstringent conditions, all EV-specific HPVs show a cross-hybridization while almost no cross-hybridization is observed with HPV-3, a type also associated with flat warts in the general population.

T. Broker and L. Chow, personal communication). For instance, at T_m −28°C, the HPV-5 genome forms heteroduplex molecules with the genomes of HPV-9 and HPV-24, the prototypes of the other EV-specific HPV groups, but not with those of HPV-1 and HPV-3. The paired regions are distributed throughout the genome, and correspond to about 33% of the genome length for HPV-5 and HPV-9 (Fig. 11a) and to about 50% for HPV-5 and HPV-24. Under the same conditions, heteroduplex molecules formed between the genomes of EV HPVs belonging to the same group show paired regions representing from about 60% of the genome length for

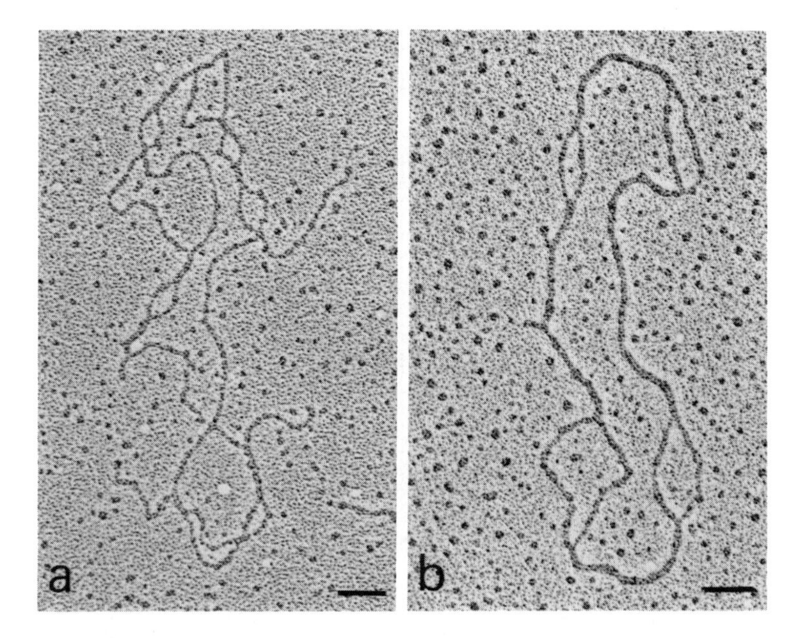

FIGURE 11. Heteroduplex molecules formed between the genomes of HPV-5 and HPV-9 (a) and the genomes of HPV-9 and HPV-15 (b) at T_m −28°C. Conditions were as previously described (Croissant *et al.*, 1982). Bars = 0.1μm.

HPV-9 and HPV-15 (Fig. 11b) to about 70% for HPV-5 and HPV-8. Regions of partial homology between HPV-1 or 3 and HPV-5, 9, and 24 have only been detected under conditions of lower stringency (T_m −41°C) (Croissant *et al.*, unpublished results).

4. Genome Organization of EV HPVs

The only available nucleotide sequence of an EV HPV is that of the HPV-8 genome (7654 bp) (Pfister *et al.*, 1985; Fuchs *et al.*, 1986). The general organization of the open reading frames (ORFs) is similar to that of the other papillomaviruses so far sequenced (Danos *et al.*, 1984; Seedorf *et al.*, 1985). The regions of greatest homology are observed in ORF L1, in the carboxy-terminal part of ORF E1, and in the amino- and carboxy-terminal parts of ORF E2, while the E2 region overlapping E4 is strikingly different. There are no ORFs E5 and E8. As for HPV-16, a classical RNA polymerase II promoter is located in front of ORF E2, and no start codon is found in ORF E4. E4 contains 4 times a direct repeat of nine base pairs, which is found 14 times in a region of the Epstein–Barr virus B95-8 genome corresponding to the EBNA 2 gene. The direct repeats are part of a sequence homology between the two genomes. The E4 protein of HPV-8 has a remarkably high proline content and shows three areas of significant homology with the EBNA 2 protein, the latter being probably involved in the immortalization of B lymphocytes (Pfister *et al.*, 1985; Fuchs *et al.*, 1986).

EV HPVs differ as to their oncogenic potential and the comparative analysis of their nucleotide sequences should contribute to the understanding of the molecular bases of their distinct biological properties. Heteroduplex analysis has revealed the existence of regions of conserved sequences, distributed throughout the whole genomes. The mapping of these regions of homology has allowed the alignment of the physical maps of the genomes of HPV types 1, 5, 8, 9, 14, 15, 17, and 24, and permitted the identification of the regions homologous to the ORFs localized on the HPV-1 and the HPV-8 genomes, as illustrated for HPV-5 (Fig. 9) (Croissant *et al.*, unpublished results). This has shown that the most variable regions of EV HPV genomes correspond to the noncoding region, ORFs E6 and E7, and, to a lesser extent, ORF E4 and the 3′ half of ORF L2.

C. HPV Infection in EV Patients

EV-specific HPVs have been cloned from lesions of patients throughout the world (Table I). The prevalence of EV-associated HPVs may be estimated from the study of 60 patients from Europe, Africa, and South America suffering from the typical EV form (Orth, 1986; G. Orth, S. Jablonska, L. A. Rueda, M. Lutzner, C. Blanchet-Bardon, W. Jacyk, P. Van Voorst Vader, M. F. Avril, S. Obalek, M. Favre, and O. Croissant, unpublished results). About one-third of the patients had familial EV or were the children of consanguineous parents, and half of them had or had previously had skin cancers. Most patients were found to be infected by more than one EV-specific HPV type (up to ten types) (Fig. 2), in agreement with previous studies (Orth *et al.*, 1978c, 1980; Kremsdorf *et al.*, 1982a, 1983, 1984; Ostrow *et al.*, 1983b; Pfister *et al.*, 1983; Gassenmaier *et al.*, 1984). The most frequently found EV HPVs were types 5, 8, 20, and 17, identified in 35, 20, 17, and 14 patients, respectively. The worldwide distribution of these HPV types is further shown by the identification of HPV types 17, 20, and 5 in 3, 2, and 2 patients, respectively, among 8 patients from a single district in Japan (M. Yutsudo, personal communication). HPV-36 also seems to be rather prevalent, since it has been detected in 7 of 18 patients (Kawashima *et al.*, 1986b). In contrast, some EV HPVs (HPV types 12, 15, 21, 23) have been detected in only one or two patients of this series. HPV-3, which is associated with flat warts in the general population, was detected in the flat wartlike lesions of 8 patients also infected with EV-specific HPVs. A double infection with HPV-3 and HPV-8 has also been reported in four familial EV cases (Androphy *et al.*, 1985). Three patients had HPV-2-associated common warts. None of the other HPVs associated with skin warts or with genital lesions were detected (Orth, 1986; G. Orth *et al.*, unpublished results).

Some patients have been diagnosed as suffering from EV on the basis of long-standing, disseminated flat warts (Jablonska and Orth, 1985). In a series of eight such patients, including the familial case discussed above (patient D, Fig. 2), none had or had previously had skin cancers, and none were found to be infected with an EV-specific HPV. HPV-3 was found in five patients, HPV-10 in one patient, and as yet uncharacterized HPVs related to both HPV-3 and HPV-10 in two patients (Orth *et al.*, 1978c, 1979, 1980; Jablonska *et al.*, 1979b; Kremsdorf *et al.*, 1983; Jablonska and Orth, 1985; Orth, 1986). A similar case has been reported by Ostrow *et al.* (1983b).

From these data, it is clear that EV patients display an abnormal susceptibility to a specific set of HPV types and, sometimes, to types associated with flat warts in the general population.

VII. ROLE OF HPVs IN EV CARCINOMAS

A. HPV Types Found in EV Cancers

According to the published reports, one-third of all EV patients develop multiple skin carcinomas (Lutzner, 1978) but, as mentioned earlier, this is probably an underestimation. Early studies had suggested that patients infected with HPV-5 or HPV-5-related types had a much higher risk of developing cancer than patients infected with HPV-3 or HPV-3-related types (Orth *et al.*, 1978c, 1979). The detection of HPV-5 genomes in two carcinomas of a Polish patient (Orth *et al.*, 1980), in a primary carcinoma and a metastatic tumor of two Japanese patients (Ostrow *et al.*, 1982), and in a primary carcinoma of a Turkish patient (Pfister *et al.*, 1983) strongly suggested a specific role of this HPV type in the development of EV carcinomas.

The association of some other HPV types with EV carcinomas has been suggested by the detection of HPV-3-related sequences (most probably HPV-10) in a vulvar bowenoid *in situ* carcinoma (Green *et al.*, 1982) and, more recently, that of HPV-17 genomes in a skin cancer (Yutsudo *et al.*, 1985). However, a study of 28 carcinomas, including a metastatic tumor, obtained from 14 EV patients from Europe, Africa, and South America, has confirmed the predominant association of EV carcinomas with HPV-5 (Table II) (Orth, 1986; G. Orth *et al.*, unpublished results). Eight of the patients were shown to be infected with at least three HPV types. The types most frequently found in benign lesions were HPV-5 (12 patients), HPV-8 and HPV-20 (5 patients), HPV-17, HPV-14, and HPV-24 (3 patients). HPV genomes were found in all but one tumor: HPV-5 in 21 cancers (12 patients) including the metastatic tumor, HPV-8 in 5 cancers (2 patients), and HPV-14b in 1 cancer. One patient had a tumor associated with HPV-5 and another one associated with HPV-8. The negative tumor was from a patient infected with six HPV types, including HPV-5 and HPV-8, who also had a cancer containing HPV-5 genomes (Orth, 1986; G. Orth *et al.*, unpublished results). It should be stressed that HPV types 5, 8, and 14 present cross-hybridization levels of 10–15% and belong to the same group of EV HPVs (Kremsdorf *et al.*, 1984). In contrast to the multiplicity of HPVs infecting the patients, only trace amounts of a second HPV type were occasionally found in the tumors (Pfister *et al.*, 1983; Orth, 1986; G. Orth *et al.*, unpublished results).

B. Physical State of HPV Genomes

HPV genomes are usually detected in EV cancers in high copy numbers (100–300 copies per diploid amount of host cell DNA) (Orth *et al.*,

TABLE II. HPV Types Associated with EV Carcinomas

References[a]	No. of cases	No. of HPV types per patient	No. of tumors	No. of tumors with DNA sequences of HPV-5	HPV-8	HPV-10	HPV-14	HPV-17	Negative tumors
1	14	1 to 11	28	21	5	—	1	—	1
2	1	1	1	—	—	1	—	—	—
3	2	1	2	2	—	—	—	—	—
4	1	5	1	1	—	—	—	—	—
5	1	1	1	—	—	—	—	1	—

[a] 1, Orth *et al.* (1980), Orth (1986), Van Voorst Vader (1986), G. Orth *et al.* (unpublished results); 2, Green *et al.* (1982); 3, Ostrow *et al.* (1982); 4, Pfister *et al.* (1983); 5, Yutsudo *et al.* (1985).

1980; Pfister *et al.*, 1983; Yutsudo *et al.*, 1985; Orth, 1986). *In situ* hybridization experiments indicate that a vegetative replication of HPV-5 DNA may occur in a few cells (Orth *et al.*, 1980; Orth, 1986), and this may account for the high average copy number of HPV genomes in cancers. HPV-5 and HPV-8 genomes have been detected as free molecules, mostly as monomers like in benign lesions, but with variable proportions of oligomers (Fig. 12) (Orth *et al.*, 1980; Ostrow *et al.*, 1982; Pfister *et al.*, 1983; Yutsudo *et al.*, 1985; Orth, 1986; G. Orth *et al.*, unpublished results). Blot hybridization experiments, carried out after one- or two-dimensional agarose gel electrophoresis of total cellular DNA, either untreated or treated with a mixture of restriction endonucleases that do not cleave the viral DNA, have shown no evidence for the integration of viral sequences into cellular DNA, with the exception of one tumor associated with HPV-14 where the viral sequences migrated with the bulk of the cellular DNA and were probably integrated in it (Orth, 1986). HPV-5 and HPV-8 genomes showing deletions have been detected in variable proportions in some primary tumors (Fig. 12) (Ostrow *et al.*, 1982; Orth, 1986; G. Orth *et al.*, unpublished results). HPV-5 genomes with deletions

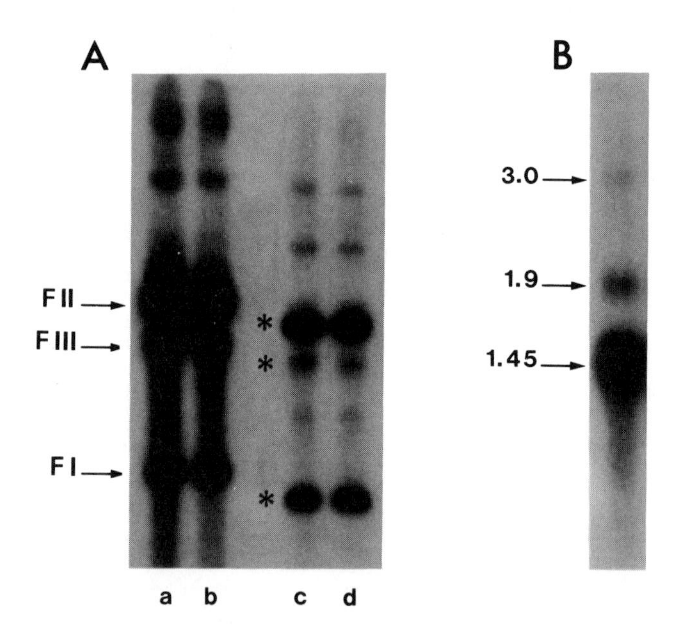

FIGURE 12. Physical state and transcription of HPV-5 genome in EV carcinomas, as analyzed by blot hybridization (G. Orth *et al.*, unpublished results). (A) HPV-5 genomes were detected in untreated total cell DNA obtained from two EV carcinomas as free monomeric molecules, intact (FI, FII, FIII) (a) or deleted (*) (c), and as bands with slower migrations. The latter bands correspond to free oligomers since their migration is unchanged after treatment with a mixture of no-cut enzymes for HPV-5 DNA (*Pvu*I, *Sma*I) (b, d). (B) HPV-5-specific transcripts were detected in a total RNA preparation obtained from an EV carcinoma. Sizes are expressed in kilobases.

have also been observed in two metastatic tumors (Ostrow *et al.*, 1982; G. Orth *et al.*, unpublished results), as the only species of HPV-5 DNA molecules in one case (G. Orth *et al.*, unpublished results). In the latter tumors, the deletions have been shown to involve a region of the HPV-5 genome corresponding to the ORFs of the late region (Watts *et al.*, 1985; G. Orth *et al.*, unpublished results).

C. Expression of HPV Genomes

Viral particles have only been irregularly detected in early malignant EV lesions and never observed in invasive EV cancers (Aaronson and Lutzner, 1967; Jablonska *et al.*, 1970; Ruiter and Van Mullem, 1970a; Grupper *et al.*, 1971; Delescluse *et al.*, 1972; Yabe and Sadakane, 1975; Rueda and Rodriguez, 1976; Yabe *et al.*, 1978; Orth *et al.*, 1980; Lutzner *et al.*, 1984). Vegetative replication of HPV-5 DNA has been demonstrated in a few cells of EV carcinomas by light and electron microscopy, after *in situ* hybridization (Orth *et al.*, 1980; Orth, 1986). No viral particles have been observed in the nuclei where the hybrids were detected by electron microscopy (Orth *et al.*, 1980). This indicates that viral replication may be initiated in a few carcinoma cells but remains abortive, probably because of an altered terminal differentiation process (Orth *et al.*, 1977b).

Evidence has recently been obtained for the transcription of HPV-5 genomes in EV carcinomas. HPV-5 transcripts, essentially two species of 1450 and 1900 bp (Fig. 12), have been demonstrated in a metastatic carcinoma and a primary carcinoma arising later in the same EV patient. Both transcripts hybridized with a subgenomic DNA fragment of HPV-5 corresponding to ORFs E6 and 7 (G. Orth *et al.*, unpublished results).

This suggests that the continuous expression of the HPV-5 genome is required during the whole process of malignant transformation of HPV-5-infected keratinocytes in EV patients. The putative transforming protein(s) remains to be identified.

VIII. HPVs AND SKIN NEOPLASIA IN NON-EV PATIENTS

The association of HPV-5 and HPV-5-related types with intraepithelial and invasive carcinomas which develop in EV patients raises the question of the role of these HPVs in the origin of premalignant and malignant skin lesions in patients known to be at risk for skin cancer and in the general population.

A. Immunosuppression or Immunodepression

Patients with genetic or acquired depression or suppression of CMI have an increased incidence of warts and skin cancers (Penn, 1980; Jablonska *et al.*, 1982a; Lutzner, 1985). HPV-5-related sequences have been found in the lesions of a patient suffering from common variable immunodeficiency and showing a generalized lichenoid papular eruption (Goldes *et al.*, 1984). HPV-17, an EV-specific HPV, and HPV-38, a type cross-hybridizing with some EV HPVs, have been found in a malignant melanoma of a patient suffering from chronic lymphatic leukemia and under immunosuppressive therapy, but were not detected in 34 other melanomas (Scheurlen *et al.*, 1986). An HPV type unrelated to EV HPVs, HPV-26, was characterized first in the common warts of an immunodeficient patient with a history of multiple squamous cell carcinomas (Ostrow *et al.*, 1984), but no further information has been available on the pathogenicity of HPV-26.

The best documented example of patients at risk for warts and premalignant or malignant lesions of the skin are renal allograft recipients who have been iatrogenically immunosuppressed to prevent rejection of kidney transplants. The frequency of warts has been reported to be 43% in a series of patients surveyed from 3 months to 9 years after transplantation (Koranda *et al.*, 1974), and 64% after continuous immunosuppression for 6 to 13 years (Spencer and Andersen, 1979). The skin cancers in kidney transplant recipients are often multiple and usually located in light-exposed areas, including the vermilion border of the lips. They are more aggressive than in the general population and form metastases more readily (Hardie *et al.*, 1980; Penn, 1980). The tumor frequency varies with the amount of sun exposure (Penn, 1980; Boyle *et al.*, 1984) and with the duration of the immunosuppression (Marshall, 1974; Hardie *et al.*, 1980). The overall incidence of skin cancers in renal transplant recipients has been reported to vary between 1.4% in a temperate region of America (Hoxtell *et al.*, 1977) and 8.6% for Caucasians living in a subtropical climate (Hardie *et al.*, 1980). This represents a 7- to 20-fold increase over the incidence in the general population (Hoxtell *et al.*, 1977; Hardie *et al.*, 1980). In a series with a 5% overall incidence of skin cancers, 17% of the patients were involved after 4 years of treatment and 28% showed premalignant keratoses (Marshall, 1974). In a series of patients living in a subtropical climate, the incidence of skin cancers increased by 5% per year after the first year, with a cumulative risk of 44% after 9 years (Hardie *et al.*, 1980). The increased incidence occurs mainly, or exclusively, among squamous cell carcinomas (Mullen *et al.*, 1976; Hoxtell *et al.*, 1977; Kinlen *et al.*, 1979; Hardie *et al.*, 1980), with a reported 5-fold (Kinlen *et al.*, 1979) to 36-fold (Hoxtell *et al.*, 1977) higher incidence. The ratio of basal to squamous cell carcinomas has been reported to reverse from 4 : 1 in the general population to 1 : 1.7 (Hardie *et al.*, 1980) or 1 : 11 (Mullen *et al.*, 1976) in renal allograft recipients. Some of these patients have widespread warts and skin cancers and may be considered as presenting an EV-like syndrome (Koranda *et al.*, 1974; Mullen *et al.*, 1976; Spencer and Andersen, 1979; Lutzner *et al.*, 1980, 1983).

There is little evidence for the infection of renal allograft recipients by EV HPVs. The warts observed in such patients are usually flat warts or common warts, which have been found associated with the same HPV types found in similar lesions in the general population (Pfister *et al.*, 1979, 1985; Rudlinger *et al.*, 1986; S. Jablonska, S. Obalek, M. Favre, and G. Orth, unpublished results). EV HPV antigens have been detected in a wartlike lesion of a patient (Lutzner *et al.*, 1980), and EV HPV-related sequences in a common wart of a patient in another series (Pfister *et al.*, 1985). Pityriasis versicolorlike macules, resembling EV lesions, have been reported in two renal transplant recipients and contained EV HPV antigens and/or HPV-5 genomes (Lutzner *et al.*, 1980, 1983; Rudlinger *et al.*, manuscript submitted). HPV-5 DNA sequences have also been detected in a carcinoma *in situ* and in a microinvasive squamous cell carcinoma of one of these patients (Lutzner *et al.*, 1983). This constitutes the only available evidence for a role of EV HPVs in the cutaneous squamous cell carcinomas of renal allograft recipients.

B. General Population

Several studies have been aimed at evaluating the role of HPVs in the etiology of benign neoplasms of the skin (keratoacanthomas), premalignant skin lesions (solar keratosis, Bowen's disease), and cutaneous basal and squamous cell carcinomas in the general population (Green *et al.*, 1981; Pfister *et al.*, 1985; Kawashima *et al.*, 1986a,b; Scheurlen *et al.*, 1986; M. Kawashima, S. Jablonska, M. Favre, S. Obalek and G. Orth, unpublished results).

Keratoacanthomas are rapidly growing, benign neoplasms, originating from hair follicles, that may be self-limited but can also cause local destruction (Lever and Schaumburg-Lever, 1975; Sanderson and Mackie, 1979). EV HPV-related DNA sequences have been detected in 2 of a series of 5 keratoacanthomas (Pfister *et al.*, 1985). The genomes of HPV-9, an EV-specific HPV, and HPV-37, a type related to HPV types 9, 15, 17, have both been found associated with 1 case, in a series of 6 keratoacanthomas (Scheurlen *et al.*, 1986). No HPV DNA sequences were detected in another series of 21 lesions (Kawashima *et al.*, 1986a,b; M. Kawashima *et al.*, unpublished results).

Basal cell carcinomas or basaliomas are slow-growing, locally invasive, rarely metastasizing tumors, derived from basal cells of the epidermis or of the epidermal appendages (Lever and Schaumburg-Lever, 1975; Sanderson and Mackie, 1979). The genome of an EV HPV, HPV-20, has been detected in a single case, in a series of 44 biopsies of basaliomas (M. Kawashima *et al.*, unpublished results).

Solar (actinic) keratosis and Bowen's disease of the skin are considered as precursors of squamous cell carcinomas (Lever and Schaumburg-Lever,

1975; Sanderson and Mackie, 1979). A possible role of HPVs in the genesis of solar keratosis has been suggested by the observation of papillomavirus-like particles in 1 of 4 cases (Spradbrow *et al.*, 1983). EV HPV-related sequences have been detected in solar keratoses of 2 of 30 patients (Kawashima *et al.*, 1986b). One of these HPV genomes has further been characterized as HPV-36, which shows 30% DNA cross-hybridization with HPV-5. HPV-36 was subsequently found in the benign lesions of 7 of 18 EV patients and should be considered as an EV-specific HPV (Kawashima *et al.*, 1986b). Lesions of Bowen's disease of the skin are *in situ* squamous cell carcinomas with characteristic histological features (Lever and Schaumburg-Lever, 1975). Although EV carcinomas usually have bowenoid features (Rueda and Rodriguez, 1976, 1983), no evidence has been found for the association of EV HPVs with Bowen's disease of the skin. HPV-16 DNA sequences, usually found in genital Bowen's disease and genital carcinomas (Dürst *et al.*, 1983; Ikenberg *et al.*, 1983; Beaudenon *et al.*, 1986), have been detected in 2 of 5 cases of extragenital Bowen's disease (Ikenberg *et al.*, 1983). The DNA of HPV-2, a type usually associated with common warts (Orth *et al.*, 1977a), has been demonstrated in a single case (Pfister and Haneke, 1984). In another series, HPV DNA sequences, weakly related to HPV-16, have been detected in only 1 of 35 specimens, and characterized as HPV-34, a type subsequently found in one case of penile bowenoid papulosis (Kawashima *et al.*, 1986a; M. Kawashima *et al.*, unpublished results). This suggests that Bowen's disease of the skin and genital Bowen's disease have different causes.

Squamous cell carcinomas are true invasive carcinomas of the epidermis (Lever and Schaumburg-Lever, 1975; Sanderson and Mackie, 1979). HPV-16 DNA was detected in 1 of 10 tumors (Pfister *et al.*, 1985), and HPV DNA sequences, cross-hybridizing weakly with HPV-5 and HPV-16 under nonstringent conditions, were found in 1 of 31 tumor samples (M. Kawashima *et al.*, unpublished results).

In conclusion, the available data bring little support to the role of HPVs in the genesis of premalignant and malignant lesions of the skin in the general population, pointing to the role of genetic factors in the control of HPV infection in EV. It cannot be ruled out that in the few lesions where they have been detected, EV HPVs or related types were travelers rather than causative agents. This is suggested by the lack of detection of HPV-36 genomes in five of seven actinic keratoses and a squamous cell carcinoma of the patient where this HPV type was first identified (Kawashima *et al.*, 1986b). HPV-37 was not detected in a basalioma of the patient from whom it was isolated (Scheurlen *et al.*, 1986). Among the factors which could be involved in the origin of nonmelanoma skin cancers (Dunham, 1972), a major role is assigned to UV radiation in the sunburn spectrum (290–320 nm UVB) (Urbach, 1982; Boyle *et al.*, 1984). However, it remains possible that premalignant and malignant cutaneous lesions may be associated with as yet unrecognized HPVs which may not have been detected with the available HPV DNA probes.

IX. *IN VITRO* STUDIES ON EV HPVs

The available data on the pathogenicity of EV HPVs point to a strict epitheliotropism, restricted to the epidermis. The expression of the late events of the life cycle of EV HPVs (vegetative viral DNA replication, synthesis of viral structural components) depends on the expression of the terminal differentiation of the host cell, and the expression of their oncogenic potential involves, most probably, a multistep process requiring many years. This may explain why the *in vitro* replication of EV HPVs and the expression of their oncogenic potential in cultured keratinocytes have not yet been obtained. This is a methodological constraint common to all HPVs (Taichman *et al.*, 1984). Furthermore, cell lines derived from EV cancers have not been obtained.

An alternative approach has been to introduce EV HPV genomes into cell lines of nonhuman origin and not deriving from epidermis. Some evidence has been presented for the maintenance and the expression of the viral genomes in such cells (Brackmann *et al.*, 1983; Watts *et al.*, 1984). Cotransformation experiments of mouse Ltk^- cells have been performed with cloned HPV-3 or HPV-9 genomes and the herpesvirus thymidine kinase gene taken as a selectable marker. Transformed tk^+ cells were shown to contain multiple high-molecular-weight inserts of HPV DNA in high copy numbers and HPV-3-specific transcripts have been demonstrated (Brackmann *et al.*, 1983). Molecularly cloned HPV-5 DNA has been reported to induce a morphological transformation of mouse C127 cells. The transformed cells contained persistent episomal copies of HPV-5 DNA in low copy numbers, and showed anchorage-independent growth and tumorigenicity in nude mice (Watts *et al.*, 1984). Naturally occurring mutant HPV-5 genomes with deletions in the late region retained the ability to transform C127 cells and to persist as episomes (Watts *et al.*, 1985). Further studies should confirm the relevance of this approach to localize the transforming functions within the genome of potentially oncogenic EV HPVs and to identify the products of EV HPV transforming genes.

X. QUESTIONS RAISED BY EV

EV raises general questions about the plurality of HPVs, the factors controlling HPV infection, those conditioning the expression of their oncogenic potential, and the molecular mechanisms involved in the development of HPV-associated cancers.

A. Plurality of HPVs

EV provides the best illustration of the plurality of HPVs, since at least 15 of the HPV types identified are specifically associated with the disease. This great diversity does not reflect a genetic instability since the HPV types, subtypes, and variants associated with EV are well-defined, stable entities found worldwide (Orth *et al.*, 1980; Pfister *et al.*, 1981; Kremsdorf *et al.*, 1982b, 1983, 1984; Ostrow *et al.*, 1982; Yutsudo *et al.*, 1982, 1985; Gassenmaier *et al.*, 1984). There is no clear evidence that the diversity of EV HPVs is generated by recombination between genomes of different types (Kremsdorf *et al.*, 1984; O. Croissant *et al.*, unpublished results). The accumulation of point mutations and DNA rearrangements, such as deletions, insertions, or duplications, are more likely to have caused the diversification of papillomaviruses (Danos *et al.*, 1984; Danos, 1986). The genome of HPV-8, the only EV HPV DNA sequenced so far (Fuchs *et al.*, 1986), appears more related to the genomes of epidermotropic PVs than to those of genital HPVs and animal PVs inducing fibropapillomas (Fuchs *et al.*, 1986; Danos, 1986). The elucidation of the nucleotide sequences of other EV HPV genomes may shed some light on the molecular basis of their specific biological properties. An intriguing question is whether EV condition favored the emergence of a specific set of related HPVs, and, if so, which mechanisms are involved.

B. Control of HPV Infection

The EV HPVs seem harmless in the general population and their reservoir remains unknown. The detection of EV HPVs and related types in single cases of premalignant and malignant skin lesions indicates that infection with these HPV types occurs in non-EV patients (Lutzner *et al.*, 1983; Pfister *et al.*, 1985; Kawashima *et al.*, 1986b; Scheurlen *et al.*, 1986). Infection may be inapparent or may provoke inconspicuous macular lesions. Such lesions could be overlooked by patients and only exceptionally recognized and reported by dermatologists (Johnson *et al.*, 1962; Cornelius *et al.*, 1968; Lutzner *et al.*, 1980). Reservoirs other than keratinocytes may be considered in view of the detection of HPV-17 in one melanoma (Scheurlen *et al.*, 1986) and of HPV-5-related sequences in the semen of an EV patient and his healthy son (Ostrow *et al.*, 1986). Although EV patients show widespread infection and must shed large amounts of viral particles, their spouses have never been reported to be infected (Jablonska and Orth, 1985). Worth mentioning, however, is the report of the onset of EV in two patients allegedly after their brother rubbed a crushed wart removed from his finger into their skin (Jablonska and Milewski, 1957), successful heteroinoculation experiments using extracts from EV lesions (Lutz, 1946; Jablonska and Formas, 1959), and the report of a hand wart with histological features of EV lesions in two individuals who had handled biopsies from an EV patient (Duverne *et al.*, 1976).

EV patients can be considered as having an abnormal genetic predisposition to infection by specific HPV types and, in some instances, by HPVs associated with flat warts in the general population. The abnormal alleles of the recessively transmitted "EV genes" could be responsible for (1) an abnormal susceptibility of keratinocytes to infection by specific HPV types, (2) a specific alteration of the systemic or local immune response of the host, (3) a higher risk of malignant transformation of infected keratinocytes. The abnormal products of the recessive "EV genes" have not been discovered yet. The analysis of restriction-fragment-length polymorphisms as genetic markers of the disease (Botstein *et al.*, 1980) in well-documented EV families should help to better define the genetics of EV.

Another example of a genetic predisposition to HPV infection is focal epithelial hyperplasia of the oral mucosa, specifically associated with HPV-13 and HPV-32 (Beaudenon *et al.*, 1987). The disease is observed with a high frequency among Eskimos and Amerindians but only rarely in the rest of the world population. A possible role of genetic factors in other HPV infections should be considered, in particular for HPV types which constitute a risk factor for the development of genital neoplasia.

C. Cofactors in HPV Oncogenesis

The predominant association of EV carcinomas with HPV-5 (and, to a lesser extent, with some other EV HPVs) contrasts with the great diversity of HPVs observed in benign lesions. This points to this specific HPV type as a major risk factor for the development of EV carcinomas. Most EV patients have been reported to have an impaired CMI but it is not clear whether this deficiency is a primary one or is secondary to the chronic HPV infection. Anyway, deficient cutaneous or systemic immune functions are likely to play a role in the lifelong HPV-5 infection of EV patients, and in the uncontrolled growth of EV carcinomas. The transformation of HPV-5-associated lesions is a rare event which probably involves additional factors. The delay between the onset of infection and the development of invasive carcinomas, through carcinomas *in situ*, points to a multistage process. The preferential localization of cancers in light-exposed sites suggests that the UV radiations of the sun, a known important factor in skin carcinogenesis (Urbach, 1982), play a role in the malignant conversion of EV lesions (Schellander and Fritsch, 1970; Ruiter, 1973). Whether or not EV patients have defective DNA repair mechanisms remains controversial (Moreno and Pruniéras, 1974; Hammar *et al.*, 1976; Kienzler *et al.*, 1979; Proniewska and Jablonska, 1980). Based on the study of rodent models, UV radiations in addition to their known mutagenic effect, may also favor the growth of skin cancers by an immunosuppressive effect mediated by the induction of specific suppressor T cells (Fisher and Kripke, 1982; Streilen, 1983, 1985; Kripke and Morrison, 1985). However, EV cancers may sometimes appear in nonexposed areas, like the scrotum and the palm (Lutzner *et al.*, 1984), and this suggests that factors other than sunlight are involved in the progression of EV lesions.

The association with specific HPV types, the long delay required for tumor progression, and the involvement of cofactors are common features to the other known examples of genital and extragenital HPV-associated carcinomas (zur Hausen and Schneider, this volume).

D. Mechanisms of Tumor Progression

The mechanisms involved in the development of EV carcinomas are far from being understood. Tumor-derived cell lines, which proved useful for the study of genital HPVs (Schwarz *et al.*, 1985; Yee *et al.*, 1985), are not available for EV HPVs. The best model for EV is the Shope papillomatosis of the rabbit. As in EV, genetic, immune, and extrinsic factors play a role in the persistence of the skin warts induced by the Shope cottontail rabbit papillomavirus (CRPV), and in their progression toward non-virus-producing carcinomas (Rous and Beard, 1935; Rous and Friedewald, 1944; Rogers and Rous, 1951; Syverton, 1952). In both EV and Shope carcinomas, the viral genomes are generally maintained as monomeric or oligomeric episomes in high copy numbers, and integration in cellular DNA is rather infrequent (Orth *et al.*, 1980; Ostrow *et al.*, 1982; Wettstein and Stevens, 1982; Pfister *et al.*, 1983; Georges *et al.*, 1984; Orth, 1986). The expression of ORFs E6 and E7 of the CRPV genome seems to be required for both the formation of a wart and the development of an invasive cancer (Georges *et al.*, 1984; Nasseri and Wettstein, 1984; Danos *et al.*, 1985). HPV-5-specific transcripts containing E6 and E7 sequences have been identified in EV carcinomas (G. Orth *et al.*, unpublished results).

EV carcinomas differ from carcinomas of the uterine cervix in that integration of HPV-16 and HPV-18 DNA sequences into the cellular genome appears to be the rule in the latter case (Boshart *et al.*, 1984; Dürst *et al.*, 1985; Schwarz *et al.*, 1985). The insertion site is usually located in a region of the viral DNA corresponding to ORFs E1 and E2 and ORFs E6 and E7 may be expressed as transcripts containing both viral and cellular sequences (Dürst *et al.*, 1985; Schwarz *et al.*, 1985; Beaudenon *et al.*, 1986). This suggests that the mechanisms involved in malignant conversion depend on the type of HPV and/or on the tissue involved. Evidence has recently been obtained for the activation of the oncogene c-*myc*, at least, in the late stages of the development of cervical carcinomas (Riou *et al.*, 1985). No data are as yet available on the nature of the cellular genes activated or inactivated during the development of EV carcinomas.

In conclusion, EV, which is a rare experiment of nature, provides an outstanding model for the study of the interactions between potentially oncogenic HPVs and genetic, immunological, and extrinsic factors in the production of human cancers.

ACKNOWLEDGMENTS. I wish to dedicate this review to Professor Stefania Jablonska. The contribution of many dermatologists, especially Dr. L. A.

Rueda, Dr. C. Blanchet-Bardon, Dr. M. Lutzner, Dr. W. Jacyk, and Dr. P. Van Voorst Vader, is acknowledged. I am grateful to Dr. H. Pfister for the communication of sequence data prior to publication. I want to thank particularly Odile Croissant, Michel Favre, and Françoise Breitburd for their help and fruitful discussions and B. Rubat du Mérac for her expert assistance during the preparation of this manuscript.

REFERENCES

Aaronson, C. M., and Lutzner, M. A., 1967, Epidermodysplasia verruciformis and epidermoid carcinoma: Electron microscopic observations, *J. Am. Med. Assoc.* **201:**775–777.

Androphy, E. J., Dvoretzky, I., Maluish, A. E., Wallace, H. J., and Lowy, D. R., 1984, Response of warts in epidermodysplasia verruciformis to treatment with systemic and intralesional alpha interferon, *J. Am. Acad. Dermatol.* **11:**197–202.

Androphy, E. J., Dvoretzky, I., and Lowy, D. R., 1985, X-linked inheritance of epidermodysplasia verruciformis: Genetic and virologic studies of a kindred, *Arch. Dermatol.* **121:**864–868.

Baker, H., 1968, Epidermodysplasia verruciformis with electron microscopic demonstration of virus, *Proc. R. Soc. Med.* **61:**7–8.

Beaudenon, S., Kremsdorf, D., Croissant, O., Jablonska, S., Wain-Hobson, S., and Orth, G., 1986, A new type of human papillomavirus associated with genital neoplasias, *Nature* **321:**246–249.

Beaudenon, S., Praetorius, F., Kremsdorf, D., Lutzner, M., Worsaae, N., Pehau-Arnaudet, G., and Orth, G., 1987, A new type of human papillomavirus associated with oral focal epithelial hyperplasia, *J. Invest. Dermatol.* **88** (in press).

Bilancia, A., 1961, Considerazioni su una sindrome dermatologica molto discussa: la epidermodisplasia verruciforme di Lewandowsky e Lutz, *Arch. Ital. Dermatol.* **32:**425–450.

Blanchet-Bardon, C., and Lutzner, M. A., 1985, Interferon and retinoid treatment of warts, *Clin. Dermatol.* **3**(4):195–199.

Blanchet-Bardon, C., Puissant, A., Lutzner, M., Orth, G., Nutini, M. T., and Guesry, P., 1981, Interferon treatment of skin cancer in patients with epidermodysplasia verruciformis, *Lancet* **1:**274.

Bloc, G., Faye, I., Basset, A., Privat, Y., Ruscher, H., Camain, R., and Sarrat, H., 1971, Epidermodysplasie verruciforme de Lewandowsky–Lutz chez des Noirs Africains, *Bull. Soc. Med. Afr. Noire Lang. Fr.* **16:**558–590.

Boshart, M., Gissmann, L., Ikenberg, H., Kleinheinz, A., Scheurlen, W., and zur Hausen, H., 1984, A new type of papillomavirus DNA, its presence in genital cancer biopsies and in cell lines derived from cervical cancer, *EMBO J.* **3:**1151–1157.

Botstein, D., White, R. L., Skolnick, M., and Davis, R. W., 1980, Construction of a genetic linkage map using restriction fragment length polymorphisms, *Am. J. Hum. Genet.* **32:**314–331.

Boyle, J., Mackie, R. M., Briggs, J. D., Junor, B. J. R., and Aitchison, T. C., 1984, Cancer, warts, and sunshine in renal transplant patients: A case-control study, *Lancet* **1:**702–705.

Brackmann, K. H., Green, M., Wold, W. S. M., Rankin, A., Loewenstein, P. M., Cartas, M. A., Sanders, P. R., Olson, K., Orth, G., Jablonska, S., Kremsdorf, D., and Favre, M., 1983, Introduction of cloned human papillomavirus genomes into mouse cells and expression at the RNA level, *Virology* **129:**12–24.

Bundino, S., Zina, A. M., and Bernengo, M. G., 1978, Dyskeratosis congenita with epidermodysplasia verruciformis of Lewandowsky and Lutz, *Dermatologica* **156:**15–23.

Califano, A., and Caputo, R., 1968, La microscopia elettronica nello studio delle precancerosi epiteliali cutanee, *Minerva Dermatol.* **43:**567–616.

Claudy, A. L., Touraine, J. L., and Mitanne, D., 1982, Epidermodysplasia verruciformis induced by a new human papillomavirus (HPV-8): Report of a case without immune dysfunction. Effect of treatment with an aromatic retinoid, *Arch. Dermatol. Res.* **274:**213–219.

Coggin, J. R., Jr., and zur Hausen, H., 1979, Workshop on papillomaviruses and cancer, *Cancer Res.* **39:**545–546.

Cornelius, C. E., Witkowski, J. A., and Wood, M. G., 1968, Viral verruca, human papovavirus infection: Epidermodysplasia verruciformis, vacuolar degeneration of the epidermis, *Arch. Dermatol.* **98:**377–384.

Croissant, O., Testanière, V., and Orth, G., 1982, Mise en évidence et localisation de régions conservées dans les génomes du papillomavirus humain la et du papillomavirus bovin 1 par analyse de molécules "hétéroduplexes" au microscope électronique, *C. R. Acad. Sci.* **294:**581–586.

Croissant, O., Breitburd, F., and Orth, G., 1985, Specificity of cytopathic effect of cutaneous human papillomaviruses, *Clin. Dermatol.* **3**(4)**:**43–55.

Danos, O., 1986, Architecture et fonctionnement du génome des papillomavirus, Thèse de Doctorat, Université Paris 7.

Danos, O., Giri, I., Thierry, F., and Yaniv, M., 1984, Papillomavirus genomes: Sequences and consequences, *J. Invest. Dermatol.* **83:**7s–11s.

Danos, O., Georges, E., Orth, G., and Yaniv, M., 1985, Fine structure of the cottontail rabbit papillomavirus mRNAs expressed in the transplantable VX2 carcinoma, *J. Virol.* **53:**735–741.

Delescluse, C., Pruniéras, M., Régnier, M., Arouete, J., and Grupper, C., 1970, Incorporation de thymidine tritiée dans les verrues vulgaires, les papillomes cornés, et l'épidermodysplasie verruciforme, *Ann. Dermatol. Syphiligr.* **97:**525–533.

Delescluse, C., Pruniéras, M., Régnier, M., Moreno, G., and Arouete, J., 1972, Epidermodysplasia verruciformis. I. Electron microscope autoradiography and tissue culture studies, *Arch. Dermatol. Forsch.* **242:**202–215.

Delescluse, C., Régnier, M., and Pruniéras, M., 1973, Epidermodysplasia verruciformis. II. Uridine incorporation, *Arch. Dermatol. Forsch.* **247:**89–97.

Depaoli, M., 1971, Epidermodisplasia verruciforme con degenerazione bowenoide et epiteliomatosa multipla, baso e spinocellulare, *Minerva Dermatol.* **46:**407–417.

Dunham, L. J., 1972, Cancer in man at site of prior benign lesion of skin or mucous membrane: A review, *Cancer Res.* **32:**1359–1374.

Dürst, M., Gissmann, L., Ikenberg, H., and zur Hausen, H., 1983, A papillomavirus DNA from a cervical carcinoma and its prevalence in cancer biopsy samples from different geographic regions, *Proc. Natl. Acad. Sci. USA* **80:**3812–3815.

Dürst, M., Kleinheinz, A., Hotz, M., and Gissmann, L., 1985, The physical state of human papillomavirus type 16 DNA in benign and malignant genital tumors, *J. Gen. Virol.* **66:**1515–1522.

Duverne, J., Pruniéras, M., Mazauric, F. X., and Vial, A. M., 1973, Epidermodysplasie verruciforme: étude clinique, histologique (microscopie électronique) et autoradiographique à propos d'un cas, *Lyon Med.* **229:**567–571.

Duverne, J., Pruniéras, M., Mazauric, F. X., and Vial, A. M., 1976, Quelques réflexions à propos d'un cas d'épidermodysplasie verruciforme, *Bull. Soc. Fr. Dermatol. Syphiligr.* **83:**304–306.

Fisher, M. S., and Kripke, M. L., 1982, Suppressor T lymphocytes control the development of primary skin cancers in ultraviolet irradiated mice, *Science* **216:**1133–1134.

Fuchs, P. G., Iftner, T., Weninger, J., and Pfister, H., 1986, Epidermodysplasia verruciformis-associated human papilloma virus 8: Genomic sequence and comparative analysis, *J. Virol.* **58:**626–634.

Gassenmaier, A., Lammel, M., and Pfister, H., 1984, Molecular cloning and characterization of the DNAs of human papillomaviruses 19, 20, and 25 from a patient with epidermodysplasia verruciformis, *J. Virol.* **52:**1019–1023.

Gavanou, J., Gambert, I., Lambert, D., Laurent, R., Agache, P., and Chapuis, J. L., 1976, Aspects cliniques d'épidermodysplasie verruciforme avec déficit congénital en IgM, *Bull. Soc. Fr. Dermatol. Syphiligr.* **83**:298–304.

Georges, E., Croissant, O., Bonneaud, N., and Orth, G., 1984, Physical state and transcription of the cottontail rabbit papillomavirus genome in warts and transplantable VX2 and VX7 carcinomas of domestic rabbits, *J. Virol.* **51**:530–538.

Gissmann, L., Pfister, H., and zur Hausen, H., 1977, Human papilloma viruses (HPV): Characterization of four different isolates, *Virology* **76**:569–580.

Glinski, W., Jablonska, S., Langner, A., Obalek, S., Haftek, M., and Proniewska, M., 1976, Cell-mediated immunity in epidermodysplasia verruciformis, *Dermatologica* **153**:218–227.

Glinski, W., Obalek, S., Jablonska, S., and Orth, G., 1981, T-cell defect in patients with epidermodysplasia verruciformis due to human papillomavirus types 3 and 5, *Dermatologica* **162**:141–147.

Goldes, J. A., Filipovitch, A. H., Neudorf, S. M., Bender, M. E., Ostrow, R. S., Faras, A., and Goltz, R. W., 1984, Epidermodysplasia verruciformis in a setting of common variable immunodeficiency, *Pediatr. Dermatol.* **2**:136–139.

Green, M., Orth, G., Wold, W. S. M., Sanders, P. R., McKay, J. K., Favre, M., and Croissant, O., 1981, Analysis of human cancers, normal tissues and verrucae plantares for DNA sequences of human papillomavirus types 1 and 2, *Virology* **110**:176–184.

Green, M., Brackmann, K. H., Sanders, P. R., Loewenstein, P. M., Freel, J. H., Eisinger, M., and Switlyk, S. A., 1982, Isolation of a human papillomavirus from a patient with epidermodysplasia verruciformis: Presence of related viral DNA genomes in human urogenital tumors, *Proc. Natl. Acad. Sci. USA* **79**:4437–4441.

Grupper, C., Pruniéras, M., Delescluse, C., Arouete, J., and Garelly, E., 1971, Epidermodysplasie verruciforme: Etudes ultrastructurale et autoradiographique, *Ann. Dermatol. Syphiligr.* **98**:33–48.

Guilhou, J. J., Malbos, S., Barneon, S., Habib, A., Baldet, P., and Meynadier, J., 1980, Epidermodysplasie verruciforme (2 observations): Etude immunologique, *Ann. Dermatol. Venereol.* **107**:611–619.

Haftek, M., Jablonska, S., and Orth, G., 1985, Specific cell-mediated immunity in patients with epidermodysplasia verruciformis and plane warts, *Dermatologica* **170**: 213–220.

Hammar, H., Hammar, L., Lambert, B., and Ringborg, U., 1976, A case report including EM and DNA repair investigations in a dermatosis associated with multiple skin cancers: Epidermodysplasia verruciformis, *Acta Med. Scand.* **200**:441–446.

Hardie, I. R., Strong, R. W., Hartley, L. C. J., Woodruff, P. W. H., and Clunie, G. J. A., 1980, Skin cancer in Caucasian renal allograft recipients living in a subtropical climate, *Surgery* **87**:177–183.

Heilman, C. A., Law, M. F., Israel, M. A., and Howley, P. M., 1980, Cloning of human papilloma virus genomic DNAs and analysis of homologous polynucleotide sequences, *J. Virol.* **36**:395–407.

Herberman, R. B. (ed.), 1983, *Basic and Clinical Tumor Immunology*, Nijhoff, The Hague.

Hoffmann, E., 1926, Über verallgemeinerte Warzenerkrankung (Verrucosis generalisata) und ihre Beziehung zur Epidermodysplasia verruciformis (Lewandowsky), *Dermatol. Z.* **48**:241–266.

Hoxtell, E. O., Mandel, J. S., Murray, S. S., Schuman, L. M., and Goltz, R. W., 1977, Incidence of skin carcinoma after renal transplantation, *Arch. Dermatol.* **113**:436–438.

Ikenberg, M., Gissmann, L., Gross, G., Grussendorf-Conen, E. I., and zur Hausen, H., 1983, Human papillomavirus type 16-related DNA in genital Bowen's disease and in Bowenoid papulosis, *Int. J. Cancer* **32**:563–565.

Jablonska, S., and Formas, I., 1959, Weitere positive Ergebnisse mit Auto- und Heteroinokulation bei Epidermodysplasia verruciformis Lewandowsky-Lutz, *Dermatologica* **118**:86–93.

Jablonska, S., and Milewski, B., 1957, Zur Kenntnis der Epidermodysplasia verruciformis, *Dermatologica* **115:**1–22.

Jablonska, S., and Orth, G., 1985, Epidermodysplasia verruciformis, *Clin. Dermatol.* **3**(4)**:**83–96.

Jablonska, S., Fabjanska, L., and Formas, I., 1966, On the viral aetiology of epidermodysplasia verruciformis, *Dermatologica* **132:**369–385.

Jablonska, S., Biczysko, W., Jakubowicz, K., and Dabrowski, J., 1968, On the viral etiology of epidermodysplasia verruciformis Lewandowsky-Lutz: Electron microscope studies, *Dermatologica* **137:**113–125.

Jablonska, S., Biczysko, W., Jakubowicz, K., and Dabrowski, J., 1970, The ultrastructure of transitional states to Bowen's disease and invasive Bowen's carcinoma in epidermodysplasia verruciformis, *Dermatologica* **140:**186–194.

Jablonska, S., Dabrowski, J., and Jakubowicz, K., 1972, Epidermodysplasia verruciformis as a model in studies on the role of papovaviruses in oncogenesis, *Cancer Res.* **32:**583–589.

Jablonska, S., Orth, G., Jarzabek-Chorzelska, M., Glinski, W., Obalek, S., Rzesa, G., Croissant, O., and Favre, M., 1979a, Twenty-one years of follow-up studies of familial epidermodysplasia verruciformis, *Dermatologica* **158:**309–327.

Jablonska, S., Orth, G., Jarzabek-Chorzelska, M., Rzesa, G., Obalek, S., Glinski, W., Favre, M., and Croissant, O., 1979b, Epidermodysplasia verruciformis versus disseminated varrucae planae: Is epidermodysplasia verruciformis a generalized infection with wart virus? *J. Invest. Dermatol.* **72:**114–119.

Jablonska, S., Orth, G., and Lutzner, M. A., 1982a, Immunopathology of papillomavirus-induced tumors in different tissues, *Springer Semin. Immunopathol.* **5:**33–62.

Jablonska, S., Obalek, S., Orth, G., Haftek, M., and Jarzabek-Chorzelska, M., 1982b, Regression of the lesions of epidermodysplasia verruciformis, *Br. J. Dermatol.* **107:**109–116.

Jablonska, S., Orth, G., Obalek, S., and Croissant, O., 1985, Cutaneous warts: Clinical, histologic, and virologic correlations, *Clin. Dermatol.* **3**(4)**:**71–82.

Jacyk, W. K., and Lechner, W., 1984, Epidermodysplasia verruciformis in lepromatous leprosy: Report of 2 cases, *Dermatologica* **168:**202–205.

Jacyk, W. K., and Subbuswamy, S. G., 1979, Epidermodysplasia verruciformis in Nigerians, *Dermatologica* **159:**256–265.

Johnson, S. A., Montgomery, H., Tuura, J. L., and Winkelmann, R. K., 1962, Two cases of unusual vacuolar degeneration of the epidermis, *Arch. Dermatol.* **85:**93–104.

Kaminski, M., Pawinska, M., Jablonska, S., Szmurlo, A., Majewski, S., and Orth, G., 1985, Increased natural killer activity in patients with epidermodysplasia verruciformis, *Arch. Dermatol.* **121:**84–86.

Kaufmann, J., Meves, C., and Ott, F., 1978, Die Epidermodysplasia verruciformis Lewandowsky-Lutz in licht-und elektronenoptischen Vergleich mit den übrigen Papova-Virus-Acanthomen, *Arch. Dermatol. Res.* **261:**39–54.

Kawashima, M., Jablonska, S., Favre, M., Obalek, S., Croissant, O., and Orth, G., 1986a, Characterization of a new type of human papillomavirus found in a lesion of Bowen's disease of the skin, *J. Virol.* **57:**688–692.

Kawashima, M., Favre, M., Jablonska, S., Obalek, S., and Orth, G., 1986b, Characterization of a new type of human papillomavirus (HPV) related to HPV5 from a case of actinic keratosis, *Virology* **54:**389–394.

Kienzler, J. L., Laurent, R., Coppey, J., Favre, M., Orth, G., Coupez, L., and Agache, P., 1979, Epidermodysplasie verruciforme: Données ultrastructurales, virologiques et photobiologiques; à propos d'une observation, *Ann. Dermatol. Venereol.* **106:**549–563.

Kienzler, J. L., Lemoine, M. T., Orth, G., Jibard, N., Blanc, D., Laurent, R., and Agache, P., 1983, Humoral and cell-mediated immunity to human papillomavirus type 1 (HPV-1) in human warts, *Br. J. Dermatol.* **108:**665–672.

Kinlen, L. J., Sheil, A. G. R., Peto, J., and Doll, R., 1979, Collaborative United Kingdom–Australasian study of cancer in patients treated with immunosuppressive drugs, *Br. Med. J.* **2:**1461–1466.

Kogoj, F., 1926, Die Epidermodysplasia verruciformis, *Acta Derm. Venereol.* **7:**170–179.

Koranda, F. C., Dehmel, E. M., Kahn, G., and Penn, I., 1974, Cutaneous complications in immunosuppressed renal homograft recipients, *J. Am. Med. Assoc.* **229**:419–424.
Kremsdorf, D., 1980, Clonage et construction d'une carte physique du génome de trois papillomavirus humains associés à la forme grave de l'épidermodysplasie verruciforme, D. E. A. Microbiol. Université Paris 6.
Kremsdorf, D., Jablonska, S., Favre, M., and Orth, G., 1982a, Biochemical characterization of two types of human papillomaviruses associated with epidermodysplasia verruciformis, *J. Virol.* **43**:436–447.
Kremsdorf, D., Jablonska, S., Favre, M., and Orth, G., 1982b, Abstracts of papers presented at the Cold Spring Harbor Meeting on Papilloma Viruses, Cold Spring Harbor, N. Y., pp. 91–96.
Kremsdorf, D., Jablonska, S., Favre, M., and Orth, G., 1983, Human papillomaviruses associated with epidermodysplasia verruciformis. II. Molecular cloning and biochemical characteriation of human papillomavirus 3a, 8, 10, and 12 genomes, *J. Virol.* **48**:340–351.
Kremsdorf, D., Favre, M., Jablonska, S., Obalek, S., Rueda, L. A., Lutzner, M. A., Blanchet-Bardon, C., Van Voorst Vader, P. C., and Orth, G., 1984, Molecular cloning and characterization of the genomes of nine newly recognized human papillomavirus types associated with epidermodysplasia verruciformis, *J. Virol.* **52**:1013–1018.
Kripke, M. L., and Morrison, W. L., 1985, Modulation of immune function by UV radiation, *J. Invest. Dermatol.* **85**:62s–66s.
Langner, A., Jablonska, S., and Darzynkiewicz, Z., 1968, Autoradiographic study of DNA synthesis by epidermal cells in epidermodysplasia verruciformis, *Acta Derm. Venereol.* **48**:501–506.
Laurent, R., Coume-Marquet, S., Kienzler, J. L., Lambert, D., and Agache, P., 1978, Comparative electron microscopic study of clear cells in epidermodysplasia verruciformis and flat warts, *Arch. Dermatol. Res.* **263**:1–12.
Lazzaro, C., Giardina, A., and Randazzo, S. D., 1966, Sulla epidermodisplasia verruciforme di Lewandowski–Lutz, *Minerva Dermatol.* **41**:4–13.
Lever, W. F., and Schaumburg-Lever, G. A., 1975, Tumors and cysts of the epidermis, in: *Histopathology of the Skin*, 5th ed., pp. 450–497, Lippincott, Philadelphia.
Lewandowsky, F., and Lutz, W., 1922, Ein Fall einer bisher nicht beschriebenen Hauterkrankung (Epidermodysplasia verruciformis), *Arch. Dermatol. Syphilol.* **141**:193–203.
Lutz, W. A., 1946, A propos de l'épidermodysplasie verruciforme, *Dermatologica* **92**:30–43.
Lutzner, M. A., 1978, Epidermodysplasia verruciformis: An autosomal recessive disease characterized by viral warts and skin cancer. A model for viral oncogenesis, *Bull. Cancer* **65**:169–182.
Lutzner, M. A., 1985, Papillomavirus lesions in immunodepression and immunosuppression, *Clin. Dermatol.* **3**(4):165–169.
Lutzner, M., Croissant, O., Ducasse, M. F., Kreis, H., Crosnier, J., and Orth, G., 1980, A potentially oncogenic human papillomavirus (HPV-5) found in two renal allograft recipients, *J. Invest. Dermatol.* **75**:353–356.
Lutzner, M. A., Orth, G., Dutronquay, V., Ducasse, M. F., Kreis, H., and Crosnier, J., 1983, Detection of human papillomavirus type 5 DNA in skin cancers of an immunosuppressed renal allograft recipient, *Lancet* **2**:422–424.
Lutzner, M. A., Blanchet-Bardon, C., and Orth, G., 1984, Clinical observations, virologic studies, and treatment trials in patients with epidermodysplasia verruciformis, a disease induced by specific human papillomaviruses, *J. Invest. Dermatol.* **83**:18s–25s.
Marill, F. G., and Grould, P., 1967, Epidermodysplasie verruciforme de Lewandowsky–Lutz avec transformation épitheliomateuse, *Bull. Soc. Fr. Dermatol. Syphiligr.* **74**:213–215.
Marshall, V., 1974, Premalignant and malignant skin tumours in immunosuppressed patients, *Transplantation* **17**:272–275.
Maschkilleisson, L. N., 1931, Ist die Epidermodysplasia verruciformis (Lewandowsky–Lutz) eine selbständige Dermatose? Ihre Beziehungen zur Verrucositas, *Dermatol. Wochenschr.* **92**:569–578.

Midana, A., 1949, Sulla questione dei rapporti tra epidermodysplasia verruciformis e verrucosi generalizzata: Osservazioni su 4 casi di E. V. a carattere famigliare, *Dermatologica* **99:**1–23.

Moreno, G., and Pruniéras, M., 1974, Epidermodysplasia verruciformis. III. Ultraviolet micro-irradiation of cells in culture, *Arch. Dermatol. Forsch.* **249:**247–254.

Mullen, D. L., Silverberg, S. G., Penn, I., and Hammond, W. S., 1976, Squamous cell carcinoma of the skin and lip in renal homograft recipients, *Cancer* **37:**729–734.

Nash, A. A., 1985, Tolerance and suppression in viral diseases, *Br. Med. Bull.* **41:**41–45.

Nasseri, M., and Wettstein, F. O., 1984, Viral transcripts in cottontail rabbit papillomavirus-induced tumors: Differences exist between benign and malignant as well as between non-virus-producing and virus-producing tumors, *J. Virol.* **51:**706–712.

Oehlschlaegel, G., Röckl, H., and Müller, E., 1966, Die Epidermodysplasia verruciformis, ein Sog. Praecancerose, *Hautarzt* **10:**450–458.

Okamoto, T., Yabe, Y., and Ohmori, S., 1970, Virus-like particles in rhabdomyosarcoma with epidermodysplasia verruciformis, *Dermatologica* **141:**309–314.

Orth, G., 1986, Epidermodysplasia verruciformis, a model for understanding the oncogenicity of human papillomaviruses, *Ciba Found. Symp.* **120:**157–174.

Orth, G., and Favre, M., 1985, Human papillomaviruses: Biochemical and biologic properties, *Clin. Dermatol.* **3**(4):27–42.

Orth, G., Favre, M., and Croissant, O., 1977a, Characterization of a new type of human papillomavirus that causes skin warts, *J. Virol.* **24:**108–120.

Orth, G., Breitburd, F., Favre, M., and Croissant, O., 1977b, Papilloma viruses: Possible role in human cancer, *Cold Spring Harbor Conf. Cell Prolif.* **4:**1043–1068.

Orth, G., Favre, M., Jablonska, S., Brylak, K., and Croissant, O., 1978a, Viral sequences related to a human skin papillomavirus in genital warts, *Nature* **275:**334–336.

Orth, G., Jablonska, S., Breitburd, F., Favre, M., and Croissant, O., 1978b, The human papillomaviruses, *Bull. Cancer* **65:**151–164.

Orth, G., Jablonska, S., Favre, M., Croissant, O., Jarzabek-Chorzelska, M., and Rzesa, G., 1978c, Characterization of two types of human papillomaviruses in lesions of epidermodysplasia verruciformis, *Proc. Natl. Acad. Sci. USA* **75:**1537–1541.

Orth, G., Jablonska, S., Jarzabek-Chorzelska, M., Rzesa, G., Obalek, S., Favre, M., and Croissant, O., 1979, Characteristics of the lesions and risk of malignant conversion as related to the type of the human papillomavirus involved in epidermodysplasia verruciformis, *Cancer Res.* **39:**1074–1082.

Orth, G., Favre, M., Breitburd, F., Croissant, O., Jablonska, S., Obalek, S., Jarzabek-Chorzelska, M., and Rzesa, G., 1980, Epidermodysplasia verruciformis: A model for the role of papillomaviruses in human cancer, *Cold Spring Harbor Conf. Cell Prolif.* **7:**259–282.

Orth, G., Jablonska, S., Favre, M., Croissant, O., Obalek, S., Jarzabel-Chorzelska, M., and Jibard, N., 1981, Identification of papillomaviruses in butchers' warts, *J. Invest. Dermatol.* **76:**97–102.

Ostrow, R. S., Bender, M., Niimura, M., Seki, T., Kawashima, M., Pass, F., and Faras, A. J., 1982, Human papillomavirus DNA in cutaneous primary and metastasized squamous cell carcinoma from patients with epidermodysplasia verruciformis, *Proc. Natl. Acad. Sci. USA* **79:**1634–1638.

Ostrow, R., Zachow, K., Watts, S., Bender, M., Pass, F., and Faras, A., 1983a, Characterization of two HPV3-related papillomaviruses from common warts that are distinct clinically from flat warts or epidermodysplasia verruciformis, *J. Invest. Dermatol.* **80:**436–440.

Ostrow, R. S., Watts, S., Bender, M., Niimura, M., Seki, T., Kawashima, M., Pass, F., and Faras, A. J., 1983b, Identification of three distinct papillomavirus genomes in a single patient with epidermodysplasia verruciformis, *J. Am. Acad. Dermatol.* **8:** 398–404.

Ostrow, R. S., Zachow, K. R., Thompson, O., and Faras, A. J., 1984, Molecular cloning and characterization of a unique type of human papillomavirus from an immune deficient patient, *J. Invest. Dermatol.* **82:**362–366.

Ostrow, R. S., Zachow, K. R., Niimura, M., Okagaki, T., Muller, S., Bender, M., and Faras, A. J., 1986, Detection of papillomavirus DNA in human semen, *Science* **231**:731–733.

Pass, F., Reissig, M., Shah, K. V., Eisinger, M., and Orth, G., 1977, Identification of an immunologically distinct papillomavirus from lesions of epidermodysplasia verruciformis, *J. Natl. Cancer Inst.* **59:**1107–1112.

Penn, I., 1980, Immunosuppression and skin cancer, *Clin. Plast. Surg.* **7:**361–368.

Pfister, H., and Haneke, E., 1984, Demonstration of human papilloma virus type 2 DNA in Bowen's disease, *Arch. Dermatol. Res.* **276:**123–125.

Pfister, H., Gross, G., and Hagedorn, M., 1979, Characterization of human papillomavirus 3 in warts of a renal allograft patient, *J. Invest. Dermatol.* **73:**349–353.

Pfister, H., Nürnberger, F., Gissmann, L., and zur Hausen, H., 1981, Characterization of a human papillomavirus from epidermodysplasia verruciformis lesions of a patient from Upper Volta, *Int. J. Cancer* **27:**645–650.

Pfister, H., Gassenmaier, A., Nürnberger, F., and Stüttgen, G., 1983, Human papillomavirus 5-DNA in carcinoma of an epidermodysplasia verruciformis patient infected with various human papillomavirus types, *Cancer Res.* **43:**1436–1441.

Pfister, H., Iftner, T., and Fuchs, P. G., 1985, Papillomaviruses from epidermodysplasia verruciformis patients and renal allograft recipients, in: *Papillomaviruses: Molecular and Clinical Aspects* (P. M. Howley and T. R. Broker, eds.), pp. 85–100, Liss, New York.

Prawer, S. E., Pass, F., Vance, J. C., Greenberg, L. J., Yunis, E. J., and Zelickson, A. S., 1977, Depressed immune function in epidermodysplasia verruciformis, *Arch. Dermatol.* **113:**495–499.

Proniewska, M., and Jablonska, S., 1980, UV-induced DNA repair synthesis in patients with epidermodysplasia verruciformis, *Dermatologica* **160:**289–296.

Pruniéras, M., 1975, Ultrastructure of the epidermis in epidermodysplasia verruciformis (EV), in: *Biomedical Aspects of Human Wart Infection* (M. Pruniéras, ed.), pp. 165–170, Fondation Mérieux, Lyon.

Prystowsky, S. O., Herndon, J. H., and Freeman, R. G., 1976, Epidermodysplasia verruciformis, *Am. J. Dis. Child.* **130:**437–440.

Pyrhönen, S., Jablonska, S., Obalek, S., and Kuismanen, E., 1980, Immune reactions in epidermodysplasia verruciformis, *Br. J. Dermatol.* **102:**247–254.

Rajagopalan, K., Bahru, J., Loo, D. S., Tay, C. H., Chin, K. N., and Tan, K. K., 1972, Familial epidermodysplasia verruciformis of Lewandowsky and Lutz, *Arch. Dermatol.* **105:**73–78.

Relias, A., Sakellariou, G., and Tsoïtis, G., 1967, Epidermodysplasie verruciforme de Lewandowsky–Lutz à multiples épithéliomas, *Ann. Dermatol. Syphiligr.* **94:**501–514.

Riou, G., Barrois, M., Dutronquay, V., and Orth, G., 1985, Presence of papillomavirus DNA sequences, amplification of c-*myc* and c-Ha-*ras* oncogenes and enhanced expression of c-*myc* in carcinomas of the uterine cervix, in: *Papillomaviruses: Molecular and Clinical Aspects* (P. M. Howley and T. R. Broker, eds.), pp. 47–56, Liss, New York.

Rogers, S., and Rous, P., 1951, Joint action of a chemical carcinogen and a neoplastic virus to induce cancer in rabbits, *J. Exp. Med.* **93:**459–487.

Rous, P., and Beard, J. W., 1935, The progression to carcinoma of virus-induced rabbit papillomas (Shope), *J. Exp. Med.* **62:**523–548.

Rous, P., and Friedewald, W. F., 1944, The effect of chemical carcinogens on virus-induced rabbit papillomas, *J. Exp. Med.* **79:**511–538.

Rüdlinger, R., Smith, I. W., Bunney, M. H., and Hunter, J. A. A., 1986, Human papillomavirus infections in a group of renal transplant patients, *Br. J. Dermatol.* (in press).

Rüdlinger, R., Bunney, M. H., Smith, I. W., and Hunter, J. A. A., 1987, Detection of human papillomavirus type 5 DNA in a renal allograft patient from Scotland, *Arch. Dermatol.* (manuscript submitted).

Rueda, L. A., and Rodriguez, G., 1972, Comparacion di la virogenesis en la epidermodisplasia verruciforme y en las verrugas planas, *Med. Cutan. Iber. Lat. Am.* **6:**451–457.

Rueda, L. A., and Rodriguez, G., 1976, Verrugas humanas por virus papova: Correlacion clinica, histologica y ultraestructural, *Med. Cutan. Iber. Lat. Am.* **2:**113–136.

Rueda, L. A., and Rodriguez, G., 1983, Histopathology and ultrastructure of epidermodysplasia verruciformis, in: *Proceedings of the XVIth International Congress of Dermatology* (A. Kukita and M. Seiji, eds.), pp. 617–620, University of Tokyo Press, Tokyo.
Ruiter, M., 1969, Malignant degeneration of skin lesions in epidermodysplasia verruciformis, *Acta Derm. Venereol.* **49:**309–313.
Ruiter, M., 1973, On the histomorphology and origin of malignant cutaneous changes in epidermodysplasia verruciformis, *Acta Derm. Venereol.* **53:**290–298.
Ruiter, M., and Van Mullem, P. J., 1966, Demonstration by electronmicroscopy of an intranuclear virus in epidermodysplasia verruciformis, *J. Invest. Dermatol.* **47:**247–252.
Ruiter, M., and Van Mullem, P. J., 1970a, Behavior of virus in malignant degeneration of skin lesions in epidermodysplasia verruciformis, *J. Invest. Dermatol.* **54:**324–331.
Ruiter, M., and Van Mullem, P. J., 1970b, Further histological investigations on malignant degeneration of cutaneous lesions in epidermodysplasia verruciformis, *Acta Derm. Venereol.* **50:**205–211.
Sanderson, K. V., and Mackie, R. M., 1979, Tumours of the skin, in: *Textbook of Dermatology* (A. J. Rook, D. S. Wilkinson, and F. J. Ebling, eds.), 3rd ed., Vol. 2, pp. 2129–2231, Blackwell, Oxford.
Sayag, J., Muratore, R., Olmer, J., and Hadida, E., 1966, Epidermodysplasie verruciforme et réticulose ganglionnaire maligne, *Bull. Soc. Fr. Dermatol. Syphiligr.* **73:**313–315.
Schellander, F., and Fritsch, P., 1970, Epidermodysplasia verruciformis: Neue Aspekte zur Symptomatologie und Pathogenese, *Dermatologica* **140:**251–263.
Scheurlen, W., Gissmann, L., Gross, G., and zur Hausen, H., 1986, Molecular cloning of two new HPV-types (HPV37 and HPV38) from a keratoacanthoma and a malignant melanoma, *Int. J. Cancer* **37:**505–510.
Schwarz, E., Freese, U. K., Gissmann, L., Mayer, W., Roggenbuck, B., Stremlau, A., and zur Hausen, H., 1985, Structure and transcription of human papillomavirus sequences in cervical carcinoma cells, *Nature* **314:**111–114.
Seedorf, K., Krämmer, G., Dürst, M., Suhai, S., and Röwekamp, W. G., 1985, Human papillomavirus type 16 DNA sequence, *Virology* **145:**181–185.
Spencer, E. S., and Andersen, H. K., 1979, Viral infections in renal allograft recipients treated with long-term immunosuppression, *Br. Med. J.* **2:**829–830.
Spradbrow, P. B., Beardmore, G. L., and Francis, J., 1983, Virions resembling papillomaviruses in hyperkeratotic lesions from sun-damaged skin, *Lancet* **1:**189.
Streilen, J. W., 1983, Skin-associated lymphoid tissues (SALT): Origins and functions, *J. Invest. Dermatol.* **80:**12s–16s.
Streilen, J. W., 1985, Circuits and signals of the skin-associated lymphoid tissues (SALT), *J. Invest. Dermatol.* **85:**10s–13s.
Sullivan, M., and Ellis, F. A., 1939, Epidermodysplasia verruciformis (Lewandowsky and Lutz), *Arch. Dermatol. Syphilol.* **40:**422–432.
Syverton, J. T., 1952, The pathogenesis of the rabbit papilloma-to-carcinoma sequence, *Ann. N. Y. Acad. Sci.* **54:**1126–1140.
Tagami, H., Aiba, S., and Rokugo, M., 1985a, Regression of flat warts and common warts, *Clin. Dermatol.* **3**(4)**:**170–178.
Tagami, H., Oku, T., and Iwatsuki, K., 1985b, Primary tissue culture of spontaneously regressing flat warts, *Cancer* **55:**2437–2441.
Taichman, L. B., Breitburd, F., Croissant, O., and Orth, G., 1984, The search for a culture system for papillomavirus, *J. Invest. Dermatol.* **83:**2s–6s.
Tsoitis, G., Destombes, P., Ravisse, P., and Ancelle, G., 1973, Epidermodysplasie verruciforme de Lewandowsky–Lutz observée chez un Camerounais, *Bull. Soc. Pathol. Exot.* **66:**411–426.
Tsumori, T., Yutsudo, M., Nakano, Y., Tanigaki, T., Kitamura, H., and Hadura, A., 1983, Molecular cloning of a new human papillomavirus isolated from epidermodysplasia verruciformis lesions, *J. Gen. Virol.* **64:**967–969.

Urbach, F., 1982, Epidemiology and etiology of skin cancer in man, in: *Cancer Campaign*, Vol. 6, *Cancer Epidemiology* (E. Grundmann, ed.), pp. 213–227, Gustav Fischer Verlag, Stuttgart/New York.

Van Voorst Vader, P. C., Orth, G., Dutronquay, V., Driessen, L. H. H. M., Eggink, H. F., Kallenberg, L. G. M., and The, T. H., 1986, Epidermodysplasia verruciformis: Skin carcinoma containing human papillomavirus type 5 DNA sequences and primary hepatocellular carcinoma associated with chronic hepatitis B virus infection in a patient, *Acta Derm. Venereol.* **66**:231–236.

Vasily, D. B., Miller, O. F., Fudenberg, H. H., Goust, J. M., and Wilson, G. B., 1984, Epidermodysplasia verruciformis: Response to therapy with dialyzable leukocyte extract (transfer factor) derived from household contacts, *J. Clin. Lab. Immunol.* **14**:49–57.

Vogel, F., and Motulsky, A. G., 1982, *Human Genetics: Problems and Approaches*, Springer-Verlag, Berlin.

Waisman, M., and Montgomery, H., 1942, Verruca plana and epithelial nevus, including a study of epidermodysplasia verruciformis, *Arch. Dermatol. Syphilol.* **45**:259–279.

Watts, S. L., Phelps, W. C., Ostrow, R. S., Zachow, K. R., and Faras, A. J., 1984, Cellular transformation by human papillomavirus DNA in vitro, *Science* **225**:634–636.

Watts, S. L., Chow, L. T., Ostrow, R. S., Faras, A. J., and Broker, T. R., 1985, Localization of HPV-5 transforming functions, in: *Papillomaviruses: Molecular and Clinical Aspects* (P. M. Howley and T. R. Broker, eds.), pp. 501–511, Liss, New York.

Wettstein, F. O., and Stevens, J. G., 1982, Variable-sized free episomes of Shope papilloma virus DNA are present in all non-virus-producing neoplasms and integrated episomes are detected in some, *Proc. Natl. Acad. Sci. USA* **79**:790–794.

Wojtulewicz-Kurkus, J., Glinski, W., Jablonska, S., Podobinska, I., and Obalek, S., 1985, Identification of HLA antigens in familial and non-familial epidermodysplasia verruciformis, *Dermatologica* **170**:53–58.

Yabe, Y., and Sadakane, H., 1975, The virus of epidermodysplasia verruciformis: Electron microscopic and fluorescent antibody studies, *J. Invest. Dermatol.* **65**:324–330.

Yabe, Y., Yasui, M., Yoshino, N., Fujiwara, T., Ohkuma, N., and Nohara, N., 1978, Epidermodysplasia verruciformis: Viral particles in early malignant lesions, *J. Invest. Dermatol.* **71**:225–228.

Yee, C., Krishnan-Hewlett, I., Baker, C. C., Schlegel, R., and Howley, P. M., 1985, Presence and expression of human papillomavirus sequences in human cervical carcinoma cell lines, *Am. J. Pathol.* **119**:361–366.

Yutsudo, M., Tanigaki, T., Tsumori, T., Watanabe, S., and Hakura, A., 1982, New human papilloma virus isolated from epidermodysplasia verruciformis lesions, *Cancer Res.* **42**:2440–2443.

Yutsudo, M., Shimakage, T., and Hakura, A., 1985, Human papillomavirus type 17 DNA in skin carcinoma tissue of a patient with epidermodysplasia verruciformis, *Virology* **144**:295–298.

zur Hausen, H., 1982, Human genital cancer: Synergism between two virus infections or synergism between a virus infection and initiating events? *Lancet* **1**:1370–1372.

CHAPTER 9

The Role of Papillomaviruses in Human Anogenital Cancer

HARALD ZUR HAUSEN AND ACHIM SCHNEIDER

I. INTRODUCTION AND HISTORICAL ASPECTS

Since 1842 (Rigoni-Stern), numerous epidemiological studies point to an infectious event in the etiology of human cervical cancer (reviewed by: Rotkin, 1973; Kessler, 1977; zur Hausen, 1977). Venereal transmission of a carcinogenic factor with a long latency period has been postulated in many of these studies. Sexual promiscuity, early onset of sexual activity, and poor hygienic conditions appear to be the prime risk factors for cervical cancer.

More recently, additional observations revealed a correlation between the incidences of cervical and penile cancer (Martinez, 1969; Graham *et al.*, 1979; Cartwright and Sinson, 1980; Maggregor and Innes, 1980; Smith *et al.*, 1980). In a number of countries studied to date, the incidence rates of penile cancer are approximately 20-fold lower than those of cervical cancer (Waterhouse *et al.*, 1982). Accordingly, areas with high rates of cervical cancer have relatively elevated incidence rates for penile cancer, although at an approximately 20-fold reduced rate. This suggests that the same etiological factors may be operative in both types of carcinomas.

Very little information exists on the epidemiology of vulvar cancer. The incidence data are usually combined with those for vaginal cancer (Waterhouse *et al.*, 1982). The available results suggest, however, that

HARALD ZUR HAUSEN • Deutsches Krebsforschungszentrum, 6900 Heidelberg, Federal Republic of Germany. ACHIM SCHNEIDER • Sektion für Gynäkologische Zytologie und Histologie, Universitäts-Frauenklinik, 7900 Ulm, Federal Republic of Germany.

vulvar cancer occurs at a slightly elevated frequency when compared to penile cancer and reveals a similar geographic distribution.

Cancer at external genital sites is seen in different age groups than those for cervical cancer. The latter occurs on average about 10–15 years earlier than penile or vulvar cancer.

The early hints on a possible involvement of an infectious event in the etiology of cervical cancer led to a number of attempts to identify known genital pathogens as causative agents of these carcinomas (see review by Rotkin, 1973). The analysis of infections by bacteria (e.g., syphilis, gonorrhea), protozoa (*Trichomonas*), and chlamydia did not result in conclusive evidence. The discovery of several tumor viruses in the 1950s and 1960s (for review see Gross, 1983) initiated research on the possible role of viruses in cervical cancer.

Since 1968, genital infections by herpes simplex virus (HSV) type 2 have been regarded as promising candidates (Rawls *et al.*, 1968; Nahmias *et al.*, 1969). Seroepidemiological studies revealed higher antibody titers against HSV-2 antigens in cervical cancer patients as compared to age-matched controls. In addition, the percentage of seroreactive women was higher in the cancer group than in control groups matched for various biological and social parameters (reviewed in Rawls *et al.*, 1980; zur Hausen, 1983).

The possible involvement of HSV received experimental support from studies of Duff and Rapp (1973) who demonstrated that partially inactivated HSV effectively transforms rodent cells *in vitro*.

A number of attempts were made to trace HSV nucleic acids or proteins in cervical cancer cells. The results are still controversial (see review by zur Hausen, 1983). The available data permit the interpretation that HSV DNA, RNA, or proteins are not regularly present in cervical cancer tissue. The occasional demonstration of fragments of HSV DNA in some tumors (Frenkel *et al.*, 1972; Park *et al.*, 1983; Galloway and McDougall, 1983) or of HSV RNA demonstrated by *in situ* hybridization awaits further clarification. Recently, host-cell sequences cross-hybridizing with cytomegalovirus DNA even under conditions of elevated stringency have been demonstrated (Rüger *et al.*, 1984). The existence of similar cross-reactivity between HSV DNA and host-cell sequences had been shown previously (Maitland *et al.*, 1981) and deserves further attention.

In summary, molecular biological studies do not provide convincing evidence for an involvement of HSV-2 in the etiology of cervical cancer. Similarly, a large prospective study recently carried out by Vonka and his colleagues in Prague (1984a,b) failed to support an involvement of HSV-2 infections in cervical cancer or preneoplastic lesions.

The first investigations on a possible involvement of a different virus group, the papillomaviruses, were reported in 1974 (zur Hausen *et al.*, 1974). The background for speculations arguing in favor of a relationship of papillomavirus infections to genital cancer has been summarized by zur Hausen (1975, 1976, 1977). In brief, papillomaviruses were known to induce mainly benign, occasionally also malignant tumors. Genital warts (condylomata acuminata) are venereally transmitted. Numerous anecdotal reports reveal a malignant conversion, usually of long-persisting condylomata acuminata (reviewed by zur Hausen, 1977). In addition, there existed indications that the incidence of genital warts of vulva and vagina coincides with dysplastic lesions of the cervix (Rutledge and Sinclair, 1968; Jagella and Stegner, 1974; Meisels and Fortin, 1976; Meisels *et al.*, 1977; Purola and Savia, 1977). In 1976 and 1977, Meisels and his co-workers reported cytological changes in cervical dysplasias revealing characteristic features of papillomavirus infection. They interpreted the presence of "koilocytotic" cells (Koss and Durfee, 1956) as indicators for this type of infection. Meisels identified what he labeled as cervical flat condyloma as a morphological entity. Other groups as well as Meisels and his colleagues subsequently demonstrated that part of cervical dysplasias and specifically koilocytotic cells contained typical papillomavirus particles as shown by electron microscopy (Laverty *et al.*, 1978; Della Torre *et al.*, 1978; Meisels *et al.*, 1981). Additional findings supported the view that papillomavirus infections may lead to cervical dysplasia: Jenson and Shah and their co-workers developed a group-specific antiserum reacting with internal capsid proteins of a wide variety of animal and human papillomaviruses (Jenson *et al.*, 1980; Shah *et al.*, 1980). By using this serum it was demonstrated that close to 50% of cervical dysplasias reveal papillomavirus group-specific antigens in superficial layers and koilocytotic cells.

Initial attempts to reveal papillomavirus DNA in cervical cancer biopsies by using radioactively labeled complementary RNA from human plantar wart virus as probe in nucleic acid hybridization experiments failed to provide evidence for persisting HPV DNA (zur Hausen *et al.*, 1974). In view of the plurality of papillomavirus types known today, retrospectively, this result is by no means surprising.

The isolation and characterization of papillomaviruses from genital warts turned out to be difficult in view of the scarcity of viral particles in these tumors (de Lap *et al.*, 1976; Orth *et al.*, 1978). Their characterization became possible in 1980 (Gissmann and zur Hausen, 1980), leading to their molecular cloning (de Villiers *et al.*, 1981) and to the analysis of their nucleic acid sequences (Schwarz *et al.*, 1983; Seedorf *et al.*, 1985; Dartmann *et al.*, 1986). The prevalent virus type in condylomata acuminata is designated as HPV-6. The availability of this probe led to the isolation of a closely related by distinct type from laryngeal papillomas and condylomata acuminata, HPV-11 (Gissmann *et al.*, 1983).

Radioactively labeled HPV-11 DNA was used under hybridization conditions of low stringency (Heilman *et al.*, 1980) to detect cross-reacting sequences in human cervical cancer. This approach led to the identification of two new papillomavirus DNAs now labeled HPV-16 (Dürst *et al.*, 1983) and HPV-18 (Boshart *et al.*, 1984). At present, HPV-16 and 18 are the most prevalent types of papillomavirus DNAs found in human genital cancer, accounting for approximately 70% of all tumors investigated. Individual tumors reveal occasionally HPV-11 (Gissmann *et al.*, 1983) and also HPV-10 DNA (Green *et al.*, 1982). Thus, specific types of papillomaviruses are rather regularly found in human genital cancer and seem to deserve special interest as putative etiological agents for this malignant condition. Additional evidence supporting their role in malignant genital tumors will be discussed below.

In recent years a number of studies identified extensive smoking over periods of time as risk factor for the development of genital cancer (Winkelstein, 1977; Wigle, 1980; Clarke *et al.*, 1982; Vonka *et al.*, 1984a,b). The available data suggest an increase in relative risk for cervical cancer development among heavy smokers by a factor of 2–12. This appears to complicate the straightforward assumption that infectious agents are etiologically involved in human genital cancer. However, it should be noted that particularly specific papillomavirus infections appear to interact synergistically or cooperatively with chemical and physical carcinogens in malignant conversion (reviewed by zur Hausen, 1977, 1983). The mechanism of this interaction is at present not understood, although enhancement of viral DNA integration by initiators may facilitate the development of malignant cell clones. It has been postulated (zur Hausen, 1982) that initiating events connected with the uptake of tobacco condensates during smoking or associated with HSV infections may synergistically interact with infections by specific papillomaviruses in the induction of human genital cancer. On the basis of these speculations, it is likely that any other initiating events occurring within the genital tract due to poor hygienic conditions (e.g., bacterial metabolites, mutagenic degradation product) would also contribute to the development of genital cancer. The probability of such processes may be substantially different for the sites cervix, vulva, or penis, respectively.

II. ISOLATION OF GENITAL PAPILLOMAVIRUSES AND THEIR CHARACTERIZATION

As outlined before, low particle production prevented isolation of papillomaviruses from genital warts for a number of years. The isolation of HPV-6 from a condyloma acuminatum (Gissmann and zur Hausen, 1980) and the subsequent characterization of its DNA (de Villiers *et al.*, 1981) led quickly to analysis of additional genital papillomavirus types (Gissmann *et al.*, 1982; Dürst *et al.*, 1983; Boshart *et al.*, 1984). The presently analyzed genital papillomavirus types and their association with genital and other proliferations are listed in Table I. HPV-6 and 11 represent closely related viruses. DNA from both virus types has been entirely sequenced and reveals an overall nucleotide homology of 82% (Schwarz *et al.*, 1983; Dartmann *et al.*, 1986). Under hybridization conditions of high stringency, they cross-hybridize by about 25% (Gissmann

TABLE I. Human Genital Papillomavirus Types and Their Association with Proliferative Diseases

HPV type	Tumor	Approximate frequency of association	Reference
HPV-6	Condyloma acuminatum	60%	Gissmann *et al.* (1983)
	Cervical koilocytotic dysplasia	Rare	Gissmann *et al.* (1983)
	Laryngeal papilloma	30–50%	Mounts *et al.* (1982)
HPV-11	Condyloma acuminatum	30%	Gissmann *et al.* (1983)
	Cervical koilocytotic dysplasia	30%	Schneider *et al.* (1985)
	Cervical cancer	3%	Gissmann *et al.* (1983)
	Laryngeal papilloma	50%	Gissmann *et al.* (1983)
HPV-16	Condyloma acuminatum	Rare	Dürst *et al.* (1983)
	Moderate + severe cervical dysplasia	30–50%	Schneider *et al.* (1985)
	Cervical cancer	40–60%	Dürst *et al.* (1983)
	Bowenoid lesions	80%	Ikenberg *et al.* (1983)
HPV-18	Condyloma acuminatum	Rare	Boshart *et al.* (1984)
	Cervical dysplasia	?	—
	Bowenoid lesions	?	—
	Cervical, vulvar, and penile cancer	20%	Boshart *et al.* (1984)
HPV-31[a]	Condyloma acuminatum	?	Lorincz *et al.* (1985)
	Cervical dysplasias	Frequent	Lorincz *et al.* (1985)
(CD 6)	Bowenoid lesions	?	—
	Cervical cancer	Rare	Lorincz *et al.* (1985)

[a]Tentative designation.

et al., 1982), thus representing individual types according to the present definition (Coggin and zur Hausen, 1979). Structural features of their genomes correspond to those reported for BPV and HPV-1 (Chen *et al.*, 1982; Danos *et al.*, 1982).

Besides condylomata acuminata and cervical dysplastic lesions, HPV-6 or 11 DNA is regularly found in an invasively growing variant of genital warts: These tumors, designated as Buschke–Löwenstein tumors or non-metastasizing verrucous carcinomas, are characterized by their exuberant exophytic growth and their invasive properties. They very rarely metastasize (Buschke and Löwenstein, 1931). In 14 biopsies analyzed to date, HPV-6 DNA was demonstrated in 12 and HPV-11 DNA in 2 samples (de Villiers *et al.*, 1986; Boshart and zur Hausen, 1986).

HPV-6 or 11 DNA is also regularly demonstrated in laryngeal papillomas and in occasional papillomas of the oral cavity, pointing to oral infections with these genital viruses. HPV-16 and 18 are only distantly related to HPV-6 and 11 and to each other. DNA from both viruses has been isolated from cervical cancer biopsies (Dürst *et al.*, 1983; Boshart *et al.*, 1984) containing sequences cross-hybridizing with HPV-11 DNA under conditions of low stringency. HPV-16 DNA has been sequenced entirely. Again it reveals a genome organization similar to that of other papillomaviruses with the possible exception of a stop codon within the E1 open reading frame (Seedorf *et al.*, 1985). Partial sequencing of HPV-18 DNA also suggests a typical genome structure corresponding to that of other papillomaviruses (Schwarz *et al.*, 1985). One isolate, initially designated CD 6 and now labeled HPV-31 (Lorincz *et al.*, 1985), was isolated and cloned from a koilocytic cervical dysplasia. It appears to differ substantially from the isolates described previously. Although the distribution of this agent has not been sufficiently studied to date, it seems to be most frequently associated with mild and moderate cervical dysplasias. It is rarely found in genital cancer and in other genital lesions.

Recently, additional types of papillomaviruses have been cloned from genital biopsies. Very little is known as yet concerning their occurrence in benign and malignant genital tumors.

The available data indicate that more than five distinct types of papillomaviruses infect the human genital tract. At least two of them are regularly found in malignant genital tumors.

III. BIOLOGY OF GENITAL PAPILLOMAVIRUS INFECTIONS

The transmission of genital papillomavirus infections seems to occur venereally (Teokharov, 1962; Oriel, 1971; zur Hausen *et al.*, 1975; Gross *et al.*, unpublished data; Schneider *et al.*, unpublished data). It appears that the availability of a cell which is still capable of dividing is a prerequisite for a successful papillomavirus infection. Microlesions at genital sites exposing cells of the basal cell layer are thus predisposing to these infections. In addition, the availability of proliferating cells at the surface at the squamo-columnar junction of the cervix probably accounts for the fact that about 90% of cervical dysplasias originate at this site.

Figure 1 schematically outlines the natural history of genital HPV infections. It appears that at least two modes of interactions are characteristic features of infections with specific types of HPVs: one seems to be represented by infections with types 6, 11, and 31, resulting in

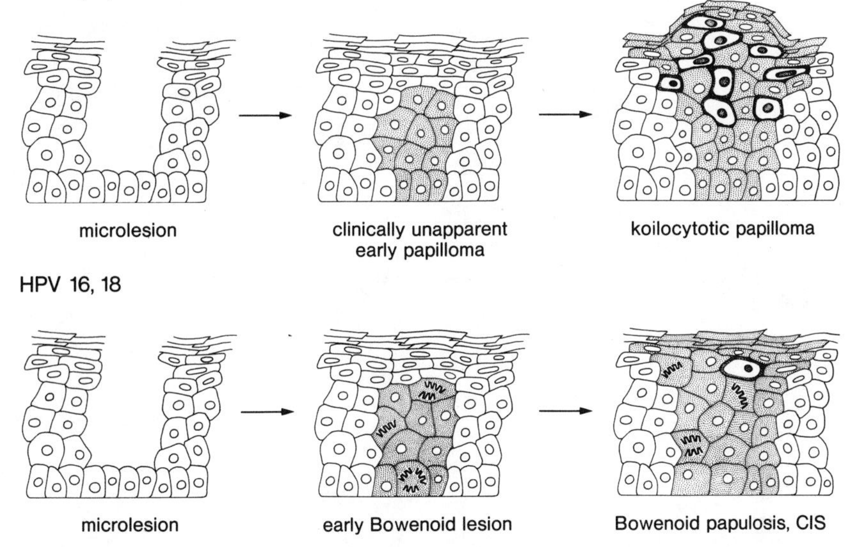

FIGURE 1. Natural history of HPV infections. Viral infections primarily appear to occur via microlesions or in proliferating cells exposed to the surface at the transformation zone. HPV-6, 11, and 31 infections result most frequently in koilocytotic dysplasias at cervical sites, whereas HPV-16 and 18 infections lead to atypical mitotic figures, bowenoid changes, and carcinoma *in situ.*

proliferations which still preserve characteristics of the regular differentiation pattern. The nuclei are regularly shaped, the DNA content is tetraploid or diploid (Fu *et al.,* 1981), and extensive koilocytosis appears to correlate to some extent with virus antigen (Jenson *et al.,* 1980) and viral particle synthesis (Meisels *et al.,* 1981).

A second type of infection, represented by types 16 and 18, is characterized by marked nuclear atypia, regular and atypic mitotic figures even in the upper third of the epithelial layer, an aneuploid karyotype, and a low number of koilocytotic cells which may even be entirely absent (Crum *et al.,* 1985). In advanced lesions the pattern corresponds to severe dysplasia or carcinoma *in situ.* Whereas the first type of infections particularly when induced by HPV types 6 and 11 reveals a more exophytic growth pattern, HPV-16 and 18 infections usually lead to flat discrete lesions which are macroscopically inconspicuous, although histologically they frequently exhibit features of a carcinoma *in situ.* At external genital sites these infections result in characteristic features of bowenoid papulosis or Bowen's disease (Ikenberg *et al.,* 1983). Macroscopic and histological characteristics of these infections of external and internal genital sites are demonstrated in Figs. 2–4.

Papillomavirus-induced proliferations in the human genital tract are characterized by long duration and frequent episodes of recurrence even

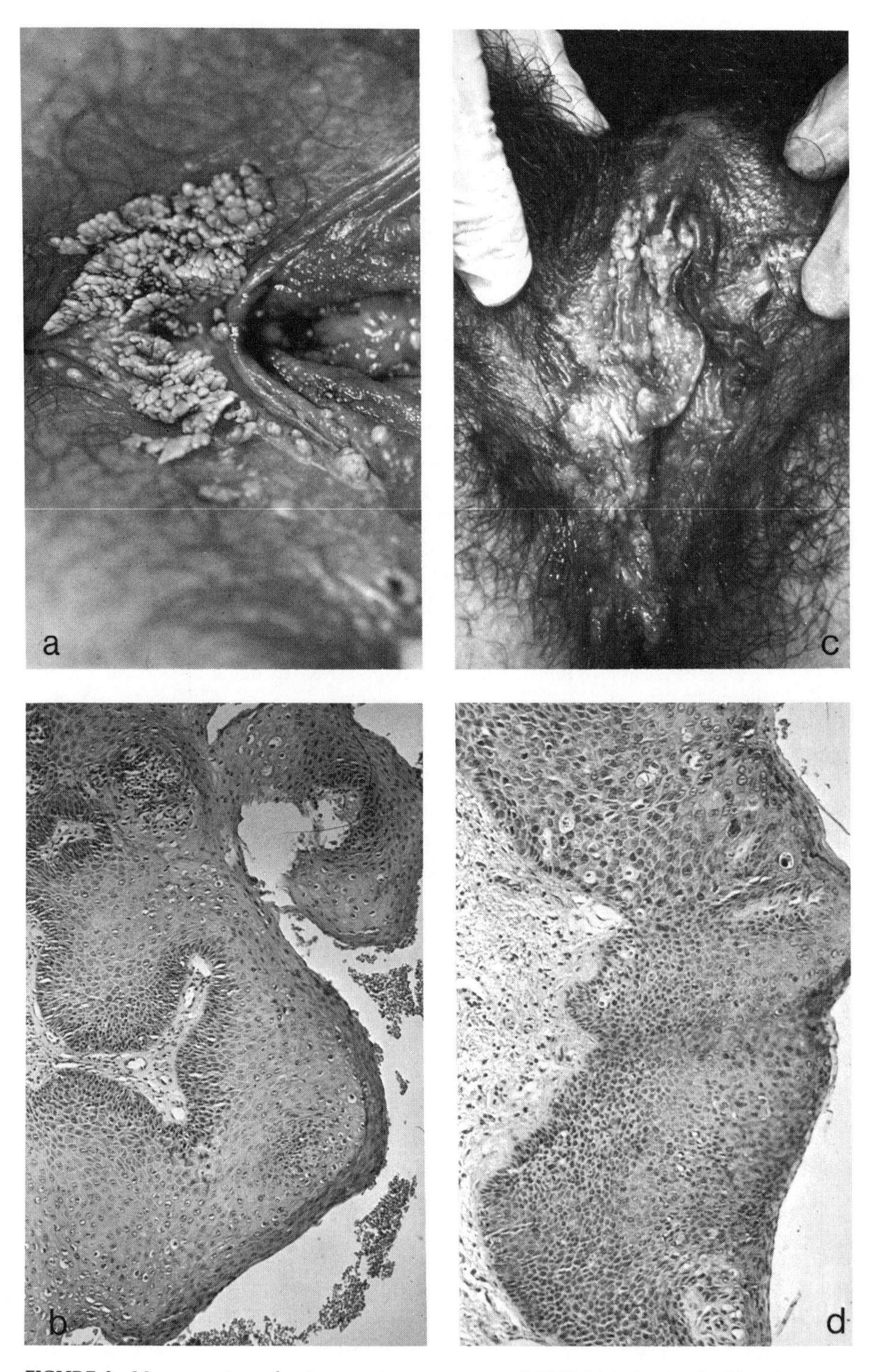

FIGURE 2. Macroscopic and microscopic appearance of HPV-11 infection (a, b) and HPV-16 infection (c, d) of the vulva. Note the nuclear atypia in (d).

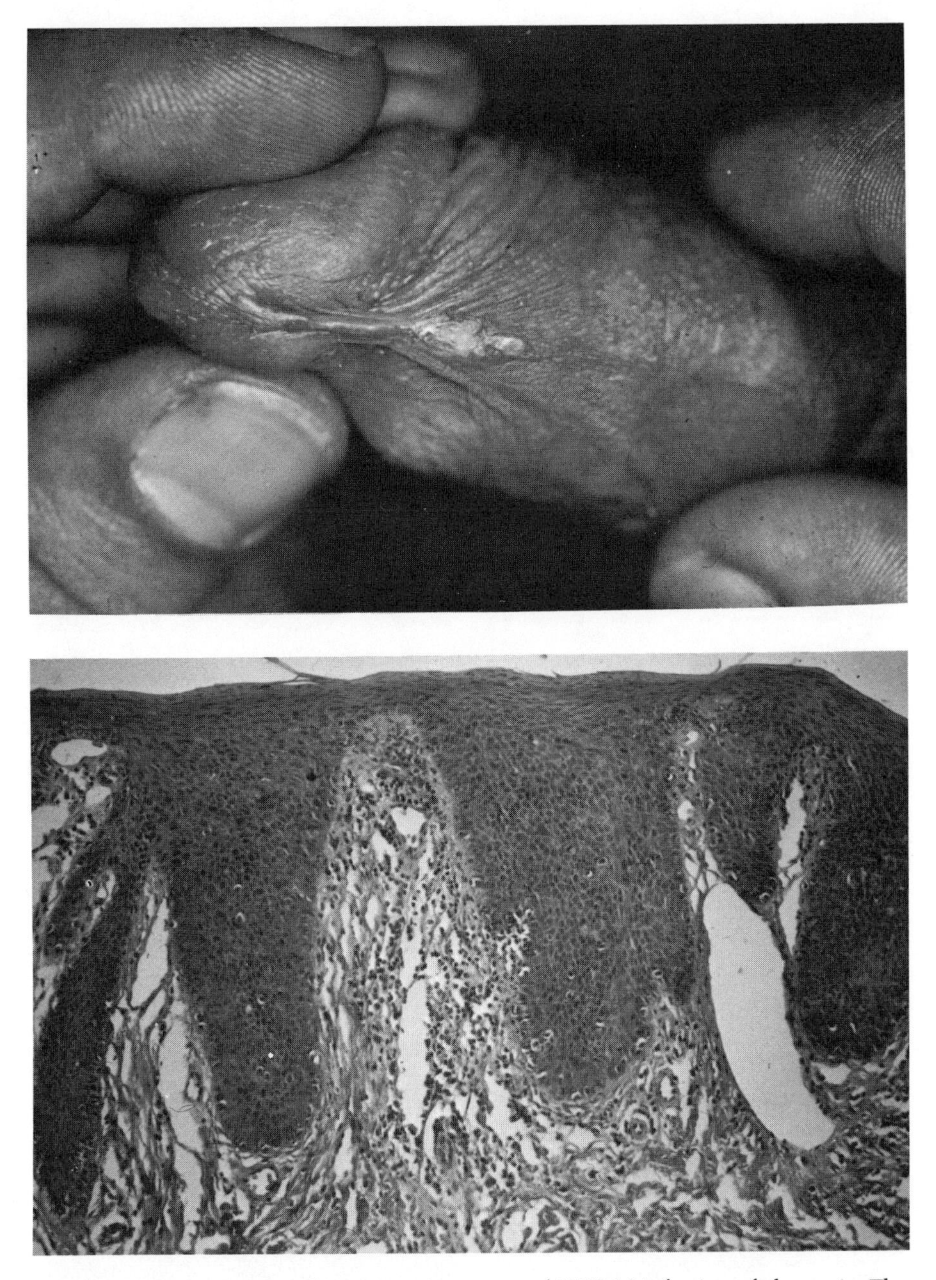

FIGURE 3. Macroscopic and microscopic pattern of HPV-16 infection of the penis. The macroscopically discrete lesion reveals histological features of a carcinoma *in situ*.

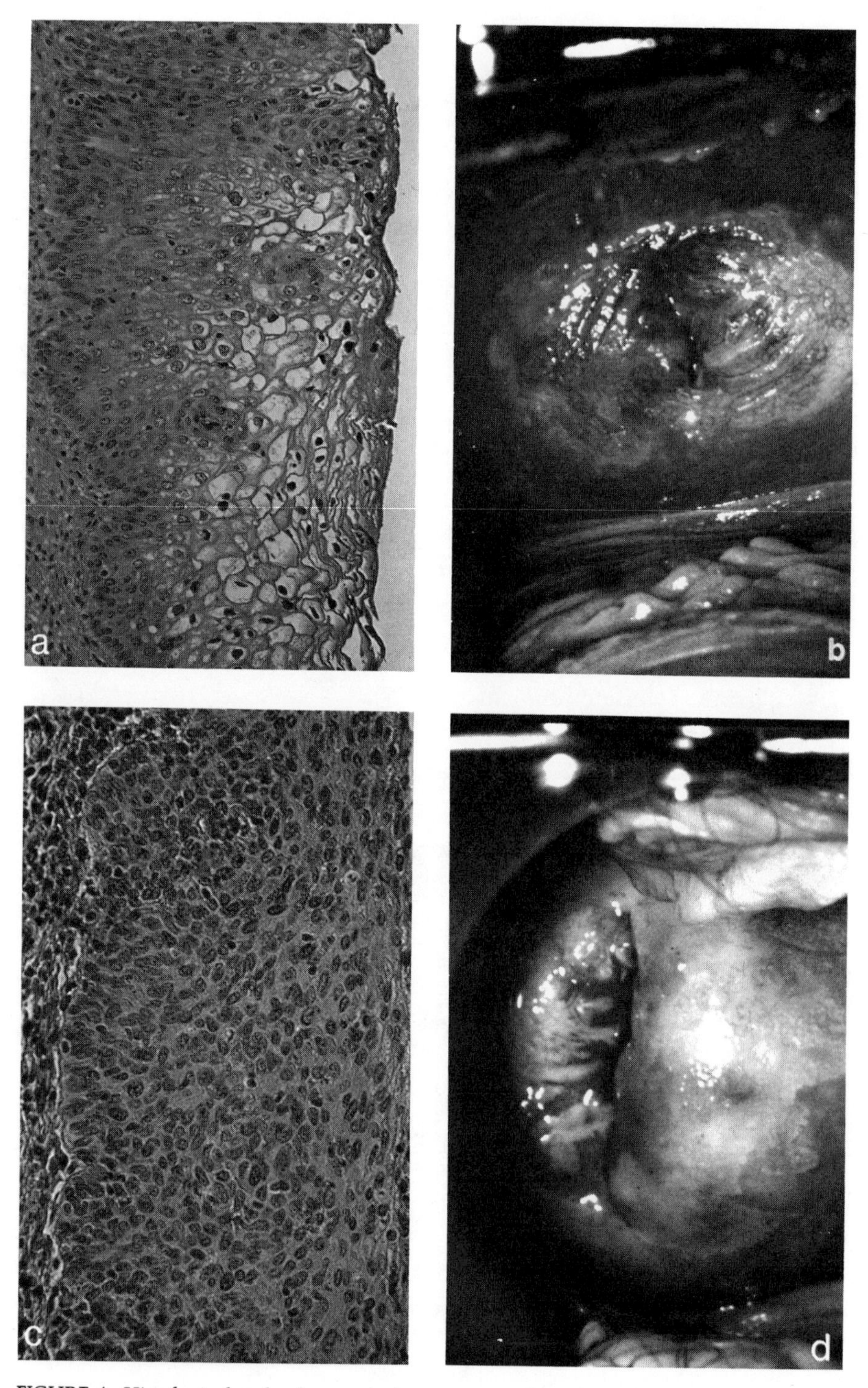

FIGURE 4. Histological and colposcopical appearance of HPV-11 infection (a, b) and HPV-16 infection of the cervix (c, d).

after surgical removal. Accurate data on average persistence of condylomata acuminata and cervical dysplasias are not available. Regression appears to occur less frequently when compared to common warts and after longer duration of the clinical symptoms. HPV-6 and 11 infections seem to regress more frequently than infections with HPV-16 or 18. Recent data indicate a substantial percentage of subclinical HPV-16 and HPV-18 infections (Schneider *et al.*, 1985; de Villiers *et al.*, unpublished data). Clinical observations support the view that the majority of all these lesions eventually disappear spontaneously, although a certain percentage of HPV-16- and HPV-18-positive proliferations progress into invasive cancer (Syrjänen *et al.*, 1985).

In view of the relatively low rates of penile and vulvar cancer in comparison to the high incidence rates of papillomavirus-induced lesions at these sites (de Villiers and zur Hausen, unpublished data), spontaneous regression or long-time persistence without malignant conversion must by far exceed the rare penile or vulvar lesions which convert into malignant tumors.

IV. HPV DNA SEQUENCES IN CELL LINES

Due to the lack of an *in vitro* system for papillomavirus replication and persistence, previous studies on human papillomaviruses entirely depended on the availability of suitable biopsy materials. Recently, several human cell lines have been identified, all derived from genital cancer, containing HPV-16 or 18 DNA (Boshart *et al.*, 1984; Schwarz *et al.*, 1985 and unpublished results; Yee *et al.*, 1985). A characterization of the positive lines is shown in Table II. All of these lines contain the viral DNA in an integrated form (Schwarz *et al.*, 1985 and unpublished data).

TABLE II. Cell Lines Containing HPV-16 or HPV-18 DNA Sequences

Cell line	Origin	Type of HPV DNA	Reference
HeLa	Cervical adenocarcinoma (Gey *et al.*, 1952)	HPV-18	Boshart *et al.* (1984), Schwarz *et al.* (1985)
SW 756	Squamous cell carcinoma cervix (Freedman *et al.* 1982)	HPV-18	Boshart *et al.* (1984), Schwarz *et al.* (1985)
C4-1	Squamous cell carcinoma cervix (Auersperg, 1964)	HPV-18	Boshart *et al.* (1984), Schwarz *et al.* (1985)
SiHa	Squamous cell carcinoma cervix (Friedl *et al.*, 1979)	HPV-16	Yee *et al.* (1985), Schwarz *et al.* (unpublished)
Caski	Squamous cell carcinoma cervix (Pattillo, 1977)	HPV-16	Yee *et al.* (1985), Schwarz *et al.* (unpublished)

It is of substantial interest that the originally circular viral DNA molecules show a remarkable specificity in their integration sites. In all HPV-18-positive lines (HeLa, C4-1, SW 756) the viral DNA molecules integrate within the E1, E2 ORFs (Schwarz *et al.*, 1985). Tentative data suggest that the viral DNA in the two other lines has a similar integration pattern (Schwarz *et al.*, unpublished results). Except for C4-1 and SiHa, all lines contain multiple genome copies. In HeLa and C4-1 cells, substantial parts of E2 and and L2 are deleted.

The arrangement of integrated viral genomes was analyzed in HeLa and SW 756 cells. The data reveal amplification of the integrated viral sequences with flanking host-cell DNA (Schwarz *et al.*, 1985). In HeLa cells, more than one integration site exists; one of them appears to result from a further deletion of HPV-18 DNA, being transposed with the 5′ flanking host-cell DNA (Schwarz *et al.*, unpublished data).

In all five lines the persisting viral DNA is actively transcribed (Schwarz *et al.*, 1985; Howley, personal communication). Analysis of transcription in HeLa, C4-1, and SW 756 cells showed that the E6/E7 region codes for RNA which is read through into the adjacent host-cell DNA. No evidence is yet available on similar integration sites within specific domains of host-cell DNA.

V. STATE OF VIRAL DNA IN BENIGN AND MALIGNANT LESIONS

The integrated state of HPV-16 and 18 DNA in cervical cancer cell lines contrasts reports on BPV and HPV nonintegration in other types of lesions (Orth *et al.*, 1980; Amtmann *et al.*, 1980; Ostrow *et al.*, 1982). Thus, it is of considerable interest to analyze the state of HPV-16 or 18 DNA in genital precursor lesions and directly in cervical cancer biopsies. An analysis of six nonmalignant HPV-16 lesions (bowenoid papulosis and two HPV-16-positive condylomatous lesions) revealed no demonstrable integrated viral DNA in these proliferations. In contrast, at least eight HPV-16- or HPV-18-positive malignant tumors contained integrated viral sequences, some in addition to episomally persisting viral DNA (Dürst *et al.*, 1985). In all tumors in which viral and flanking host-cell sequences have been analyzed, at least some viral molecules were interrupted within the E1 or E2 ORFs.

Thus, integration of viral DNA seems to occur rather regularly in the course of malignant conversion or during tumor progression. It will be interesting to see whether the so far consistent specificity of integration within the E1, E2 ORFs after analysis of a larger number of biopsies will be a specific feature of malignant tumors.

At least in some of the HPV-16 DNA-containing cervical cancer biopsies the persisting viral DNA is transcribed (Schwarz *et al.*, 1985). A detailed analysis of the transcripts is presently not available.

The presence of specific papillomavirus DNA sequences in malignant genital tumors and their precursor lesions, a difference in the state of viral DNA between precursor lesions and malignant tumors, a probably specific integration site within the viral genome, and transcription of a putative transforming gene in cell lines and at least several tumor biopsies appear to be strong arguments for a role of HPV infections in the etiology of human genital cancer.

VI. INTERACTION OF PAPILLOMAVIRUS INFECTIONS WITH INITIATORS IN MALIGNANT CONVERSION

Numerous observations in specific animal and human papillomavirus-induced proliferations which subsequently undergo malignant conversion stress a role of chemical or physical initiating factors in this sequence of events (see reviews by zur Hausen, 1977; Orth *et al.*, 1980; zur Hausen *et al.*, 1984). This is well demonstrated in epidermodysplasia verruciformis in man. Malignant conversion is noted almost exclusively in warts at sun-exposed sites. All converting papillomas contain only specific types of papillomaviruses, usually HPV types 5 or 8. A similar situation appears to account for other animal and human papillomatoses (see reviews cited above), justifying the questions whether this interaction is a general rule for papillomavirus-induced malignant tumors.

Are there initiating events affecting the human genital tract which in combination with HPV-16 and HPV-18 infections result in malignant conversion?

Indeed, in several recent studies it has been shown that heavy and prolonged smoking represents a significant risk factor for cervical cancer (Winkelstein, 1977; Wigle, 1980; Clarke *et al.*, 1982). It is likely that either initiating or promoting components of tobacco smoke reach the cervical cells via bloodstream and tissue diffusion and nicotine and cotinine have been demonstrated in vaginal fluids (Hoffmann *et al.*, 1985).

HSV infections, incriminated by seroepidemiological and experimental studies as playing a role in the induction of cervical cancer (see review by zur Hausen, 1983), were recently shown to reveal initiatorlike functions in infected cells (Schlehofer and zur Hausen, 1982; Schlehofer *et al.*, 1983; Matz *et al.*, 1984).

Besides smoking and HSV infections, initiating events may also originate from bacterial, protozoal, or chlamydial metabolites, exposing cervical sites to a much higher risk of initiating events than vulvar and penile locations. This could account for the approximately 20-fold higher risk of acquiring cervical cancer in comparison to vulvar and penile cancer. Chemical and physical initiators (carcinogens) appear to share two properties: they are usually mutagenic, although not all mutagens are carcinogens (MacMahon and Sugimura, 1984), and induce selective DNA amplification (SDA) in a wide variety of cells. SDA involves specific stretches of DNA and seems to be particularly important for initiation (Schlehofer *et al.*, 1983; Heilbronn *et al.*, 1985). It results from a cascade of events triggered by modifications at the DNA level. Functionally, SDA appears to lead to adaptation to modified microenvironmental conditions (Heilbronn *et al.*, 1984) but may also involve growth-promoting cellular sequences not expressed in the respective state of differentiation.

How can we reconcile these events following initiation with the epidemiology of human genital cancer and particularly with the specific HPV infections found in these tumors? Two possible models can be considered:

1. Recombinatory events induced by carcinogens could facilitate integration of viral DNA. In instances where integration specifically disrupts the regulation of early sequences, this would lead to a malignant cell clone. This mode of interaction would permit the prediction that mutations or deletions disrupting transactivation of the early region (Spalholz *et al.*, 1985) or occurring within the regulatory region could result in the same consequences. Thus, there would be no definite requirement for integration.
2. Alternatively, stabilization of mRNA or protein product of the E6–E7 region by fusion transcripts from adjacent host-cell DNA and/or functional modification may also lead to malignant conversion.

In view of the suspected role of SDA in initiation, it may be even more tempting to speculate on a switch-on of growth-promoting host-cell sequences by specific types of HPV infections amplified in rare instances by initiating events. Although this model would not immediately imply a need for viral DNA integration, increased stability of the transcribed product or additional functional modifications due to the fusion process could lead to a higher effectiveness of this process.

The availability of models explaining this interesting interaction between papillomavirus infections and truly carcinogenic events should stimulate further research to evaluate these synergistic or cooperative functions which are potentially very important for human carcinogenesis.

VII. ASSOCIATION OF MALIGNANT TUMORS WITH GENITAL PAPILLOMAVIRUS INFECTIONS AT EXTRAGENITAL SITES

Several HPV-16-positive carcinomas have been identified at extragenital sites: they have been recorded in the oral–respiratory region or at ano-perianal sites.

The presence of HPV-16 DNA in two of seven squamous cell carcinomas of the tongue is of specific interest since a third of these tumors contained an HPV-2-related papillomavirus (de Villiers *et al.*, 1986). This relatively high percentage of HPV-positive tongue carcinomas deserves special attention.

Among a high number of laryngeal carcinomas tested, one tumor contained HPV-16 DNA (Stremlau *et al.*, 1985; Scheurlen *et al.*, 1986), whereas a new papillomavirus (tentatively designated as HPV-30) was isolated from another laryngeal cancer biopsy (Kahn *et al.*, 1986).

One small cell carcinoma of the lung also contained HPV-16 DNA (Stremlau *et al.*, 1985). This tumor originated from a patient who was operated on 9 years earlier for cervical cancer. It is not possible to rule out the development of a late metastasis in this patient.

Some anal and perianal carcinomas also contain HPV-16 sequences (Beckmann *et al.*, 1985; Scheurlen *et al.*, unpublished data). The contribution of these viral infections to these types of tumors still awaits further clarification.

These isolates point to the importance of further studies in tumors of the respiratory and digestive tract for papillomavirus infections. Since many of these tumors are clearly related to other environmental carcinogens (e.g., smoking), they may have originated from a similar interaction between HPV infection and initiators as described before.

REFERENCES

Amtmann, E., Müller, H., and Sauer, G., 1980, Equine connective tissue tumors contain unintegrated bovine papilloma virus DNA, *J. Virol.* **35:**962–964.

Auersperg, N., 1964, Long-term cultivation of hypodiploid human tumor cells, *J. Natl. Cancer Inst.* **32:**135–163.

Beckmann, A. M., Daling, J. R., and McDougall, J. K., 1985, Human papillomavirus DNA in anogenital carcinomas, *J. Cell Biochem. Suppl.* **9c:**68.

Boshart, M., and zur Hausen, H., 1986, Human papillomavirus (HPV) in Buschke-Lowenstein tumors: Physical state of the DNA and identification of a tandem duplication in the non-coding region of a HPV-6 subtype, *J. Virol.* **58:**963–966.

Boshart, M., Gissmann, L., Ikenberg, H., Kleinheinz, A., Scheurlen, W., and zur Hausen, H., 1984, A new type of papillomavirus DNA, its presence in genital cancer biopsies and in cell lines derived from cervical cancer, *EMBO J.* **3:**1151–1157.

Buschke, A., and Löwenstein, L., 1931, Über carcinomähnliche Condylomata acuminata des Penis, *Arch. Dermatol. Syph.* **163:**30–46.

Cartwright, R. A., and Sinson, J. D., 1980, Carcinoma of penis and cervix, *Lancet* **1:**97.

Chen, E. Y., Howley, P. M., Levinson, A. D., and Seeburg, P. H., 1982, The primary structure and genetic organization of the bovine papillomavirus type 1 genome, *Nature* **299:**529–534.

Clarke, E. A., Morgan, R. W., and Newman, A. M., 1982, Smoking as a risk factor in cancer of the cervix: Additional evidence from a case control study, *Am. J. Epidemiol.* **115:**59–66.

Coggin, J. R., and zur Hausen, H., 1979, Workshop on papillomaviruses and cancer, *Cancer Res.* **39:**545–546.

Crum, C. P., Nagai, N., Mitao, M., Levine, R. U., and Silverstein, S. J., 1985, Histological and molecular analysis of early cervical neoplasia, *J. Cell Biochem. Suppl.* **9c:**70.

Danos, O., Katinka, M., and Yaniv, M., 1982, Human papillomavirus 1a complete DNA sequence: A novel type of genome organization among papovaviridae, *EMBO J.* **1:**231–236.

Dartmann, K., Schwarz, E., Gissmann, L., and zur Hausen, H., 1986, The nucleotide sequence and genome organization of human papilloma virus type 11, *Virology* **151:**124–130.

deLap, R., Friedman-Kien, A., and Rush, M. G., 1976, The absence of human papilloma viral sequences in condylomata acuminata, *Virology* **74:**268–272.

Della Torre, G., Pilotti, S., de Palo, G., and Rilke, F., 1978, Viral particles in cervical condylomatous lesions, *Tumori* **64:**456–463.

de Villiers, E.-M., Gissmann, L., and zur Hausen, H., 1981, Molecular cloning of viral DNA from human genital warts, *J. Virol.* **40:**932–935.

de Villiers, E.-M., Schneider, A., Gross, G., and zur Hausen, H., 1986, Analysis of benign and malignant urogenital tumors for human papillomavirus infection by labeling cellular DNA, *Med. Microbiol. Immunol.* **174:**281–286.

Duff, R., and Rapp, F., 1973, Oncogenic transformation of hamster embryo cells after exposure to inactivated herpes simplex virus type 1, *J. Virol.* **12:**209–217.

Dürst, M., Gissmann, L., Ikenberg, H., and zur Hausen, H., 1983, A papillomavirus DNA from a cervical carcinoma and its prevalence in cancer biopsy samples from different geographic regions, *Proc. Natl. Acad. Sci. USA* **80:**3812–3815.

Dürst, M., Kleinheinz, A., Hotz, M., and Gissmann, L., 1985, The physical state of human papillomavirus type 16 DNA in benign and malignant genital tumors, *J. Gen. Virol.* **66:**1515–1522.

Freedman, R. S., Bowen, J. M., Leibovitz, A., Pethak, S., Siciliano, M. J., Gallagher, H. S., and Giovanella, B. C., 1982, Characterization of a cell line (SW 756) derived from a human squamous carcinoma of the uterine cervix, *In Vitro* **18:**719–726.

Frenkel, N., Roizman, B., Cassai, E., and Nahmias, A., 1972, A DNA fragment of herpes simplex 2 and its transcription in human cervical cancer tissue, *Proc. Natl. Acad. Sci. USA* **69:**3784–3789.

Friedl, F., Kimura, I., and Osato, T., 1979, Studies on a new human cell line (SiHa) derived from carcinoma of uterus. I. Its establishment and morphology, *Proc. Soc. Exp. Biol. Med.* **135:**543–545.

Fu, Y. S., Reagan, J. W., and Richart, R. M., 1981, Definition of precursors, *Gynecol. Oncol.* **12:**220–231.

Galloway, D. A., and McDougall, J. K., 1983, The oncogenic potential of herpes simplex viruses: Evidence for a "hit-and-run" mechanism, *Nature* **302:**21–24.

Gey, G. O., Coffman, W. D., and Kubicek, M. T., 1952, Tissue culture studies of the proliferative capacity of cervical carcinoma and normal epithelium, *Cancer Res.* **12:**264–265.

Gissmann, L., and zur Hausen, H., 1980, Partial characterization of viral DNA from human genital warts (condylomata acuminata), *Int. J. Cancer* **25:**605–609.

Gissmann, L., Diehl, V., Schultz-Coulon, H.-J., and zur Hausen, H., 1982, Molecular cloning and characterization of human papilloma virus DNA derived from a laryngeal papilloma, *J. Virol.* **44:**393–400.

Gissmann, L., Wolnik, L., Ikenberg, H., Koldovsky, U., Schnürch, H. G., and zur Hausen, H., 1983, Human papillomavirus types 6 and 11 DNA sequences in genital and laryngeal papillomas and in some cervical cancers, *Proc. Natl. Acad. Sci. USA* **80:**560–563.

Graham, S., Priore, R., Graham, M., Browne, R., Burnett, W., and West, D., 1979, Genital cancer in wives of penile cancer patients, *Cancer* **44:**1870–1974.

Green, M., Brackmann, K. H., Sanders, P. R., Loewenstein, P. M., Freel, J. H., Eisinger, M., and Switlyk, S. A., 1982, Isolation of a human papillomavirus from a patient with epidermodysplasia verruciformis: Presence of related viral DNA in human urogenital tumors, *Proc. Natl. Acad. Sci. USA* **79:**4437–4441.

Gross, L., 1983, *Oncogenic Viruses,* Pergamon Press, Elmsford, N.Y.

Heilbronn, R., Schlehofer, J. R., and zur Hausen, H., 1984, Selective killing of carcinogen-treated SV-transformed Chinese hamster cells by a defective parvovirus, *Virology* **136:**439–441.

Heilbronn, R., Schlehofer, J. R., Yalkinoglu, A. Ö., and zur Hausen, H., 1985, Selective DNA-amplification induced by carcinogens (initiators): Evidence for a role of proteases and DNA polymerase alpha, *Int. J. Cancer* **36:**85–91.

Heilman, C. A., Law, M.-F., Israel, M. A., and Howley, P. M., 1980, Cloning of human papilloma virus genomic DNAs and analysis of homologous polynucleotide sequence, *J. Virol.* **36**:395–407.

Hoffmann, D., Hecht, S. S., Haley, N. J., Brunnemann, K. D., Adams, J. D., and Wynder, E. L., 1985, Turmorigenic agents in tobacco products and their uptake by chewers, smokers and nonsmokers, *J. Cell Biochem. Suppl.* **9C**:33.

Ikenberg, H., Gissmann, L., Gross, G., Grussendorf, E.-I., and zur Hausen, H., 1983, Human papillomavirus type 16-related DNA in genital Bowen's disease and in Bowenoid papulosis, *Int. J. Cancer* **32**:563–565.

Jagella, H. P., and Stegner, H. E., 1974, Zur Dignität der Condylomata acuminata: Klinische, histopathologische und cytophotometrische Befunde, *Arch. Gynaekol.* **216**:119–132.

Jenson, A. B., Rosenthal, J. D., Olson, C., Pass, F., Lancaster, W. D., and Shah, K., 1980, Immunologic relatedness of papillomaviruses from different species, *J. Natl. Cancer Res.* **64**:495–500.

Kahn, T., Schwarz, E., and zur Hausen, H., 1986, Molecular cloning and characterization of the DNA of a new human papillomavirus (HPV 30) from a laryngeal carcinoma, *Int. J. Cancer* **37**:61–65.

Kessler, I., 1977, Venereal factors in human cervical cancer: Evidence from marital clusters, *Cancer* **39**:1912–1919.

Koss, L. G., and Durfee, G. R., 1956, Unusual patterns of squamous epithelium of the uterine cervix: Cytologic and pathologic study of koilocytotic atypia, *Ann. N.Y. Acad. Sci.* **63**:1245–1261.

Laverty, C. R., Russel, P., Hills, E., and Booth, N., 1978, The significance of the non-condylomatous wart virus infection of the cervical transformation zone: A review with discussion of two illustrative cases, *Acta Cytol.* **22**:195–201.

Lorincz, A. T., Lancaster, W. D., and Temple, G. F., 1985, Detection and characterization of a new type of human papilloma virus, *J. Cell Biochem. Suppl.* **9c**:75.

MacMahon, B., and Sugimura, T. (eds.), 1984, *Coffee and Health*, Banbury Report 17, Cold Spring Harbor Laboratory, Cold Spring Harbor, N.Y.

Maggregor, J. E., and Innes, G., 1980, Carcinoma of penis and cervix, *Lancet* **1**:1246–1247.

Maitland, N. J., Kinross, J. H., Buttusil, A., Ludgate, S. M., Smart, G. E., and Jones, K. W., 1981, The detection of DNA tumour virus-specific RNA sequences in abnormal human cervical biopsies by in situ hybridization, *J. Gen. Virol.* **55**:123–137.

Martinez, J., 1969, Relationship of squamous cell carcinoma of the cervix uteri to squamous cell carcinoma of the penis, *Cancer* **24**:777–780.

Matz, B., Schlehofer, J. R., and zur Hausen, H., 1984, Identification of a gene function of herpes simplex virus type 1 essential for amplification of simian virus 40 DNA sequences in transformed hamster cells, *Virology* **134**:328–337.

Meisels, A., and Fortin, R., 1976, Condylomatous lesions of the cervix and vagina. I. Cytologic patterns, *Acta Cytol.* **20**:505–509.

Meisels, A., Fortin, R., and Roy, M., 1977, Condylomatous lesions of the cervix. II. Cytologic, colposcopic and histopathologic study, *Acta Cytol.* **21**:379–390.

Meisels, A., Roy, M., Fortier, M., Morin, C., Casas-Cordero, M., Shah, K. V., and Turgeon, H., 1981, Human papilloma virus infection of the cervix: The atypical condyloma, *Acta Cytol.* **25**:7–16.

Mounts, P., Shah, K. V., and Kashima, H., 1982, Viral etiology of juvenile- and adult onset squamous papilloma of the larynx, *Proc. Natl. Acad. Sci. USA* **79**:5425–5429.

Nahmias, A. J., Josey, W. E., Naib, Z. M., Luce, C. F., and Guest, B. A., 1969, Antibodies to herpes virus hominis types 1 and 2 in humans. II. Women with cervical cancer, *Am. J. Epidemiol.* **91**:547–552.

Oriel, J. D., 1971, Anal warts and anal coitus, *Br. J. Vener. Dis.* **47**:373–376.

Orth, G., Favre, N., Jablonska, S., Brylak, K., and Croissant, O., 1978, Viral sequences related to a human skin papilloma virus in genital warts, *Nature* **275**:334–336.

Orth, G., Favre, M., Breitburd, F., Croissant, O., Jablonska, S., Obalek, S., Jarzabek-Chorzelska, M., and Rzesa, G., 1980, Epidermodysplasia verruciformis: A model for the role of papillomaviruses in human cancer, in: *Viruses in Naturally Occurring Cancers* (M. Essex, G. Todaro, and H. zur Hausen, eds.), pp. 259–282, Cold Spring Harbor Laboratory, Cold Spring Harbor, N.Y.

Ostrow, R. S., Bender, M., Niimura, M., Seki, T., Kawashima, M., Pass, F., and Faras, A. J., 1982, Human papillomavirus DNA in cutaneous primary and metastasized squamous cell carcinomas from patients with epidermodysplasia verruciformis, *Proc. Natl. Acad. Sci. USA* **79:**1634–1638.

Park, M., Kitchener, H. C., and Macnab, J. C. M., 1983, Detection of herpes simplex virus type 2 DNA restriction fragments in human cervical carcinoma tissue, *EMBO J.* **2:**1029–1034.

Pattillo, R. A., 1977, Tumor antigen and human gonadotropin in Caski cells: A new epidermoid cervical cancer cell line, *Science* **196:**1456–1458.

Purola, E., and Savia, E., 1977, Cytology of gynecologic condyloma acuminatum, *Acta Cytol.* **21:**26–31.

Rawls, W. E., Tompkins, W. A. F., Figueroa, M. E., and Melnick, J. L., 1968, Herpes simplex virus type 2: Association with carcinoma of the cervix, *Science* **161:**1255–1256.

Rawls, W. E., Clarke, A., Smith, K. O., Docherty, J. J., Gilman, S. C., and Graham, S., 1980, Specific antibodies to herpes simplex virus type 2 among women with cervical cancer, in: *Viruses in Naturally Occurring Cancers* (M. Essex, G. Todaro, and H. zur Hausen, eds.), pp. 117–133, Cold Spring Harbor Laboratory, Cold Spring Harbor, N.Y.

Rigoni-Stern, D., 1842, Fatti statistici relativi alle malatie cancerose, *Giorn. Service Prog. Pathol. Terap.* **2:**507–517.

Rotkin, I. D., 1973, A comparison review of key epidemiological studies in cervical cancer related to current searches for transmissible agents, *Cancer Res.* **33:**1353–1367.

Rüger, R., Bornkamm, G. W., and Fleckenstein, B., 1984, Human cytomegalovirus DNA sequences with homologies to the cellular genome, *J. Gen. Virol.* **65:**1351–1364.

Rutledge, F., and Sinclair, M., 1968, Treatment of intraepithelial carcinoma of the vulva by skin excision and graft, *J. Obstet. Gynecol.* **102:**806–818.

Scheurlen, W., Stremlau, A., Gillmann, L., Höhn, D., Zenner, H.-P. and zur Hausen, H., 1986, Rearranged HPV 16 molecules in an anal and in a laryngeal carcinoma, *Int. J. Cancer* **38:**671–676.

Schlehofer, J. R., and zur Hausen, H., 1982, Induction of mutations within the host genome by partially inactivated herpes simplex virus type 1, *Virology* **122:**471–475.

Schlehofer, J. R., Gissmann, L., and zur Hausen, H., 1983, Herpes simplex virus induced amplification of SV40 sequences in transformed Chinese hamster embryo cells, *Int. J. Cancer* **32:**99–103.

Schneider, A., Kraus, H., Schuhmann, R., and Gissmann, L., 1985, Papillomavirus infection of the lower genital tract: Detection of viral DNA in gynecological swabs, *Int. J. Cancer* **35:**443–448.

Schwarz, E., Dürst, M., Demankowski, C., Lattermann, O., Zech, R., Wolfsperger, E., Suhai, S., and zur Hausen, H., 1983, DNA sequence and genome organization of genital human papillomavirus type 6b, *EMBO J.* **2:**2341–2348.

Schwarz, E., Freese, U. K., Gissmann, L., Mayer, W., Roggenbuck, B., Stremlau, A., and zur Hausen, H., 1985, Structure and transcription of human papillomavirus sequences in cervical carcinoma cells, *Nature* **314:**111–114.

Seedorf, K., Krämmer, G., Dürst, M., Suhai, S., and Röwekamp, W. G., 1985, Human papillomavirus type 16 sequence, *Virology* **145:**181–186.

Shah, K. H., Lewis, M. G., Jenson, A. B., Kurman, R. J., and Lancaster, W. D., 1980, Papillomavirus and cervical dysplasia, *Lancet* **2:**1190.

Smith, P. G., Kinley, L. J., White, G. C., Adelstein, A. M., and Fox, A. J., 1980, Mortality of wives of men dying with cancer of the penis, *Br. J. Cancer* **41:**422–428.

Spalholz, B. A., Yang, Y.-C., and Howley, P. M., 1985, Trans-activation of a BPV-I enhancer, *J. Cell Biochem. Suppl* **9c:**89.

Stremlau, A., Gissmann, L., Ikenberg, H., Stark, M., Bannasch, P., and zur Hausen, H., 1985, Human papillomavirus type 16-related DNA in an anaplastic carcinoma of the lung, *Cancer* **55:**737–740.

Syrjänen, K., de Villiers, E.-M., Saarikoski, S., Castren, O., Väyrynen, M., Mäntyjärvi, R., and Parkkinnen, S., 1985, Cervical papillomavirus infection progressing to invasive cancer in less than three years, *Lancet* **1**:510–511.

Teokharov, B. A., 1962, On the problem of the nature and epidemiology of acuminate condylomatosis, *Vestn. Dermatol. Venerol.* **36**:51–56.

Vonka, V., Kanka, J., Jelinek, J., Subrt, I., Sucharek, A., Havrankova, A., Vachal, M., Hirsch, I., Domorazkova, E., Zavadova, H., Richterova, V., Naprstkova, J., Dvorakova, V., and Svoboda, B., 1984a, Prospective study on the relationship between cervical neoplasia and herpes simplex type-2 virus. I. Epidemiological characteristics, *Int. J. Cancer* **33**:49–60.

Vonka, V., Kanka, J., Hirsch, I., Zavadova, H., Krcmar, M., Suchankova, A., Rezakova, D., Broucek, J., Press, M., Domorazkova, E., Svoboda, B., Havrankova, A., and Jelinek, J., 1984b, Prospective study on the relationship between cervical neoplasia and herpes simplex type-2 virus. II. Herpes simplex type-2 antibody presence in sera taken at enrollment, *Int. J. Cancer* **33**:61–66.

Waterhouse, J., Muir, C., Shanmugaratnam, K., and Powell, J. (eds.), 1982, *Cancer Incidence in Five Continents,* Vol. 4, Int. Agric. Res. Cancer, Lyons.

Wigle, D. T., 1980, Re: 'Smoking and cancer of the cervix: Hypothesis,' *Am. J. Epidemiol.* **111**:125–127.

Winkelstein, W., Jr., 1977, Smoking and cancer of the uterine cervix, *Am. J. Epidemiol.* **106**:257–259.

Yee, C., Krishnan-Hewlett, I., Baker, C. C., Schlegel, R., and Howley, P., 1985, Presence and expression of human papillomavirus sequences in human cervical carcinoma cell lines, *Am. J. Pathol.* **119**:361–366.

zur Hausen, H., 1975, Oncogenic herpesvirus, *Biochim. Biophys. Acta* **417**:25–53.

zur Hausen, H., 1976, Condylomata acuminata and human genital cancer, *Cancer Res.* **36**:794.

zur Hausen, H., 1977, Human papillomaviruses and their possible role in squamous cell carcinomas, *Curr. Top. Microbiol. Immunol.* **78**:1–30.

zur Hausen, H., 1982, Human genital cancer: Synergism between two virus infections or synergism between a virus infection and initiating events? *Lancet* **2**:1370–1372.

zur Hausen, H., 1983, Herpes simplex virus in human genital cancer, *Int. Rev. Exp. Pathol.* **25**:307–326.

zur Hausen, H., Meinhof, W., Scheiber, W., and Bornkamm, G. W., 1974, Attempts to detect virus-specific DNA in human tumors. I. Nucleic acid hybridizations with complementary RNA of human wart virus, *Int. J. Cancer* **13**:650–656.

zur Hausen, H., Gissmann, L., Steiner, W., Dippold, W., and Dregger, I., 1975, Human papilloma virus and cancer, *Bibl. Haematol. (Basel)* **43**:569–571.

zur Hausen, H., Gissmann, L., and Schlehofer, J. R., 1984, Viruses in the etiology of human genital cancer, *Prog. Med. Virol.* **30**:170–186.

CHAPTER 10

Laryngeal Papillomas
Clinical Aspects and *in Vitro* Studies

BETTIE M. STEINBERG

I. INTRODUCTION

Laryngeal papillomas are rapidly growing, noninvasive epithelial tumors caused by human papillomaviruses (HPV). They were first described as "warts in the throat" in the 17th century, differentiated clinically from other laryngeal masses and called papilloma in 1871. They are usually located on the true vocal cords and epiglottis, but can involve all of the larynx, the trachea, bronchial tubes, and even the lungs.

The actual incidence of the disease in the United States is not known, since it is not classed as a mandatory reportable disease by any of the State Health Departments or the National Communicable Disease Center. Estimates place the figure at approximately 1 per 100,000 people. In England, there is one case of laryngeal papillomas to 1000 to 6000 admissions to children's hospitals (Hajek, 1956). There is a generally perceived idea that the worldwide frequency seems to differ according to locality, with South America being the highest.

In spite of its relatively low incidence, laryngeal papillomatosis is of major concern to clinicians and interest to molecular biologists. This is in part because it has a high degree of morbidity and mortality, and managing the disease is of significant cost to the patient and to society. In this chapter we will discuss various aspects of laryngeal papillomatosis: clinical description, treatment, pathology, and molecular and cellular biology.

BETTIE M. STEINBERG • Department of Otolaryngology, Long Island Jewish Medical Center, New Hyde Park, New York 11042.

II. CLINICAL DESCRIPTION

Clinical symptoms usually start as hoarseness or abnormal cry, with subsequent respiratory distress, stridor, and aphonia. If not treated, these lesions can become extensive enough to cause death. In this respect they are quite different from benign papillomas in other parts of the body, which can be disfiguring, unpleasant, and associated with significant morbity, but are not usually life threatening unless they convert to a malignancy. The serious nature of laryngeal papillomas results from massive amounts of tissue in the airway, which has a restricted size and can tolerate very little obstruction. The appearance of a normal larynx, with the vocal cords partially open, is shown in Fig. 1A. A large papilloma on the epiglottis can be seen in Fig. 1B. It is obvious that a papilloma of that size would significantly block the airway if it were located on the vocal cords.

The most serious form of laryngeal papillomatosis involves extension into the trachea, bronchial tubes, and lungs. Weiss and Kashima (1983) have reported that, in their studies, 26% of the patients had tracheal involvement, 5% had bronchial lesions, and 8% (3 of 39 patients) had pulmonary involvement. These numbers are higher than those reported by others (Singer *et al.*, 1966; Strong *et al.*, 1976) which ranged from 2% to 17% for tracheal extension. All studies agree that the incidence of tracheal disease is greater than 50% in those patients who have undergone a tracheotomy to establish and maintain an airway. Pulmonary disease, while rare, is extremely difficult to manage. It is frequently fatal, even in patients receiving medical treatment.

The most notable characteristic of laryngeal papillomas is their tendency to recur following surgical removal. It has been frequently stated that recurrent disease is restricted to children, and it is often referred to as juvenile laryngeal papillomatosis. However, this concept is not accurate. Better than 30 years ago, Holinger *et al.* (1950) reported that among their patients were nearly equal numbers of juvenile and adult-onset disease. The favored term today is recurrent laryngeal papillomatosis.

Table I summarizes some of the clinical data for 45 patients at Long Island Jewish Medical Center (courtesy of Dr. Allan Abramson). The pattern is typical for this disease. Equal numbers of children and adults presented as patients. It is unusual for papillomas to present in the adolescent years and there were none in this series. Sex distribution in juvenile-onset patients is equal, while adult-onset is more common in males (15:7). The need for repeated surgery is seen in both groups. Only one of the juvenile-onset and four of the adult-onset patients were free of disease after one operation (data not shown). More common is the pattern shown in Table I, with the majority of patients of both groups requiring surgery at least once or twice per year. It is clear that juvenile-onset disease tends to be more aggressive, since none of the adult-onset patients averaged three or more surgical removal procedures per year over

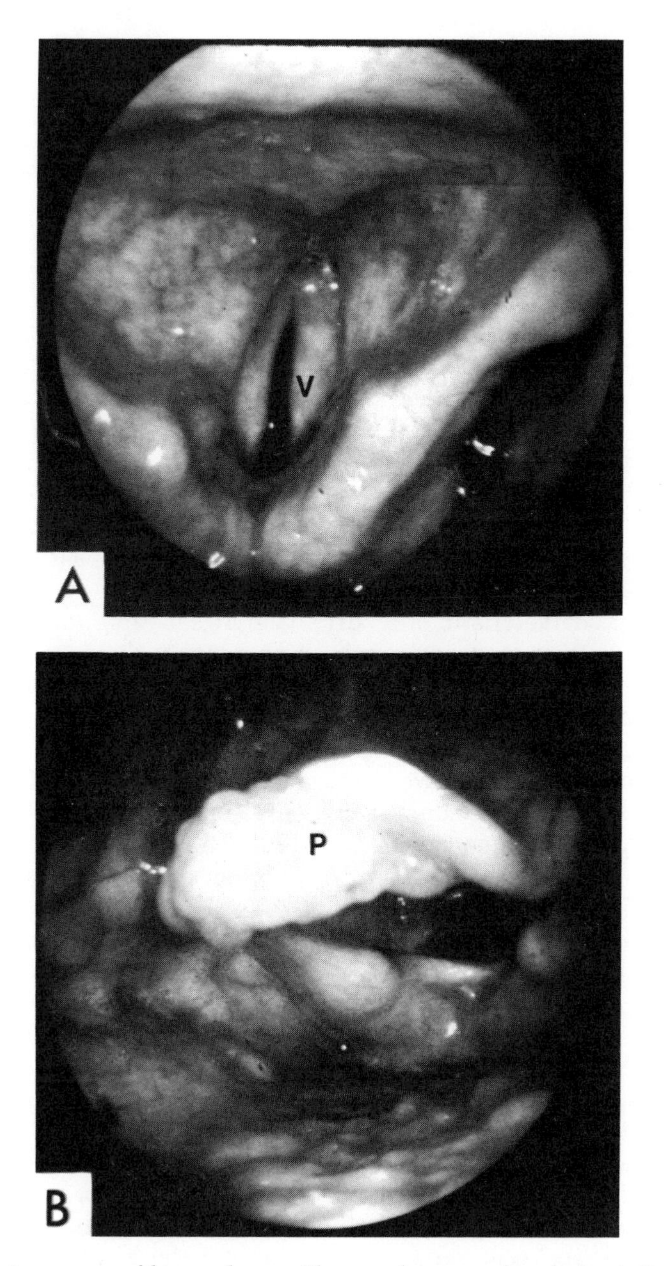

FIGURE 1. Appearance of human larynx. Photographs were taken during indirect laryngoscopy. (A) Normal larynx, clearly showing the smooth edges of the vocal cords (V). (B) A large papilloma (P) on the epiglottis prevents visualization of the vocal cords. Reprinted by permission of *Clinics in Dermatology* (S. Jablonska, G. Orth, and L. Parish, eds.), Oct.–Dec., 1985.

TABLE I. Clinical Pattern of Recurrent Laryngeal Papillomas

	Sex		Frequency of surgery (percent of patients)[a]		
Age of onset	M	F	3 or more/year	1–2/year	Less than 1/year
0–10	11	12	33	44	23
20–60+	15	7	0	88	22

[a] Averaged over total duration of active disease to date.

the total course of their illness. This does not, however, mean that the adult-onset papilloma is always slow growing. One recent patient, not included in these data since she has had papilloma less than 1 year, has had three surgical removals in less than 8 months. She can be compared to two children in this series who had 25 operations in 4 years and 26 operations in 6 years, respectively. These patients have particularly severe disease, but they typify the serious nature of recurrent laryngeal papillomatosis.

Another characteristic of recurrent laryngeal papillomas, like other papillomas, is their tendency to regress spontaneously in the majority of patients. At the present time, 62% of the juvenile-onset patients and 82% of the adult-onset patients shown in Table I are free of symptoms. The period of active disease ranged from 1 year to 22 years before remission was achieved. Conversely, two of the patients with juvenile-onset disease have been symptomatic for 24 years and show no signs of remission. Remission can be temporary, or can last for the lifetime of the patient. Since there is always the possibility of recurrence, we talk only about remission, not cure.

It is very possible that hormonal status plays some role in the course of laryngeal papillomatosis. There is a 2:1 preponderance of males in adult-onset disease, which is not seen in juvenile-onset (Table I; Holinger *et al.*, 1968a). It is generally believed that juvenile-onset disease regresses at or before puberty. Like other "rules" regarding laryngeal papillomas, this is frequently not observed. In the series of patients reviewed in Table I, 9 of the juvenile-onset patients still have active disease. Of these 9, 4 are now adults. If hormonal changes at puberty play a role in inducing remission, it does not happen in every patient. In support of the "rule," however, the average age of those juvenile-onset patients who did enter remission was 10.2 years, and only 3 of the 14 patients were in their teens when remission began (data not shown).

Hormonal changes during pregnancy may also affect the clinical course of laryngeal papillomas. There are reports in the literature of laryngeal papillomas becoming both worse and better during pregnancy (Holinger *et al.*, 1950, 1968a). The patients described in Table I include seven who become pregnant. Of the three who were already in remission, one had a reactivation of disease and two did not. Three of the four with active disease required increased frequency of surgery during pregnancy, and one was unchanged. In all cases where pregnancy exacerbated the disease, frequency of surgery returned to previous levels as soon as the babies were born. It is safe to say that pregnancy tends to enhance the growth of laryngeal papillomas.

The epidemiology of laryngeal papillomas is only partially resolved. Cook *et al.* (1973) and Quick *et al.* (1980) have data which suggested that there is a link between genital condyloma and laryngeal papillomas. Two-thirds of the patients in the study by Quick *et al.* (1980) had a positive maternal history of condyloma, and those with a definite negative history were primarily adult-onset patients. The source of infection in the adult-onset patients is not known. It is postulated that infants born to women with condyloma are infected during birth. However, there are instances where newborns, who were delivered by cesarean section, developed laryngeal papillomas during the first weeks of life (Allan Abramson, personal communication). Thus, it is possible that the fetus can be infected *in utero*.

Laryngeal papillomatosis is a serious and potentially life-threatening disease even though the tumors are usually benign. It is, however, possible for them to become malignant. There are reports in the literature documenting malignant transformation, either spontaneously or after x-irradiation (Ogilvie, 1953; Altman *et al.*, 1955; Galloway *et al.*, 1960; Justus *et al.*, 1970; Zehnder and Lyons, 1975; zur Hausen, 1977). The incidence of spontaneous transformation is very low. Irradiation with x-rays markedly increases the risk of malignant conversion. In a large study of patients at the Mayo Clinic from 1914 to 1960, 14% of the 43 treated with radiation therapy developed squamous cell carcinoma, while a control group of 58 papilloma patients treated with surgery alone showed no malignant conversion (Majoros *et al.*, 1963).

This description of the clinical pattern of laryngeal papillomatosis can be summarized as follows: It is a serious disease with a high degree of morbidity and a great deal of unpredictability.

III. TREATMENT

Many different types of treatment have been employed to control laryngeal papillomas, with varying degrees of success. These regimens can be classified as physical, surgical, medical, and immunological. The very fact that so many modalities have been used attests to the difficulty in managing patients with this disease, and that none of the treatments has successfully cured or induced remission consistently in the majority of patients. In almost every case, preliminary reports of a new therapy look promising, and engender a great deal of interest. Application of the procedure to additional patients by other investigators is usually disappointing. The various classes of therapy will be briefly discussed.

A. Physical Treatment

Ultrasound therapy had been the subject of enthusiastic reports but it was cautioned that we must await long-term follow-up to evaluate the side effects on neural tissue and growth centers of the larynx (Holinger *et al.*, 1968a), and has fallen into disfavor. Cryosurgery using liquid nitrogen was used in the 1960s but without too much success. Cauterization with corrosives and electrocautery have been used, but repeated application has resulted in excessive scarring. The local application of nitric and trichloracetic acid, and formaldehyde have a significant sclerosing effect which can lead to laryngeal stenosis. This, in turn, can be severe enough to require tracheotomy. Podophyllum, a caustic resin which has been extremely successful in treating venereal papillomas, has also been used for laryngeal papillomas. Hollingsworth *et al.* (1950) initially recommended it. In 1982, Dedo and Jackler reported using 25% podophyllum in conjunction with CO_2 laser surgery on 109 patients, a very large series. However, their success correlated well with that of Strong *et al.* (1976) who only used the CO_2 laser, so that the effectiveness of the podophyllum is questionable. As previously discussed, radiation therapy has been shown to greatly increase the possibility of malignant conversion, and is no longer being used (Zehnder and Lyons, 1975).

B. Surgical Treatment

Repeated surgical removal at proper intervals is at present the most effective method of protecting the larynx from airway obstruction by papilloma. Traditionally, this was done using forceps to strip the papillomas from the laryngeal surfaces. Recently, the CO_2 laser has been used in conjunction with the operating microscope to remove the papillomas in a safe and accurate method (Jako, 1972; Strong *et al.*, 1976). Tissue is vaporized to a controlled depth depending on power setting (Stern *et al.*, 1980). Bleeding is controlled, and edema following this procedure is much less than with cup forceps removal. Although laser surgery, like conventional surgery, is primarily intended to control papillomatosis, the frequency of remission appears to be higher in patients operated on in this way and fewer tracheostomies are required (Strong *et al.*, 1976).

C. Medical Treatment

The clinical pattern of active disease followed by abrupt spontaneous remissions, which was discussed previously, has made evaluation of medical treatment difficult. Large numbers of patients must be evaluated, with a significant number of controls, to draw valid conclusions. Until very recently this was not done. Rather, the literature contains many reports of therapies used in small numbers of patients. Hormones, especially estradiol, have been taken orally, applied topically, and injected into the papilloma with varying success (Broyles, 1940; Szpunar, 1977). Steroids have been used without significant benefit. Antibiotics such as tetracycline, penicillin, and streptomycin have been used as an adjunct to forceps removal without apparent effect (Holinger *et al.*, 1950).

Chemotherapeutic drugs normally used for treatment of malignancies have also been used to treat laryngeal papillomas, including methotrexate, bleomycin, and 5-fluorouracil (Holinger *et al.*, 1968a; Smith *et al.*, 1980). These are very toxic agents, and their use has not been widespread. The very fact that they have been tried indicates the difficulty in managing this disease.

Among the newest drug treatments is the use of retinoids. These analogues of vitamin A have been successfully used to treat acne. They function to alter the differentiation of epithelial cells, with clear differences in the responses of different epithelial tissues (Sherman, 1961). It is premature to state whether retinoids will be effective against laryngeal papillomas.

D. Immunotherapy

An entirely different approach to treating laryngeal papillomas has been the use of immunological therapy. Because of the suspected viral etiology of human laryngeal papilloma, attempts have been made to immunize patients against their own virus, using autogenous vaccine. Holinger *et al.* (1968b) used this type of vaccine in 37 patients and found that the annual number of operations per patient decreased. Strome (1969) was unable to produce similar favorable results in 7 cases. Transfer factor, mumps skin test antigen, and Levamisole have been used either alone or in combination therapy with laryngeal surgery to alter the course of the disease (Holinger *et al.*, 1968b; DeRosa, 1977; Greensher, 1980). These too have been disappointing, although they are still being used occasionally (Borkowsky, 1984).

The one nonsurgical treatment which appears to hold great promise is the use of interferon. This might be classed as either medical or immunological therapy. α-Interferon, the type which has been most thoroughly investigated in this context, is produced naturally by and purified from human lymphocytes in response to viral infection.

Haglund *et al.* (1981) were the first to report on the use of interferon for this disease. Three million units of interferon were given to seven patients three times weekly by intramuscular injection, for periods of time ranging from 9 to 18 months. At that time, with complete regression of tumors, the frequency of injections was reduced to twice weekly, then once weekly, and if they remained tumor free, finally discontinued. Within 2 months, five of the seven patients had such severe recurrence of papillomas that interferon therapy had to be renewed or increased. Again, there was complete regression of tumors. Four of these cases remained tumor free after the second course of treatment, while one required the initiation of a third course.

Similar protocols, in conjunction with initial removal of papillomas using the CO_2 laser, have subsequently been used by investigators in Europe and the United States with comparable results for relatively small numbers of patients (Schouten *et al.*, 1982; Goepfert *et al.*, 1982; McCabe and Clark, 1983). Most of these investigators used α-interferon, but Schouten *et al.* (1982) compared α- and β-interferon on three cases of laryngeal papilloma. They found that α-interferon was more effective in controlling growth of the tumors.

A very large clinical trial is presently under way, under the direction of Dr. Gerald Healy (Children's Hospital, Boston), to determine more precisely the effectiveness of interferon in pediatric patients. Under the sponsorship of the National Institutes of Health, patients are being enlisted from a large number of participating hospitals throughout the United States. Patients are assigned to one of two groups: one receives interferon, the other serves as controls. A second large study, sponsored by Burroughs–Wellcome Company, is being conducted from Johns Hopkins University, Baltimore, under the direction of Dr. Haskins Kashima. In this study, each patient serves as both control and experimental subject, receiving interferon for 6 months and a placebo for 6 months. These studies should finally result in a definitive statement regarding the effectiveness of interferon in either controlling or curing laryngeal papillomatosis. The preliminary results are most encouraging. It has been stated that if interferon was a drug looking for a disease to treat, perhaps laryngeal papillomatosis is the disease.

IV. PATHOLOGY

The laryngeal cavity is a part of the respiratory system, and the mucous membrane is contiguous with that of the mouth and pharynx, and the trachea below. Most of the laryngeal membrane, like the trachea, is composed of ciliated columnar epithelium. However, most of the epiglottis and the false cord and all of the true cord are covered by stratified squamous epithelium (Hast, 1984). It is the stratified squamous epithelium which develops papillomas following HPV infection.

Laryngeal papillomas appear as white or pink-red, glistening wartlike masses composed of vascular connective tissue cores covered by a hyperplastic stratified squamous epithelium. There are no distinctive histological differences between the adult and juvenile-onset recurring papillomas. However, Batsakis *et al.* (1983) do distinguish these from solitary adult "papillomas" which they believe are not true papillomas.

The differences between the normal stratified squamous epithelium of the larynx and laryngeal papilloma are mostly but not completely quantitative (Incze *et al.*, 1977). The structural composition of the basement membrane region is usually normal. However, it may occasionally be discontinuous, with the cytoplasmic processes from the basal cells extending into the connective tissue (Lundquist *et al.*, 1975). In the submucosal and epithelial layer, polymorphonuclear leukocytes and lymphocytes can sometimes be identified. Electron microscopy reveals the papilloma cell membrane to have umbilicated microvilli, increased numbers of desmosomes, and increased amounts of fibrillar material in dilated intercellular spaces (Scully *et al.*, 1983).

The epithelial portion of the lesion is characteristically composed of a much thickened spinous layer, with increased nucleoli, and with variation in the size and shape of the cells. Cross sections of a normal vocal cord and vocal cord papilloma are shown in Fig. 2. Note the connective tissue cores with open blood vessels and the very thickened spinous layer in the papilloma.

The normal membrane covering the vocal cords is classified as nonkeratinizing stratified squamous epithelium. This means that the final differentiation of the cells does not result in a "stratum corneum" but the cells, like all epithelial cells, do make cytokeratins. The absence of a cornified layer is clearly seen in Fig. 2A. Typical laryngeal papillomas also do not form a cornified layer (Fig. 2B). In fact, maturation of the superficial epithelial layer appears delayed, with less flattening of the cells. Conversely, some of the solitary adult-onset papillomas are hyperkeratinized. Again, this suggests that they are a different entity. Koilocytes, large cells with an irregularly shaped nucleus and a perinuclear halo (Koss and Durfee, 1956), can often be seen in laryngeal papillomas. These cells probably reflect a cytopathic effect of the HPV. They are commonly seen in other HPV infections as well. Figure 3, an enlargement of one area from Fig. 2B, clearly shows the presence of koilocytes in this papilloma.

The normal tissue has patchy areas which stain with an antibody directed against filaggrin, a differentiation-specific protein found in the granular layer of stratified squamous epithelium (Fig. 4). Recurrent papillomas appear to make little or no filaggrin. The papilloma shown in Fig. 5 had only two cells which contained filaggrin, while the normal tissue (Fig. 4) had several hundred positive cells in a section of similar size. This also suggests that differentiation is delayed in papillomas.

Quick *et al.* (1979) histologically classified laryngeal papillomas into two groups, typical and atypical according to the relative width of the proliferative zone of the epithelium and to the pleomorphism and lack of polarization of the cells. Typical papillomas fit the description given above: thickened spinous layer, intact basement membrane, and a normal-appearing basal–parabasal layer. Atypical papillomas were classified as mild, moderate, or severe, based primarily on the thickness of the basal–parabasal layer. Severe atypia was also characterized by abnormal

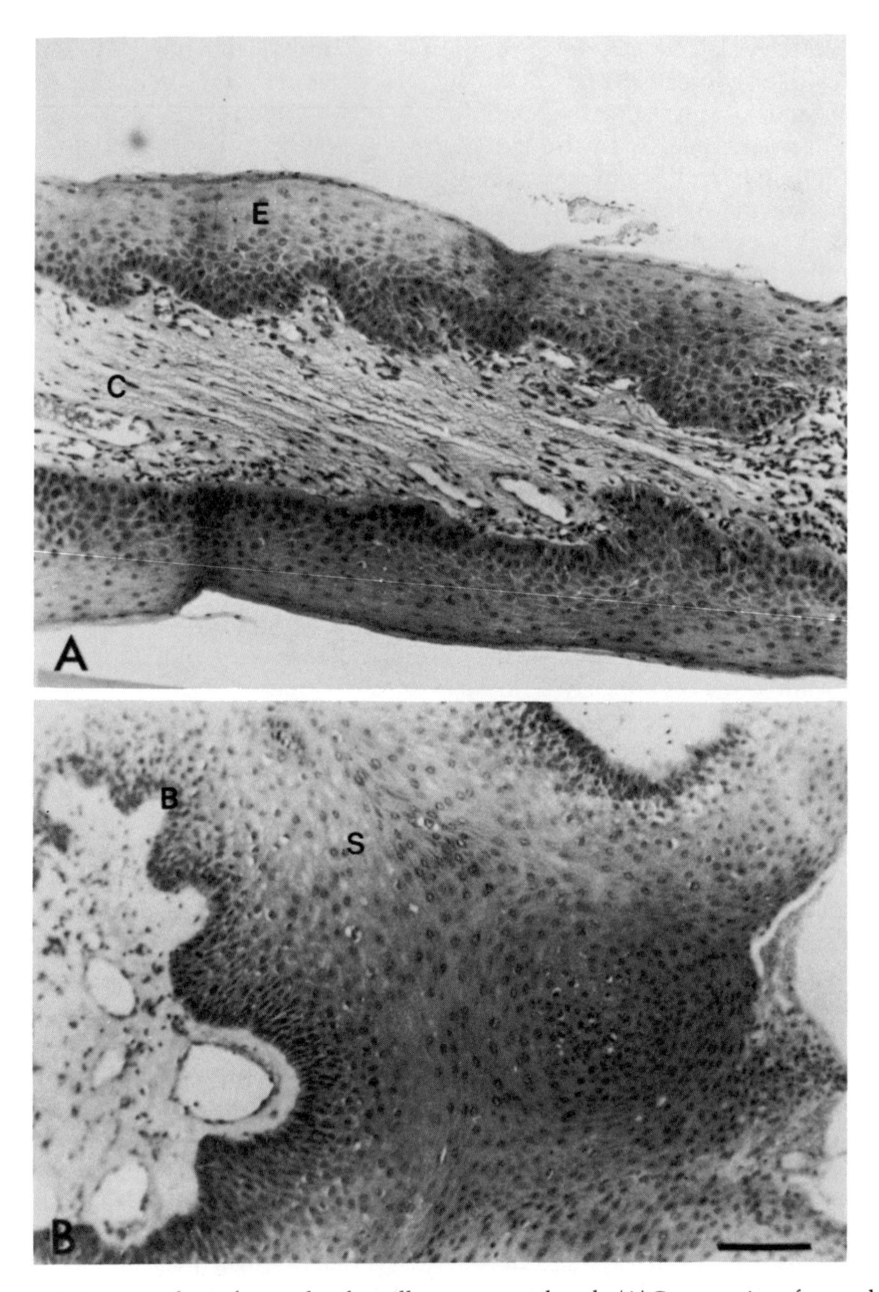

FIGURE 2. Histology of normal and papillomatous vocal cords. (A) Cross section of normal cord, showing relatively thin epithelium (E) with center core of connective tissue (C). (B) Cross section through one portion of papilloma. Note very thickened spinous layers above basal cells (B) surrounding one of the many connective tissue cores. H & E; bar = 100 μm. Reprinted by permission of *Clinics in Dermatology* (S. Jablonska, G. Orth, and L. Parish, eds.), Oct.–Dec., 1985.

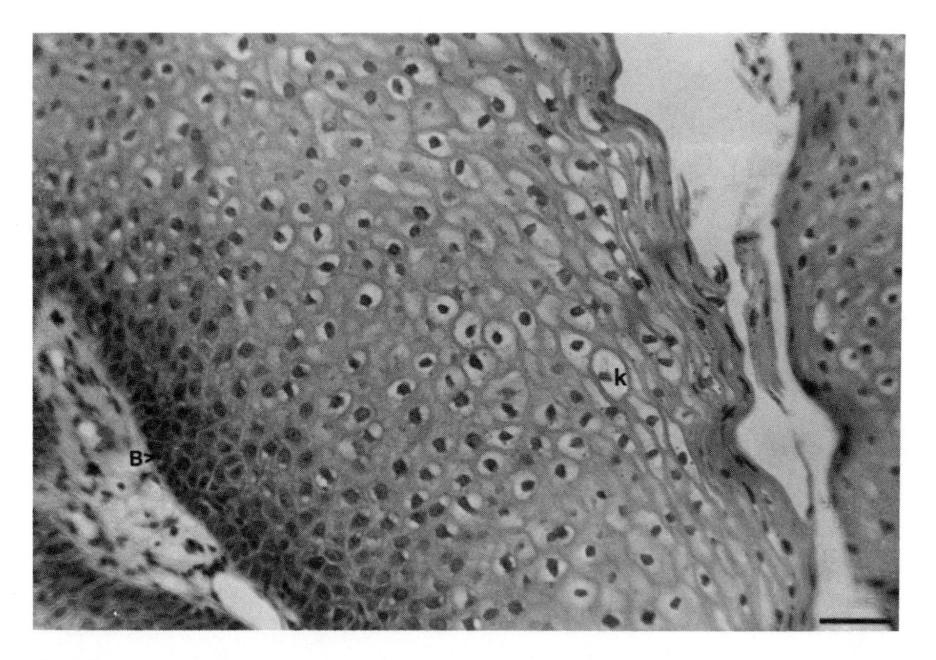

FIGURE 3. Presence of koilocytes in laryngeal papillomas. Cross section of a laryngeal papilloma. "K" marks one of the numerous koilocytotic cells located in the spinous layer of the tissue. These cells are not present in the basal (B) and immediate superbasal layer. H & E; bar = 20 μm

mitotic figures. Quick *et al.* (1979) suggested a correlation between grade of atypia and frequency of clinical recurrence, with the moderate and severe atypia requiring more frequent surgery. It still remains uncertain whether the atypia found is a true neoplastic alteration or whether it reflects the rapid but benign growth of the lesion.

V. MOLECULAR AND CELLULAR BIOLOGY

A. Etiology

The first evidence that laryngeal papilloma was caused by a virus was provided by Ullman (1923). He prepared a sterile, cell-free extract from a papilloma, injected it into his own arm, and developed a skin wart. However, attempts to visualize viral particles in the nuclei of these papillomas have been only marginally successful. Svoboda *et al.* (1963) reported the observation of viruslike particles by electron microscopy. A more definite description of viral crystalline assays was reported by Boyle *et al.* (1973). However, Cook *et al.* (1973) reported negative results, and Spoendlin and Kistler (1978) were able to find viral particles in only 1 of 12 patients. The conclusion to be drawn from these reports is that some

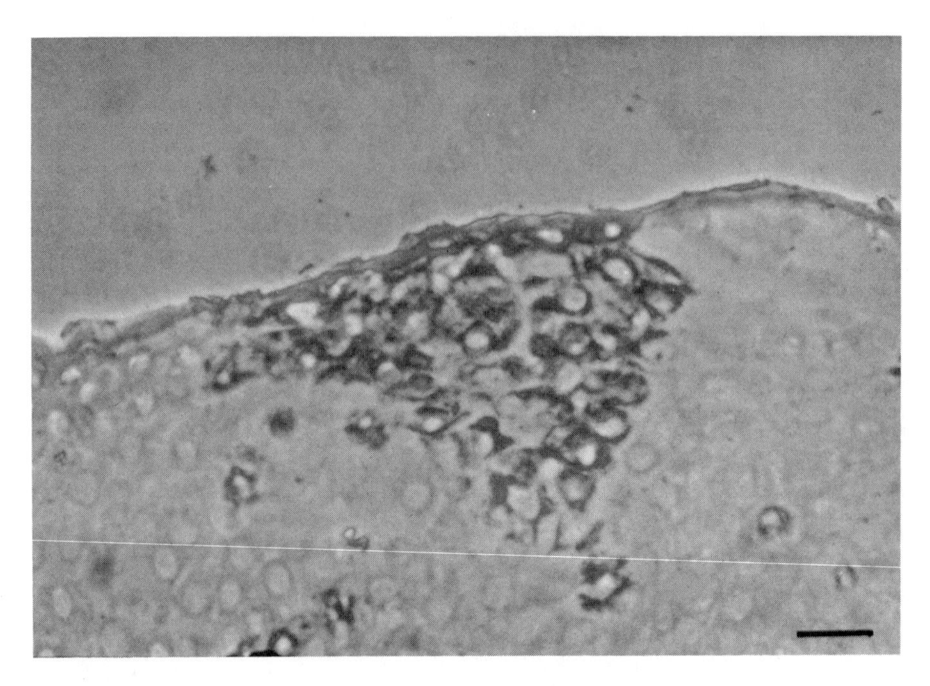

FIGURE 4. Differentiation of normal vocal cord epithelium. Five-micrometer tissue sections were incubated with antibody (generously provided by Dr. Beverly Dale, Seattle) against human filaggrin, washed with phosphate-buffered saline, and detected with the indirect immunoperoxidase ABC procedure (Vector Laboratories). Filaggrin, a protein involved with aggregation of intermediate filaments in the more differentiated cells, is located in scattered individual cells and in small patches of superficial cells. Bar = 20 μm.

small fraction of the cells in laryngeal papillomas do contain viral particles, but most do not.

Another approach to determining the presence of viruses in laryngeal papillomas involved the use of papilloma-specific antibodies for immunohistochemical staining. In this procedure, fixed and embedded tissue sections are deparaffined, incubated with antiserum against the viral capsid antigens, washed, and then incubated with a second antibody which is directed against the first immunoglobulin. The second antibody is either conjugated directly to peroxidase enzyme or to biotin which links to an avidin–biotin peroxidase complex. Once the peroxidase is localized in the tissue, a histochemical stain is used which produces a dark precipitate at the site of the enzyme, thereby localizing the HPV antigen. This technique is much quicker than electron microscopy, and also provides information regarding the type of virus present in the cells. Lack *et al.* (1980) found HPV antigens in laryngeal papillomas from 26 of 35 patients (74%) using a genus-specific antiserum prepared in rabbits against SDS-disrupted virus isolated from pooled human plantar warts. Similar findings have subsequently been reported using antiserum to SDS-disrupted BPV (Lancaster and Jenson, 1981) or HPV (Costa *et al.*, 1981; Braun *et al.*, 1982; Mounts *et al.*, 1982), with biopsy samples from 16% to 75% of juvenile-onset papilloma patients positive for viral antigens. Interestingly, Mounts *et al.* (1982) were the first to report viral capsid antigen in adult-onset disease (2 of 8 patients). The previous studies had been negative with adult-onset papillomas.

In all cases, staining tissue sections for capsid antigen revealed individual cells or small clusters of cells which were positive. These were located in the most superficial layers of the papillomas. It was often necessary to stain several sections taken from more than one biopsy before a positive result was seen for any individual patient. This observation, like the electron microscopy studies, suggests that very few cells in a papilloma make viral particles, since only a small fraction contain significant amounts of capsid antigen. In part, this is probably a reflection of the degree to which laryngeal papillomas differentiate. HPV virion production normally occurs in the upper layers of stratified squamous epithelium, primarily in the granular layer and stratum corneum. Since the "typical" laryngeal papilloma is not keratinized, it is not surprising that there is little production of viral capsid antigen.

Transformation of epithelial cells with HPV, resulting in a papilloma, presumably requires that basal cells be infected since these are the di-

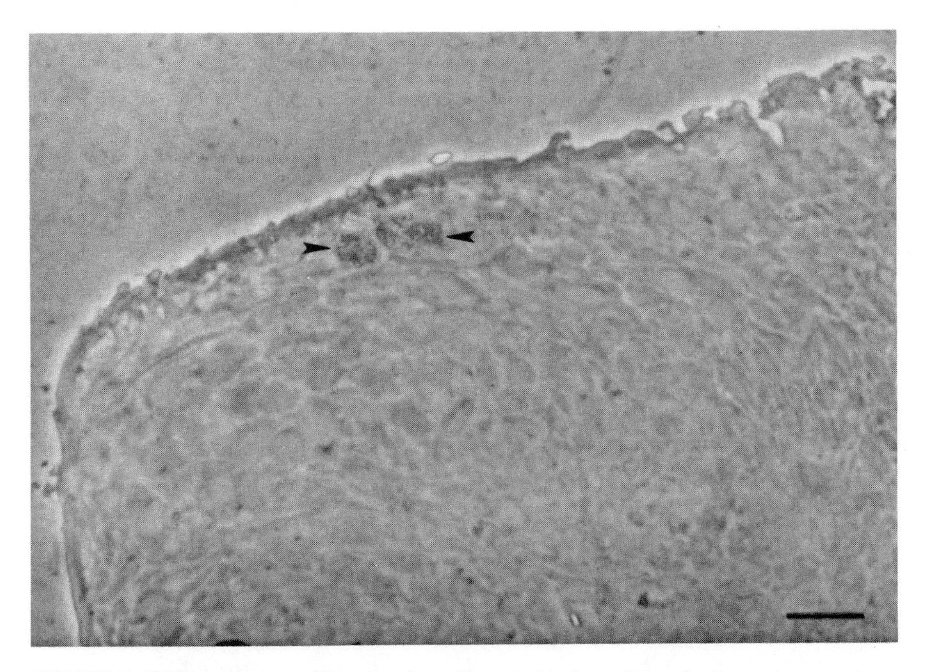

FIGURE 5. Differentiation of laryngeal papilloma. Sections through the projections on a papilloma were stained for filaggrin as described in Fig. 4. Only two faintly positive cells (arrowheads) were found in four sections of this papilloma. Bar = 20 μm.

viding cells which are the source of all the more differentiated suprabasal cells. Therefore, absence of viral capsid antigen and absence of viral particles should not indicate absence of viral DNA. Faras *et al.* (1980) first reported that HPV-related sequences were present in DNA isolated from three laryngeal papillomas. Using ^{32}P-labeled HPV DNA isolated from cutaneous warts as a probe, they were able to detect form I, form II, and form III DNA on Southern blots. Shortly thereafter, Lancaster and Jenson (1981) reported that they were able to detect HPV DNA in two of four laryngeal papillomas using a BPV-1 probe. Both of these studies used nonstringent conditions for hybridization so that heterologous HPVs could be detected. Heilman *et al.* (1980) have shown that varying formamide concentration in the hybridization solution permitted the heterologous hybrization of different HPVs, and the formation of HPV–BPV hybrids. Optimal cross-hybridization occurs at approximately $T_m - 50°C$, whereas stringent or homologous hybridization is usually best at $T_m - 28°C$. The conditions used by Lancaster and Jenson (1981) were somewhat more stringent than used by Faras *et al.* (1980). The increased stringency probably explains the inability to detect HPV DNA in two of the four papillomas, especially since one of the HPV DNA-negative tissues was positive for HPV capsid antigen. This could reflect either differences in HPV type or copy number.

Neither of the hybridization studies described above provided any information regarding the specific type of HPV found in laryngeal papillomas. Quick *et al.* (1980) reported that HPV-2 hybridized to two of nine samples of laryngeal papilloma under stringent conditions. This finding was the first to suggest that laryngeal papillomas, like skin warts and other HPV-induced lesions, can be caused by more than one type of HPV. It also suggested that HPV-2 can cause laryngeal papillomas.

Mounts *et al.* (1982) used Southern blot hybridization to analyze 12 juvenile-onset and 8 adult-onset laryngeal papillomas for the presence of HPV DNA. The probe was HPV-6 which they had isolated from a genital condyloma and cloned into the plasmid pBR322. Hybridization was done under stringent conditions ($T_m - 28°C$), and all 20 samples were positive for HPV-6-related DNA. Digestion of the DNAs with a battery of restriction enzymes (*Eco*RI, *Hpa*I, *Pst*I, *Bam*HI, and *Hin*dIII) resulted in four different cleavage patterns. Mounts *et al.* (1982) concluded that there are at least four different HPV-6 subtypes in laryngeal papillomas. The subtype did not correlate with the sex of the patient or the age of onset of disease. Presumably, none of their tissues contained HPV-2, since there is no evidence for more than one type of HPV within a single laryngeal lesion.

It is possible that at least one of the HPV-6 subtypes described by Mounts *et al.* (1982) is actually another type of closely related HPV. Gissmann *et al.* (1982) cloned and characterized an HPV from a patient with severe recurrent laryngeal papilloma. This HPV, subsequently called HPV-11, shares a great deal of homology with HPV-6. The two viral types have sufficient regions of homology to cross-hybridize under stringent conditions. However, they can be distinguished if the filter is washed at $T_m - 4°C$. It is very possible that HPV-6c of Mounts *et al.* (1982) is actually HPV-11, since the restriction pattern with *Pst*I is very much like that of HPV-11. The relationship between the other HPV subtypes of Mounts *et al.* (1982) and HPV-11 is not at all clear. Gissmann *et al.* (1982) found that four of nine additional laryngeal papillomas in their study hybridized to HPV-11. Negative samples could reflect either lack of sufficient DNA for analysis or true absence of HPV. Positive samples most probably did not contain HPV-6, even though HPV-6 and HPV-11 cross-hybridize, since there was no significant loss of label relative to the standards when the filter was washed at $T_m - 4°C$. Therefore, it would seem that a minimum of 50% of the laryngeal papillomas sampled by Gissmann *et al.* (1982) contained HPV-11. Mounts *et al.* (1982) found HPV-6-related DNA in all of their laryngeal papilloma patients, Gissmann *et al.* (1982) found HPV-11 in half of theirs, and Quick *et al.* (1980) found HPV-2 in 22% of their patients. It would therefore appear that multiple types of HPV can cause laryngeal papillomas. There might well be a tissue preference for HPV-6 or HPV-11, which does not absolutely exclude HPV-2 or other HPVs. Conversely, the high incidence of HPV-6 and HPV-11 in the larynx might reflect the usual source of the infection (the genitalia) and not be related to tissue specificity. In any case, heterogeneity of HPV DNA is common in laryngeal papillomas.

The heterogeneity of HPVs in laryngeal papillomas is illustrated in Fig. 6. Biopsies were taken from ten patients and DNA prepared from the 12 different biopsies was digested with the restriction enzyme *Hin*cII, which cuts HPV-11 to generate one large and one tiny fragment and cuts HPV-6b to generate four fragments. The digested DNA was then analyzed by Southern blot hybridization using the HPV-6b probe of de Villiers *et al.* (1981). All of the papillomas contained HPV sequences which hybridized to the probe in 30% formamide ($T_m - 35°C$). When hybridization was done in 50% formamide ($T_m - 20°C$), HPV DNA was barely detectable in lane 9, and not detectable in lane 14. The papillomas in these two patients apparently contained HPVs which were not closely related to HPV-6b or HPV-11, and which were different from the HPVs found in the other eight papilloma patients.

Those papillomas which did show hybridization to HPV-6b under stringent conditions (50% formamide) also had some heterogeneity in their genomes. Based solely on the restriction pattern with *Hin*cII, one can see differences on the Southern blot. This could reflect different HPV-6 subtypes, the presence of HPV-11 in some of the papillomas, or simply variability in a single *Hin*cII site. No comparison can be made to the four HPV-6 subtypes described by Mounts *et al.* (1982), since the DNAs shown in Fig. 6 were digested with a different enzyme than those used by Mounts *et al.* (1982). It is clear from Fig. 6 that the observed variability is between different patients, not between multiple papillomas from the same pa-

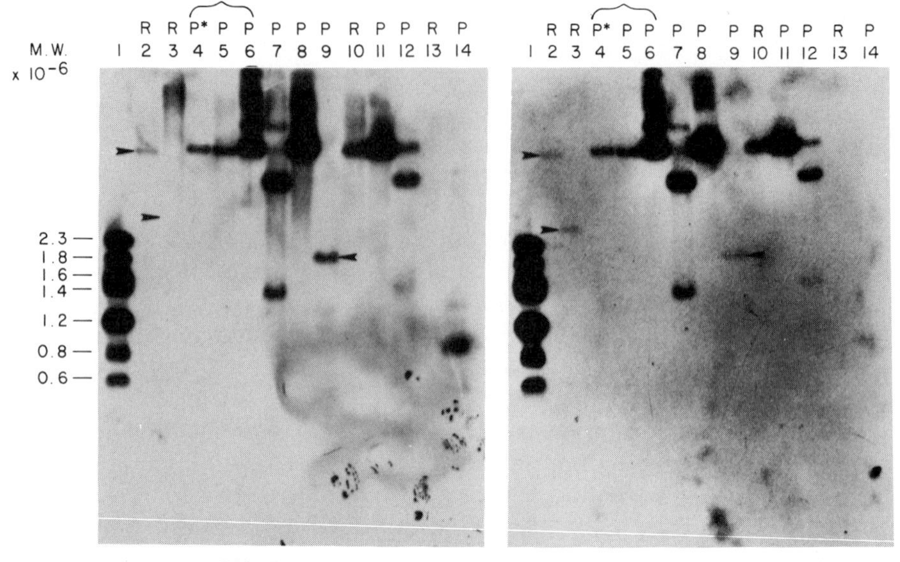

FIGURE 6. Heterogeneity of HPV DNAs in laryngeal papillomas. DNA was extracted from tissues or cultured cells, digested with *Hin*cII, and probed with HPV-6b as described by Steinberg *et al.* (1983). R, patients in remission; P, papilloma; asterisk, cells grown in culture. Braces group samples obtained from one patient during a single surgical procedure. Arrowheads locate fragments which are faint or show differences between nonstringent (30% formamide) or stringent (50% formamide) conditions. Reprinted by permission of *The New England Journal of Medicine* **308:**1261–1264, 1983.

tient. Lanes 4, 5, and 6 contained DNA from a vocal cord papilloma grown in culture for 2 months, the original papilloma before culture, and a uvula papilloma, respectively. All showed a similar pattern. The tissues in lanes 7 and 12 were removed from the same patient in two separate surgical procedures 4 months apart. The constancy of the *Hin*cII pattern suggest that the HPV infecting a patient does not vary with time, and that the variation we see is probably not due to minor genomic changes occurring during vegetative replication of the HPV in the replicating epithelial cells. This in contrast to reports of deletion mutants and heterogeneous HPVs within a single skin wart (Yoshiike and Defendi, 1977; Orth *et al.*, 1977), or genital condyloma (Schneider *et al.*, 1984).

There is a possible relationship between viral types or viral subtype and the clinical course of laryngeal papillomatosis. Mounts and Kashima (1984) reported that there appeared to be a correlation between HPV-6 subtype and severity of disease, as measured both by frequency of surgery and by extension of papillomas below the larynx into the trachea, bronchus, and lungs. They classified their HPV-6s into four subtypes: HPV-6c, HPV-6d, HPV-6e, and HPV-6f. All of the patients requiring surgery four or more times per year, and all of those with papillomas in or below the trachea contained HPV-6c. However, HPV-6c was found in 13 of their 21 patients, while 4 had HPV-6d and only 2 each had HPV-6e and HPV-

6f. It is certainly suggestive that the most severe disease was found only in patients containing HPV-6c, but additional cases with the other HPV-6 subtypes must be studied before firm conclusions can be drawn.

One additional piece of information regarding the role of HPV in laryngeal papillomas can be derived from Fig. 6. DNA from three patients who had been in remission for more than 18 months still contained sequences related to HPV-6 (lanes 2, 3, 10). This was the first evidence that the recurrent nature of laryngeal papillomas might be due to a reactivation of a latent viral infection.

The presence of HPV sequences in additional biopsies of "normal" laryngeal tissue is seen in Fig. 7. DNA from three additional patients in remission was positive for HPV when hybridized under nonstringent conditions (lanes 1, 6, 10–12). This result confirms the previous findings. Figure 7 also shows the results of hybridizing the HPV-6b probe to DNA from morphologically normal laryngeal tissues of two patients with active papillomas. One patient was an adult with a 24-year history of juvenile-onset disease (lanes 7–9a), the other an adult with a 2-year history of adult-onset disease (lanes 2b and 3b). Biopsy sites were: papillomas of the true vocal cord (lanes 7a and 2b), epiglottis (8a), false cord (9a), and opposite true cord (3b). It is apparent, even from this small sample, that HPV can be found in several different laryngeal tissues, and that the extent of infected tissue is much greater than the extent of active papilloma growth.

Presence of HPV DNA in morphologically normal tissue (Figs. 6 and 7) helps to explain the clinical pattern of laryngeal papillomatosis. The multifocal nature of the disease is due to the extensive area of infection. The frequent regrowth following surgical removal is probably not due to any seeding of virus during surgery followed by new infection as was previously believed. Rather, it may reflect the transforming activity of HPV which has been present in the laryngeal tissues since the initial viral infection. In order to completely remove the infected tissue, all of the laryngeal mucosa would have to be excised, not just the papilloma. This is obviously not possible, and it is for this reason that surgery can only help control the symptoms of laryngeal papillomas, not cure the disease.

Finding HPV sequences in the tissues of patients in remission explains the recurrence of papillomas after years of remission. Again, one no longer must look for the source of new infection. The regrowth of papillomas following remission induced by successful interferon therapy may reflect a similar mechanism. Studies are currently in progress in our laboratory to quantify the amount of HPV DNA present in laryngeal biopsy samples taken during interferon therapy by the Boston study. Concurrently, similar assays are being done with the Baltimore study (P. Mounts, personal communication). Preliminary results suggest that viral DNA persists during interferon therapy, and that induction of clinical remission precedes any detectable reduction in HPV copy number (Steinberg *et al.*, 1985). This finding is not too surprising. Tissue culture studies have been done with mouse cells infected with BPV-1 (Turek *et al.*, 1982).

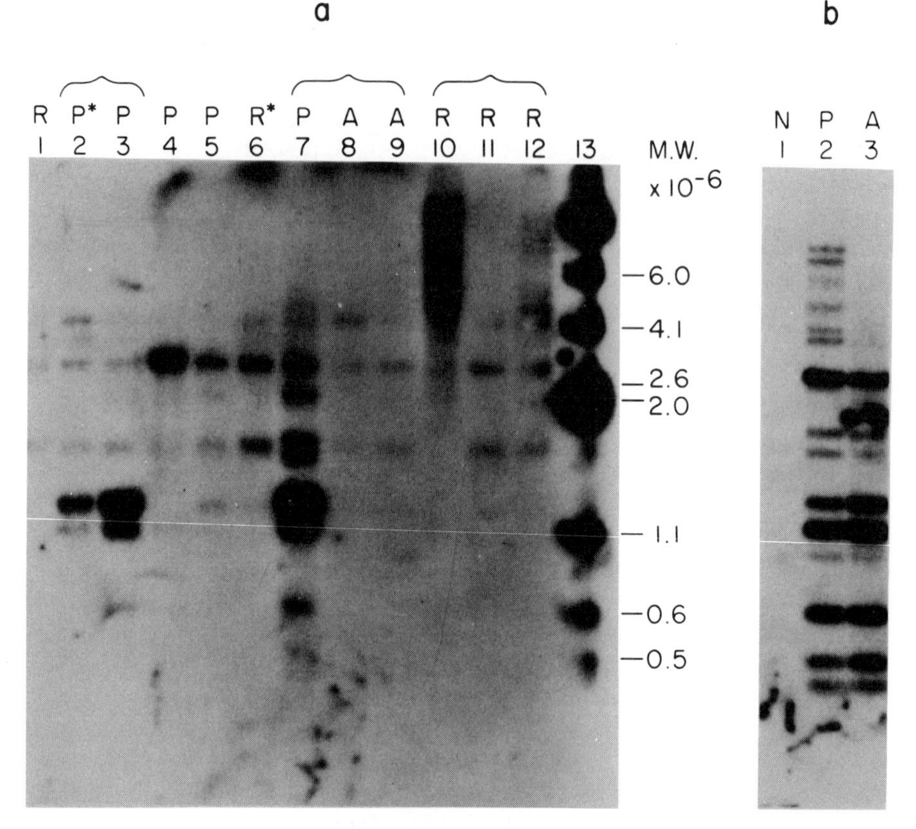

FIGURE 7. Ability of HPV DNA to exist in latent infections. DNA was digested with a mixture of *Eco*RI, *Bam*I, and *Pst*I as described by Steinberg *et al.* (1983). HPV DNA appears to be present both in adjacent sites and in tissues from patients in remission. (a) Twelve samples analyzed in one gel; (b) three samples analyzed in a second gel. R, patient in remission; P, papilloma; A, tissue adjacent to a papilloma; N, normal tissue from patient with base of tongue carcinoma; asterisks, cells grown in culture. Braces, group samples obtained from one patient during a single surgical procedure. Reprinted by permission of *The New England Journal of Medicine* **308:**1261–1264, 1983.

Interferon treatment of these cells results in a very gradual decrease in the number of BPV genomes per cell, presumably because the replication of BPV is slowed relative to the replication of the cells themselves. As the cultured cells continue to divide, the viral DNA is slowly diluted out. It would take a long time to significantly reduce the amount of HPV in patients on interferon by this mechanism. Removal of interferon from the cultured cells was followed by a rapid return of the BPV copy number to its original level. If a similar result is found in papilloma patients, it might partially explain the recurrence of disease when interferon therapy is discontinued.

B. Tissue Culture Studies

Very little is known about the biology of HPVs. This is primarily because propagation of HPV *in vitro* has not been accomplished. Butel (1972) described her extensive, unsuccessful attempts. Sporadic reports of successful culture leading to virion production (Eisenger *et al.*, 1975) have not been successfully reproduced with any consistency in other laboratories (see Chapter 4). In order to study the cell biology of laryngeal papillomatosis, we developed a way to grow epithelial cells derived from these papillomas in cell culture (Steinberg *et al.*, 1982). The primary cells are grown from tissue fragments embedded in a hydrated collagen gel which contains a nutrient mixture plus 15% fetal calf serum. The cultured papilloma cells do not differ markedly in morphology from normal laryngeal cells in culture. This is not surprising, since the morphology of a single basal or spinous cell in a laryngeal papilloma is not very different from a normal epithelial cell. The extensive thickening of the spinous layer in a papilloma is not reproduced in tissue culture. Both normal and papilloma cells grow to only a three- to four-cell thickness before the superficial layer is sloughed into the medium.

Primary cells grown by the method of Steinberg *et al.* (1982) can be maintained in culture for many months before they die. They can only be passaged three to four times before the culture senesces, probably because the plating efficiency is very low (approximately 1%) and four passages represents 25–30 doublings. Attempts to establish a permanent cell line from these cells have been unsuccessful. The cells grown in our laboratory (Steinberg *et al.*, 1982) do not stain with the genus-specific anti-HPV antiserum discussed earlier in this chapter. They do contain HPV DNA. Positive hybridization to the HPV-6b probe can be seen with DNA isolated from subcultured cells grown *in vitro* for 2–3 months (Fig. 6, lane 4; Fig. 7, lanes 2a and 6a).

Establishment of a permanent laryngeal papilloma line containing HPV DNA could facilitate studies of the biology of papillomaviruses. Immortality of a human cell line is generally associated with the transformation of the cells to a malignant phenotype *in vivo*, and therefore such cells would not be representative of benign papillomas. For this reason, our laboratory has chosen to study the properties of first-passage cells derived from laryngeal papillomas and from normal laryngeal epithelium. Results of some of these ongoing experiments are described below.

1. Cytoskeletal Studies

Both types of cells (normal and papilloma) were grown on glass coverslips, fixed, and stained with antibodies against various cytoskeletal proteins (Romani *et al.*, 1986). These included keratin, tubulin, vimentin, desmin, and actin. Both normal and papilloma cells were negative for desmin, normally found in muscle cells. They were positive for keratin, an epithelial intermediate-filament protein, and vimentin, a mesenchymal intermediate-filament protein. Many epithelial cells in culture express the vimentin gene. There was no obvious difference in the microtubule organization of the two cell types, as visualized with antitubulin.

Distribution of keratin in intermediate filaments was different in the papilloma and normal cells. Most of the papilloma cells showed a very clumped perinuclear area of antikeratin stain, with only a few fibers at the periphery of the cell. Normal cells had many keratin fibers throughout the cytoplasm, often radiating from a heavy circular perinuclear band.

Actin-containing microfilament distribution was also different in the two cell types. The normal cells had very little actin fiber which could be visualized by indirect immunofluorescence. There was some positive staining which outlined the edges of the cells. This was especially evident in the most superficial layer of the cultured cells. The papilloma cells showed much heavier actin staining. Not only were edges of the cells well stained, but actin fibers could be clearly seen traversing the cell and extending past the cell membrane to apparently terminate on the glass coverslip. This pattern of heavier actin staining in cells from benign tumors, relative to normal epithelial cells, has also been noted in colonic tumors (Friedman *et al.*, 1985).

The effects of calcium concentration of the cytoskeletal fibers are currently being investigated. Using antiactin and antikeratin antibodies, and immunohistochemical stains, we are localizing these proteins in the cytoskeleton at both the light and electron microscope level. Special attention is being focused on the interaction of the fibers with the cell membrane and cell–cell junctions.

We have also determined the molecular weights of the cytokeratins present in normal and papilloma cells, as well as normal and papilloma biopsy samples, using SDS–polyacrylamide gel electrophoresis. All of the samples contained keratins with molecular weights between 40K and 58K. Neither normal nor papilloma cells or tissues contained the 66K and 68K keratins which are specific for keratinizing epithelium. The papilloma cells and tissues contained 46K and 53K keratins which were not present in the normal cells and tissues. These keratins may serve as markers for papilloma cells.

2. Effects of Interferon on Cultured Cells

Because interferon seems effective in controlling the symptoms of laryngeal papillomatosis (see Section III.D), we have tested its effects on cultured papilloma cells. There is very little difference in response between normal cells and papilloma cells (Steinberg *et al.*, 1985). For each experiment, cells were treated with α-interferon at 0, 10, 100, 500, and 1000 units/ml medium. *In vivo* levels of interferon usually reach a peak of 100–200 units/ml blood approximately 6 hr after an injection, and decline to a level of approximately 10 units/ml blood within 24 hr.

Rate of outgrowth of cells from tissue fragments, a composite of migration and cell division, was not appreciably affected until the interferon concentration reached 500 units/ml. Incorporation of [^{3}H]thymidine into acid-insoluble macromolecules, during 24 hr of continuous labeling, was decreased by the presence of interferon. The maximum effect was seen at 100 units/ml interferon, and both types of cells had very similar response, with approximately a two-third reduction in [^{3}H]thymidine incorporation. The relative reduction was the same, whether total acid-precipitable cpm per microgram cell protein or percentage of labeled nuclei following autoradiography was measured. Interferon either reduces the fraction of cells which go through the S phase of the cell cycle in 24 hr, or reduces the fraction of cells in the population which incorporate exogenous thymidine during DNA synthesis. There was no effect on protein synthesis as measured by tritiated amino acid incorporation.

Growth of primary cells in the presence of interferon for 1 month, a rather short period of time, did not have any appreciable effect on the number of HPV genomes per cell. We also did not detect any reduction in the copy number of HPV in tissues from a patient on interferon therapy for 3 months. There was a definite positive response to treatment by this patient, since the amount of regrowth of papilloma in the larynx was much less than during previous 3-month intervals.

As mentioned earlier, we are continuing our studies of the effects of clinical use of interferon on HPV copy number in laryngeal papillomas. We are also continuing our investigations into the *in vitro* response of papillomas to interferon. We hope to be able to determine whether the interferon used clinically to induce remission is acting primarily as a direct antitumor agent, an antivirus agent, or an enhancer of the immune system.

3. Effects of Retinoic Acid on Cultured Cells

Retinoids have long been known to regulate epithelial differentiation *in vivo* (for review see Lotan, 1980). Since laryngeal papillomas appear to have an altered ability to differentiate completely, it is reasonable that they might respond differently to retinoids. A few investigators have begun to treat laryngeal papilloma patients with 13-*cis*-retinoic acid, in an attempt to control the disease, even though nothing is known about the response of papilloma cells to increased retinoid concentrations. A great deal is known about its effects on other epithelia. *In vivo,* excess retinoids block the differentiation of stratified squamous epithelium and suppress keratinization (Lotan, 1980) and similar changes have been seen *in vitro* (Fuchs and Green, 1981).

We have evaluated the *in vitro* effects of retinoids on normal and papilloma cells. Both the all-*trans* retinoic acid and 13-*cis*-retinoic acid were used in our studies, in concentrations ranging from 10^{-10} M to 10^{-5} M. These experiments were done with cells growing in the completely defined serum-free medium MCDB 153, developed by Tsao *et al.* (1982), since the standard medium with 15% FCS contains approximately 6×10^{-8} M retinoid.

Outgrowth of both types of primary cells was enhanced four- to sixfold by the addition of retinoid. There was only a small difference between the two types of retinoic acid, with the 13-*cis*-retinoic acid slightly more stimulating than the *trans* retinoic acid. Optimal concentration of retinoid for the normal cells was approximately 10^{-9} M, and for the papilloma cells was between 10^{-7} and 10^{-8} M. Growth of secondary cultures of both types was also enhanced by the addition of retinoids at the same concentrations. Retinoid at 10^{-5} M was toxic for both types of cells. Adding retinoid certainly did not block the growth of papilloma cells!

Keratin production was not markedly changed by the addition of retinoic acid. SDS–polyacrylamide gel electrophoresis of keratins synthesized by secondary cells growing in varying concentrations of retinoic acid showed very similar keratin patterns, with the addition of a minor band at 43.5K in the presence of retinoids. Neither type of cell synthesized the high-molecular-weight keratins characteristic of keratinized epithelium in the absence of retinoic acid.

One measure of differentiation is the ability to form cornified envelopes upon suspension of cells in the presence of a calcium ionophore (Rice and Green, 1979). Cornified envelopes are highly insoluble structures composed of the protein involucrin, which is cross-linked by the enzyme transglutaminase in the presence of high levels of calcium. These envelopes are found just beneath the cell membrane in cells of the uppermost layers of stratified squamous epithelium. Retinoids block the formation of these envelopes in cultured cells (Yupsa *et al.*, 1982). We have found that both normal and papilloma cells can produce cornified envelopes in culture. The two cell types are equally sensitive to the effects of retinoic acid on this process, with greater than 90% inhibition of envelope formation when the cells are cultured with 10^{-9} M retinoic acid. Again, the response was the same with both the all-*trans* and 13-*cis*-retinoic acid.

We have concluded from these studies that there is no significant difference between normal and papilloma cells in their *in vitro* response to retinoids. Both show enhanced growth and suppressed differentiation. If anything, the optimal retinoid concentration for the growth of papilloma cells is higher than for normal cells.

VI. LARYNGEAL PAPILLOMAS AND LARYNGEAL CANCER

Irradiation of recurrent laryngeal papillomas with x-rays increases the probability of transformation from a benign papilloma to a squamous cell carcinoma (Galloway *et al.*, 1960; Majoros *et al.*, 1963). The frequency of transformation of recurrent papillomas following radiation was 14% in one large study (Majoros *et al.*, 1963) with conversion to carcinoma occurring after a lapse of 1–15 years. In that same study, with 58 control papilloma patients, there was no spontaneous malignant conversion. Clearly, there is some interaction between the HPV-induced lesions and ionizing radiation which results in carcinoma.

Those adult-onset papillomas which spontaneously convert to carcinoma may represent, at least in part, a different phenomenon. The conversion rate in the solitary papillomas has been reported to be as high as 20% (zur Hausen, 1977), and it has been suggested that they represent a different premalignant lesion. Batsakis *et al.* (1983) commented that in contrast to papillomas in children, squamous "papillomas" of the larynx in adults have included a variety of papillary or exophytic lesions which include papillary and verrucous carcinomas, and even vocal nodules. They state that one type of adult papilloma is nonkeratinizing, and identical to juvenile-onset papillomas in both pathology and biological behavior; the other is a keratinizing lesion which is almost always solitary. These latter lesions might have a different etiology and a different prognosis.

The relationship of HPV to laryngeal carcinoma in patients with no previous history of laryngeal papillomas has recently been under investigation. Syrjanen and Syrjanen (1981) first reported that 49 of 116 cases of laryngeal squamous cell carcinoma had morphological similarities to genital condyloma, including the presence of koilocytes which are usually associated with HPV infection. Subsequently, they found HPV-specific capsid antigens in 36% of these condylomatous-like lesions, using antiserum against SDS-disrupted HPV (Syrjanen *et al.*, 1982).

HPV-16 was isolated from a cervical cancer biopsy and cloned into pBR322 (Dürst *et al.*, 1983). A high percentage of genital cancers are associated with HPV-16, suggesting that this type of HPV may be able to induce malignant transformation under the proper conditions. We have recently detected HPV-16-related DNA in several cases of verrucous carcinoma of the larynx (Brandsma *et al.*, 1986). Using the cloned HPV-16 generously provided by Dürst *et al.* (1983), we had positive hybridization to verrucous carcinoma DNA on both Southern blots and "dot blots" at T_m −35°C, which was not completely removed when the filters were washed at approximately T_m −25°C. A small percentage of benign laryngeal lesions were also positive for HPV-16-related DNA.

We do not know whether those laryngeal papillomas which convert to carcinomas after radiotherapy contain HPVs related to HPV-16 or the HPVs usually associated with benign papillomas (i.e., HPV-6, HPV-11). Certainly, our data showing that HPV-16-related sequences can be present in some benign lesions of the larynx are consistent with the results of Dürst *et al.* (1983) for genital condylomas.

VII. MAJOR UNANSWERED QUESTIONS

Most of the major questions which need to be answered in order to understand laryngeal papillomas must also be answered for other papillomavirus-induced neoplasms. We will briefly discuss them here within the context of laryngeal papillomatosis.

The epidemiology of laryngeal papillomas is not at all well understood. We do not know if all infections are contracted from genital condylomas. Assuming that juvenile-onset disease is contracted only by an infant born to a mother with condyloma, is the infection always, usually, rarely, or never intrauterine? Some anecdotal data suggest that intrauterine infection can occur. However, its frequency is not know. If expectant mothers with condyloma are at risk of having children who develop laryngeal papillomas, why is the incidence of laryngeal papillomas very low (0.001%) while the current estimates of condyloma in young women is approximately 2%? Are we about to see an "epidemic" of laryngeal papillomatosis? Doctors who treat this disease are not reporting any marked increase in the number of patients. Are there other significant risk factors which we do not recognize? Are these presumptive risk factors related to acquiring a laryngeal HPV infection, or related to the expression of infection resulting in papillomas severe enough to be symptomatic? It is possible that the number of patients with this disease is only a fraction of those harboring a latent infection, since we know that latent HPV can exist in the larynx of papilloma patients.

Similar questions can be asked about adult-onset laryngeal papillomatosis. Is this the expression of a latent infection acquired from the mother, or a new acquired infection? Again, route of transmission and other determining risk factors are unknown.

We do not know what host or viral factors affect the expression of laryngeal HPV: what induces a remission, what causes the reactivation of disease following remission? Because the biology of the papillomaviruses is still poorly understood, we do not know how transcription and translation of these viruses is regulated. If this were better understood, treatment procedures could possibly be developed which would be much more effective than those currently in use.

The role of the immune system in the expression of this disease is also not understood. Are subtle failures of the immune system part of the presumptive risk factors? To date, no one has reported a correlation between laryngeal papillomas and immune system defects.

The relationship between laryngeal HPV infection and laryngeal carcinoma is also poorly understood. We do know that irradiation of papillomas can increase the risk of conversion to carcinoma, and that the risk of spontaneous conversion is extremely low. We also know that heavy use of tobacco and alcohol are major risk factors for laryngeal carcinoma. Does the presence of latent HPV in conjunction with the alcohol and tobacco use increase this risk?

All of the questions raised above deal with the epidemiology or clinical expression of laryngeal HPV infections in one way or another. Another set of questions are related to the mechanism of transformation of laryngeal epithelial cells, which results in a papilloma. We presently have very little understanding of this whole phenomenon. Development of tissue culture systems which permit this type of transformation *in vitro* will hopefully be accomplished in the near future. Until this is possible, it will be very difficult to truly understand the biology of the papillomaviruses.

REFERENCES

Altman, F., Basek, M., and Stout, A. P., 1955, Papillomatosis of the larynx with intraepithelial anaplastic changes, *Arch. Otolaryngol.* **62:**478–485.

Batsakis, J. G., Raymond, A. K., and Rice, D. H., 1983, The pathology of head and neck tumors: Papillomas of the upper aerodigestive tracts, *Head Neck Surg.* **5:**332–344.

Borkowsky, W., Martin, D., and Lawrence, S., 1984, Juvenile laryngeal papillomatosis with pulmonary spread: Regression following transfer factor therapy, *Am. J. Dis. Child.* **138:**667–669.

Boyle, W. F., Riggs, G. B., Oshiro, L. S., and Lenette, E. H., 1973, Electron microscopic identification of papova virus in laryngeal papilloma, *Laryngoscope* **83:**1102–1108.

Brandsma, J. L., Steinberg, B. M., Abramson, A. L., and Winkler, B., 1986, Presence of human papillomavirus type 16 related sequences in verrucous carcinoma of the larynx, *Cancer Res.* **46:**2185–2188.

Braun, L., Kashima, H., Eggleston, J., and Shaw, K., 1982, Demonstration of papillomavirus antigen in parraffin sections of laryngeal papillomas, *Laryngoscope* **92:**640–643.

Broyles, F. N., 1940, Treatment of papilloma in children with estrogenic hormone, *Bull. Johns Hopkins Hosp.* **66:**319–322.

Butel, J., 1972, Studies with human papilloma virus modeled after known papovavirus systems, *J. Natl. Cancer Inst.* **48:**285–299.

Cook, T. A., Brunschwig, J. P., Butel, J. S., Cohen, A. M., Goepfert, H., and Rawls, W., 1973, Laryngeal papilloma: Etiologic and therapeutic considerations, *Ann. Otol. Rhinol. Laryngol.* **82:**649–655.

Costa, J., Howley, P. M., Bowling, M. C., Howard, R., and Bauer, W. C., 1981, Presence of human papilloma viral antigens in juvenile multiple laryngeal papilloma, *Am. J. Clin. Pathol.* **75:**194–197.

Dedo, H. H., and Jackler, R. K., 1982, Laryngeal papilloma: Results of treatment with the CO_2 laser and podophyllum, *Ann. Otol. Rhinol. Laryngol.* **91:**425–430.

DeRosa, E., 1977, Laryngeal papillomas: New treatment with levamisole, *Trans. Am. Acad. Ophthalmol. Otolaryngol.* **84:**75.

de Villiers, E.-M., Gissmann, L., and zur Hausen, H., 1981, Molecular cloning of viral DNA from human genital warts, *J. Virol.* **40:**932–935.

Dürst, M., Gissmann, L., Ikenberg, H., and zur Hausen, H., 1983, A papillomavirus from a cervical carcinoma and its prevalence in cancer biopsy samples from different geographical regions, *Proc. Natl. Acad. Sci. USA* **80:**3812–3815.

Eisenger, M., Kucarova, O., Sarkar, N. H., and Good, R. A., 1975, Propagation of human wart virus in tissue culture, *Nature* **256:**432–434.

Faras, A. J., Krzyzek, R. A., Ostrow, R. S., Watts, S. L., Smith, D. M., Anderson, D. L., Quick, C. A., and Pass, F., 1980, Genetic variation among papillomavirus, *Ann. N.Y. Acad. Sci.* **354:**60–79.

Friedman, E., Verderame, M., Winawer, S., and Pollack, R., 1985, Altered actin cytoskeletal patterns in two premalignant stages in human colon carcinoma development, *Cancer Res.* **45:**3236–3242.

Fuchs, E., and Green, H., 1981, Regulation of terminal differentiation of cultured human keratinocytes by vitamin A, *Cell* **25:**617–625.

Galloway, T. C., Soper, G. R., and Elsen, J., 1960, Carcinoma of the larynx after irradiation for papilloma, *Arch. Otolaryngol.* **72:**289–294.

Gissmann, L., Diehl, V., Schultz-Coulon, H. J., and zur Hausen, H., 1982, Molecular cloning and characterization of human papilloma virus DNA derived from a laryngeal papilloma, *J. Virol.* **44**:393–400.

Goepfert, H., Sessions, R. B., Gutterman, J., Cangir, A., Diehlel, W., and Sulek, M., 1982, Leukocyte interferon in patients with juvenile laryngeal papillomatosis, *Ann. Otol. Rhinol. Laryngol.* **91**:431–436.

Greensher, A., 1980, Treatment of laryngeal papillomas with mumps skin test antigens, *Lancet* **2**:920.

Haglund, S., Lundquist, P. G., Cantell, K., and Strander, H., 1981, Interferon therapy in juvenile laryngeal papillomatosis, *Arch. Otolaryngol.* **107**:327–332.

Hajek, E., 1956, Contribution to the etiology of laryngeal papilloma in children, *J. Laryngol. Otol.* **70**:160–168.

Hast, M., 1986, Anatomy of the larynx, in: *Otolaryngology*, Vol. 3, *Diseases of the Larynx, Pharynx, and Upper Respiratory Tract* (G. M. English, ed.), pp. 1–16, Harper and Row, Philadelphia.

Heilman, C. A., Law, M.-F., Israel, M. A., and Howley, P. M., 1980, Cloning of human papillomavirus genomic DNAs and analysis of homologous polynucleotide sequences, *J. Virol.* **36**:395–407.

Holinger, P. H., Johnston, K. C., and Anison, G. C., 1950, Papilloma of larynx: Review of 109 cases with preliminary report of aureomycin therapy, *Ann. Otol. Rhinol. Laryngol.* **59**:547–563.

Holinger, P. H., Schild, J. A., and Maurizi, D. G., 1968a, Laryngeal papilloma: Review of etiology and therapy, *Laryngoscope* **78**:1462–1467.

Holinger, P. H., Shipkowitz, N. L., Holper, J. C., and Worland, M. C., 1968b, Studies of laryngeal papilloma and an autogenous papilloma vaccine, *Acta Oto-Laryngol.* **65**:63.

Hollingsworth, J. B., Kohlmoos, H. W., and McNaught, R. C., 1950, Treatment of juvenile papilloma of larynx with resin of podophyllum, *Arch. Otolaryngol.* **52**:82–87.

Incze, J., Strong, M. S., Vaughan, C. W., and Clemente, M. P, 1977, The morphology of human papillomas of the upper respiratory tract, *Cancer* **39**:1634–1646.

Jako, G. J., 1972, Laser surgery of the vocal cords, *Laryngoscope* **82**:2204–2216.

Justus, J., Baerthold, W., and Pierbish-Effenberger, R., 1970, Juvenile Larynx-papillomatose mit Ausbreitung uber das Tracheobronchial system und maligne Entartung ohne Strohlen therapie, *Hans Nas. Ohrenheilkd.* **18**:349–354.

Koss, L. G., and Durfee, G. R., 1956, Unusual patterns of squamous epithelium of the uterine cervix: Cytological and pathologic study of koilocytotic atypia, *Ann. N.Y. Acad. Sci.* **63**:1245–1261.

Lack, E. E., Jenson, A. B., Smith, H. G., Healy, G. B., Pass, F., and Vawter, G. F., 1980, Immunoperoxidase localization of human papillomavirus in laryngeal papillomas, *Intervirology* **14**:148–154.

Lancaster, W. D., and Jenson, A. B., 1981, Evidence for papillomavirus genus-specific antigens and DNA in laryngeal papilloma, *Intervirology* **15**:204–212.

Lotan, R., 1980, Effects of vitamin A and its analogs (retinoids) on normal and neoplastic cells, *Biochim. Biophys. Acta* **605**:33–91.

Lundquist, P., Frithiof, L., and Wersall, J., 1975, Ultrastructural features of human juvenile laryngeal papillomas, *Acta Oto-Laryngol.* **80**:137–149.

McCabe, B. F., and Clark, K. F., 1983, Interferon and laryngeal papillomatosis: The Iowa experience, *Ann. Otol. Rhinol. Laryngol.* **92**:2–7.

Majoros, M., Parkhill, E. M., and Devine, K. D., 1963, Malignant transportation of benign laryngeal papillomas in children after radiation therapy, *Surg. Clin. North Am.* **43**:1019–1061.

Mounts, P., and Kashima, H., 1984, Association of human papillomatosis subtype and clinical course in respiratory papillomatosis, *Laryngoscope* **94**:28–33.

Mounts, P., Shaw, K. V., and Kashima, H., 1982, Viral etiology of juvenile and adult onset squamous papilloma of the larynx, *Proc. Natl. Acad. Sci. USA* **79**:5425–5429.

Ogilvie, O. E., 1953, Multiple papillomas of trachea with malignant degeneration, *Arch. Otolaryngol.* **58:**10–18.
Orth, G., Jablonska, S., and Croissant, O., 1977, Characterization of a new type of human papillomavirus that causes skin warts, *J. Virol.* **24:**108–120.
Quick, C. A., Foucar, E., and Dehiver, L. P., 1979, Frequency and significance of epithelial atypia in laryngeal papillomatosis, *Laryngoscope* **89:**550–560.
Quick, C. A., Watts, S. L., Krzyzek, R. A., and Faras, A. J., 1980, Relationship between condylomata and laryngeal papillomata, *Ann. Otol. Rhinol. Laryngol.* **89:**467–471.
Rice R. H., and Green, H., 1979, Presence in human epidermal cells of a soluble protein precursor of the cross-linked envelope: Activation of the cross linking by calcium ions, *Cell* **18:**681–694.
Romani, V. G., Abramson, A. L., and Steinberg, B. M., 1986, Laryngeal papilloma cells in culture have an altered cytoskeleton, *Acta Otolaryngol.* (in press).
Schneider, P. S., Krumholz, B. A., Topp, W. C., Steinberg, B. M., and Abramson, A. L., 1984, Molecular heterogeneity of female genital wart (condylomata acuminata) papilloma viruses, *Int. J. Gynecol. Pathol.* **2:**329–336.
Schouten, T. J., Weimar, W., Bos, J. H., Bos, C. E., Cremers, C. W. R., Jr., and Schellekins, H., 1982, Treatment of juvenile laryngeal papillomatosis with two types of interferon, *Laryngoscope* **92:**686–688.
Scully, P., Stratton, C., and Brophy, J., 1983, Stereoscanning electron microscopy of argon laser excised laryngeal papilloma, *Laryngoscope* **93:**188–195.
Sherman, B. S., 1961, The effect of vitamin A on epithelial mitosis *in vitro* and *in vivo, J. Invest. Dermatol.* **37:**469–480.
Singer, D. B., Greenbert, S. D., and Harrison, G. M., 1966, Papillomatosis of the lung, *Am. Rev. Respir. Dis.* **94:**777–781.
Smith, H. G., Healy, G. B., Vaughan, C. W., and Strong, H. S., 1980, Topical chemotherapy of recurrent respiratory papillomatosis, a preliminary report, *Ann. Otol. Rhinol. Laryngol.* **89:**472–478.
Spoendlin, H., and Kistler, G., 1978, Papova virus in human laryngeal papillomas, *Ann. Otol. Rhinol. Laryngol.* **218:**289–292.
Steinberg, B. M., Abramson, A. L., and Meade, R. P., 1982, Culture of human laryngeal papilloma cells *in vitro, Otolaryngol. Head Neck Surg.* **90:**728–735.
Steinberg, B. M., Topp, W. C., Schneider, P. S., and Abramson, A. L., 1983, Laryngeal papillomavirus infection during clinical remission, *N. Engl. J. Med.* **308:**1261–1264.
Steinberg, B. M., Abramson, A. L., and Horowitz, B., 1985, Effects of alpha-interferon on cultured laryngeal papilloma cells, in: *Papillomaviruses: Molecular and Clinical Aspects* (P. M. Howley and T. R. Broker, eds.), pp. 413–426, Liss, New York.
Stern, L. S., Abramson, A. L., and Grimes, G. W., 1980, Qualitative and morphometric evaluation of vocal cord lesions produced by the carbon dioxide laser, *Laryngoscope* **90:**792–808.
Strome, M., 1969, Analysis of an autogenous vaccine in the treatment of juvenile papillomatosis of the larynx, *Laryngoscope* **79:**272–279.
Strong, M. S., Vaughn, C. W., Healy, G. B., Cooperband, S. R., and Clemente, M. A. C. P., 1976, Recurrent respiratory papillomatosis—Management with the CO_2 laser, *Ann. Otol. Rhinol. Laryngol.* **85:**508–516.
Svoboda, D. J., Kirchner, F. R., and Proud, G. O., 1963, Electron microscopic study of human laryngeal papillomatosis, *Cancer Res.* **23:**1084–1089.
Syrjanen, K. J., and Syrjanen, S. M., 1981, Histological evidence for the presence of condylomatous epithelial lesions in association with laryngeal squamous cell carcinoma, *J. Otorhinolaryngol. Relat. Spec.* **43:**181–194.
Syrjanen, K., Syrjanen, S., and Pyrhonen, S., 1982, Human papilloma virus (HPV) antigens in lesions of laryngeal squamous cell carcinomas, *J. Otorhinolaryngol. Relat. Spec.* **44:**323–334.
Szpunar, J., 1977, Juvenile laryngeal papillomatosis, *Otolaryngol. Clin. North Am.* **10:**67–70.

Tsao, M. C., Walthall, B. J., and Ham, R. G., 1982, Clonal growth of normal human epidermal keratinocytes in a defined medium, *J. Cell. Physiol.* **10:**219–229.

Turek, L. P., Byrne, J. C., Lowy, D., Dvoretzky, I., Friedman, R. M., and Howley, P. M., 1982, Interferon inhibits bovine papilloma virus transformation of mouse cells and induces reversion of established transformants, *UCLA Symp. Mol. Biol.* **25:**181–191.

Ullman, E. V., 1923, On the etiology of laryngeal papilloma, *Acta Otolaryngol.* **5:**317–338.

Weiss, M. D., and Kashima, H. K., 1983, Tracheal involvement in laryngeal papillomatosis, *Laryngoscope* **93:**45–48.

Yoshiike, K., and Defendi, V., 1977, Presence of deletion molecules in human wart virus DNA, *J. Virol.* **21:**415–418.

Yupsa, S. H., Ben, T., and Steinert, P., 1982, Retinoic acid induces transglutaminase activity but inhibits cornification of cultured epidermal cells, *J. Biol. Chem.* **257:**9906–9908.

Zehnder, P. R., and Lyons, G. D., 1975, Carcinoma and juvenile papillomatosis, *Ann. Otol. Rhinol. Laryngol.* **84:**614–618.

zur Hausen, H., 1977, Human papilloma viruses and their possible role in squamous cell carcinomas, *Curr. Top. Microbiol. Immunol.* **78:**1–30.

CHAPTER 11

Papillomavirus Cloning Vectors

DANIEL DIMAIO

I. INTRODUCTION

The papillomaviruses are proving to be valuable tools for transferring genes into living mammalian cells because the viral DNA can be used to establish permanent cell lines containing the transferred genes as unrearranged, extrachromosomal DNA molecules. This use of these viruses may help elucidate the function and regulation of eukaryotic genes and may enable the production of gene products that would otherwise be difficult to obtain. Bovine papillomavirus type 1 is the only papillomavirus to be exploited so far as a cloning vector, but it is likely that similar vectors will also be derived from DNA of other papillomaviruses.

A number of techniques have been developed for introducing purified genes into cells in tissue culture. The gene of interest is often covalently joined to a cloning vehicle (or vector) to facilitate identification of cells that have taken up the DNA or to ensure its propagation or expression. Cloning vectors have been developed from a number of animal viruses including SV40 and polyoma virus, adenoviruses, herpesviruses, poxviruses, and retroviruses (Gluzman, 1982). Infection by most of these viruses culminates in cell death, so it is often not possible to establish permanent cell lines for long-term biochemical and genetic manipulation. Methods that do generate stable cell lines almost invariably result in rearrangement of the transferred DNA and its joining to diverse uncharacterized segments of DNA (Perucho *et al.*, 1980). Consequently, quantitative com-

DANIEL DIMAIO • Department of Human Genetics, Yale University School of Medicine, New Haven, Connecticut 06510.

parisons of gene expression in different cell lines are difficult to make. In an attempt to circumvent these problems, a new class of eukaryotic cloning vectors has been developed from BPV-1 DNA. This chapter describes these vectors, summarizes the results obtained by using them, and discusses their potential applications.

II. TRANSFORMATION BY BOVINE PAPILLOMAVIRUS

BPV-1 DNA can oncogenically transform susceptible rodent fibroblasts in culture. This process and the properties of the transformed cells are discussed in detail in an earlier chapter in this volume, but the relevant points will be reiterated here. (1) About 1 to 2 weeks after purified viral DNA is applied to certain lines of mouse (C127, NIH 3T3) or rat (FR 3T3) fibroblasts, dense foci of cells appear on the monolayer of normal cells. Viral DNA cloned in *E. coli* retains this focus-forming ability. (2) An intact, unit-length viral genome is not required for transformation because a 5500-bp subgenomic fragment of viral DNA is sufficient to transform cells. (3) Papillomavirus transformation does not result in cell death; rather, the cells stably acquire a set of altered growth properties including anchorage independence and the ability to form tumors in nude mice. (4) The viral DNA does not appear to integrate into the cellular genome. Instead, up to several hundred apparently identical copies of the viral DNA are usually maintained as circular DNA molecules in the nuclei of transformed cells. DNA of BPV-2 and BPV-4 (Moar *et al.*, 1981; Campo and Spadidos, 1983), the cottontail rabbit papillomavirus (Watts *et al.*, 1983), the deer fibromavirus (Groff *et al.*, 1983), the European elk papillomavirus (Stenlund *et al.*, 1983b), and human papillomavirus types 1 and 5 (Watts *et al.*, 1984) have also been reported to transform rodent cells in an analogous manner.

Most gene transfer experiments using BPV-1 vectors take the following form. Recombinant molecules are constructed that contain the gene of interest, a transforming segment of BPV-1 DNA, and a bacterial replicon. The availability of the complete BPV-1 DNA sequence greatly facilitates the construction of these molecules (Chen *et al.*, 1982). The prokaryotic sequences are included to allow the easy preparation of large amounts of the recombinant DNA for subsequent steps. The DNA is then transferred as a calcium phosphate precipitate (usually with carrier DNA) or via protoplast fusion to susceptible rodent cells growing on plastic. Because papillomaviruses do not form infectious progeny in tissue culture, it has not been possible to encapsidate recombinant molecules in BPV-1 virions for efficient introduction into cells. About 2 weeks after exposure to DNA, individual transformed foci or colonies can be isolated and expanded into cell lines for analysis. Although only very few cells exposed to DNA or plasmid-containing bacteria become stably transformed (often less than 10^{-4}), it is possible to derive cell lines consisting entirely of transformed cells which can be propagated indefinitely or frozen in media containing 8% dimethyl sulfoxide. There have been no reports of transient expression experiments in which expression of a BPV-1-linked gene is assayed within a few days of DNA transfer. Technical details concerning the generation and analysis of BPV-1-transformed cell lines can be found in earlier brief reviews on BPV-1 cloning vectors (Howley *et al.*, 1983; Sarver *et al.*, 1983) and in the references cited in this chapter. Recently, Campo (1985) has compiled a number of relevant protocols and numerous restriction endonuclease cleavage maps of cloning vectors containing BPV-1 DNA.

In transformed cells, the transferred, BPV-1-linked DNA is usually present as an extrachromosomal DNA molecule. This is the primary attraction to using BPV-1 as a vector because it allows the establishment of permanent cell lines which contain the transferred gene in a stable, completely defined sequence environment. The advantages of studying transferred genes in this configuration are listed in Table I. It is clear that understanding the factors which influence either the efficiency of focus formation by BPV-1 plasmids or the extrachromosomal maintenance of these plasmids is crucial for the successful design of vectors. The protocols used for DNA transfer and the structure of the BPV-1 plasmids both affect transformation efficiency. NIH 3T3 cells are more readily transformed by BPV-1 DNA than are C127 cells (Lowy *et al.*, 1980), but the latter cell line has been used in most gene transfer experiments because transformed C127 cells give rise to foci which are more distinct. Primary hamster embryo cells can also be transformed by BPV-1 DNA (Morgan and Meinke, 1980), but attempts to transform human cells with BPV-1 or BPV-1 DNA have been unsuccessful (Black *et al.*, 1963; Dvoretzky *et al.*, 1980). It is the general experience that as C127 cells are passaged in culture, they become refractory to DNA-mediated transformation and are more likely to undergo spontaneous transformation. The batch of serum used for cell culture can also greatly affect the efficiency with which transformed foci are generated (Giri *et al.*, 1983), and in one set of experiments the omission of carrier DNA abolished the focus-forming activity of one of two recombinant plasmids tested (Matthias *et al.*, 1983). Although DNA has been transferred as a calcium phosphate precipitate in most experiments, it

TABLE I. Advantages of Cloning Vectors That Propagate as Plasmids

Plasmid DNA can readily be purified away from high-molecular-weight cellular DNA
Minichromosomes containing the gene of interest can readily be purified away from cellular chromosomes
DNA on all copies of a particular plasmid is in a uniform sequence environment. This may allow uniform expression levels of genes on these plasmids
The potential problem of integration of transferred DNA into inactive regions of cellular DNA is eliminated

has been reported for several plasmids that foci are more readily obtained following protoplast fusion (Schaffner, 1980; Binetruy *et al.*, 1982).

Because much of the molecular biology of BPV-1 is not understood in detail, the construction of stable BPV-1 cloning vectors is largely a matter of trial-and-error. Nevertheless, a few general principles of vector design can be outlined (see Fig. 1 for schematic diagram of an idealized BPV-1 cloning vector). The DNA segment inserted into the vector must not interrupt an essential gene or regulatory signal of the viral DNA. The finding that the 5500-bp *Hin*dIII to *Bam*HI restriction fragment of BPV-1 DNA is sufficient to transform cells indicated that this fragment may be a suitable BPV-1 vector (Lowy *et al.*, 1980; Sarver *et al.*, 1981). The realization that transformation by the 5500-bp fragment is facilitated by the remaining portion of the viral genome or by certain segments of cellular DNA suggests that one of these facilitatory segments should be included in BPV-1 vectors (DiMaio *et al.*, 1982; Sarver *et al.*, 1982; Kushner *et al.*, 1982; Karin *et al.*, 1983). For the full-length viral genome linked to bacterial plasmid DNA, the *Bam*HI rather than the *Hin*dIII permutation

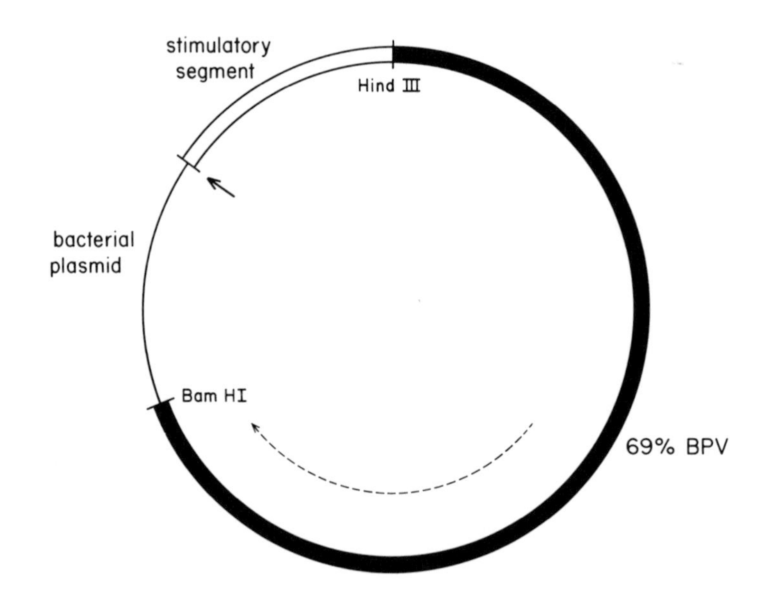

FIGURE 1. Schematic diagram of an idealized BPV-1 cloning vector. The circle represents a circular, double-stranded DNA molecule. The dark line represents the 5500-bp *Hin*dIII to *Bam*HI restriction fragment of BPV-1 DNA which is sufficient to transform mouse cells. The direction of transcription of this fragment in mouse cells is indicated by the dashed arrow. The double line represents a segment of DNA which stimulates mouse cell transformation by the BPV-1 DNA fragment. Such segments include the remaining 2500 bp of the viral genome and any of a number of DNA fragments from cellular genomes. The single line represents a bacterial replicon containing sequences that allow selection and amplification of the plasmid in bacteria. Additional DNA segment can be inserted at the position indicated by the solid arrow. Vectors of this general design include pBPV-142-6 and pBPV-BV1 (Sarver *et al.*, 1982; Zinn *et al.*, 1983).

seems preferable because it has been reported that the former is more stable (Sarver *et al.*, 1982).

The influence of linkage of BPV-1 DNA to bacterial plasmids deserves further comment because there are conflicting reports in the literature. A number of investigators have observed that the efficiency of focus induction by BPV-1 DNA is markedly reduced by covalent linkage to the bacterial plasmid pBR322 (Lowy *et al.*, 1980; Sarver *et al.*, 1981, 1982; DiMaio *et al.*, 1982; Binetruy *et al.*, 1982; Kushner *et al.*, 1982; Richards *et al.*, 1984). One group finds that this inhibition remains when pBR322 is replaced by pML2 (Binetruy *et al.*, 1982), a derivative of pBR322 lacking specific sequences which inhibit replication of linked SV40 DNA in monkey cells (Lusky and Botchan, 1981), but others find that insertion of the full-length *Bam*HI-cut BPV-1 genome into this deleted plasmid is compatible with efficient focus formation (Sarver *et al.*, 1982). In fact, other groups see no inhibition by bacterial plasmid sequences (Campo and Spandidos, 1983; Giri *et al.*, 1983). Numerous factors could account for these discrepancies. These include differences in viral isolates, host mammalian cells, transformation or cell culture protocols, DNA modifications introduced during propagation in different bacterial hosts, and exact bacterial plasmid sequences present in the recombinants. Since a systematic evaluation of the effect of these parameters has not been reported, it seems prudent to follow precisely the protocols of the laboratory from which each particular vector originated. In any case, a number of recombinant plasmids have been constructed which under certain relatively well-defined conditions efficiently give rise to transformed foci and which are maintained with minimal DNA rearrangement in the transformed cells.

III. DEVELOPMENT OF BPV-1 VECTORS

Following initial reports that bacterial plasmid sequences inhibit transformation by BPV-1 DNA, early gene transfer experiments using BPV-1 included cleavage of the recombinant plasmid with restriction endonucleases to dissociate the bacterial sequences from the DNA to be transferred. BPV-1 was first employed as a cloning vector by Sarver *et al.* (1981) who used the 5500-bp fragment of BPV-1 DNA to transfer the rat preproinsulin I gene into C127 cells. Prior to transformation, pBR322 sequences which had been used to amplify this DNA in bacteria were removed. Cells that had acquired the transferred DNA were identified on the basis of their ability to form morphologically transformed foci, and all 48 cell lines tested secreted rat proinsulin into the culture medium. These experiments established several important points about using BPV-1 as a cloning vector. In addition to demonstrating that BPV-1-induced morphological transformation could be used to select cells containing transferred genes, they also indicated that a number of appropriate eukaryotic cell-specific posttranslational modifications could occur in these cells. Importantly, the DNA was maintained as a multicopy plasmid with apparently no integration in the transformants, demonstrating that DNA linked to BPV-1 DNA is also propagated as a plasmid. In many of the cell lines, the DNA had undergone different rearrangements including acquisition of uncharacterized segments of DNA. This variability in DNA structure among different cell lines may have accounted for the variability that was also observed in gene expression. Similar experiments have since been performed with a number of BPV-1-linked genes excised from bacterial plasmid sequences.

There are three reasons why it is unsatisfactory to separate bacterial plasmid sequences from BPV-1 DNA prior to transformation. First, the requirement for enzymatic cleavage places constraints on acceptable locations of restriction endonuclease recognition sites in the recombinant plasmids. Second, linear BPV-1 DNA often undergoes rearrangements at its ends during recircularization in cells. Finally, the separation of eukaryotic and prokaryotic replicons prevents the facile transfer of plasmids between bacteria and mouse cells. These problems were overcome to a large extent by the second generation of BPV-1 vectors, plasmids which transformed cells without prior removal of bacterial plasmid sequences. A stable prokaryotic/eukaryotic plasmid replicon containing BPV-1 and bacterial plasmid DNA was first reported by DiMaio *et al.* (1982) who found that a 7600-bp DNA fragment containing the human β-globin gene caused a several hundred-fold stimulation of focus formation in C127 cells by the 5500-bp segment of BPV-1 DNA inserted in a bacterial plasmid. The transferred DNA in the transformants was often present initially as about 30 unrearranged, monomeric plasmids per cell, although rearrangements occasionally occurred with continued passage of the cell. Plasmids could be reestablished in *E. coli* by transforming competent bacteria with low-molecular-weight DNA from mouse cells and selecting for ampicillin-resistant colonies. The methylation pattern of the DNA giving rise to these colonies was consistent with its having replicated in animal cells, indicating that it was not contaminating DNA, and colonies generated in this manner contained plasmid DNA indistinguishable from the molecules originally used to transform mouse cells (Fig. 2). These experiments demonstrated that it was possible to introduce an intact plasmid into mouse cells where it was stably maintained as a monomeric, circular DNA molecule, and to reestablish the plasmid in bacteria without further enzymatic manipulation.

The mechanism by which the globin DNA stimulates transformation is unknown, but analysis of mutants with deletions in the globin region indicate that 2700 bp of human DNA is sufficient to cause this effect (Zinn *et al.*, 1983). Transformation of C127 cells by the subgenomic segment of BPV-1 DNA is stimulated by other cellular DNA fragments including the rat and human growth hormone genes (Kushner *et al.*, 1982; Sarver *et al.*, 1982), human metallothionein II_A gene (Karin *et al.*, 1983), an intergenic DNA segment from the rat fibrinogen gene cluster (Sarver

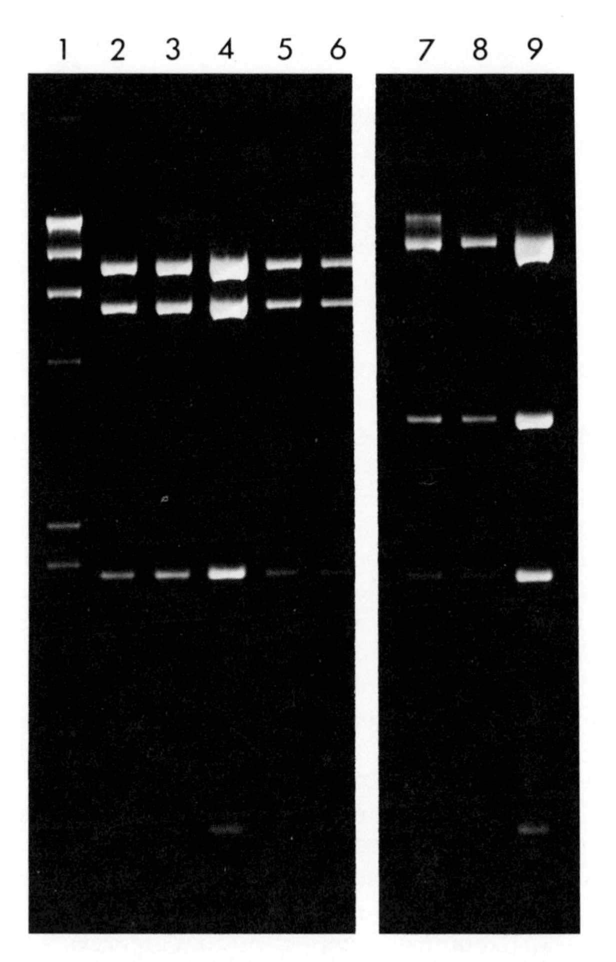

FIGURE 2. Recovery of BPV-1 plasmids from transformed mouse cells. C127 cells were transformed with pBPV-β1 and pBPV-β3 (DiMaio *et al.*, 1982). Low-molecular-weight DNA was isolated from cell lines established from individual foci of transformed cells and transferred into bacteria. DNA of the input plasmids was compared to DNA from individual ampicillin-resistant bacterial colonies after digestion with *Bam*HI and agarose gel electrophoresis. Lane 1: molecular weight markers. Lanes 2, 3, 5, 6: plasmids rescued from mouse cells transformed with pBPV-β3. Lane 4: pBPV-β3. Lanes 7, 8: plasmids rescued from mouse cells transformed with pBPV-β1. Lane 9: pBPV-β1.

et al., 1982), and the rat preproinsulin gene (Sarver *et al.*, 1982). Except for the human metallothionein gene which also stimulates transformation of NIH 3T3 cells, it is not known if these sequences are active in other cell types. Inactive DNA fragments include the human β-interferon gene (Zinn *et al.*, 1982), the rat γ-fibrinogen gene (Sarver *et al.*, 1983), the mouse metallothionein I gene (Sarver *et al.*, 1983), a human class I histocompatibility antigen heavy chain gene (DiMaio *et al.*, 1984), and the human metallothionein I gene (Richards *et al.*, 1984). Cells can also be efficiently transformed by certain full-length BPV-1–pML2 plasmids without additional DNA segments (Sarver *et al.*, 1982). These plasmids are also stably maintained in the transformants, and they can be rescued in bacteria. It should be possible to derive useful vectors from any of these plasmids unless the insertion of additional DNA interferes with focus formation or plasmid stability. Stable plasmid vectors containing either the full-length *Bam*HI-generated viral genome or the 5500-bp transforming fragment plus a stimulatory segment of cellular DNA have been successfully used to transfer additional DNA segments (Law *et al.*, 1983; Zinn *et al.*, 1983; Schenborn *et al.*, 1985; Sambrook *et al.*, 1985).

The utility of such vectors would probably be greatly increased if the recombinant plasmids contained selectable markers in addition to the BPV-1 transforming function. Experiments using the herpes simplex virus thymidine kinase gene (Lusky *et al.*, 1983; Sekiguchi *et al.*, 1983), the *E. coli* gpt gene (Law *et al.*, 1982; Giri *et al.*, 1983), or a mouse dihydrofolate reductase gene (Breathnach, 1984) were unsatisfactory because the BPV-1 DNA in cells containing the selected gene had frequently undergone rearrangement and/or integration into cellular DNA. More recently, both the neomycin analogue G418 and heavy metals have been used to select for extrachromosomal BPV-1-linked genes. Most mammalian cells are killed by G418, but resistance can be conferred by a gene isolated from Tn5, a bacterial transposon (Colbere-Garapin *et al.*, 1981; Southern and Berg, 1982). Several groups have constructed recombinant plasmids containing BPV-1 DNA and the G418 resistance gene under the control of eukaryotic regulatory signals and have demonstrated that after transfer of these recombinants and selection of drug-resistant cells, the DNA is maintained as a multicopy plasmid in the resistant mouse or rat cells (Law *et al.*, 1983; Matthias *et al.*, 1983; Lusky and Botchan, 1984; Meneguzzi *et al.*, 1984). Some of these plasmids are relatively unstable, but there appears to be less plasmid rearrangement in cell lines selected for drug resistance rather than for morphological transformation (Matthias *et al.*, 1983; Meneguzzi *et al.*, 1984). Similarly, resistance to the toxic heavy metals cadmium and zinc can be conferred by BPV-1-linked, extrachromosomal cloned mouse or human metallothionein genes (Karin *et al.*, 1983; Pavlakis and Hamer, 1983a,b).

If insertion of additional DNA is still compatible with selection and does not result in plasmid instability, these vectors may allow the introduction of extrachromosomal, BPV-1-linked genes into cell types not susceptible to oncogenic transformation by the papillomaviruses. Moreover, Lusky and Botchan (1984) have found that autonomous replication can proceed in the absence of some BPV-1 functions required for cell transformation. One can thus envision a class of transformation-defective BPV-1 vectors that still propagate as plasmids and can be selected biochemically. For many studies of gene expression, it may be preferable to use such vectors to eliminate the effects of oncogenic transformation.

IV. EXPRESSION OF GENES CLONED ON BPV-1

The ultimate goal of most gene transfer experiments is the examination of the expression of the transferred gene in the host cell. The nucleotide sequences controlling the structure and expression of a gene product can be identified and characterized by transferring genes carrying defined mutations. The effect of a foreign gene on the cell can also be assessed, as can the effects of an altered gene product or altered levels of a normal gene product. BPV-1 vectors are especially suited for studying genes isolated from mammals because the rodent host cells contain much of the cellular machinery necessary for accurate expression of these genes. In fact, not only are numerous posttranscriptional and posttranslational modifications correctly carried out in BPV-1-transformed cells (Some of which are listed in Table II), but certain levels of gene control are also retained in the transformants.

Using BPV-1 vectors to study gene regulation is not without its hazards. Events occurring in cells oncogenically transformed by BPV-1 may not necessarily reflect the situation in normal cells. For example, it is known that transforming proteins of other DNA viruses can activate transferred cellular and viral genes (Imperiale *et al.*, 1983; Green *et al.*, 1983). The multicopy nature of the BPV-1 system poses additional com-

TABLE II. Posttranslational Modifications Reported for Products of BPV-1-Linked Genes

Proteolytic cleavage
Rat preproinsulin I
Human growth hormone
Glycosylation
Hepatitis B virus surface antigen
Vesicular stomatitis virus G protein
Influenza virus hemagglutinin
Membrane insertion
Vesicular stomatitis virus G protein
Human histocompatibility antigen heavy chain
Influenza virus hemagglutinin
Secretion
Rat preproinsulin I
Hepatitis B virus surface antigen
Human β-interferon
Human growth hormone
Vesicular stomatitis virus G protein (mutant)
Minute virus of mice capsid proteins
Influenza virus hemagglutinin (mutant)
Assembly in macromolecular aggregates
Hepatitis B virus surface antigen
Human histocompatibility antigen heavy chain
Minute virus of mice capsid proteins

plications because the proportion of transcriptionally active genomes has not been established. Therefore, transcriptional induction could be due either to an increased rate of transcription of each gene or to recruitment of more templates into the active fraction. Although the assumption is made that the expression detected results from transcription of extrachromosomal templates, in most of these gene transfer experiments it has not been ruled out that the transformants contain a small amount of integrated exogenous DNA. It is particularly difficult to distinguish between large oligomeric plasmids and tandemly arrange integrated molecules. It is conceivable that in some experiments a small number of very active, integrated or rearranged templates may account for the observed expression.

The sections below describe the experience with a number of specific genes introduced into cells on BPV-1 vectors.

A. Human β-Globin Gene

The globin genes code for the polypeptide subunits of hemoglobin, the major soluble protein and oxygen-carrying component of red blood cells. Because of their important physiological role and because of the ease of obtaining large amounts of purified material, the globin genes and their products have been intensively studied for many years. The globin DNA fragment which stimulates transformation by the 5500-bp fragment of BPV-1 DNA contains an intact human β-globin gene which is actively transcribed in the mouse fibroblasts transformed by these recombinants (DiMaio *et al.*, 1982). RNA mapping experiments using nuclease S1 and primer extension with reverse transcriptase indicate that the β-globin gene in the BPV-1 transformants is transcribed from the promoter normally active in adult red cell precursors (although this promoter is normally inactive in fibroblasts), that both intervening sequences are faithfully removed, and that a poly(A) tail is affixed at the correct site. It was estimated that between 0.1 and 1% of the mRNA in the transformed cells is globin-specific; this is at least two orders of magnitude higher than the amount of globin RNA in mouse L cells containing a transferred human β-globin gene (Charney *et al.*, 1984). Such high expression is somewhat surprising in light of several reports that the levels of BPV-1-specific RNA are exceedingly low in transformed cells (Heilman *et al.*, 1982; Amtmann and Sauer, 1982; Engel *et al.*, 1983), but numerous other cloned genes on BPV-1 vectors are expressed at a very high level. In addition to globin RNA with the correct 5′ end, a small amount of RNA initiated farther upstream is produced when the globin gene is in the same transcriptional orientation as the BPV-1 genes, but not when it is in the opposite orientation.

The generation of stable cell lines producing large amounts of globin mRNA prompted Treisman *et al.* (1983) to use this system to investigate the transcripts produced from mutant globin genes isolated from patients with the hereditary hemoglobin disorder, beta thalassemia. Mouse cell lines were generated with BPV-1 plasmids containing these mutant genes in place of the normal one. Nucleotide substitutions within the globin gene often resulted in the appearance of incorrectly processed RNA molecules which were unable to produce normal globin polypeptides. In one case, a single base change resulted in the utilization of a normally cryptic splice site several hundred nucleotides removed from the location of the mutation. This approach of using BPV-1 vectors to generate cell lines producing large amounts of RNA for structural analysis should be widely applicable. However, it should be noted that gross rearrangement of the transferred DNA can be accompanied by the utilization of abnormal splicing and polyadenylation signals as has been reported for an integrated, rearranged BPV-1-linked dihydrofolate reductase gene (Breathnach, 1984).

B. Metallothioneins

The metallothioneins are small, cysteine-rich polypeptides which bind and detoxify heavy metal ions. The expression of the metallothionein genes is normally stimulated by heavy metals or by the glucocorticoid, dexamethasone. The concentration of RNA transcribed from the extrachromosomal mouse metallothionein I, the human metallothionein II_A, and the human metallothionein II_B promoters is increased two- to tenfold after administration of zinc or cadmium, indicating that the genes do not have to be located in a specific position in a cellular chromosome for this response (Pavlakis and Hamer, 1983a; Karin *et al.*, 1983; Richards *et al.*, 1984). When these genes are carried on a stable plasmid vector, there is remarkable uniformity in basal and induced expression levels in independently derived cell lines (Karin *et al.*, 1983).

Dexamethasone does not induce the metallothionein genes on the BPV-1 recombinants although this treatment does induce the endogenous cellular genes. The differential response of the BPV-1-linked but not the endogenous genes to metals and glucocorticoids indicates that dexamethasone does not normally induce by increasing intracellular levels of heavy metal. The same human metallothionein II_A promoter used in these BPV-1 experiments has also been stably incorporated into the chromosomal DNA of human cells, and in some of these cell lines it is induced by dexamethasone (Richards *et al.*, 1984). Therefore, the lack of induction in the BPV-1 system cannot be caused by the absence of a signal from the metallothionein gene fragment and may instead be due to the extrachromosomal state of the DNA or to the use of the heterologous NIH 3T3 host cell.

As mentioned in a previous section, it is possible to select BPV-1-transformed mouse cells containing extrachromosomal mouse and human metallothionein genes. Moreover, to facilitate the expression of other genes that would normally not be actively transcribed, Pavlakis and Hamer have developed a set of cloning vectors where foreign DNA can be inserted in such a manner as to come under the control of the mouse metallothionein I promoter which is expressed at a high level in cells transformed by the BPV-1–metallothionein recombinants (Pavlakis and Hamer, 1983a,b). Using this sort of vector, 2–6 $\times$ 10^8 molecules/cell of growth hormone are secreted daily by some cell lines containing a BPV-1-linked human growth hormone gene fused to the metallothionein promoter. This level of expression is about tenfold higher than in cells infected with the human growth hormone gene on a lytic SV40-based vector (Hamer and Walling, 1982).

C. Hepatitis B Surface Antigen

Infection by hepatitis B virus affects hundreds of millions of individuals worldwide. A safe and effective hepatitis B vaccine is being produced from hepatitis B surface antigen (HBsAg) purified from the serum of chronic carriers, but the cost of the vaccine prohibits its widespread use in developing countries and the possibility that it contains as-yet-unrecognized infectious agents must be considered. Furthermore, the requirement for extensive posttranslational processing of the primary translation product of HBsAg mRNA makes the antigen an unlikely candidate for economic, large-scale production in bacteria, which cannot carry out these modifications. These considerations have led a number of groups to use the HBsAg gene linked to BPV-1 DNA to generate stable mouse cell lines that secrete large amounts of mature HBsAg. Although the experimental details differ, the general experience has been that most cell lines selected for morphological transformation by the HBsAg-linked BPV-1 DNA produce detectable and often very high levels of HBsAg (Stenlund *et al.*, 1983a; Wang *et al.*, 1983; Hsiung *et al.*, 1984). The glycosylated antigen is secreted into the culture medium as lipoprotein particles with buoyant density, electron microscopic appearance, polypeptide composition, and immunogenicity indistinguishable from comparable 22-nm particles purified from the serum of chronic hepatitis carriers.

To increase levels of expression, Hsiung *et al.* (1984) placed the gene under the control of regulatory signals derived from the mouse metallothionein I gene. In addition to obtaining high levels of gene expression, they found that they could extend the life of the cell cultures at confluence by an alternating regimen of serum-containing and serum-free culture media. In this manner they were able to maintain high-expressing cell lines in roller bottle culture for 60 days without passaging. Such *in vitro* manipulation of recombinants to maximize expression levels and improved cell culture techniques to increase yield may allow large-scale production of commercially important polypeptides. Although the level of gene expression achieved so far with BPV-1 vectors has probably not

TABLE III. Production of Hepatitis B Virus Surface Antigen in Mammalian Cell Cultures

Vector	Host	Amount[a]	Duration[b]	Reference
—	PLC/PRF/5 (human hepatoma cell line)	0.2–0.3	—	Alexander *et al.* (1976), Stratowa *et al.* (1982), Wang *et al.* (1983)
Viral DNA[c]	L tk⁻ (mouse)	1–2	7 days	Dubois *et al.* (1980)
Viral DNA[d]	NIH 3T3 (mouse)	8–10	30 cell passages	Christman *et al.* (1982)
Moloney murine sarcoma virus	NIH 3T3 (mouse)	4.5	1 week	Stratowa *et al.* (1982)
SV40	CV-1 (monkey)	1.25	3 days	Moriarty *et al.* (1981)
SV40	CV-1 (monkey)	38	2 weeks	Liu *et al.* (1982)
SV40	COS-1 (monkey)	45	2–3 weeks	Crowley *et al.* (1983)
SV40	COS-1 (monkey)	16–20	60 hr	Siddiqui (1983)
BPV-1	C127 (mouse)	6	1 month	Stenlund *et al.* (1983a)
BPV-1	NIH 3T3 (mouse)	6	6 months	Wang *et al.* (1983)
BPV-1[e]	C127 (mouse)	10–20	5 months	Hsiung *et al.* (1984)

[a] μg hepatitis B surface antigen/10^7 cells per day.
[b] Length of time cell cultures produced antigen.
[c] Hepatitis B virus DNA was cotransferred with herpes simplex virus thymidine kinase gene.
[d] Hepatitus B virus DNA was cotransferred with a dihydrofolate reductase gene, and expression was assayed after gene amplification in methotrexate.
[e] Surface antigen gene was transcribed from a mouse metallothionein I promoter.

been optimized, it compares favorably with the HBsAg yields reported for a variety of other host–vector systems (Table III).

D. Human β-Interferon Gene

The interferons are secreted proteins that play an important role in the defense against viral infection. Because of their therapeutic potential and because they represent a class of inducible genes, the interferon (IFN) genes have been studied using several types of gene transfer experiments. The human β (or fibroblast) IFN gene has been transferred into mouse C127 cells using two different types of BPV-1 vectors. In the earlier experiments, the IFN-β gene linked to the subgenomic transforming fragment of BPV-1 DNA was excised from pBR322 sequences prior to transformation (Zinn *et al.*, 1982; Mitrani-Rosenbaum *et al.*, 1983). Many cell lines selected for morphological transformation secreted human IFN into the culture medium. The secreted IFN activity and IFN RNA levels (but not BPV-1 RNA levels) were increased by treatment of the cells with either poly(rI · rC) or Newcastle disease virus, two well-studied inducing agents of the IFN-β genes. Induction is also caused by cycloheximide treatment (Maroteaux *et al.*, 1983). Although the DNA appeared extrachromosomal in the transformants, there was considerable DNA rearrangement among different cell lines. Perhaps as a consequence of these diverse genomic configurations, the expression levels and extent of induction varied widely among different cell lines established with the same DNA fragment.

The results were more satisfactory when the same gene was transferred using a stable plasmid vector containing the 5500-bp segment of BPV-1 DNA and the stimulatory segment of human globin DNA (Zinn *et al.*, 1983). With this recombinant, most transformed cell lines contained plasmid DNA identical to the input molecule. Moreover, the basal and induced level of human IFN RNA and secreted IFN activity were essentially identical in all cell lines even though the latter were not subcloned from single transformed cells (Table IV). In contrast, other methods of establishing cell lines stably expressing a transferred gene are characterized by substantial variability of expression among different cell lines. For example, several laboratories have used biochemical selection to establish cell lines containing a transferred integrated human IFN-β gene; within each set of cell lines the extent of inducibility varied at least tenfold (Canaani and Berg, 1982; Ohno and Taniguchi, 1982; Pitha *et al.*, 1982; Hauser *et al.*, 1982).

In the BPV-1 system, induction of the wild-type IFN gene with poly(rI · rC) resulted in an approximately 400-fold increase in the level of correctly initiated human IFN-β, but in only a 40-fold increase in human IFN activity. The endogenous mouse fibroblast IFN activity was induced at least 2000-fold. Low levels of expression of the transferred IFN-β gene were also seen with the unstable linearized BPV-1 vectors and with other non-BPV-1 vector systems (Pitha *et al.*, 1982; Ohno and Taniguchi, 1982; Hauser *et al.*, 1982; Canaani and Berg, 1982). The factors responsible for this suboptimal regulatory response have not been defined.

The lack of variability in gene expression among cell lines enabled

TABLE IV. Interferon Induction in Mouse Cells Transformed with pBV-IFΔ3[a,b]

Cell line	Human IFN activity (units/ml)		Mouse IFN activity (units/ml)		Induced/uninduced human IFN activity
	−	+	−	+	
3-1	200	2000	<20	40,000	10
3-5	400	4000	ND	ND	10
3-6	300	4000	<20	40,000	13
3-7	200	2000	<20	40,000	10
3-8	200	2000	<20	40,000	10
3-9	400	4000	<20	40,000	10
3-10	400	4000	ND	ND	10

[a] From Zinn *et al.* (1983) with permission.

[b] Several independent cell lines were established with pBV-IFΔ3, a plasmid containing a deletion in the 5′ flanking region of the human IFN-β gene. Secreted IFN activity was measured for each cell line in the absence of induction (−) or after treatment with poly(I·C) (+). IFN secretion by this mutant is induced about 10-fold; the wild-type gene is induced about 40-fold.

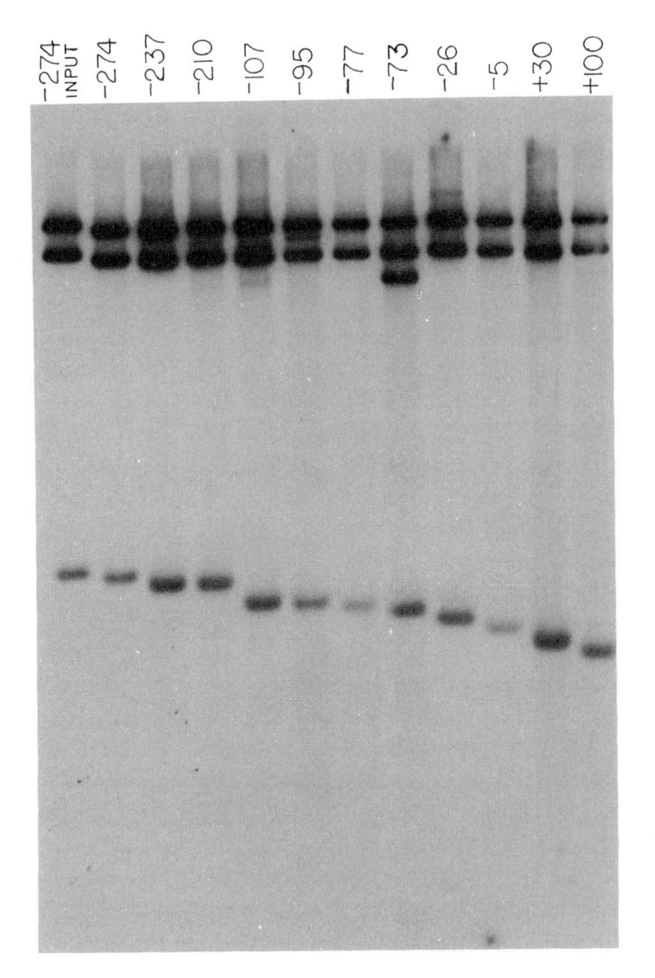

FIGURE 3. Structure of interferon gene deletion mutants in transformed mouse cells. C127 cells were transformed with a series of interferon gene deletion mutants, and individual foci were expanded into cell lines from which low-molecular-weight DNA was isolated. After digestion with restriction endonucleases, the DNA fragments were separated by agarose gel electrophoresis and the transferred DNA was detected by hybridization. The numbers above each lane indicate the mutant plasmid used for transformation, and the lowest band in each lane is the DNA fragment which contains the constructed deletion. (From Zinn *et al.*, 1983, with permission.)

Zinn *et al.* (1983) to assess the effect of a series of mutations in the putative regulatory region of the human IFN gene. Deletions in the 5′ end of the gene were constructed *in vitro* and transferred individually into cells. Several transformed cell lines were generated from each mutant and were analyzed for DNA content and for IFN gene expression. In most of the transformants, the transferred DNA had the same structure as the input molecule (Fig. 3). Analysis of IFN expression confirmed the lack of variability within each set of cell lines and indicated that there are two regulatory regions adjacent to the 5′ end of the human IFN gene. One region, between 107 and 210 bp 5′ from the site of transcription initiation, plays a role in maintaining a low level of transcription in the absence of induction. Genes with deletions of this region also exhibit altered induction kinetics. The second region, located between 19 and 77 bp 5′ from the gene, is required for increased accumulation of IFN RNA after induction. Recently, it has been found that a 40-base segment normally located between 37 and 77 bp 5′ to the IFN gene is sufficient to confer high levels of inducibility to heterologous BPV-1-linked promoters (Goodbourn *et al.*, 1985). Similar analysis of the effects of mutations and isolated regulatory segments on other genes will probably be one of the major uses of BPV-1 vectors in the near future. BPV-1-linked human IFN-γ and IFN-α genes have been expressed in mouse cells as cDNA clones driven from the SV40 early promoter (Fukunaga *et al.*, 1984).

E. Mouse Mammary Tumor Virus

Transcription of the mouse mammary tumor virus (MMTV) is stimulated by glucocorticoids, and gene transfer experiments have localized the hormonally responsive element to the long terminal repeat of this retrovirus (reviewed in Ringold, 1983). To develop a completely defined model system to investigate this response, Ostrowski *et al.* (1983) constructed BPV-1 recombinants containing the MMTV LTR and introduced them into C127 cells after removing bacterial plasmid sequences. Between 50 and 200 copies of the transferred DNA were maintained extrachromosomally in the transformants, but in contrast to the common experience with BPV-1 vectors excised from pBR322, there was little or no evidence of rearrangement. RNA analysis indicated that there was a great deal of variability in gene expression in the absence of induction and that dexamethasone treatment of cells caused the accumulation of increased amounts of RNA initiated both at the MMTV promoter and at sites farther upstream (but caused no change in the number of DNA templates). The extent of this increase varied from 1.3- to 100-fold among different cell lines, perhaps reflecting either minor, undetected DNA rearrangements or an inherent property of the MMTV promoter. The latter possibility is suggested by the variability of transcriptional induction in cells containing MMTV proviruses (Ringold, 1983)

Nuclei were prepared from transformed cells grown in the presence or the absence of dexamethasone, and recombinant minichromosomes were extracted using a procedure that resulted in approximately a 500-fold enrichment of extrachromosomal molecules. Nascent transcripts were then elongated *in vitro* and the newly synthesized RNA was analyzed. The results of this experiment indicate that hormonal induction in this system takes place at the level of transcription initiation. Importantly, these experiments also demonstrate convincingly that induction is occurring on extrachromosomal templates.

F. Polypeptide Hormones

Genes for several polypeptide hormones have been introduced into mouse cells on BPV-1 vectors; in all cases the hormones are secreted from the transformed cells. The results with the rat preproinsulin gene (Sarver *et al.*, 1981) were discussed above (see Section III). The BPV-1-linked rat growth hormone gene is also expressed in transformed cells (Kushner *et al.*, 1982). The cell lines reported in this study are not suitable for studies of some aspects of the regulation of growth hormone gene expression because transcription of the transferred gene does not respond to dexamethasone, an inducer of the endogenous gene. Moreover, the growth hormone RNA in the transformants initiated upstream from the initiation site of bona fide rat growth hormone mRNA.

As summarized in Section IV.B, BPV-1 vectors have been used to produce high levels of human growth hormone (Pavlakis and Hamer, 1983b). Using a similar vector, Ramabhandran *et al.* (1985) reported high-level production of bovine growth hormone able to specifically bind growth hormone receptors. The gene encoding the common α subunit of the human glycoprotein hormones has also been introduced into mouse cells, and the synthesis, glycosylation, and secretion of the protein have been described (Ramabhandran *et al.*, 1984).

G. Other Genes

BPV-1 vectors have been used to introduce a number of other genes into cells. Stable mouse cell lines have been generated which express the vesicular stomatitis virus glycoprotein (a protein which mediates membrane fusion at low pH) or a mutant version lacking the COOH-terminal membrane anchor sequence (Florkiewicz *et al.*, 1983; Florkiewicz and Rose, 1984). Transcription of these genes was initiated at a linked SV40 promoter. At least 95% of the transformed cells synthesize the glycoprotein, but the level of expression varied greatly from cell to cell. Acidification of the medium caused the transformed cells to fuse, indicating that the viral protein was biologically active. Comparison of expression of the wild-type and mutant genes demonstrated that the mutant protein was secreted from the cells rather than anchored into the membrane. Moreover, the attachment of complex oligosaccharides and transit from the rough endoplasmic reticulum to the Golgi apparatus were delayed for the mutant. Similar experiments have been performed with a BPV-1-linked influenza virus hemagglutinin gene (Sambrook *et al.*, 1985). The authors suggest a novel use of cell lines expressing the viral hemagglutinin: the ability of red cell ghosts to bind these cells and to disgorge their contents may allow the efficient delivery of macromolecules into the cytoplasm of cells.

A cellular surface antigen has been expressed using BPV-1 vectors as well. A human class I histocompatibility antigen heavy chain polypeptide is synthesized at a very high rate in C127 cells transformed with its gene linked to BPV-1 DNA (DiMaio *et al.*, 1984). The heavy chain associates with β_2-microglobulin at the cell surface of these transformants, but the protein seems less stable than the corresponding protein in human lymphoblastoid cells.

C127 cells have also been transformed with BPV-1 recombinants containing the capsid protein genes of the minute virus of mice (MVM), an autonomous parvovirus, and these cells secrete MVM capsids (Pintel *et al.*, 1984). Immunofluorescence experiments demonstrated a great deal of variability in both the amount of capsid proteins expressed and the intracellular localization of these proteins. These results were obtained even though the cells were cloned several times, and the molecular basis of this variability is obscure. A human U1 RNA gene has been introduced into mouse cells on a variety of BPV-1 vectors (Schenborn *et al.*, 1985). As much as 15% of the total U1 RNA in the transformed cells was encoded by the transferred gene, and no more than several hundred base pairs of flanking DNA sequence were required for efficient synthesis.

V. STABILITY OF BPV-1 RECOMBINANTS

The precise structure of BPV recombinants seems to play a major role in determining plasmid stability. One of the globin DNA-containing vectors, pBPV-BV1, appeared more stable than several other vectors when linked to the influenza virus hemagglutinin gene (Sambrook *et al.*, 1985), but it is premature to generalize from a single set of recombinants. When the goal of a gene transfer experiment is the establishment of permanent, stable cell lines, it is essential that the DNA is maintained without rearrangement. Although these conditions are often met with BPV-1 vectors, examples of plasmid instability are numerous. Transformation with linear BPV-1 molecules almost invariably results in circularization with rearrangement of the DNA molecule at its ends and acquisition of new DNA sequences (Law *et al.*, 1981; Sarver *et al.*, 1981; Zinn *et al.*, 1982). BPV-1 vectors that transform cells as intact plasmid molecules eliminate this problem, but with continued passage of the resulting cell lines, deletions can occur even in these plasmids (Fig. 4) (DiMaio *et al.*, 1982). These deletions often appear first in the bacterial plasmid segments of the recombinants, suggesting that these segments may be selected against during plasmid replication. If the ability to rescue the recombinant from mouse cells is not a prerequisite of an experiment, it may be advisable to remove the bacterial plasmid sequences and circularize the BPV-1/foreign gene segment *in vitro* prior to transformation. Apparent oligomerization of both unrearranged and deleted plasmids has also been reported (Binetruy *et al.*, 1982), but segregation of BPV-1 plasmids from transformed cells appears to occur very infrequently (Turek *et al.*, 1982).

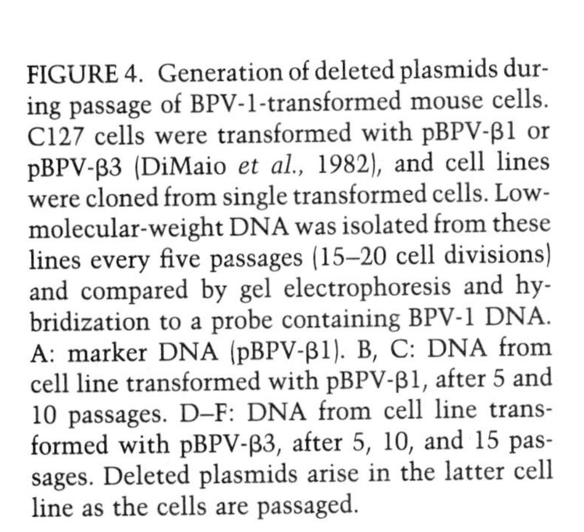

FIGURE 4. Generation of deleted plasmids during passage of BPV-1-transformed mouse cells. C127 cells were transformed with pBPV-β1 or pBPV-β3 (DiMaio *et al.*, 1982), and cell lines were cloned from single transformed cells. Low-molecular-weight DNA was isolated from these lines every five passages (15–20 cell divisions) and compared by gel electrophoresis and hybridization to a probe containing BPV-1 DNA. A: marker DNA (pBPV-β1). B, C: DNA from cell line transformed with pBPV-β1, after 5 and 10 passages. D–F: DNA from cell line transformed with pBPV-β3, after 5, 10, and 15 passages. Deleted plasmids arise in the latter cell line as the cells are passaged.

Inclusion or removal of specific DNA sequences can also result in instability. When a murine retroviral transcriptional enhancer is inserted into MMTV/BPV-1 recombinants, integration and rearrangement occur in most cell lines (Ostrowski *et al.*, 1983). When a portion of the stimulatory segment of human globin DNA is removed from a plasmid containing the 5500-bp fragment of BPV-1 DNA, the resulting molecule still transforms efficiently and is maintained as a plasmid; but the transferred DNA is frequently rearranged (Zinn *et al.*, 1983). Although most deletions in the 5′ portion of the human IFN-β gene are compatible with stable plasmid maintenance, a few deletion mutants generated exclusively cell lines that were unstable in terms of either the structure of the plasmid DNA or IFN gene expression (Zinn *et al.*, 1983). The position of the human DNA insert influenced the stability of BPV plasmids containing the human U1 RNA gene (Schenborn *et al.*, 1985). DNA segments could cause instability by inhibiting the production of virus-encoded *trans*-acting factors required for plasmid replication or segregation or by interfering with viral sequences that are required in *cis* for these phenomena. Because only the *cis*-acting sequences need be physically joined to the foreign DNA to ensure its extrachromosomal propagation, perhaps it will be possible to supply *trans*-acting replication factors from a second viral genome in the cell, thus eliminating the possibility that a linked foreign DNA segment will adversely affect the production of proteins required for replication. Recombination between multiple plasmids in a single cell and segregation of a plasmid which is not maintained by selection are potential pitfalls of this approach, but Lusky and Botchan (1984) have reported the stable co-maintenance of a wild-type BPV-1 genome supplying *trans* replication factors and a plasmid containing only a small segment of BPV-1 DNA, which presumably supplies the *cis* replication requirements for the plasmid. Although it has been reported that the absence of adenine methylation stabilized a BPV recombinant containing a thymidine kinase gene (Lusky *et al.*, 1983), this does not seem to be a general effect (Schenborn *et al.*, 1985).

Because viral DNA is not encapsidated in the BPV-1 system, there is no obvious size limitation on BPV-1 recombinants. Stable 16,000-bp recombinants have been reported (DiMaio *et al.*, 1982). Transfer of a 24,000-bp plasmid resulted in its conversion into a higher-molecular-weight (possibly integrated) form; it has not been determined whether this behavior was a consequence of its size or some other feature or the plasmid (DiMaio *et al.*, 1984).

As is clear from this brief discussion, there are probably numerous factors that influence plasmid stability, and they are still poorly defined. All cell lines generated with BPV-1 recombinants must be evaluated separately for the presence of DNA rearrangements and integration. The presence of a moderate amount of rearrangement may not be a serious problem for some experiments such as the production of high levels of gene products (e.g., Treisman *et al.*, 1983; Stenlund *et al.*, 1983a). Moreover, plasmids can be rescued by transforming bacteria with DNA from mouse cell lines containing predominantly integrated or oligomerized BPV-1 recombinants (Sekiguchi *et al.*, 1983; Binetruy *et al.*, 1982). As more experience is gained in the use of these vectors and as the molecular biology of the papillomaviruses is studied, perhaps the rules governing plasmid stability will emerge.

VI. FUTURE DIRECTIONS

Some of the numerous potential applications of BPV-1 vectors have already been discussed. Their use in generating cell lines synthesizing large amounts of RNA or mature gene products has been amply demonstrated. To synthesize products that are normally toxic to the cells, it may be possible to construct vectors which allow the plasmid copy number or transcription to be acutely increased after cell lines have been established. Attempts to use the mouse metallothionein I promoter in this manner have met with limited success. Only a modest level of induction has been observed for the human growth hormone structural gene fused to this promoter (Pavlakis and Hamer, 1983a); no induction was seen for an HBsAg gene fused to the same promoter (Hsiung *et al.*, 1984). Gene fusions using the MMTV or the human IFN-β promoter may be more effective.

As is shown by the mutational analysis of the human IFN-β gene (Zinn *et al.*, 1983), the uniform levels of expression of BPV-linked genes make this an ideal system for quantitative evaluation of regulatory mutants. Goodbourn (personal communication) has recently found similar levels of expression in cell lines derived from individual foci and in pooled cell lines descended by relatively few cell divisions from multiple, independent, transformed cells. Use of pooled cell lines should greatly speed the analysis of mutants; the short period of cell expansion may also allow the analysis of recombinants which are not stable enough to tolerate the extended cell culture required to generate a cell line from a single cell. Unfortunately, many regulatory phenomena cannot be studied in mouse fibroblasts. The use of biochemical selection may allow the transfer of BPV-1-linked genes into a wide range of cells, but it has not been determined whether BPV-1 DNA will replicate as a plasmid in many cell types that are resistant to morphological transformation by the virus. It is unlikely that uniform expression levels will be achieved unless the BPV-1 vector is maintained extrachromosomally.

The inability to purify specific segments of chromatin from the bulk of cellular chromatin has necessitated the use of indirect, hybridization techniques to analyze the chromatin associated with specific cellular genes. Using BPV-1 vectors, it is now possible to isolate and partially purify chromatin associated with specific genes. Such preparations can be examined biochemically or with the electron microscope, as has been done for BPV-1 chromatin (Rösl *et al.*, 1983). The relatively low copy number of BPV-1 plasmids in cells will make biochemical characterization a formidable task, but it also makes it less likely that important regulatory molecules will be titrated out by massively replicating templates, as may be the case with some other viral vectors.

There is also a great deal of interest in exploiting BPV-1 recombinants to isolate plasmids that have undergone sequence alterations during replication in mouse cells. For example, after growth of a mouse cell line containing a BPV-1 recombinant, bacteria can be transformed with plasmids purified from these cells; if the appropriate selection schemes can be devised, it should be possible to isolate altered plasmids which have undergone mutation or recombination. Plasmids that confer a selectable phenotype on the transformed mammalian cells can also be isolated (Fig. 5). In one straightforward sort of experiment, a gene cloned on BPV-1 is mutagenized *in vitro* and the pool of mutagenized molecules is transferred into mammalian cells. Selection is then imposed to isolate rare transformants that have acquired a mutant gene and therefore express a selectable phenotype. Plasmid DNA is then extracted from these transformants and used to transform bacteria to amplify the variant DNA for

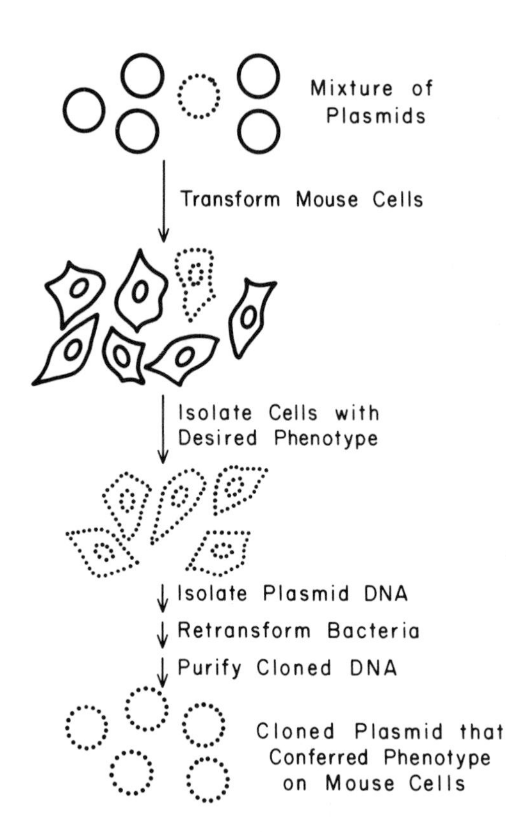

FIGURE 5. Schematic diagram of a method to rescue genes that confer a selectable phenotype on mouse cells. The circles at the top and bottom of the figure represent plasmid DNA molecules that have been amplified in bacteria. The other forms represent transformed mouse cells. Plasmids that can confer a selectable phenotype after transfer into mouse cells are represented by dotted circles; dotted forms represent cells that contain this plasmid. If the plasmid mixture initially used to transform mouse cells contains rare members that confer a selectable phenotype on mouse cells, by selecting cells with this phenotype it may be possible to isolate the responsible plasmids. (From DiMaio, 1984.)

detailed analysis. Using similar approaches, it may be possible to isolate specific genes from genomic or cDNA libraries constructed in BPV-1 vectors.

Analogous "shuttle vectors" are being developed from cosmids (Lund *et al.*, 1982; Lau and Kan, 1983), bacteriophage (Hamada *et al.*, 1983; Lindenmaier *et al.*, 1982), SV40 (Breitman *et al.*, 1982; Conrad *et al.*, 1982), retroviruses (Cepko *et al.*, 1984), and systems that require *in vitro* enzymatic manipulation (Howard *et al.*, 1983), but none offer the uniform expression levels and the ease of plasmid recovery that are potentially attainable with the BPV-1 system. Implementation of this approach obviously requires efficient selection or screening procedures that are compatible with stable extrachromosomal replication of the recombinant and with the expansion of the transformed foci into mass cultures for DNA extraction. There are numerous technical problems which must first be overcome. The conditions used to initially transform mouse cells must generate a large number of independent transformants in order to make recovery of a rare mutant or gene feasible. Some inserts may interfere with focus formation or plasmid stability. Moreover, it will probably be difficult to construct a representative library in plasmid vectors given the problem of generating many thousands of different circular DNA molecules each containing both a BPV-1 and an inserted DNA segment. The inclusion of a bacteriophage lambda *cos* site on a BPV-1 vector may facilitate construction of such a library (Matthias *et al.*, 1983), but it has not been established that a 50,000-bp BPV-1 recombinant can be stably maintained in mammalian cells or be efficiently recovered from them.

The persistence of multiple different plasmids within a clone of transformed mouse cells poses additional problems. When an equimolar mixture of two very similar but distinguishable BPV-1 plasmids is used to transform cells, both persist in cell lines derived from individual transformed foci (DiMaio, unpublished results). This indicates that selection conditions should be designed to enable isolation of the gene of interest if it coexists in cells with other plasmids. Multiple different plasmids will be recovered in bacteria if there are multiple species in the selected mouse cells, and these recovered plasmids will then have to be rescreened individually for activity in mammalian cells. Further experiments are needed to determine whether these potential problems represent an insurmountable obstacle or merely a technical challenge.

Finally, it is clear that better papillomavirus cloning vectors will be constructed as the molecular biology of these viruses becomes better understood. Factors that influence transformation efficiency, stability, and copy number have to be identified and evaluated before vectors can be designed on a rational basis. A systematic assessment of vectors derived from a number of papillomaviruses may identify a specific papillomavirus–host cell combination that is particulary compatible with stable plasmid maintenance. Conversely, studies using papillomavirus vectors will undoubtedly play an essential role in elucidating the molecular biology of these viruses.

ACKNOWLEDGMENTS. The author wishes to thank the many colleagues who contributed material for inclusion in this chapter. In addition, the support of the Edward J. Mallinckrodt Foundation is gratefully acknowledged.

REFERENCES

Alexander, J., Bey, E. M., Geddes, E. W., and Lecatsas, G., 1976, Establishment of a continuously growing cell line from primary carcinoma of the liver, *S. Afr. Med. J.* **50:**2124–2128.

Amtmann, E., and Sauer, G., 1982, Bovine papillomavirus transcription: Polyadenylated RNA species and assessment of the direction of transcription, *J. Virol.* **43:**59–66.

Binetruy, B., Meneguzzi, G., Breathnach, R., and Cuzin, F., 1982, Recombinant DNA molecules comprising bovine papilloma virus type 1 DNA linked to plasmid DNA are maintained in a plasmidial state both in rodent fibroblasts and in bacterial cells, *EMBO J.* **1:**621–628.

Black, P. H., Hartley, J. W., Rowe, W. P., and Heubner, R. J., 1963, Transformation of bovine tissue cells by bovine papillomavirus, *Nature* **199:**1016–1018.

Breathnach, R., 1984, Selective amplification in methotrexate-resistant mouse cells of an artificial dihydrofolate reductase transcription unit making use of cryptic splicing and polyadenylation sites, *EMBO J.* **3:**901–908.

Breitman, M. L., Tsui, L.-C., Buchwald, M., and Siminovitch, L., 1982, Introduction and recovery of a selectable bacterial gene from the genome of mammalian cells, *Mol. Cell. Biol.* **2:**966–976.

Campo, M. S., 1985, Bovine papillomavirus DNA: A eukaryotic cloning vector, in: *DNA Cloning: A Practical Approach* (D. Glover, ed.), IRL Press.

Campo, M. S., and Spandidos, D. A., 1983, Molecularly cloned bovine papillomavirus DNA transforms mouse fibroblasts *in vitro*, *J. Gen. Virol.* **64:**549–557.

Canaani, D., and Berg, P., 1982, Regulated expression of the human beta 1 interferon gene after transduction into cultured mouse and rabbit cells, *Proc. Natl. Acad. Sci. USA* **79:**5166–5170.

Cepko, C. L., Roberts, B. E., and Mulligan, R. C., 1984, Construction and applications of a highly transmissible murine retrovirus shuttle vector, *Cell* **37:**1053–1062.

Charney, P., Treisman, R., Mellon, P., Chao, M., Axel, R., and Maniatis, T., 1984, Differences in human α and β-globin gene expression in mouse erythroleukemia cells: The role of intragenic sequences, *Cell* **38:**251–263.

Chen, C. Y., Howley, P. M., Levinson, A. D., and Seeburg, P. H., 1982, The primary structure and the genetic organization of bovine papilloma virus type 1 genome, *Nature* **299:**530–535.

Christman, J. K., Gerber, M., Price, P. M., Flordellis, C., Edelman, J., and Acs, G., 1982, Amplification of expression of hepatitis B surface antigen in 3T3 cells cotransfected with a dominant active gene and cloned viral DNA, *Proc. Natl. Acad. Sci. USA* **79:**1815–1819.

Colbere-Garapin, F., Horodniceanu, F., Kourilsky, P., and Gerapin, A.-C., 1981, A new dominant hybrid selective marker for higher eukaryotic cells, *J. Mol. Biol.* **150:**1–14.

Conrad, S. E., Liu, C.-P., and Botchan, M. R., 1982, Fragment spanning the SV40 replication origin is the only DNA sequence required in cis for viral excision, *Science* **218:**1223–1225.

Crowley, C. W., Liu, C. C., and Levinson, A., 1983, Plasmid-directed synthesis of hepatitis B surface antigen in monkey cells, *Mol. Cell. Biol.* **3:**44–55.

DiMaio, D., 1984, Eukaryotic cloning vectors based on bovine papillomaviruses, *BioEssays* **1:**23–26.

DiMaio, D., Treisman, R. H., and Maniatis, T., 1982, Bovine papilloma virus vector that propagates as a plasmid in both mouse and bacterial cells, *Proc. Natl. Acad. Sci. USA* **79:**4030–4034.

DiMaio, D., Corbin, V., Sibley, E., and Maniatis, T., 1984, High level expression of a cloned HLA heavy chain gene introduced into mouse cells on a bovine papillomavirus vector, *Mol. Cell. Biol.* **4:**340–350.

Dubois, M. F., Pourcel, C., Rousset, S., Chany, C., and Tiolais, P., 1980, Excretion of hepatitis B surface antigen particles from mouse cells transformed with cloned viral DNA, *Proc. Natl. Acad. Sci. USA* **77:**4549–4553.

Dvoretzky, I., Shober, R., Chattopadhyay, S. K., and Lowy, D. R., 1980, A quantitative *in vitro* focus forming assay for bovine papilloma virus, *Virology* **103:**369–375.

Engel, L. W., Heilman, C. A., and Howley, P. M., 1983, Transcriptional organization of bovine papillomavirus type I, *J. Virol.* **47:**516–528.

Florkiewicz, R. Z., and Rose, J. K., 1984, A cell line expressing vesicular stomatitis virus glycoprotein fuses at low pH, *Science* **225:**721–723.

Florkiewicz, R. Z., Smith, A., Bergmann, J. E., and Rose, J. K., 1983, Isolation of stable mouse cell lines that express cell surface and secreted forms of the vesicular stomatitis virus glycoprotein, *J. Cell Biol.* **97:**1381–1388.

Fukunaga, R., Sokawa, Y., and Nagata, S., 1984, Constitutive production of human interferons by mouse cells with bovine papillomavirus as a vector, *Proc. Natl. Acad. Sci. USA* **81:**5086–5090.

Giri, I., Jouanneau, J., and Yaniv, M., 1983, Comparative studies of the expression of linked *Escherichia coli* gpt gene and BPV-1 DNAs in transfected cells, *Virology* **127**:385–396.

Gluzman, Y., 1982, *Viral Vectors*, Cold Spring Harbor laboratory, Cold Spring Harbor, N. Y.

Goodbourn, S., Zinn, K., and Maniatis, T., 1985, Human beta-interferon gene expression is regulated by an inducible enhancer sequence, *Cell* **41**:509–520.

Green, M. R., Treisman, R., and Maniatis, T., 1983, Transcriptional activation of cloned human β-globin genes by viral immediate-early gene products, *Cell* **35**:137–148.

Groff, D. E., Sundberg, J. P., and Lancaster, W. D., 1983, Extrachromosomal deer fibromavirus DNA in deer fibromas and virus transformed mouse cells, *Virology* **131**:546–550.

Hamada, Y., Tsuijimoto, Y., Ishiura, M., and Susuki, Y., 1983, A vehicle for DNA transfer and for recovery of transferred genes: Charon phage–pBR322 hybrid, *Gene* **24**:245–253.

Hamer, D. H., and Walling, M., 1982, Regulation *in vivo* of a cloned mammalian gene; cadmium induces the transcription of a mouse metallothionein gene in SV40 vectors, *J. Mol. Appl. Genet.* **1**:273–288.

Hauser, H., Gross, G., Bruns, W., Hochkeppel, H.-K., Myr, U., and Collins, J., 1982, Inducibility of human beta interferon gene in mouse L-cell clones, *Nature* **296**:650–654.

Heilman, C. A., Engel, L., Lowy, D. R., and Howley, P. M., 1982, Virus-specific transcription in bovine papillomavirus transformed mouse cells, *Virology* **119**:22–34.

Howard, B. H., Auerbach, J., Wickner, S., and Gottesman, M. E., 1983, λ SV2: A plasmid cloning vector that can be stably integrated in *E. coli*, in: *Experimental Manipulation of Gene Expression* (M. Inouye, ed.), pp. 137–153, Academic Press, New York.

Howley, P. M., Sarver, N., and Law, M.-F., 1983 Eukaryotic cloning vectors derived from bovine papillomavirus DNA, *Methods Enzymol.* **101**:387–402.

Hsiung, N., Fitts, R., Wilson, S., Milne, A., and Hamer, D., 1984, Efficient production of hepatitis B surface antigen using a bovine papillomavirus–metallothionein vector, *J. Mol. Appl. Genet.* **2**:497–506.

Imperiale, M. J., Feldman, L. T., and Nevins, J. R., 1983, Activation of gene expression by adenovirus and herpesvirus regulatory genes acting in trans and by a cis-acting adenovirus enhancer element, *Cell* **35**:127–136.

Karin, M., Cathala, G., and Nguyen-Huu, M. C., 1983, Expression and regulation of a human metallothionein gene carried on an autonomously replicating shuttle vector, *Proc. Natl. Acad. Sci. USA* **80**:4040–4044.

Kushner, P. J., Levinson, B. B., and Goodman, H. M., 1982, A plasmid that replicates in both mouse and *E. coli* cells, *J. Mol. Appl. Genet.*, **1**:527–538.

Lau, Y.-F., and Kan, Y. W., 1983, Versatile cosmid vectors for the isolation, expression, and rescue of gene sequences: Studies with the human α-globin gene cluster, *Proc. Natl. Acad. Sci. USA* **80**:5225–5229.

Law, M.-F., Lowy, D. R., Dvoretzky, I., and Howley, P. M., 1981, Mouse cells transformed by bovine papillomavirus contain only extrachromosomal viral DNA sequences, *Proc. Natl. Acad. Sci. USA* **78**:2727–2731.

Law, M.-F., Howard, B., Sarver, N., and Howley, P. M., 1982, Expression of selective traits in mouse cells transformed with a BPV DNA-derived hybrid molecule containing *Escherichia coli* gpt, in: *Eukaryotic Viral Vectors* (Y. Gluzman, ed.), pp. 79–85, Cold Spring Harbor Press, Cold Spring Harbor, N.Y.

Law, M.-F., Bryne, J. C., and Howley, P. M., 1983, A stable bovine papillomavirus hybrid plasmid that expresses a dominant selective trait, *Mol. Cell. Biol.* **3**:2110–2115.

Lindenmaier, W., Hauser, H., Greiser de Wilke, I., and Schutz, G., 1982, Gene shuttling: Moving DNA into and out of eucaryotic cells, *Nucleic Acids Res.* **10**:1243–1256.

Liu, C. C., Yansura, D., and Levinson, A. D., 1982, Direct expression of hepatitis B surface antigen in monkey cells from an SV40 vector, *DNA* **1**:213–221.

Lowy, D. R., Dvoretzky, I., Shober, R., Law, M.-F., Engel, L., and Howley, P. M., 1980, *In vitro* tumorigenic transformation by a defined sub-genomic fragment of bovine papilloma virus DNA, *Nature* **287**:72–74.

Lund, T., Grosveld, F. G., and Flavell, R. A., 1982, Isolation of transforming DNA by cosmid rescue, *Proc. Natl. Acad. Sci. USA* **79**:520–524.

Lusky, M., and Botchan, M., 1981, Inhibition of SV40 replication in simian cells by specific pBR322 DNA sequences, *Nature* **293:**79–81.

Lusky, M., and Botchan, M., 1984, Characterization of the bovine papillomavirus plasmid maintenance sequences, *Cell* **36:**391–401.

Lusky, M., Berg, L., Weiher, H., and Botchan, M., 1983, Bovine papillomavirus contains an activator of gene expression at the distal end of the early transcription unit, *Mol. Cell. Biol.* **3:**1108–1122.

Maroteauz, L., Chen, L., Mitrani-Rosenbaum, S., Howley, P. M., and Revel, M., 1983, Cycloheximide induces expression of the human interferon beta 1 gene in mouse cells transformed by bovine papillomavirus–interferon beta 1 recombinants, *J. Virol.* **47:**89–95.

Matthias, P. D., Bernard, H. U., Scott, A., Brady, G., Hashimoto-Gotoh, T., and Schutz, G., 1983, A bovine papillomavirus vector with a dominant resistance marker replicates extrachromosomally in mouse and *E. coli* cells, *EMBO J.* **2:**1487–1492.

Meneguzzi, G., Binetruy, B., Grisoni, M., and Cuzin, F., 1984, Plasmidial maintenance in rodent fibroblasts of a BPV-1–pBR322 shuttle vector without immediately apparent oncogenic transformation of the recipient cells, *EMBO J.* **3:**365–371.

Mitrani-Rosenbaum, S., Maroteaux, L., Mory, Y., Revel, M., and Howley, P. M., 1983, Inducible expression of the human interferon beta 1 gene linked to a bovine papillomavirus DNA vector and maintained extrachromosomally in mouse cells, *Mol. Cell. Biol.* **3:**233.

Moar, M. H., Campo, M. S., Laird, H., and Jarrett, W. F. H., 1981, Persistence of nonintegrated viral DNA in bovine cells transformed by bovine papillomavirus type 2, *Nature* **293:**749–751.

Morgan, D. M., and Meinke, W., 1980, Isolation of clones of hamster embryo cells transformed by the bovine papillomavirus, *Curr. Microbiol.* **3:**247–251.

Moriarty, A. M., Hoyer, B. H., Shih, J. W., Gerin, J. L., and Hamer, D. H., 1981, Expression of the hepatitis B virus surface antigen gene in cell culture using a SV40 vector, *Proc. Natl. Acad. Sci. USA* **78:**2606–2610.

Ohno, S., and Taniguchi, T., 1982, Inducer-responsive expression of the cloned human $beta_1$ interferon gene introduced into cultured mouse cells, *Nucleic Acids Res.* **10:**967–977.

Ostrawski, M. C., Richard-Foy, H., Wolford, R. G., Berand, D. S., and Hager, G., 1983, Glucocorticoid regulation of transcription at an amplified episomal promoter, *Mol. Cell. Biol.* **3:**2045–2057.

Pavlakis, G. N., and Hamer, D. H., 1983a, Regulation of a metallothionein–growth hormone hybrid gene in bovine papilloma virus, *Proc. Natl. Acad. Sci. USA* **80:**397–401.

Pavlakis, G. N., and Hamer, D. H., 1983b, Expression of cloned growth hormone and metallothionein genes in heterologous cells, *Recent Prog. Horm. Res.* **39:**353–385.

Perucho, M., Hanahan, D., and Wigler, M., 1980, Genetic and physical linkage of exogenous DNA sequences in transformed cells, *Cell* **22:**309–317.

Pintel, D., Merchlinsky, M. J., and Ward, D. C., 1984, Expression of minute virus of mice (MVM) structural proteins in murine cell lines transformed by bovine papilloma–MVM plasmid chimeras, *J. Virol.* **52:**320–327.

Pitha, P. M., Ciufo, D. M., Kellum, M., Raj, N. B. K., Reyes, G. R., and Hayward, G. S., 1982, Induction of human beta interferon synthesis with poly (rI–rC) in mouse cells transfected with cloned cDNA plasmids, *Proc. Natl. Acad. Sci. USA* **79:**4337–4341.

Ramabhadran, T. V., Reitz, B. A., and Tiemeier, D. C., 1984, Synthesis and glycosylation of the common α subunit of human glycoprotein hormones in mouse cells, *Proc. Natl. Acad. Sci. USA* **81:**6701–6705.

Ramabhadran, T. V., Reitz, B. A., and Shah, D. M., 1985, Isolation and high level expression of the bovine growth hormone gene in heterologous mammalian cells, *Gene* **38:**111–125.

Richards, R. I., Heguy, A., and Karin, M., 1984, Structural and functional analysis of the human metallothionein-IA gene: Differential induction by metal ions and glucocorticoids, *Cell* **37:**263–272.

Ringold, G. M., 1983, Regulation of mouse mammary tumor virus gene expression by glucocorticoid hormones, *Curr. Top. Microbiol.* pp. 79–103.

Rösl, R., Waldeck, W., and Sauer, G., 1983, Isolation of episomal bovine papillomavirus chromatin and identification of a DNase I-hypersensitive region, *J. Virol.* **46**:567–574.

Sambrook, J., Rogers, L., White, J., and Gething, M. J., 1985, Lines of BPV-transformed murine cells that constitutively express influenza virus hemagglutinin, *EMBO J.* **4**:91–103.

Sarver, N., Gruss, P., Law, M.-F., Khoury, G., and Howley, P. M., 1981, Bovine papilloma virus deoxyribonucleic acid: A novel eucaryotic cloning vector, *Mol. Cell. Biol.* **1**:486–496.

Sarver, N., Byrne, J. C., and Howley, P. M., 982, Transformation and replication in mouse cells of a bovine papillomavirus–pML2 plasmid vector that can be rescued in bacteria, *Proc. Natl. Acad. Sci. USA* **79**:7147–7151.

Sarver, N., Mitrani-Rosenbaum, S., Law, M.-F., McAllister, W. T., Bryne, J. C., and Howley, P. M., 1983, Bovine papillomavirus shuttle vectors, in: *Gene Expression* (J. Setlow and A. Hollaender, eds.), pp. 173–190, Plenum Press, New York.

Schaffner, W., 1980, Direct transfer of cloned genes from bacteria to mammalian cells, *Proc. Natl. Acad. Sci. USA* **77**:2163–2167.

Schenborn, E. T., Lund, E., Mitchen, J. L., and Dahlberg, J. E., 1985, Expression of a human U1 RNA gene introduced into mouse cells via bovine papilloma virus DNA vectors, *Mol. Cell. Biol. 5:*1318–1326.

Sekiguchi, Y., Nishimoto, T., Kai, R., and Sekiguchi, M., 1983, Recovery of a hybrid vector, derived from bovine papillomavirus DNA, pBR322 and the HSV *tk* gene, by bacterial transformation with extrachromosomal DNA from transfected rodent cells, *Gene* **21**:267–272.

Siddiqui, A., 1983, Expression of hepatitis B virus surface antigen gene in cultured cells by using recombinant plasmid vectors, *Mol. Cell. Biol.* **3**:143–146.

Southern, P. J., and Berg, P., 1982, Transformation of mammalian cells to antibiotic resistance with a bacterial gene under control of the SV40 early region promoter, *J. Mol. Appl. Genet.* **1**:327–341.

Stenlund, A., Lamy, D., Moreno-Lopez, J., Ahola, H., Pettersson, U., and Tiollais, P., 1983a, Secretion of the hepatitis B virus surface antigen from mouse cells using an extrachromosomal eucaryotic vector, *EMBO J.* **2**:669–673.

Stenlund, A., Moreno-Lopez, J., Ahola, H., and Pettersson, U., 1983b, European elk papillomavirus: Characterization of the genome, induction of tumors in animals and transformation in vitro, *J. Virol.* **48**:370–376.

Stratowa, C., Doehmer, J., Wang, Y., and Hofschneider, P. H., 1982, Recombinant retroviral DNA yielding high expression of hepatitis B surface antigen, *EMBO J.* **1**:1573–1578.

Treisman, R., Orkin, S. H., and Maniatis, T., 1983, Specific transcription and RNA splicing defects in five cloned beta thalassaemia genes, *Nature* **302**:591–596.

Turek, L. P., Byrne, J. C., Lowy, D. R., Dvoretzky, I., Friedman, R. M., and Howley, P. M., 1982, Interferon induces morphologic reversion with elimination of extrachromosomal viral genomes in bovine papillomavirus-transformed mouse cells, *Proc. Natl. Acad. Sci. USA* **79**:7914–7918.

Wang, Y., Stratowa, C., Schaefer-Ridder, M., Doehmer, J., and Hofschneider, P. H., 1983, Enhanced production of hepatitis B surface antigen in NIH3T3 mouse fibroblasts by using extrachromosomally replicating bovine papillomavirus vector, *Mol. Cell. Biol.* **3**:1032–1039.

Watts, S. L., Ostrow, R. S., Phelps, W. C., Prince, J. T., and Faras, A. J., 1983, Free cottontail rabbit papillomavirus DNA persists in warts and carcinomas of infected rabbits and in cells in culture transformed with virus or viral DNA, *Virology* **125**:127–138.

Watts, S. L., Phelps, W. C., Ostrow, R. S., Zachow, K. R., and Faras, A. J., 1984, Cellular transformation by human papillomavirus DNA *in vitro*, *Science* **225**:634–636.

Zinn, K., Mellon, P., Ptashne, M., and Maniatis, T., 1982, Regulated expression of an extrachromosomal human beta interferon gene in mouse cells, *Proc. Natl. Acad. Sci. USA* **79**:4897–4901.

Zinn, K., DiMaio, D., and Maniatis, T., 1983, Identification of two distinct regulatory regions adjacent to the human β interferon gene, *Cell* **34**:865–879.

APPENDIX

Sequence Analysis of Papillomavirus Genomes

CARL C. BAKER

I. PAPILLOMAVIRUS SEQUENCES

The first of the papillomaviruses to be sequenced in its entirety was bovine papillomavirus type 1 (BPV-1) (Chen *et al.*, 1982). The BPV-1 genome is a double-stranded circular DNA molecule of length 7945 bp. Nucleotide 1 is defined as the first base pair of the *Hpa*I restriction endonuclease recognition site. The sequence proceeds clockwise around the genome in the direction of transcription. By convention, this orientation has been adopted for all other papillomaviruses. For the sequences not containing the *Hpa*I recognition site, nucleotide 1 is defined as the first base pair of a homologous sequence. The DNA sequences of BPV-1 (Chen *et al.*, 1982; Danos *et al.*, 1983; Stenlund *et al.*, 1985), cottontail rabbit papillomavirus (Shope, CRPV) (Giri *et al.*, 1985), human papillomavirus type 1a (HPV-1a) (Danos *et al.*, 1982, 1983; Clad *et al.*, 1982; Schwarz *et al.*, 1983), HPV-6b (Schwarz *et al.*, 1983), and HPV-16 (Seedorf *et al.*, 1985) are presented in Figs. 1–5, respectively. DNA sequences for deer papillomavirus (DPV) (Groff and Lancaster, 1985), HPV-11 (Dartmann *et al.*, 1986), and HPV-8 (Fuchs *et al.*, 1986) are now available but will not be presented in their entirety here.

CARL C. BAKER • Laboratory of Tumor Virus Biology, National Cancer Institute, Bethesda, Maryland 20892.

```
        10         20         30         40         50         60         70
GTTAACAATA ATCACACCAT CACCGTTTTT TCAAGCGGGA AAAAATAGCC AGCTAACTAT AAAAAGCTGC
CAATTGTTAT TAGTGTGGTA GTGGCAAAAA AGTTCGCCCT TTTTTATCGG TCGATTGATA TTTTTCGACG

        80         90        100        110        120        130        140
TGACAGACCC CGGTTTTCAC ATGGACCTGA AACCTTTTGC AAGAACCAAT CCATTCTCAG GGTTGGATTG
ACTGTCTGGG GCCAAAAGTG TACCTGGACT TTGGAAAACG TTCTTGGTTA GGTAAGAGTC CCAACCTAAC

       150        160        170        180        190        200        210
TCTGTGGTGC AGAGAGCCTC TTACAGAAGT TGATGCTTTT AGGTGCATGG TCAAAGACTT TCATGTTGTA
AGACACCACG TCTCTCGGAG AATGTCTTCA ACTACGAAAA TCCACGTACC AGTTTCTGAA AGTACAACAT

       220        230        240        250        260        270        280
ATTCGGGAAG GCTGTAGATA TGGTGCATGT ACCATTTGTC TTGAAAACTG TTTAGCTACT GAAAGAAGAC
TAAGCCCTTC CGACATCTAT ACCACGTACA TGGTAAACAG AACTTTTGAC AAATCGATGA CTTTCTTCTG

       290        300        310        320        330        340        350
TTTGGCAAGG TGTTCCAGTA ACAGGTGAGG AAGCTGAATT ATTGCATGGC AAAACACTTG ATAGGCTTTG
AAACCGTTCC ACAAGGTCAT TGTCCACTCC TTCGACTTAA TAACGTACCG TTTTGTGAAC TATCCGAAAC

       360        370        380        390        400        410        420
CATAAGATGC TGCTACTGTG GGGGCAAACT AACAAAAAAT GAAAAACATC GGCATGTGCT TTTTAATGAG
GTATTCTACG ACGATGACAC CCCCGTTTGA TTGTTTTTTA CTTTTTGTAG CCGTACACGA AAAATTACTC

       430        440        450        460        470        480        490
CCTTTCTGCA AAACCAGAGC TAACATAATT AGAGGACGCT GCTACGACTG CTGCAGACAT GGTTCAAGGT
GGAAAGACGT TTTGGTCTCG ATTGTATTAA TCTCCTGCGA CGATGCTGAC GACGTCTGTA CCAAGTTCCA

       500        510        520        530        540        550        560
CCAAATACCC ATAGAAACTT GGATGATTCA CCTGCAGGAC CGTTGCTGAT TTTAAGTCCA TGTGCAGGCA
GGTTTATGGG TATCTTTGAA CCTACTAAGT GGACGTCCTG GCAACGACTA AAATTCAGGT ACACGTCCGT

       570        580        590        600        610        620        630
CACCTACCAG GTCTCCTGCA GCACCTGATG CACCTGATTT CAGACTTCCG TGCCATTTCG GCCGTCCTAC
GTGGATGGTC CAGAGGACGT CGTGGACTAC GTGGACTAAA GTCTGAAGGC ACGGTAAAGC CGGCAGGATG

       640        650        660        670        680        690        700
TAGGAAGCGA GGTCCCACTA CGCCTCCGCT TTCCTCTCCC GGAAAACTGT GTGCAACAGG GCCACGTCGA
ATCCTTCGCT CCAGGGTGAT GCGGAGGCGA AAGGAGAGGG CCTTTTGACA CACGTTGTCC CGGTGCAGCT

       710        720        730        740        750        760        770
GTGTATTCTG TGACTGTCTG CTGTGGAAAC TGCGGAAAAG AGCTGACTTT TGCTGTGAAG ACCAGCTCGA
CACATAAGAC ACTGACAGAC GACACCTTTG ACGCCTTTTC TCGACTGAAA ACGACACTTC TGGTCGAGCT

       780        790        800        810        820        830        840
CGTCCCTGCT TGGATTTGAA CACCTTTTAA ACTCAGATTT AGACCTCTTG TGTCCACGTT GTGAATCTCG
GCAGGGACGA ACCTAAACTT GTGGAAAATT TGAGTCTAAA TCTGGAGAAC ACAGGTGCAA CACTTAGAGC

       850        860        870        880        890        900        910
CGAGCGTCAT GGCAAACGAT AAAGGTAGCA ATTGGGATTC GGGCTTGGGA TGCTCATATC TGCTGACTGA
GCTCGCAGTA CCGTTTGCTA TTTCCATCGT TAACCCTAAG CCCGAACCCT ACGAGTATAG ACGACTGACT

       920        930        940        950        960        970        980
GGCAGAATGT GAAAGTGACA AAGAGAATGA GGAACCCGGG GCAGGTGTAG AACTGTCTGT GGAATCTGAT
CCGTCTTACA CTTTCACTGT TTCTCTTACT CCTTGGGCCC CGTCCACATC TTGACAGACA CCTTAGACTA

       990       1000       1010       1020       1030       1040       1050
CGGTATGATA GCCAGGATGA GGATTTTGTT GACAATGCAT CAGTCTTTCA GGGAAATCAC CTGGAGGTCT
GCCATACTAT CGGTCCTACT CCTAAAACAA CTGTTACGTA GTCAGAAAGT CCCTTTAGTG GACCTCCAGA

      1060       1070       1080       1090       1100       1110       1120
TCCAGGCATT AGAGAAAAAG GCGGGTGAGG AGCAGATTTT AAATTTGAAA AGAAAAGTAT TGGGGAGTTC
AGGTCCGTAA TCTCTTTTTC CGCCCACTCC TCGTCTAAAA TTTAAACTTT TCTTTTCATA ACCCCTCAAG

      1130       1140       1150       1160       1170       1180       1190
GCAAAACAGC AGCGGTTCCG AAGCATCTGA AACTCCAGTT AAAAGACGGA AATCAGGAGC AAAGCGAAGA
CGTTTTGTCG TCGCCAAGGC TTCGTAGACT TTGAGGTCAA TTTTCTGCCT TTAGTCCTCG TTTCGCTTCT

      1200       1210       1220       1230       1240       1250       1260
TTATTTGCTG AAAATGAAGC TAACCGTGTT CTTACGCCCC TCCAGGTACA GGGGGAGGGG GAGGGGAGGC
AATAAACGAC TTTTACTTCG ATTGGCACAA GAATGCGGGG AGGTCCATGT CCCCCTCCCC CTCCCCTCCG

      1270       1280       1290       1300       1310       1320       1330
AAGAACTTAA TGAGGAGCAG GCAATTAGTC ATCTACATCT GCAGCTTGTT AAATCTAAAA ATGCTACAGT
TTCTTGAATT ACTCCTCGTC CGTTAATCAG TAGATGTAGA CGTCGAACAA TTTAGATTTT TACGATGTCA
```

FIGURE 1. DNA sequence of BPV-1.

1340 1350 1360 1370 1380 1390 1400
TTTTAAGCTG GGGCTCTTTA AATCTTTGTT CCTTTGTAGC TTCCATGATA TTACGAGGTT GTTTAAGAAT
AAAATTCGAC CCCGAGAAAT TTAGAAACAA GGAAACATCG AAGGTACTAT AATGCTCCAA CAAATTCTTA

1410 1420 1430 1440 1450 1460 1470
GATAAGACCA CTAATCAGCA ATGGGTGCTG GCTGTGTTTG GCCTTGCAGA GGTGTTTTTT GAGGCGAGTT
CTATTCTGGT GATTAGTCGT TACCCACGAC CGACACAAAC CGGAACGTCT CCACAAAAAA CTCCGCTCAA

1480 1490 1500 1510 1520 1530 1540
TCGAACTCCT AAAGAAGCAG TGTAGTTTTC TGCAGATGCA AAAAAGATCT CATGAAGGAG GAACTTGTGC
AGCTTGAGGA TTTCTTCGTC ACATCAAAAG ACGTCTACGT TTTTTCTAGA GTACTTCCTC CTTGAACACG

1550 1560 1570 1580 1590 1600 1610
AGTTTACTTA ATCTGCTTTA ACACAGCTAA AAGCAGAGAA ACAGTCCGGA ATCTGATGGC AAACACGCTA
TCAAATGAAT TAGACGAAAT TGTGTCGATT TTCGTCTCTT TGTCAGGCCT TAGACTACCG TTTGTGCGAT

1620 1630 1640 1650 1660 1670 1680
AATGTAAGAG AAGAGTGTTT GATGCTGCAG CCAGCTAAAA TTCGAGGACT CAGCGCAGCT CTATTCTGGT
TTACATTCTC TTCTCACAAA CTACGACGTC GGTCGATTTT AAGCTCCTGA GTCGCGTCGA GATAAGACCA

1690 1700 1710 1720 1730 1740 1750
TTAAAAGTAG TTTGTCACCC GCTACACTTA AACATGGTGC TTTACCTGAG TGGATACGGG CGCAAACTAC
AATTTTCATC AAACAGTGGG CGATGTGAAT TTGTACCACG AAATGGACTC ACCTATGCCC GCGTTTGATG

1760 1770 1780 1790 1800 1810 1820
TCTGAACGAG AGCTTGCAGA CCGAGAAATT CGACTTCGGA ACTATGGTGC AATGGGCCTA TGATCACAAA
AGACTTGCTC TCGAACGTCT GGCTCTTTAA GCTGAAGCCT TGATACCACG TTACCCGGAT ACTAGTGTTT

1830 1840 1850 1860 1870 1880 1890
TATGCTGAGG AGTCTAAAAT AGCCTATGAA TATGCTTTGG CTGCAGGATC TGATAGCAAT GCACGGGCTT
ATACGACTCC TCAGATTTTA TCGGATACTT ATACGAAACC GACGTCCTAG ACTATCGTTA CGTGCCCGAA

1900 1910 1920 1930 1940 1950 1960
TTTTAGCAAC TAACAGCCAA GCTAAGCATG TGAAGGACTG TGCAACTATG GTAAGACACT ATCTAAGAGC
AAAATCGTTG ATTGTCGGTT CGATTCGTAC ACTTCCTGAC ACGTTGATAC CATTCTGTGA TAGATTCTCG

1970 1980 1990 2000 2010 2020 2030
TGAAACACAA GCATTAAGCA TGCCTGCATA TATTAAAGCT AGGTGCAAGC TGGCAACTGG GGAAGGAAGC
ACTTTGTGTT CGTAATTCGT ACGGACGTAT ATAATTTCGA TCCACGTTCG ACCGTTGACC CCTTCCTTCG

2040 2050 2060 2070 2080 2090 2100
TGGAAGTCTA TCCTAACTTT TTTTAACTAT CAGAATATTG AATTAATTAC CTTTATTAAT GCTTTAAAGC
ACCTTCAGAT AGGATTGAAA AAAATTGATA GTCTTATAAC TTAATTAATG GAAATAATTA CGAAATTTCG

2110 2120 2130 2140 2150 2160 2170
TCTGGCTAAA AGGAATTCCA AAAAAAAACT GTTTAGCATT TATTGGCCCT CCAAACACAG GCAAGTCTAT
AGACCGATTT TCCTTAAGGT TTTTTTTTGA CAAATCGTAA ATAACCGGGA GGTTTGTGTC CGTTCAGATA

2180 2190 2200 2210 2220 2230 2240
GCTCTGCAAC TCATTAATTC ATTTTTTGGG TGGTAGTGTT TTATCTTTTG CCAACCATAA AAGTCACTTT
CGAGACGTTG AGTAATTAAG TAAAAAACCC ACCATCACAA AATAGAAAAC GGTTGGTATT TTCAGTGAAA

2250 2260 2270 2280 2290 2300 2310
TGGCTTGCTT CCCTAGCAGA TACTAGAGCT GCTTTAGTAG ATGATGCTAC TCATGCTTGC TGGAGGTACT
ACCGAACGAA GGGATCGTCT ATGATCTCGA CGAAATCATC TACTACGATG AGTACGAACG ACCTCCATGA

2320 2330 2340 2350 2360 2370 2380
TTGACACATA CCTCAGAAAT GCATTGGATG GCTACCCTGT CAGTATTGAT AGAAAACACA AAGCAGCGGT
AACTGTGTAT GGAGTCTTTA CGTAACCTAC CGATGGGACA GTCATAACTA TCTTTTGTGT TTCGTCGCCA

2390 2400 2410 2420 2430 2440 2450
TCAAATTAAA GCTCCACCCC TCCTGGTAAC CAGTAATATT GATGTGCAGG CAGAGGACAG ATATTTGTAC
AGTTTAATTT CGAGGTGGGG AGGACCATTG GTCATTATAA CTACACGTCC GTCTCCTGTC TATAAACATG

2460 2470 2480 2490 2500 2510 2520
TTGCATAGTC GGGTGCAAAC CTTTCGCTTT GAGCAGCCAT GCACAGATGA ATCGGGTGAG CAACCTTTTA
AACGTATCAG CCCACGTTTG GAAAGCGAAA CTCGTCGGTA CGTGTCTACT TAGCCCACTC GTTGGAAAAT

2530 2540 2550 2560 2570 2580 2590
ATATTACTGA TGCAGATTGG AAATCTTTTT TTGTAAGGTT ATGGGGGCGT TTAGACCTGA TTGACGAGGA
TATAATGACT ACGTCTAACC TTTAGAAAAA AACATTCCAA TACCCCCGCA AATCTGGACT AACTGCTCCT

2600 2610 2620 2630 2640 2650 2660
GGAGGATAGT GAAGAGGATG GAGACAGCAT GCGAACGTTT ACATGTAGCG CAAGAAACAC AAATGCAGTT
CCTCCTATCA CTTCTCCTAC CTCTGTCGTA CGCTTGCAAA TGTACATCGC GTTCTTTGTG TTTACGTCAA

FIGURE 1. (*Continued*)

```
      2670       2680       2690       2700       2710       2720       2730
GATTGAGAAA AGTAGTGATA AGTTGCAAGA TCATATACTG TACTGGACTG CTGTTAGAAC TGAGAACACA
CTAACTCTTT TCATCACTAT TCAACGTTCT AGTATATGAC ATGACCTGAC GACAATCTTG ACTCTTGTGT

      2740       2750       2760       2770       2780       2790       2800
CTGCTTTATG CTGCAAGGAA AAAAGGGGTG ACTGTCCTAG GACACTGCAG AGTACCACAC TCTGTAGTTT
GACGAAATAC GACGTTCCTT TTTTCCCCAC TGACAGGATC CTGTGACGTC TCATGGTGTG AGACATCAAA

      2810       2820       2830       2840       2850       2860       2870
GTCAAGAGAG AGCCAAGCAG GCCATTGAAA TGCAGTTGTC TTTGCAGGAG TTAAGCAAAA CTGAGTTTGG
CAGTTCTCTC TCGGTTCGTC CGGTAACTTT ACGTCAACAG AAACGTCCTC AATTCGTTTT GACTCAAACC

      2880       2890       2900       2910       2920       2930       2940
GGATGAACCA TGGTCTTTGC TTGACACAAG CTGGGACCGA TATATGTCAG AACCTAAACG GTGCTTTAAG
CCTACTTGGT ACCAGAAACG AACTGTGTTC GACCCTGGCT ATATACAGTC TTGGATTTGC CACGAAATTC

      2950       2960       2970       2980       2990       3000       3010
AAAGGCGCCA GGGTGGTAGA GGTGGAGTTT GATGGAAATG CAAGCAATAC AAACTGGTAC ACTGTCTACA
TTTCCGCGGT CCCACCATCT CCACCTCAAA CTACCTTTAC GTTCGTTATG TTTGACCATG TGACAGATGT

      3020       3030       3040       3050       3060       3070       3080
GCAATTTGTA CATGCGCACA GAGGACGGCT GGCAGCTTGC GAAGGCTGGG GCTGACGGAA CTGGGCTCTA
CGTTAAACAT GTACGCGTGT CTCCTGCCGA CCGTCGAACG CTTCCGACCC CGACTGCCTT GACCCGAGAT

      3090       3100       3110       3120       3130       3140       3150
CTACTGCACC ATGGCCGGTG CTGGACGCAT TTACTATTCT CGCTTTGGTG ACGAGGCAGC CAGATTTAGT
GATGACGTGG TACCGGCCAC GACCTGCGTA AATGATAAGA GCGAAACCAC TGCTCCGTCG GTCTAAATCA

      3160       3170       3180       3190       3200       3210       3220
ACAACAGGGC ATTACTCTGT AAGAGATCAG GACAGAGTGT ATGCTGGTGT CTCATCCACC TCTTCTGATT
TGTTGTCCCG TAATGAGACA TTCTCTAGTC CTGTCTCACA TACGACCACA GAGTAGGTGG AGAAGACTAA

      3230       3240       3250       3260       3270       3280       3290
TTAGAGATCG CCCAGACGGA GTCTGGGTCG CATCCGAAGG ACCTGAAGGA GACCCTGCAG GAAAAGAAGC
AATCTCTAGC GGGTCTGCCT CAGACCCAGC GTAGGCTTCC TGGACTTCCT CTGGGACGTC CTTTTCTTCG

      3300       3310       3320       3330       3340       3350       3360
CGAGCCAGCC CAGCCTGTCT CTTCTTTGCT CGGCTCCCCC GCCTGCGGTC CCATCAGAGC AGGCCTCGGT
GCTCGGTCGG GTCGGACAGA GAAGAAACGA GCCGAGGGGG CGGACGCCAG GGTAGTCTCG TCCGGAGCCA

      3370       3380       3390       3400       3410       3420       3430
TGGGTACGGG ACGGTCCTCG CTCGCACCCC TACAATTTTC CTGCAGGCTC GGGGGGCTCT ATTCTCCGCT
ACCCATGCCC TGCCAGGAGC GAGCGTGGGG ATGTTAAAAG GACGTCCGAG CCCCCCGAGA TAAGAGGCGA

      3440       3450       3460       3470       3480       3490       3500
CTTCCTCCAC CCCGGTGCAG GGCACGGTAC CGGTGGACTT GGCATCAAGG CAGGAAGAAG AGGAGCAGTC
GAAGGAGGTG GGGCCACGTC CCGTGCCATG GCCACCTGAA CCGTAGTTCC GTCCTTCTTC TCCTCGTCAG

      3510       3520       3530       3540       3550       3560       3570
GCCCGACTCC ACAGAGGAAG AACCAGTGAC TCTCCCAAGG CGCACCACCA ATGATGGATT CCACCTGTTA
CGGGCTGAGG TGTCTCCTTC TTGGTCACTG AGAGGGTTCC GCGTGGTGGT TACTACCTAA GGTGGACAAT

      3580       3590       3600       3610       3620       3630       3640
AAGGCAGGAG GGTCATGCTT TGCTCTAATT TCAGGAACTG CTAACCAGGT AAAGTGCTAT CGCTTTCGGG
TTCCGTCCTC CCAGTACGAA ACGAGATTAA AGTCCTTGAC GATTGGTCCA TTTCACGATA GCGAAAGCCC

      3650       3660       3670       3680       3690       3700       3710
TGAAAAAGAA CCATAGACAT CGCTACGAGA ACTGCACCAC CACCTGGTTC ACAGTTGCTG ACAACGGTGC
ACTTTTTCTT GGTATCTGTA GCGATGCTCT TGACGTGGTG GTGGACCAAG TGTCAACGAC TGTTGCCACG

      3720       3730       3740       3750       3760       3770       3780
TGAAAGACAA GGACAAGCAC AAATACTGAT CACCTTTGGA TCGCCAAGTC AAAGGCAAGA CTTTCTGAAA
ACTTTCTGTT CCTGTTCGTG TTTATGACTA GTGGAAACCT AGCGGTTCAG TTTCCGTTCT GAAAGACTTT

      3790       3800       3810       3820       3830       3840       3850
CATGTACCAC TACCTCCTGG AATGAACATT TCCGGCTTTA CAGCCAGCTT GGACTTCTGA TCACTGCCAT
GTACATGGTG ATGGAGGACC TTACTTGTAA AGGCCGAAAT GTCGGTCGAA CCTGAAGACT AGTGACGGTA

      3860       3870       3880       3890       3900       3910       3920
TGCCTTTTCT TCATCTGACT GGTGTACTAT GCCAAATCTA TGGTTTCTAT TGTTCTTGGG ACTAGTTGCT
ACGGAAAAGA AGTAGACTGA CCACATGATA CGGTTTAGAT ACCAAAGATA ACAAGAACCC TGATCAACGA

      3930       3940       3950       3960       3970       3980       3990
GCAATGCAAC TGCTGCTATT ACTGTTCTTA CTCTTGTTTT TTCTTGTATA CTGGGATCAT TTTGAGTGCT
CGTTACGTTG ACGACGATAA TGACAAGAAT GAGAACAAAA AAGAACATAT GACCCTAGTA AAACTCACGA
```

FIGURE 1. (*Continued*)

```
      4000       4010       4020       4030       4040       4050       4060
CCTGTACAGG TCTGCCCTTT TAATGCCTTT ACATCACTGG CTATTGGCTG TGTTTTTACT GTTGTGTGGA
GGACATGTCC AGACGGGAAA ATTACGGAAA TGTAGTGACC GATAACCGAC ACAAAAATGA CAACACACCT

      4070       4080       4090       4100       4110       4120       4130
TTTGATTTGT TTTATATACT GTATGAAGTT TTTTCATTTG TGCTTGTATT GCTGTTTGTA AGTTTTTTAC
AAACTAAACA AAATATATGA CATACTTCAA AAAAGTAAAC ACGAACATAA CGACAAACAT TCAAAAAATG

      4140       4150       4160       4170       4180       4190       4200
TAGAGTTTGT ATTCCCCCTG CTCAGATTTT ATATGGTTTA AGCTGCAGCA ATAAAAATGA GTGCACGAAA
ATCTCAAACA TAAGGGGGAC GAGTCTAAAA TATACCAAAT TCGACGTCGT TATTTTTACT CACGTGCTTT

      4210       4220       4230       4240       4250       4260       4270
AAGAGTAAAA CGTGCCAGTG CCTATGACCT GTACAGGACA TGCAAGCAAG CGGGCACATG TCCACCAGAT
TTCTCATTTT GCACGGTCAC GGATACTGGA CATGTCCTGT ACGTTCGTTC GCCCGTGTAC AGGTGGTCTA

      4280       4290       4300       4310       4320       4330       4340
GTGATACCAA AGGTAGAAGG AGATACTATA GCAGATAAAA TTTTGAAATT TGGGGGTCTT GCAATCTACT
CACTATGGTT TCCATCTTCC TCTATGATAT CGTCTATTTT AAAACTTTAA ACCCCCAGAA CGTTAGATGA

      4350       4360       4370       4380       4390       4400       4410
TAGGAGGGCT AGGAATAGGA ACATGGTCTA CTGGAAGGGT TGCTGCAGGT GGATCACCAA GGTACACACC
ATCCTCCCGA TCCTTATCCT TGTACCAGAT GACCTTCCCA ACGACGTCCA CCTAGTGGTT CCATGTGTGG

      4420       4430       4440       4450       4460       4470       4480
ACTCCGAACA GCAGGGTCCA CATCATCGCT TGCATCAATA GGATCCAGAG CTGTAACAGC AGGGACCCGC
TGAGGCTTGT CGTCCCAGGT GTAGTAGCGA ACGTAGTTAT CCTAGGTCTC GACATTGTCG TCCCTGGGCG

      4490       4500       4510       4520       4530       4540       4550
CCCAGTATAG GTGCGGGCAT TCCTTTAGAC ACCCTTGAAA CTCTTGGGGC CTTGCGTCCA GGGGTGTATG
GGGTCATATC CACGCCCGTA AGGAAATCTG TGGGAACTTT GAGAACCCCG GAACGCAGGT CCCCACATAC

      4560       4570       4580       4590       4600       4610       4620
AGGACACTGT GCTACCAGAG GCCCCTGCAA TAGTCACTCC TGATGCTGTT CCTGCAGATT CAGGGCTTGA
TCCTGTGACA CGATGGTCTC CGGGGACGTT ATCAGTGAGG ACTACGACAA GGACGTCTAA GTCCCGAACT

      4630       4640       4650       4660       4670       4680       4690
TGCCCTGTCC ATAGGTACAG ACTCGTCCAC GGAGACCCTC ATTACTCTGC TAGAGCCTGA GGGTCCCGAG
ACGGGACAGG TATCCATGTC TGAGCAGGTG CCTCTGGGAG TAATGAGACG ATCTCGGACT CCCAGGGCTC

      4700       4710       4720       4730       4740       4750       4760
GACATAGCGG TTCTTGAGCT GCAACCCCTG GACCGTCCAA CTTGGCAAGT AAGCAATGCT GTTCATCAGT
CTGTATCGCC AAGAACTCGA CGTTGGGGAC CTGGCAGGTT GAACCGTTCA TTCGTTACGA CAAGTAGTCA

      4770       4780       4790       4800       4810       4820       4830
CCTCTGCATA CCACGCCCCT CTGCAGCTGC AATCGTCCAT TGCAGAAACA TCTGGTTTAG AAAATATTTT
GGAGACGTAT GGTGCGGGGA GACGTCGACG TTAGCAGGTA ACGTCTTTGT AGACCAAATC TTTTATAAAA

      4840       4850       4860       4870       4880       4890       4900
TGTAGGAGGC TCGGGTTTAG GGGATACAGG AGGAGAAAAC ATTGAACTGA CATACTTCGG GTCCCCACGA
ACATCCTCCG AGCCCAAATC CCCTATGTCC TCCTCTTTTG TAACTTGACT GTATGAAGCC CAGGGGTGCT

      4910       4920       4930       4940       4950       4960       4970
ACAAGCACGC CCCGCAGTAT TGCCTCTAAA TCACGTGGCA TTTTAAACTG GTTCAGTAAA CGGTACTACA
TGTTCGTGCG GGGCGTCATA ACGGAGATTT AGTGCACCGT AAAATTTGAC CAAGTCATTT GCCATGATGT

      4980       4990       5000       5010       5020       5030       5040
CACAGGTGCC CACGGAAGAT CCTGAAGTGT TTTCATCCCA AACATTTGCA AACCCACTGT ATGAAGCAGA
GTGTCCACGG GTGCCTTCTA GGACTTCACA AAAGTAGGGT TTGTAAACGT TTGGGTGACA TACTTCGTCT

      5050       5060       5070       5080       5090       5100       5110
ACCAGCTGTG CTTAAGGGAC CTAGTGGACG TGTTGGACTC AGTCAGGTTT ATAAACCTGA TACACTTACA
TGGTCGACAC GAATTCCCTG GATCACCTGC ACAACCTGAG TCAGTCCAAA TATTTGGACT ATGTGAATGT

      5120       5130       5140       5150       5160       5170       5180
ACACGTAGCG GGACAGAGGT GGGACCACAG CTACATGTCA GGTACTCATT GAGTACTATA CATGAAGATG
TGTGCATCGC CCTGTCTCCA CCCTGGTGTC GATGTACAGT CCATGAGTAA CTCATGATAT GTACTTCTAC

      5190       5200       5210       5220       5230       5240       5250
TAGAAGCAAT CCCCTACACA GTTGATGAAA ATACACAGGG ACTTGCATTC GTACCCTTGC ATGAAGAGCA
ATCTTCGTTA GGGGATGTGT CAACTACTTT TATGTGTCCC TGAACGTAAG CATGGGAACG TACTTCTCGT

      5260       5270       5280       5290       5300       5310       5320
AGCAGGTTTT GAGGAGATAG AATTAGATGA TTTTAGTGAG ACACATAGAC TGCTACCTCA GAACACCTCT
TCGTCCAAAA CTCCTCTATC TTAATCTACT AAAATCACTC TGTGTATCTG ACGATGGAGT CTTGTGGAGA
```

FIGURE 1. (*Continued*)

```
      5330       5340       5350       5360       5370       5380       5390
TCTACACCTG TTGGTAGTGG TGTACGAAGA AGCCTCATTC CAACTCAGGA ATTTAGTGCA ACACGGCCTA
AGATGTGGAC AACCATCACC ACATGCTTCT TCGGAGTAAG GTTGAGTCCT TAAATCACGT TGTGCCGGAT

      5400       5410       5420       5430       5440       5450       5460
CAGGTGTTGT AACCTATGGC TCACCTGACA CTTACTCTGC TAGCCCAGTT ACTGACCCTG ATTCTACCTC
GTCCACAACA TTGGATACCG AGTGGACTGT GAATGAGACG ATCGGGTCAA TGACTGGGAC TAAGATGGAG

      5470       5480       5490       5500       5510       5520       5530
TCCTAGTCTA GTTATCGATG ACACTACTAC TACACCAATC ATTATAATTG ATGGGCACAC AGTTGATTTG
AGGATCAGAT CAATAGCTAC TGTGATGATG ATGTGGTTAG TAATATTAAC TACCCGTGTG TCAACTAAAC

      5540       5550       5560       5570       5580       5590       5600
TACAGCAGTA ACTACACCTT GCATCCCTCC TTGTTGAGGA AACGAAAAAA ACGGAAACAT GCCTAATTTT
ATGTCGTCAT TGATGTGGAA CGTAGGGAGG AACAACTCCT TTGCTTTTTT TGCCTTTGTA CGGATTAAAA

      5610       5620       5630       5640       5650       5660       5670
TTTTGCAGAT GGCGTTGTGG CAACAAGGCC AGAAGCTGTA TCTCCCTCCA ACCCCTGTAA GCAAGGTGCT
AAAACGTCTA CCGCAACACC GTTGTTCCGG TCTTCGACAT AGAGGGAGGT TGGGGACATT CGTTCCACGA

      5680       5690       5700       5710       5720       5730       5740
TTGCAGTGAA ACCTATGTGC AAAGAAAAAG CATTTTTTAT CATGCAGAAA CGGAGCGCCT GCTAACTATA
AACGTCACTT TGGATACACG TTTCTTTTTC GTAAAAAATA GTACGTCTTT GCCTCGCGGA CGATTGATAT

      5750       5760       5770       5780       5790       5800       5810
GGACATCCAT ATTACCCAGT GTCTATCGGG GCCAAAACTG TTCCTAAGGT CTCTGCAAAT CAGTATAGGG
CCTGTAGGTA TAATGGGTCA CAGATAGCCC CGGTTTTGAC AAGGATTCCA GAGACGTTTA GTCATATCCC

      5820       5830       5840       5850       5860       5870       5880
TATTTAAAAT ACAACTACCT GATCCCAATC AATTTGCACT ACCTGACAGG ACTGTTCACA ACCCAAGTAA
ATAAATTTTA TGTTGATGGA CTAGGGTTAG TTAAACGTGA TGGACTGTCC TGACAAGTGT TGGGTTCATT

      5890       5900       5910       5920       5930       5940       5950
AGAGCGGCTG GTGTGGGCAG TCATAGGTGT GCAGGTGTCC AGAGGGCAGC CTCTTGGAGG TACTGTAACT
TCTCGCCGAC CACACCCGTC AGTATCCACA CGTCCACAGG TCTCCCGTCG GAGAACCTCC ATGACATTGA

      5960       5970       5980       5990       6000       6010       6020
GGGCACCCCA CTTTTAATGC TTTGCTTGAT GCAGAAAATG TGAATAGAAA AGTCACCACC CAAACAACAG
CCCGTGGGGT GAAAATTACG AAACGAACTA CGTCTTTTAC ACTTATCTTT TCAGTGGTGG GTTTGTTGTC

      6030       6040       6050       6060       6070       6080       6090
ATGACAGGAA ACAAACAGGC CTAGATGCTA AGCAACAACA GATTCTGTTG CTAGGCTGTA CCCCTGCTGA
TACTGTCCTT TGTTTGTCCG GATCTACGAT TCGTTGTTGT CTAAGACAAC GATCCGACAT GGGGACGACT

      6100       6110       6120       6130       6140       6150       6160
AGGGGAATAT TGGACAACAG CCCGTCCATG TGTTACTGAT CGTCTAGAAA ATGGCGCCTG CCCTCCTCTT
TCCCCTTATA ACCTGTTGTC GGGCAGGTAC ACAATGACTA GCAGATCTTT TACCGCGGAC GGGAGGAGAA

      6170       6180       6190       6200       6210       6220       6230
GAATTAAAAA ACAAGCACAT AGAAGATGGG GATATGATGG AAATTGGGTT TGGTGCAGCC AACTTCAAAG
CTTAATTTTT TGTTCGTGTA TCTTCTACCC CTATACTACC TTTAACCCAA ACCACGTCGG TTGAAGTTTC

      6240       6250       6260       6270       6280       6290       6300
AAATTAATGC AAGTAAATCA GATCTACCTC TTGACATTCA AAATGAGATC TGCTTGTACC CAGACTACCT
TTTAATTACG TTCATTTAGT CTAGATGGAG AACTGTAAGT TTTACTCTAG ACGAACATGG GTCTGATGGA

      6310       6320       6330       6340       6350       6360       6370
CAAAATGGCT GAGGACGCTG CTGGTAATAG CATGTTCTTT TTTGCAAGGA AAGAACAGGT GTATGTTAGA
GTTTTACCGA CTCCTGCGAC GACCATTATC GTACAAGAAA AAACGTTCCT TTCTTGTCCA CATACAATCT

      6380       6390       6400       6410       6420       6430       6440
CACATCTGGA CCAGAGGGGG CTCGGAGAAA GAAGCCCCTA CCACAGATTT TTATTTAAAG AATAATAAAG
GTGTAGACCT GGTCTCCCCC GAGCCTCTTT CTTCGGGGAT GGTGTCTAAA AATAAATTTC TTATTATTTC

      6450       6460       6470       6480       6490       6500       6510
GGGATGCCAC CCTTAAAATA CCCAGTGTGC ATTTTGGTAG TCCCAGTGGC TCACTAGTCT CAACTGATAA
CCCTACGGTG GGAATTTTAT GGGTCACACG TAAAACCATC AGGGTCACCG AGTGATCAGA GTTGACTATT

      6520       6530       6540       6550       6560       6570       6580
TCAAATTTTT AATCGGCCCT ACTGGCTATT CCGTGCCCAG GGCATGAACA ATGGAATTGC ATGGAATAAT
AGTTTAAAAA TTAGCCGGGA TGACCGATAA GGCACGGGTC CCGTACTTGT TACCTTAACG TACCTTATTA

      6590       6600       6610       6620       6630       6640       6650
TTATTGTTTT TAACAGTGGG GGACAATACA CGTGGTACTA ATCTTACCAT AAGTGTAGCC TCAGATGGAA
AATAACAAAA ATTGTCACCC CCTGTTATGT GCACCATGAT TAGAATGGTA TTCACATCGG AGTCTACCTT
```

FIGURE 1. (*Continued*)

```
      6660       6670       6680       6690       6700       6710       6720
CCCCACTAAC AGAGTATGAT AGCTCAAAAT TCAATGTATA CCATAGACAT ATGGAAGAAT ATAAGCTAGC
GGGGTGATTG TCTCATACTA TCGAGTTTTA AGTTACATAT GGTATCTGTA TACCTTCTTA TATTCGATCG

      6730       6740       6750       6760       6770       6780       6790
CTTTATATTA GAGCTATGCT CTGTGGAAAT CACAGCTCAA ACTGTGTCAC ATCTGCAAGG ACTTATGCCC
GAAATATAAT CTCGATACGA GACACCTTTA GTGTCGAGTT TGACACAGTG TAGACGTTCC TGAATACGGG

      6800       6810       6820       6830       6840       6850       6860
TCTGTGCTTG AAAATTGGGA AATAGGTGTG CAGCCTCCTA CCTCATCGAT ATTAGAGGAC ACCTATCGCT
AGACACGAAC TTTTAACCCT TTATCCACAC GTCGGAGGAT GGAGTAGCTA TAATCTCCTG TGGATAGCGA

      6870       6880       6890       6900       6910       6920       6930
ATATAGAGTC TCCTGCAACT AAATGTGCAA GCAATGTAAT TCCTGCAAAA GAAGACCCTT ATGCAGGGTT
TATATCTCAG AGGACGTTGA TTTACACGTT CGTTACATTA AGGACGTTTT CTTCTGGGAA TACGTCCCAA

      6940       6950       6960       6970       6980       6990       7000
TAAGTTTTGG AACATAGATC TTAAAGAAAA GCTTTCTTTG GACTTAGATC AATTTCCCTT GGGAAGAAGA
ATTCAAAACC TTGTATCTAG AATTTCTTTT CGAAAGAAAC CTGAATCTAG TTAAAGGGAA CCCTTCTTCT

      7010       7020       7030       7040       7050       7060       7070
TTTTTAGCAC AGCAAGGGGC AGGATGTTCA ACTGTGAGAA AACGAAGAAT TAGCCAAAAA ACTTCCAGTA
AAAAATCGTG TCGTTCCCCG TCCTACAAGT TGACACTCTT TTGCTTCTTA ATCGGTTTTT TGAAGGTCAT

      7080       7090       7100       7110       7120       7130       7140
AGCCTGCAAA AAAAAAAAAA AAATAAAAGC TAAGTTTCTA TAAATGTTCT GTAAATGTAA AACAGAAGGT
TCGGACGTTT TTTTTTTTTT TTTATTTTCG ATTCAAAGAT ATTTACAAGA CATTTACATT TTGTCTTCCA

      7150       7160       7170       7180       7190       7200       7210
AAGTCAACTG CACCTAATAA AAATCACTTA ATAGCAATGT GCTGTGTCAG TTGTTTATTG GAACCACACC
TTCAGTTGAC GTGGATTATT TTTAGTGAAT TATCGTTACA CGACACAGTC AACAAATAAC CTTGGTGTGG

      7220       7230       7240       7250       7260       7270       7280
CGGTACACAT CCTGTCCAGC ATTTGCAGTG CGTGCATTGA ATTATTGTGC TGGCTAGACT TCATGGCGCC
GCCATGTGTA GGACAGGTCG TAAACGTCAC GCACGTAACT TAATAACACG ACCGATCTGA AGTACCGCGG

      7290       7300       7310       7320       7330       7340       7350
TGGCACCGAA TCCTGCCTTC TCAGCCAAAA TGAATAATTG CTTTGTTGGC AAGAAACTAA GCATCAATGG
ACCGTGGCTT AGGACGGAAG AGTCGGTTTT ACTTATTAAC GAAACAACCG TTCTTTGATT CGTAGTTACC

      7360       7370       7380       7390       7400       7410       7420
GACGCGTGCA AAGCACCGGC GGCGGTAGAT GCGGGGTAAG TACTGAATTT TAATTCGACC TATCCCGGTA
CTGCGCACGT TTCGTGGCCG CCGCCATCTA CGCCCCATTC ATGACTTAAA ATTAAGCTGG ATAGGGCCAT

      7430       7440       7450       7460       7470       7480       7490
AAGCGAAAGC GACACGCTTT TTTTTCACAC ATAGCGGGAC CGAACACGTT ATAAGTATCG ATTAGGTCTA
TTCGCTTTCG CTGTGCGAAA AAAAAGTGTG TATCGCCCTG GCTTGTGCAA TATTCATAGC TAATCCAGAT

      7500       7510       7520       7530       7540       7550       7560
TTTTTGTCTC TCTGTCGGAA CCAGAACTGG TAAAAGTTTC CATTGCGTCT GGGCTTGTCT ATCATTGCGT
AAAAACAGAG AGACAGCCTT GGTCTTGACC ATTTTCAAAG GTAACGCAGA CCCGAACAGA TAGTAACGCA

      7570       7580       7590       7600       7610       7620       7630
CTCTATGGTT TTTGGAGGAT TAGACGGGCC ACCAGTAATG GTGCATAGCG GATGTCTGTA CCGCCATCGG
GAGATACCAA AAACCTCCTA ATCTGCCCGG TGGTCATTAC CACGTATCGC CTACAGACAT GGCGGTAGCC

      7640       7650       7660       7670       7680       7690       7700
TGCACCGATA TAGGTTTGGG GCTCCCCAAG GGACTGCTGG GATGACAGCT TCATATTATA TTGAATGGGC
ACGTGGCTAT ATCCAAACCC CGAGGGGTTC CCTGACGACC CTACTGTCGA AGTATAATAT AACTTACCCG

      7710       7720       7730       7740       7750       7760       7770
GCATAATCAG CTTAATTGGT GAGGACAAGC TACAAGTTGT AACCTGATCT CCACAAAGTA CGTTGCCGGT
CGTATTAGTC GAATTAACCA CTCCTGTTCG ATGTTCAACA TTGGACTAGA GGTGTTTCAT GCAACGGCCA

      7780       7790       7800       7810       7820       7830       7840
CGGGGTCAAA CCGTCTTCGG TGCTCGAAAC CGCCTTAAAC TACAGACAGG TCCCAGCCAA GTAGGCGGAT
GCCCCAGTTT GGCAGAAGCC ACGAGCTTTG GCGGAATTTG ATGTCTGTCC AGGGTCGGTT CATCCGCCTA

      7850       7860       7870       7880       7890       7900       7910
CAAAACCTCA AAAAGGCGGG AGCCAATCAA AATGCAGCAT TATATTTTAA GCTCACCGAA ACCGGTAAGT
GTTTTGGAGT TTTTCCGCCC TCGGTTAGTT TTACGTCGTA ATATAAAATT CGAGTGGCTT TGGCCATTCA

      7920       7930       7940
AAAGACTATG TATTTTTTCC CAGTGAATAA TTGTT
TTTCTGATAC ATAAAAAAGG GTCACTTATT AACAA
```

FIGURE 1. (*Continued*)

```
        10         20         30         40         50         60         70
GCTAACAATA ATTAAGAAAC ATGTAATGGC CAGAAACCGA TATCGGTTGC TGGCACTGTA TATCTGAGAT
CGATTGTTAT TAATTCTTTG TACATTACCG GTCTTTGGCT ATAGCCAACG ACCGTGACAT ATAGACTCTA

        80         90        100        110        120        130        140
CGCATCGGTT GCTGGCACTG TGTATCTGAG ATCGCAACGC ATTGCCAGGA ATTTCTGCAT ATAAGACAAG
GCGTAGCCAA CGACCGTGAC ACATAGACTC TAGCGTTGCG TAACGGTCCT TAAAGACGTA TATTCTGTTC

       150        160        170        180        190        200        210
AAACTTAGAG CAGATGGAGA ACTGCCTGCC ACGCTCGCTA GAGAAGCTGC AGCAAATATT ACAAATATCA
TTTGAATCTC GTCTACCTCT TGACGGACGG TGCGAGCGAT CTCTTCGACG TCGTTTATAA TGTTTATAGT

       220        230        240        250        260        270        280
TTGGAGGACT TGCCGTTTGG TTGTATATTT TGCGGGAAAT TGCTTGGGGC TGCAGAAAAA CAATTGTTCA
AACCTCCTGA ACGGCAAACC AACATATAAA ACGCCCTTTA ACGAACCCCG ACGTCTTTTT GTTAACAAGT

       290        300        310        320        330        340        350
AATGCACGGG GCTATGCATT GTATGGCATA AAGGGTGGCC GTATGGGACC TGCAGAGACT GCACTGTATT
TTACGTGCCC CGATACGTAA CATACCGTAT TTCCCACCGG CATACCCTGG ACGTCTCTGA CGTGACATAA

       360        370        380        390        400        410        420
GTCTTGTGCT TTGGATCTTT ATTGTCACCT TGCTCTTACT GCTCCTGCTT TGGAGGCTGA AGCGCTGGTT
CAGAACACGA AACCTAGAAA TAACAGTGGA ACGAGAATGA CGAGGACGAA ACCTCCGACT TCGCGACCAA

       430        440        450        460        470        480        490
GGTCAGGAAA TATCTAGCTG GTTCATGCGT TGTACAGTTT GCGGAAGAAG ATTAACTATT CCAGAAAAGA
CCAGTCCTTT ATAGATCGAC CAAGTACGCA ACATGTCAAA CGCCTTCTTC TAATTGATAA GGTCTTTTCT

       500        510        520        530        540        550        560
TTGAATTAAG AGCTAGAAAT TGCACGCTTT GTTGTATTGA TAAAGGTCAA TATTTCCAGT GGAGGGGTCA
AACTTAATTC TCGATCTTTA ACGTGCGAAA CAACATAACT ATTTCCAGTT ATAAAGGTCA CCTCCCCAGT

       570        580        590        600        610        620        630
TTGCAGTTCT TGCAAACTGT CAGACCAAGG TGATTTGGGG GGCTATCCCC CGAGTCCCGG CAGTCGCTGC
AACGTCAAGA ACGTTTGACA GTCTGGTTCC ACTAAACCCC CCGATAGGGG GCTCAGGGCC GTCAGCGACG

       640        650        660        670        680        690        700
GGGGAATGTG ACGAGTGTTG CGTCCCGGAC CTGACACATC TAACTCCGGT GGATCTGGAG GAACTTGGAT
CCCCTTACAC TGCTCACAAC GCAGGGCCTG GACTGTGTAG ATTGAGGCCA CCTAGACCTC CTTGAACCTA

       710        720        730        740        750        760        770
TATATCCAGG CCCCGAAGGA ACCTATCCGG ATTTAGTTGA CCTAGGGCCA GGCGTTTTTG GGGAAGAAGA
ATATAGGTCC GGGGCTTCCT TGGATAGGCC TAAATCAACT GGATCCCGGT CCGCAAAAAC CCCTTCTTCT

       780        790        800        810        820        830        840
CGAGGAGGGG GGTGGGCTGT TTGACAGCTT CGAGGAGGAG GATCCTGGAC CCAACCAGTG TGGGTGTTTT
GCTCCTCCCC CCACCCGACA AACTGTCGAA GCTCCTCCTC CTAGGACCTG GGTTGGTCAC ACCCACAAAA

       850        860        870        880        890        900        910
TTTTGCACCA GCTATCCGTC CGGAACAGGT GATACAGATA TAAATCAGGG ACCGGCAGGA GCTGCAGGGA
AAAACGTGGT CGATAGGCAG GCCTTGTCCA CTATGTCTAT ATTTAGTCCC TGGCCGTCCT CGACGTCCCT

       920        930        940        950        960        970        980
TTGCACTGCA GTCAGATCCA GTCTGTTTCT GTGAGAATTG TATTAACTTC ACAGAATTTA GATGATAGTA
AACGTGACGT CAGTCTAGGT CAGACAAAGA CACTCTTAAC ATAATTGAAG TGTCTTAAAT CTACTATCAT

       990       1000       1010       1020       1030       1040       1050
TTTCTGCTAT CCTGTGCGCA GGGCTGCTTC TTTATCTTTT TTCTTTATAT ACTACTGTTT TTCCTTCTGT
AAAGACGATA GGACACGCGT CCCGACGAAG AAATAGAAAA AAGAAATATA TGATGACAAA AAGGAAGACA

      1060       1070       1080       1090       1100       1110       1120
ACTGGCTTTA TCGAATTCTG CAAGATGATA GGCAGAACTC CTAAGCTTAG TGAGCTGGTT TTAGGTGAAA
TGACCGAAAT AGCTTAAGAC GTTCTACTAT CCGTCTTGAG GATTCGAATC ACTCGACCAA AATCCACTTT

      1130       1140       1150       1160       1170       1180       1190
CTGCTGAAGC GCTTAGTCTG CATTGCGACG AAGCATTAGA GAATTTAAGT GATGATGATG AGGAGGATCA
GACGACTTCG CGAATCAGAC GTAACGCTGC TTCGTAATCT CTTAAATTCA CTACTACTAC TCCTCCTAGT

      1200       1210       1220       1230       1240       1250       1260
TCAAGATAGA CAGGTGTTCA TAGAAAGGCC CTATGCAGTG TCCGTGCCAT GTAAGCGCTG TAGGCAAACT
AGTTCTATCT GTCCACAAGT ATCTTTCCGG GATACGTCAC AGGCACGGTA CATTCGCGAC ATCCGTTTGA

      1270       1280       1290       1300       1310       1320       1330
ATCAGCTTCG TCTGCGTCTG TGCTCCAGAA GCCATAAGAA CCTTGAATCG ACTGCTATCC GCATCGCTTT
TAGTCGAAGC AGACGCAGAC ACGAGGTCTT CGGTATTCTT GGAACTTAGC TGACGATAGG CGTAGCGAAA
```

FIGURE 2. DNA sequence of CRPV.

```
      1340       1350       1360       1370       1380       1390       1400
CCCTGGTGTG CCCGGAGTGT TGTAACTGAA AATGGCTGAA GGTACAGACC CTTTAGATGA CTGTGGGGGG
GGGACCACAC GGGCCTCACA ACATTGACTT TTACCGACTT CCATGTCTGG GAAATCTACT GACACCCCCC

      1410       1420       1430       1440       1450       1460       1470
TTCTTAGACA CGGAAGCGGA CTGTTTAGAC TGTGACAACC TTGAGGAGGA CCTGACAGAG CTGTTTGATG
AAGAATCTGT GCCTTCGCCT GACAAATCTG ACACTGTTGG AACTCCTCCT GGACTGTCTC GACAAACTAC

      1480       1490       1500       1510       1520       1530       1540
CTGACACTGT AAGCAGTTTA CTAGATGATA CAGATCAGGT GCAGGGAAAT TCCCTGGAAC CTTTTCAGCA
GACTGTGACA TTCGTCAAAT GATCTACTAT GTCTAGTCCA CGTCCCTTTA AGGGACCTTG GAAAAGTCGT

      1550       1560       1570       1580       1590       1600       1610
TCATGAGGCG ACTGAGACCT TGAAAAGCAT AGAGCATCTC AAGAGAAAGT ATGTCGATAG TCCTGATAAG
AGTACTCCGC TGACTCTGGA ACTTTTCGTA TCTCGTAGAG TTCTCTTTCA TACAGCTATC AGGACTATTC

      1620       1630       1640       1650       1660       1670       1680
AGCCTGGGTA TCGACAACTC CGTCAATGCC TTGAGTCCAA GATTACAAGC TTTCTCACTG TCAGGACAAA
TCGGACCCAT AGCTGTTGAG GCAGTTACGG AACTCAGGTT CTAATGTTCG AAAGAGTGAC AGTCCTGTTT

      1690       1700       1710       1720       1730       1740       1750
AAAAGGCTGT TAAAAAGAGA CTTTTCGGTA CTGACGGAGA TGAAGCTGCT TCTGGTGCTG AGTCGTTACA
TTTTCCGACA ATTTTTCTCT GAAAAGCCAT GACTGCCTCT ACTTCGACGA AGACCACGAC TCAGCAATGT

      1760       1770       1780       1790       1800       1810       1820
GGTAGAATCG GGATTTGGGT CTCAACAAAG CGTATCAGAT ACACCTGTGA CTGACATTTT AAATGCAAAT
CCATCTTAGC CCTAAACCCA GAGTTGTTTC GCATAGTCTA TGTGGACACT GACTGTAAAA TTTACGTTTA

      1830       1840       1850       1860       1870       1880       1890
ACAGCAAGAG TCAAACATTT GTTGTTATTT AGGCAAGCTC ACAGTGTTAG CTTTTCGGAG CTCACCAGAA
TGTCGTTCTC AGTTTGTAAA CAACAATAAA TCCGTTCGAG TGTCACAATC GAAAAGCCTC GAGTGGTCTT

      1900       1910       1920       1930       1940       1950       1960
CATTTCAAAG TGACAAGACT ATGAGTTGGG ATTGGGTAGG TGGGCTGGCG GACATTCATG TAAGCGTGTT
GTAAAGTTTC ACTGTTCTGA TACTCAACCC TAACCCATCC ACCCGACCGC CTGTAAGTAC ATTCGCACAA

      1970       1980       1990       2000       2010       2020       2030
GGAGAGCTTG CAGACATCTC TGAGAAGTCA TTGCGTATAT GTTCAGTATG ATCTCAATTT TGCAGAGACA
CCTCTCGAAC GTCTGTAGAG ACTCTTCAGT AACGCATATA CAAGTCATAC TAGAGTTAAA ACGTCTCTGT

      2040       2050       2060       2070       2080       2090       2100
AATGCTTCAT CTCTGCTGCT GCTCCTGAGA TTTAAAGCAC AAAAATGTAG GGACGGGGTT AAAGCGCTGC
TTACGAAGTA GAGACGACGA CGAGGACTCT AAATTTCGTG TTTTTACATC CCTGCCCCAA TTTCGCGACG

      2110       2120       2130       2140       2150       2160       2170
TATCCCAATT GTTGGGAGTT CAAGATCTAA AAGTTTTATT AGAACCTCCA AAAACAAGGA GTGTCGCTGT
ATAGGGTTAA CAACCCTCAA GTTCTAGATT TTCAAAATAA TCTTGGAGGT TTTTGTTCCT CACAGCGACA

      2180       2190       2200       2210       2220       2230       2240
TGCATTGTTC TGGTACAAAA GGGCGATGGT TTCTGGGGTT TTTAGCTACG GTCCAATGCC TGAATGGATA
ACGTAACAAG ACCATGTTTT CCCGCTACCA AAGACCCCAA AAATCGATGC CAGGTTACGG ACTTACCTAT

      2250       2260       2270       2280       2290       2300       2310
ACGCAGCAGA CAAATGTTAA CCATCAAATG TTGCAGGAAA AGCCGTTTCA GTTGTCTGTC ATGGTCCAGT
TGCGTCGTCT GTTTACAATT GGTAGTTTAC AACGTCCTTT TCGGCAAAGT CAACAGACAG TACCAGGTCA

      2320       2330       2340       2350       2360       2370       2380
GGGCATATGA TAACCACCTT CAGGATGAAA GTAGTATTGC ATACAAGTAT GCAATGCTCG CTGAAACTGA
CCCGTATACT ATTGGTGGAA GTCCTACTTT CATCATAACG TATGTTCATA CGTTACGAGC GACTTTGACT

      2390       2400       2410       2420       2430       2440       2450
TGAGAATGCA AGAGCGTTTC TAGCTTCTAA TTCTCAGGCG AAGTATGTTA GGGACTGTTG CAACATGGTC
ACTCTTACGT TCTCGCAAAG ATCGAAGATT AAGAGTCCGC TTCATACAAT CCCTGACAAC GTTGTACCAG

      2460       2470       2480       2490       2500       2510       2520
AGACTCTATT TAAGAGCAGA AATGAGACAG ATGACCATGT CTGCATGGAT AAACTACAGA TTGGATGGGA
TCTGAGATAA ATTCTCGTCT TTACTCTGTC TACTGGTACA GACGTACCTA TTTGATGTCT AACCTACCCT

      2530       2540       2550       2560       2570       2580       2590
TGAACGATGA TGGGGATTGG AAGGTGGTCG TGCATTTTCT GCGGCACCAA CGAGTGGAGT TCATACCTTT
ACTTGCTACT ACCCCTAACC TTCCACCAGC ACGTAAAAGA CGCCGTGGTT GCTCACCTCA AGTATGGAAA

      2600       2610       2620       2630       2640       2650       2660
CATGGTGAAG CTGAAGGCCT TCCTAAGAGG AACACCAAAA AAAAATTGCA TGGTGTTTTA TGGGCCACCA
GTACCACTTC GACTTCCGGA AGGATTCTCC TTGTGGTTTT TTTTTAACGT ACCACAAAAT ACCCGGTGGT
```

FIGURE 2. (*Continued*)

```
      2670       2680       2690       2700       2710       2720       2730
AATAGTGGGA AGTCATATTT TTGCATGAGC CTCATAAGAT TACTTGCAGG ACGGGTCTTG TCGTTTGCAA
TTATCACCCT TCAGTATAAA AACGTACTCG GAGTATTCTA ATGAACGTCC TGCCCAGAAC AGCAAACGTT

      2740       2750       2760       2770       2780       2790       2800
ACAGCAGAAG CCATTTTTGG CTGCAACCAT TAGCAGACGC CAAGCTAGCG CTCGTGGATG ATGCTACATC
TGTCGTCTTC GGTAAAAACC GACGTTGGTA ATCGTCTGCG GTTCGATCGC GAGCACCTAC TACGATGTAG

      2810       2820       2830       2840       2850       2860       2870
CGCGTGCTGG GATTTCATTG ATACATACCT CAGAAATGCC CTTGATGGCA ATCCCATATC GGTGGACCTG
GCGCACGACC CTAAAGTAAC TATGTATGGA GTCTTTACGG GAACTACCGT TAGGGTATAG CCACCTGGAC

      2880       2890       2900       2910       2920       2930       2940
AAGCACAAGG CACCAATAGA GATTAAGTGC CCTCCCCTCC TGATAACCAC AAATGTGGAC GTCAAATCAG
TTCGTGTTCC GTGGTTATCT CTAATTCACG GGAGGGGAGG ACTATTGGTG TTTACACCTG CAGTTTAGTC

      2950       2960       2970       2980       2990       3000       3010
ATGATAGATG GAGATACTTA TTTAGTAGAA TTTGTGTGTT TAACTTTTTG CAAGAATTGC CCATTAGAAA
TACTATCTAC CTCTATGAAT AAATCATCTT AAACACACAA ATTGAAAAAC GTTCTTAACG GGTAATCTTT

      3020       3030       3040       3050       3060       3070       3080
TGGGACACCT GTGTATGAAT TAAATGATGC AAACTGGAAA TCTTTTTTTA AAAGGTTCTG GTCCACCTTA
ACCCTGTGGA CACATACTTA ATTTACTACG TTTGACCTTT AGAAAAAAAT TTTCCAAGAC CAGGTGGAAT

      3090       3100       3110       3120       3130       3140       3150
GAACTAAGCG ACCCGGAAGA CGAGGGTGAC GATGGAGGCT CTCAGCCAGC GCTTAGACTC CATACAGGAG
CTTGATTCGC TGGGCCTTCT GCTCCCACTG CTACCTCCGA GAGTCGGTCG CGAATCTGAG GTATGTCCTC

      3160       3170       3180       3190       3200       3210       3220
GAACTTCTCA GTCTCTATGA GAAGGAGAGC ACGAGTTTGG AGTCCCAGCT ACAGCACTGG AACTTACTAA
CTTGAAGAGT CAGAGATACT CTTCCTCTCG TGCTCAAACC TCAGGGTCGA TGTCGTGACC TTGAATGATT

      3230       3240       3250       3260       3270       3280       3290
GAAAAGAACA GGTCCTTTTA CATTTCTGTA AAAAACACGG GATCAGGCAA CTGGGCTACA CGCCTGTCCC
CTTTTCTTGT CCAGGAAAAT GTAAAGACAT TTTTTGTGCC CTAGTCCGTT GACCCGATGT GCGGACAGGG

      3300       3310       3320       3330       3340       3350       3360
GTCTCTTCTT ACCTCACAGG AATGTGCAAA GCAAGCCATA GAAATGGTGC TGTACATTGA AAGCCTACTC
CAGAGAAGAA TGGAGTGTCC TTACACGTTT CGTTCGGTAT CTTTACCACG ACATGTAACT TTCGGATGAG

      3370       3380       3390       3400       3410       3420       3430
AGGTCCCCGT ATTCAGATGA GCCATGGACA TTGCAGGATA CCAGTAGAGA AAGGTTCGAA AGCCCTCCGC
TCCAGGGGCA TAAGTCTACT CGGTACCTGT AACGTCCTAT GGTCATCTCT TTCCAAGCTT TCGGGAGGCG

      3440       3450       3460       3470       3480       3490       3500
AAAAGACATT CAAAAAGAAC CCAGCTATTG TTGAGGTTTA CTATGATGGT GACAGAGGGA ACAACAATGA
TTTTCTGTAA GTTTTTCTTG GGTCGATAAC AACTCCAAAT GATACTACCA CTGTCTCCCT TGTTGTTACT

      3510       3520       3530       3540       3550       3560       3570
ATACACACTG TGGGGTATAT TTATTATTGG GAACGCTGAT GGGGAGTGGG TTAAGACTGA AAGTGGAGTG
TATGTGTGAC ACCCCATATA AATAATAACC CTTGCGACTA CCCCTCACCC AATTCTGACT TTCACCTCAC

      3580       3590       3600       3610       3620       3630       3640
GACTATAGAG GGATTTATTA TGTGGACTCT GAAGGAAACT ATGTGTATTA TGTGGACTTC TCAACCGACG
CTGATATCTC CCTAAATAAT ACACCTGAGA CTTCCTTTGA TACACATAAT ACACCTGAAG AGTTGGCTGC

      3650       3660       3670       3680       3690       3700       3710
CGGGACGTTT TGCTGCTAAT GGACACTATG ACGTGGTGTT TCAAAACATG CGCCTCTCTT CTTCTGTCAC
GCCCTGCAAA ACGACGATTA CCTGTGATAC TGCACCACAA AGTTTTGTAC GCGGAGAGAA GAAGACAGTG

      3720       3730       3740       3750       3760       3770       3780
CAGCTCCCCC CAGCCGCTGG TCAGTGCCCC TGAAGACACC GTCCCCGAAG AGGCCCCCGA CAGTGCAGTG
GTCGAGGGGG GTCGGCGACC AGTCACGGGG ACTTCTGTGG CAGGGGCTTC TCCGGGGGCT GTCACGTCAC

      3790       3800       3810       3820       3830       3840       3850
CCCGCCGCTC AAAAGAAAAC AGGGCCCAAA ACCACGCGTA CACTGGGCAG ACGAAGGTCA AGGTCACCAG
GGGCGGCGAG TTTTCTTTTG TCCCGGGTTT TGGTGCGCAT GTGACCCGTC TGCTTCCAGT TCCAGTGGTC

      3860       3870       3880       3890       3900       3910       3920
GGGTGCAACG AAGGCCGGCA AAGCAACGAA AACAGGCCGC CCCGGACGAA GCGGATTCTG CTGCCGGGGA
CCCACGTTGC TTCCGGCCGT TTCGTTGCTT TTGTCCGGCG GGGCCTGCTT CGCCTAAGAC GACGGCCCCT

      3930       3940       3950       3960       3970       3980       3990
CATCAGACCG CCTGCTCCAG AGGACGTTGG ACGAAGAACT ACGACGGTTG GAAGAACACC TCCCGGGCGG
GTAGTCTGGC GGACGAGGTC TCCTGCAACC TGCTTCTTGA TGCTGCCAAC CTTCTTGTGG AGGGCCCGCC
```

FIGURE 2. (*Continued*)

```
      4000       4010       4020       4030       4040       4050       4060
AATAGACGGC TTCGCGAGCT TATAACAGAA GCTAGCGATC CGCCCGTGAT TTGCTTGAAA GGGGGGCACA
TTATCTGCCG AAGCGCTCGA ATATTGTCTT CGATCGCTAG GCGGGCACTA AACGAACTTT CCCCCCGTGT

      4070       4080       4090       4100       4110       4120       4130
ACCAGCTTAA GTGCTTAAGG TATCGCCTTA AAAGCAAGCA CTCCTCACTA TTCGACTGCA TAAGCACTAC
TGGTCGAATT CACGAATTCC ATAGCGGAAT TTTCGTTCGT GAGGAGTGAT AAGCTGACGT ATTCGTGATG

      4140       4150       4160       4170       4180       4190       4200
TTGGAGCTGG GTTGACACAA CGAGCACATG CAGGCTAGGT AGCGGGCGCA TGCTTATAAA GTTTGCGGAC
AACCTCGACC CAACTGTGTT GCTCGTGTAC GTCCGATCCA TCGCCCGCGT ACGAATATTT CAAACGCCTG

      4210       4220       4230       4240       4250       4260       4270
TCTGAGCAGC GCGATAAGTT TCTTAGCAGG GTCCCACTCC CATCAACAAC GCAGGTGTTT TTAGGGAATT
AGACTCGTCG CGCTATTCAA AGAATCGTCC CAGGGTGAGG GTAGTTGTTG CGTCCACAAA AATCCCTTAA

      4280       4290       4300       4310       4320       4330       4340
TTTATGGGCT TTAGTGACGT GTATGCATGT AACCCATTCC CATCAGCAGC TTTTGTAACG CAACGTTTTT
AAATACCCGA AATCACTGCA CATACGTACA TTGGGTAAGG GTAGTCGTCG AAAACATTGC GTTGCAAAAA

      4350       4360       4370       4380       4390       4400       4410
TTGTACCAAT AAATCTTGCA CATACGCAAA AGGTGTCATG GTTGCACGGT CACGAAAACG CAGGGCTGCA
AACATGGTTA TTTAGAACGT GTATGCGTTT TCCACAGTAC CAACGTGCCA GTGCTTTTGC GTCCCGACGT

      4420       4430       4440       4450       4460       4470       4480
CCACAAGACA TTTATCCAAC ATGCAAAATT GCTGGCAATT GCCCAGCTGA CATACAGAAT AAATTTGAAA
GGTGTTCTGT AAATAGGTTG TACGTTTTAA CGACCGTTAA CGGGTCGACT GTATGTCTTA TTTAAACTTT

      4490       4500       4510       4520       4530       4540       4550
ACAAAACAAT TGCAGATAAA ATTCTGCAGT ATGGCAGCCT TGGTGTTTTC TTTGGAGGAC TTGGAATCAG
TGTTTTGTTA ACGTCTATTT TAAGACGTCA TACCGTCGGA ACCACAAAAG AAACCTCCTG AACCTTAGTC

      4560       4570       4580       4590       4600       4610       4620
TAGTGCCGGA GGTTCTGGGG GTCGACTAGG GTACACCCCA TTATCTGGAG GGGGGGGACG TGTCATAGCA
ATCACGGCCT CCAAGACCCC CAGCTGATCC CATGTGGGGT AATAGACCTC CCCCCCCTGC ACAGTATCGT

      4630       4640       4650       4660       4670       4680       4690
GCAGCCCCAG TAAGACCTCC CATAACAACA GAATCTGTAG GCCCTCTAGA TATAGTGCCT GAGGTAGCTG
CGTCGGGGTC ATTCTGGAGG GTATTGTTGT CTTAGACATC CGGGAGATCT ATATCACGGA CTCCATCGAC

      4700       4710       4720       4730       4740       4750       4760
ATCCTGGGGG TCCTACTCTA GTGTCACTAC ATGAACTGCC TGCAGAAACA CCATATGTAT CAAGCACAAA
TAGGACCCCC AGGATGAGAT CACAGTGATG TACTTGACGG ACGTCTTTGT GGTATACATA GTTCGTGTTT

      4770       4780       4790       4800       4810       4820       4830
TGTTACAGGG GATGGCGCAG CAGAGCCCCT TCCAGCTGGT CATGGGGGAA GCCAGATTTC AGACGTCACA
ACAATGTCCC CTACCGCGTC GTCTCGGGGA AGGTCGACCA GTACCCCCTT CGGTCTAAAG TCTGCAGTGT

      4840       4850       4860       4870       4880       4890       4900
TCTGGTACAT CCGGCACAGT GTCCAGAACA CACATTAATA ACCCTGTATT CGAGGCTCCA ATGACCGGTG
AGACCATGTA GGCCGTGTCA CAGGTCTTGT GTGTAATTAT TGGGACATAA GCTCCGAGGT TACTGGCCAC

      4910       4920       4930       4940       4950       4960       4970
ATCAGGATGT CTCGGATGTG CATGTATTTG CTCACTCTGA AAGTAGTATA ACTATCAACC AAACAGAAAA
TAGTCCTACA GAGCCTACAC GTACATAAAC GAGTGAGACT TTCATCATAT TGATAGTTGG TTTGTCTTTT

      4980       4990       5000       5010       5020       5030       5040
CACGGGCGGA GAGCTAATAG AGATGGTCCC CCTCAGACAC CCCCCTCGCA GTGAGGGAGA TTTCAGGGAA
GTGCCCGCCT CTCGATTATC TCTACCAGGG GGAGTCTGTG GGGGGAGCGT CACTCCCTCT AAAGTCCCTT

      5050       5060       5070       5080       5090       5100       5110
ACATCCTTCA GCACAAGCAC ACCAATCCCT GATAGATCTG CGTTGCGATC TATAAACGTA GCTAGCAGAA
TGTAGGAAGT CGTGTTCGTG TGGTTAGGGA CTATCTAGAC GCAACGCTAG ATATTTGCAT CGATCGTCTT

      5120       5130       5140       5150       5160       5170       5180
GATATCAGCA GGTACAAGTA GAAAACCCTG CTTTCCTGAA CAGGCCCAGG GAACTGGTGC AATTCGAAAA
CTATAGTCGT CCATGTTCAT CTTTTGGGAC GAAAGGACTT GTCCGGGTCC CTTGACCACG TTAAGCTTTT

      5190       5200       5210       5220       5230       5240       5250
CACATTTGAC AATCCTGCTT TTGTGGATGA TGAGCAACTA AGCCTCCTTT TTGAACAGGA TCTAGACACC
GTGTAAACTG TTAGGACGAA AACACCTACT ACTCGTTGAT TCGGAGGAAA AACTTGTCCT AGATCTGTGG

      5260       5270       5280       5290       5300       5310       5320
GTAGTTGCAA CTCCAGATCC TGCGTTCCAG GATGTTGTGC GTTTAAGTAG GCCTAGTTTC ACTCAATCCA
CATCAACGTT GAGGTCTAGG ACGCAAGGTC CTACAACACG CAAATTCATC CGGATCAAAG TGAGTTAGGT
```

FIGURE 2. (*Continued*)

```
      5330       5340       5350       5360       5370       5380       5390
GAGCTGGTAG GGTTCGCGTC AGCCGCCTGG GGAGGACGCT AACAATGCAA ACACGCAGTG GTAAGGCCTT
CTCGACCATC CCAAGCGCAG TCGGCGGACC CCTCCTGCGA TTGTTACGTT TGTGCGTCAC CATTCCGGAA

      5400       5410       5420       5430       5440       5450       5460
TGGGCCTGCC AAACACTTCT ACTATGAGCT TTCAAGCATA GCAGAAGGTC CCGAGCCAGA CATCCTCATC
ACCCGGACGG TTTGTGAAGA TGATACTCGA AAGTTCGTAT CGTCTTCCAG GGCTCGGTCT GTAGGAGTAG

      5470       5480       5490       5500       5510       5520       5530
CCCGAATCAG AACAGGAAAC ATCATTCACA GATGCCACAT CTAAAGACAC ACAACAGGAA GCAGAAGTGT
GGGCTTAGTC TTGTCCTTTG TAGTAAGTGT CTACGGTGTA GATTTCTGTG TGTTGTCCTT CGTCTTCACA

      5540       5550       5560       5570       5580       5590       5600
ATGCAGATGG TTCAACCCTG GAAACTGACA CATCGGCAGA TGAAAACTTG ACACTTGTCT TTTCAGACAG
TACGTCTACC AAGTTGGGAC CTTTGACTGT GTAGCCGTCT ACTTTTGAAC TGTGAACAGA AAAGTCTGTC

      5610       5620       5630       5640       5650       5660       5670
AGGCAGGGGT CAGGGTTCAC ATGTACCTAT TCCAGGCAAG TCCACAATTG GGGGTCCTGT AAATATTGGG
TCCGTCCCCA GTCCCAAGTG TACATGGATA AGGTCCGTTC AGGTGTTAAC CCCCAGGACA TTTATAACCC

      5680       5690       5700       5710       5720       5730       5740
GACAGCAAAT ACTATACTCT GAACCCTGGA GAAACTACAA GCTTTGAAGC AGATGTAATT TCACCTGTTT
CTGTCGTTTA TGATATGAGA CTTGGGACCT CTTTGATGTT CGAAACTTCG TCTACATTAA AGTGGACAAA

      5750       5760       5770       5780       5790       5800       5810
TCATATTTGA GGGTAACGCA GATGGCACTT ACTATCTAGA GGAACCTCTA CGAAAGAAAA GACGCAAATC
AGTATAAACT CCCATTGCGT CTACCGTGAA TGATAGATCT CCTTGGAGAT GCTTTCTTTT CTGCGTTTAG

      5820       5830       5840       5850       5860       5870       5880
TATCTTTTTA CTTGCAGATG GCAGTGTGGC TGTCTACGCA GAATAAGTTT TACCTGCCGC CTCAGCCTGT
ATAGAAAAAT GAACGTCTAC CGTCACACCG ACAGATGCGT CTTATTCAAA ATGGACGGCG GAGTCGGACA

      5890       5900       5910       5920       5930       5940       5950
CACAAAGATA CCTAGCACGG ATGAATACGT TACTCGGACA AACGTCTTTT ATTATGCATC CAGTGACCGG
GTGTTTCTAT GGATCGTGCC TACTTATGCA ATGAGCCTGT TTGCAGAAAA TAATACGTAG GTCACTGGCC

      5960       5970       5980       5990       6000       6010       6020
CTACTCACAG TGGGACATCC ATACTATGAA ATACGTGATA AAGGCACCAT GCTTGTTCCA AAGGTTTCTC
GATGAGTGTC ACCCTGTAGG TATGATACTT TATGCACTAT TTCCGTGGTA CGAACAAGGT TTCCAAAGAG

      6030       6040       6050       6060       6070       6080       6090
CAAACCAATA CAGAGTATTT AGAATCAAAC TCCCTGACCC TAACAAGTTT GCATTTGGTG ACAAGCAACT
GTTTGGTTAT GTCTCATAAA TCTTAGTTTG AGGGACTGGG ATTGTTCAAA CGTAAACCAC TGTTCGTTGA

      6100       6110       6120       6130       6140       6150       6160
ATATGATCCA GAGAAGGAAC GGCTTGTGTG GTGCCTTAGA GGTATTGAGG TCAATCGAGG CCAGCCTCTA
TATACTAGGT CTCTTCCTTG CCGAACACAC CACGGAATCT CCATAACTCC AGTTAGCTCC GGTCGGAGAT

      6170       6180       6190       6200       6210       6220       6230
GGAGTCAGTG TCACAGGAAA CCCTATCTTT AATAAATTTG ATGATGTCGA GAATCCCACA AAGTATTACA
CCTCAGTCAC AGTGTCCTTT GGGATAGAAA TTATTTAAAC TACTACAGCT CTTAGGGTGT TTCATAATGT

      6240       6250       6260       6270       6280       6290       6300
ATAACCATGC AGACCAGCAA GACTACAGAA AAAGCATGGC GTTCGACCCC AAGCAAGTGC AGCTGTTAAT
TATTGGTACG TCTGGTCGTT CTGATGTCTT TTTCGTACCG CAAGCTGGGG TTCGTTCACG TCGACAATTA

      6310       6320       6330       6340       6350       6360       6370
GCTTGGATGC GTCCCTGCCA CAGGAGAACA CTGGGCTCAG GCAAAGCAGT GTGCAGAGGA TCCACCACAA
CGAACCTACG CAGGGACGGT GTCCTCTTGT GACCCGAGTC CGTTTCGTCA CACGTCTCCT AGGTGGTGTT

      6380       6390       6400       6410       6420       6430       6440
CAGACCGACT GTCCCCCCAT TGAACTAGTG AACACTGTTA TAGAAGATGG GGACATGTGT GAAATAGGCT
GTCTGGCTGA CAGGGGGGTA ACTTGATCAC TTGTGACAAT ATCTTCTACC CCTGTACACA CTTTATCCGA

      6450       6460       6470       6480       6490       6500       6510
TTGGGGCAAT GGACCATAAA ACATTGCAGG CCAGTTTATC AGAGGTTCCC CTTGAGTTAG CACAGTCAAT
AACCCCGTTA CCTGGTATTT TGTAACGTCC GGTCAAATAG TCTCCAAGGG GAACTCAATC GTGTCAGTTA

      6520       6530       6540       6550       6560       6570       6580
CAGCAAGTAT CCAGACTATC TAAAAATGCA AAAAGATCAG TTTGGGGATT CTATGTTCTT TTATGCTAGA
GTCGTTCATA GGTCTGATAG ATTTTTACGT TTTTCTAGTC AAACCCCTAA GATACAAGAA AATACGATCT

      6590       6600       6610       6620       6630       6640       6650
AGAGAGCAGA TGTATGCTAG ACATTTCTTC AGCAGGGCAG GAGGGGACAA GGAAAATGTG AAGAGCAGGG
TCTCTCGTCT ACATACGATC TGTAAAGAAG TCGTCCCGTC CTCCCCTGTT CCTTTTACAC TTCTCGTCCC
```

FIGURE 2. (*Continued*)

```
      6660       6670       6680       6690       6700       6710       6720
CCTACATAAA ACGCACACAG ATGCAGGGAG AGGCAAATGC CAACATTGCA ACTGACAATT ACTGCATCAC
GGATGTATTT TGCGTGTGTC TACGTCCCTC TCCGTTTACG GTTGTAACGT TGACTGTTAA TGACGTAGTG

      6730       6740       6750       6760       6770       6780       6790
ACCTAGTGGA TCTTTGGTCT CTAGCGATTC ACAGGTTTTT AACCGTGCAT ATTGGCTCCA GAAAGCTCAA
TGGATCACCT AGAAACCAGA GATCGCTAAG TGTCCAAAAA TTGGCACGTA TAACCGAGGT CTTTCGAGTT

      6800       6810       6820       6830       6840       6850       6860
GGCATGAACA ATGGAGTTTG CTGGGACAAT CAAATTTTTG TGACTGTGGT AGATAACACC AGGGGTACAA
CCGTACTTGT TACCTCAAAC GACCCTGTTA GTTTAAAAAC ACTGACACCA TCTATTGTGG TCCCCATGTT

      6870       6880       6890       6900       6910       6920       6930
TATTAAGTCT TGTCACAAAA TCCAAGGAGC AAATCAAGAA GACCCATGGA AAAACAGTAC ATTTTTCTTC
ATAATTCAGA ACAGTGTTTT AGGTTCCTCG TTTAGTTCTT CTGGGTACCT TTTTGTCATG TAAAAAGAAG

      6940       6950       6960       6970       6980       6990       7000
CTATCTAAGG CATGTGGAGG AGTATGAACT GCAATTTGTG CTCCAGCTAT GTAAGGTCAA GTTAACACCC
GATAGATTCC GTACACCTCC TCATACTTGA CGTTAAACAC GAGGTCGATA CATTCCAGTT CAATTGTGGG

      7010       7020       7030       7040       7050       7060       7070
GAAAACCTAT CATACCTACA TAGCATGCAC CCAACAATCA TAGATAATTG GCAATTGTCA GTGTCAGCTC
CTTTTGGATA GTATGGATGT ATCGTACGTG GGTTGTTAGT ATCTATTAAC CGTTAACAGT CACAGTCGAG

      7080       7090       7100       7110       7120       7130       7140
AGCCCAGTGG AACGCTAGAA GACCAGTACA GATACCTGCA GTCCATTGCA ACCAAATGTC CACCCCCAGA
TCGGGTCACC TTGCGATCTT CTGGTCATGT CTATGGACGT CAGGTAACGT TGGTTTACAG GTGGGGGTCT

      7150       7160       7170       7180       7190       7200       7210
ACCTCCCAAA GAAAACACTG ACCCATATAA AAACTATAAG TTTTGGGAAG TAGATTTGTC TGAGAAGCTA
TGGAGGGTTT CTTTTGTGAC TGGGTATATT TTTGATATTC AAAACCCTTC ATCTAAACAG ACTCTTCGAT

      7220       7230       7240       7250       7260       7270       7280
TCTGATCAGC TAGATCAGTA TCCACTTGGC AGAAAGTTCC TAAATCAAAG TGGCCTGCAA AGAATTGGTA
AGACTAGTCG ATCTAGTCAT AGGTGAACCG TCTTTCAAGG ATTTAGTTTC ACCGGACGTT TCTTAACCAT

      7290       7300       7310       7320       7330       7340       7350
CAAAAAGACC TGCACCTGCA CCTGTTAGTA TTGTGAAATC ATCTAAACGC AAGAGACGTA CTTAATTGTA
GTTTTTCTGG ACGTGGACGT GGACAATCAT AACACTTTAG TAGATTTGCG TTCTCTGCAT GAATTAACAT

      7360       7370       7380       7390       7400       7410       7420
TATTATTGGA TGTGTATTTA ATACCTCATA ACTTATAAAT ATACTAAATA AAGTTTAAAA CCGATGCATA
ATAATAACCT ACACATAAAT TATGGAGTAT TGAATATTTA TATGATTTAT TTCAAATTTT GGCTACGTAT

      7430       7440       7450       7460       7470       7480       7490
CTAGTATTAC TTCTTTTTTG TGAAGCTCTT TGCGACCGCA CCCGGTGGCA TGTGATGACA CATCACACAG
GATCATAATG AAGAAAAAAC ACTTCGAGAA ACGCTGGCGT GGGCCACCGT ACACTACTGT GTAGTGTGTC

      7500       7510       7520       7530       7540       7550       7560
TAAGCCTTGA GAGAACGCCA GAGGTTGTAA ACATCCTGTT CCGGCTGGCA CTGAGCCTGA TCGCAGGCAG
ATTCGGAACT CTCTTGCGGT CTCCAACATT TGTAGGACAA GGCCGACCGT GACTCGGACT AGCGTCCGTC

      7570       7580       7590       7600       7610       7620       7630
TGATCTTAAC CGTAAGCGTT CTCATTAATT CTGGGGTATT TCCAAGAAAA ACGTTATCGC TTTTGGCCAC
ACTAGAATTG GCATTCGCAA GAGTAATTAA GACCCCATAA AGGTTCTTTT TGCAATAGCG AAAACCGGTG

      7640       7650       7660       7670       7680       7690       7700
CGTACTCGGT GTCAGCAAAT ACCAAGCCAA GATCTTTTTT GGTGCCAAAA ATCTAAGCAA CCGCTCCCGG
GCATGAGCCA CAGTCGTTTA TGGTTCGGTT CTAGAAAAAA CCACGGTTTT TAGATTCGTT GGCGAGGGCC

      7710       7720       7730       7740       7750       7760       7770
TCCCGGCAAC CGCTCCCGGT CTTGCAACCG CCCACGTTCG ACCGCTCCCG GTCTTGCAAC CGCCCACGTT
AGGGCCGTTG GCGAGGGCCA GAACGTTGGC GGGTGCAAGC TGGCGAGGGC CAGAACGTTG GCGGGTGCAA

      7780       7790       7800       7810       7820       7830       7840
CGACCGCTCC CGGTCTTGCA ACCGCCAGGT GTGCATGACT GTAAGTATTG ACTCACACCG CATGCGGTGT
GCTGGCGAGG GCCAGAACGT TGGCGGTCCA CACGTACTGA CATTCATAAC TGAGTGTGGC GTACGCCACA

      7850       7860
TATGCACTTT GATTTAATGA TGGTTGTT
ATACGTGAAA CTAAATTACT ACCAACAA
```

FIGURE 2. (*Continued*)

```
         10         20         30         40         50         60         70
GTTAACTACC ATCATTCATT ATTCTAGTTA CAACAAGAAC CTAGGAGTTA TATGCCAGAA GTAAGCCTAT
CAATTGATGG TAGTAAGTAA TAAGATCAAT GTTGTTCTTG GATCCTCAAT ATACGGTCTT CATTCGGATA

         80         90        100        110        120        130        140
AAAATACACA GGTAAGACTC TGCACAGGAC CAGATGGCGA CACCAATCCG GACCGTCAGA CAGCTTTCCG
TTTTATGTGT CCATTCTGAG ACGTGTCCTG GTCTACCGCT GTGGTTAGGC CTGGCAGTCT GTCGAAAGGC

        150        160        170        180        190        200        210
AAAGCCTCTG TATCCCATAT ATTGATGTTT TATTGCCTTG TAATTTTTGT AATTATTTTT TGTCTAATGC
TTTCGGAGAC ATAGGGTATA TAACTACAAA ATAACGGAAC ATTAAAAACA TTAATAAAAA ACAGATTACG

        220        230        240        250        260        270        280
TGAGAAGCTG CTTTTTGATC ATTTTGATTT GCATCTTGTC TGGAGAGACA ATTTGGTGTT TGGATGCTGT
ACTCTTCGAC GAAAAACTAG TAAAACTAAA CGTAGAACAG ACCTCTCTGT TAAACCACAA ACCTACGACA

        290        300        310        320        330        340        350
CAAGGGTGTG CTAGAACTGT TAGCCTATTG GAGTTTGTTT TATATTATCA GGAGTCTTAT GAGGTACCGG
GTTCCCACAC GATCTTGACA ATCGGATAAC CTCAAACAAA ATATAATAGT CCTCAGAATA CTCCATGGCC

        360        370        380        390        400        410        420
AAATAGAAGA AATTTTGGAC AGACCTTTAT TGCAAATTGA ACTCCGTTGT GTTACATGCA TAAAAAAACT
TTTATCTTCT TTAAAACCTG TCTGGAAATA ACGTTTAACT TGAGGCAACA CAATGTACGT ATTTTTTTGA

        430        440        450        460        470        480        490
GAGTGTTGCT GAAAAATTGG AGGTTGTGTC AAACGGAGAA AGAGTGCATA GAGTTAGAAA CAGACTTAAA
CTCACAACGA CTTTTTAACC TCCAACACAG TTTGCCTCTT TCTCACGTAT CTCAATCTTT GTCTGAATTT

        500        510        520        530        540        550        560
GCAAAGTGTA GTTTGTGTCG CTTGTATGCT ATATAACAAT GGTGGGCGAA ATGCCAGCAC TAAAGGACCT
CGTTTCACAT CAAACACAGC GAACATACGA TATATTGTTA CCACCCGCTT TACGGTCGTG ATTTCCTGGA

        570        580        590        600        610        620        630
GGTTCTTCAA CTTGAACCAA GCGTCCTAGA TTTAGATCTT TATTGTTACG AGGAGGTGCC TCCTGATGAC
CCAAGAAGTT GAACTTGGTT CGCAGGATCT AAATCTAGAA ATAACAATGC TCCTCCACGG AGGACTACTG

        640        650        660        670        680        690        700
ATAGAGGAGG AGTTAGTGTC GCCTCAGCAA CCTTATGCTG TCGTTGCTTC CTGTGCCTAT TGCGAGAAAC
TATCTCCTCC TCAATCACAG CGGAGTCGTT GGAATACGAC AGCAACGAAG GACACGGATA ACGCTCTTTG

        710        720        730        740        750        760        770
TGGTTCGATT GACCGTCCTC GCGGATCACA GCGCCATTAG ACAGCTGGAG GAACTCCTTC TGCGATCTTT
ACCAAGCTAA CTGGCAGGAG CGCCTAGTGT CGCGGTAATC TGTCGACCTC CTTGAGGAAG ACGCTAGAAA

        780        790        800        810        820        830        840
GAACATCGTG TGCCCACTGT GCACCCTACA GCGACAGTAA AATGGCAGAT AATAAAGGTA CTGAAAACGA
CTTGTAGCAC ACGGGTGACA CGTGGGATGT CGCTGTCATT TTACCGTCTA TTATTTCCAT GACTTTTGCT

        850        860        870        880        890        900        910
TTGGTTTTTG GTGGAGGCGA CAGATTGTGA GGAAACGTTA GAGGAAACCT CACTTGGTGA CCTAGATAAT
AACCAAAAAC CACCTCCGCT GTCTAACACT CCTTTGCAAT CTCCTTTGGA GTGAACCACT GGATCTATTA

        920        930        940        950        960        970        980
GTTTCTTGTG TTAGCGACTT ATCTGATTTA TTAGACGAGG CGCCGCAAAG CCAGGGGAAT TCCCTGGAAT
CAAAGAACAC AATCGCTGAA TAGACTAAAT AATCTGCTCC GCGGCGTTTC GGTCCCCTTA AGGGACCTTA

        990       1000       1010       1020       1030       1040       1050
TGTTCCACAA GCAAGAATCG CTGGAAAGCG AACAGGAACT TAATGCTTTA AAACGAAAGT TACTTTACAG
ACAAGGTGTT CGTTCTTAGC GACCTTTCGC TTGTCCTTGA ATTACGAAAT TTTGCTTTCA ATGAAATGTC

       1060       1070       1080       1090       1100       1110       1120
TCCTCAGGCG AGAAGCGCGG ACGAAACAGA CATTGCTAGC ATTAGTCCTA GATTAGAAAC TATTTCTATT
AGGAGTCCGC TCTTCGCGCC TGCTTTGTCT GTAACGATCG TAATCAGGAT CTAATCTTTG ATAAAGATAA

       1130       1140       1150       1160       1170       1180       1190
ACAAAGCAAG ACAAAAAAAG GTATCGAAGG CAACTGTTTT CTCAGGATGA TAGTGGTTTA GAGCTATCGC
TGTTTCGTTC TGTTTTTTTC CATAGCTTCC GTTGACAAAA GAGTCCTACT ATCACCAAAT CTCGATAGCG

       1200       1210       1220       1230       1240       1250       1260
TGCTTCAGGA TGAAACTGAA AATATTGATG AATCGACACA GGTAGATCAA CAGCAGAAAG AACATACTGG
ACGAAGTCCT ACTTTGACTT TTATAACTAC TTAGCTGTGT CCATCTAGTT GTCGTCTTTC TTGTATGACC

       1270       1280       1290       1300       1310       1320       1330
GGAAGTTGGG GCCGCTGGGG TGAACATTTT GAAAGCTAGT AATATCCGCG CCGCATTATT AAGCAGATTT
CCTTCAACCC CGGCGACCCC ACTTGTAAAA CTTTCGATCA TTATAGGCGC GGCGTAATAA TTCGTCTAAA
```

FIGURE 3. DNA sequence of HPV-1a.

```
      1340       1350       1360       1370       1380       1390       1400
AAAGATACGG CTGGCGTCAG TTTTACAGAC CTGACGCGGT CGTACAAGAG CAACAAAACC TGTTGTGGAG
TTTCTATGCC GACCGCAGTC AAAATGTCTG GACTGCGCCA GCATGTTCTC GTTGTTTTGG ACAACACCTC

      1410       1420       1430       1440       1450       1460       1470
ATTGGGTTTT GGCAGTTTGG GGTGTCCGTG AAAATTTAAT TGACAGTGTA AAAGAATTAT TGCAAACCCA
TAACCCAAAA CCGTCAAACC CCACAGGCAC TTTTAAATTA ACTGTCACAT TTTCTTAATA ACGTTTGGGT

      1480       1490       1500       1510       1520       1530       1540
TTGTGTGTAT ATTCAATTGG AACATGCAGT AACTGAAAAA AATAGATTTT TATTTTTATT GGTACGATTT
AACACACATA TAAGTTAACC TTGTACGTCA TTGACTTTTT TTATCTAAAA ATAAAAATAA CCATGCTAAA

      1550       1560       1570       1580       1590       1600       1610
AAAGCCCAGA AAAGTAGAGA GACTGTGATA AAACTTATAA CCACAATTCT TCCAGTTGAT GCTAGCTATA
TTTCGGGTCT TTTCATCTCT CTGACACTAT TTTGAATATT GGTGTTAAGA AGGTCAACTA CGATCGATAT

      1620       1630       1640       1650       1660       1670       1680
TTTTGTCTGA GCCTCCAAAA TCAAGAAGTG TGGCTGCTGC ATTATTTTGG TATAAAAGAT CTATGTCTTC
AAAACAGACT CGGAGGTTTT AGTTCTTCAC ACCGACGACG TAATAAAACC ATATTTTCTA GATACAGAAG

      1690       1700       1710       1720       1730       1740       1750
AACTGTTTTT ACATGGGGTA CAACTTTGGA GTGGATTGCA CAGCAAACCC TTATTAATCA TCAGTTAGAT
TTGACAAAAA TGTACCCCAT GTTGAAACCT CACCTAACGT GTCGTTTGGG AATAATTAGT AGTCAATCTA

      1760       1770       1780       1790       1800       1810       1820
TCCGAAAGTC CCTTTGAGCT TTGTAAAATG GTTCAGTGGG CCTATGATAA TGGACATACA GAAGAGTGTA
AGGCTTTCAG GGAAACTCGA AACATTTTAC CAAGTCACCC GGATACTATT ACCTGTATGT CTTCTCACAT

      1830       1840       1850       1860       1870       1880       1890
AAATTGCATA TTATTATGCT GTTTTAGCAG ATGAGGATGA AAATGCAAGG GCATTTCTAA GCTCTAATTC
TTTAACGTAT AATAATACGA CAAAATCGTC TACTCCTACT TTTACGTTCC CGTAAAGATT CGAGATTAAG

      1900       1910       1920       1930       1940       1950       1960
ACAGGAAAAA TATGTGAAAG ACTGTGCACA AATGGTAAGA CACTATTTAC GTGCTGAGAT GGCACAAATG
TGTCCTTTTT ATACACTTTC TGACACGTGT TTACCATTCT GTGATAAATG CACGACTCTA CCGTGTTTAC

      1970       1980       1990       2000       2010       2020       2030
TCTATGTCAG AGTGGATTTT TAGAAAACTA GATAATGTAG AAGGTTCTGG TAATTGGAAA GAAATTGTAA
AGATACAGTC TCACCTAAAA ATCTTTTGAT CTATTACATC TTCCAAGACC ATTAACCTTT CTTTAACATT

      2040       2050       2060       2070       2080       2090       2100
GATTTTTAAG ATTTCAAGAA GTTGAATTTA TAAGCTTTAT GATTGCATTT AAAGATTTGT TATGTGGTAA
CTAAAAATTC TAAAGTTCTT CAACTTAAAT ATTCGAAATA CTAACGTAAA TTTCTAAACA ATACACCATT

      2110       2120       2130       2140       2150       2160       2170
GCCAAAGAAA AACTGTTTGT TAATATTTGG ACCTCCAAAT ACAGGAAAAT CAATGTTTTG TACAAGTTTA
CGGTTTCTTT TTGACAAACA ATTATAAACC TGGAGGTTTA TGTCCTTTTA GTTACAAAAC ATGTTCAAAT

      2180       2190       2200       2210       2220       2230       2240
TTAAAGTTGT TAGGAGGGAA AGTGATTTCA TACTGTAACA GTAAAAGTCA GTTTTGGTTG CAGCCTCTGG
AATTTCAACA ATCCTCCCTT TCACTAAAGT ATGACATTGT CATTTTCAGT CAAAACCAAC GTCGGAGACC

      2250       2260       2270       2280       2290       2300       2310
CTGATGCTAA GATAGGGCTA TTAGATGATG CAACAAAGCC ATGTTGGGAT TATATGGACA TTTATATGAG
GACTACGATT CTATCCCGAT AATCTACTAC GTTGTTTCGG TACAACCCTA ATATACCTGT AAATATACTC

      2320       2330       2340       2350       2360       2370       2380
AAATGCATTG GATGGTAACA CTATTTGTAT TGATTTAAAA CATAGAGCTC CTCAACAAAT TAAATGCCCA
TTTACGTAAC CTACCATTGT GATAAACATA ACTAAATTTT GTATCTCGAG GAGTTGTTTA ATTTACGGGT

      2390       2400       2410       2420       2430       2440       2450
CCTTTACTTA TTACTAGTAA TATTGATGTT AAATCAGATA CCTGTTGGAT GTATTTGCAT AGTAGAATAT
GGAAATGAAT AATGATCATT ATAACTACAA TTTAGTCTAT GGACAACCTA CATAAACGTA TCATCTTATA

      2460       2470       2480       2490       2500       2510       2520
CAGCTTTTAA ATTTGCTCAT GAGTTTCCAT TTAAAGACAA TGGTGATCCA GGATTTTCCT TAACAGACGA
GTCGAAAATT TAAACGAGTA CTCAAAGGTA AATTTCTGTT ACCACTAGGT CCTAAAAGGA ATTGTCTGCT

      2530       2540       2550       2560       2570       2580       2590
AAATTGGAAA TCTTTCTTTG AAAGGTTTTG GCAACAGTTA GAATTAAGTG ACCAAGAAGA CGAGGGAAAC
TTTAACCTTT AGAAAGAAAC TTTCCAAAAC CGTTGTCAAT CTTAATTCAC TGGTTCTTCT GCTCCCTTTG

      2600       2610       2620       2630       2640       2650       2660
GATGGAAAAC CTCAGCAGTC GCTTAGACTT ACTGCAAGAG CAGCTAATGA ACCTATATGA ACAGGACAGT
CTACCTTTTG GAGTCGTCAG CGAATCTGAA TGACGTTCTC GTCGATTACT TGGATATACT TGTCCTGTCA
```

FIGURE 3. (*Continued*)

```
      2670       2680       2690       2700       2710       2720       2730
AAATTGATAG AAGATCAAAT TAAGCAGTGG AATCTAATTA GACAAGAACA AGTTCTTTTC CATTTCGCCA
TTTAACTATC TTCTAGTTTA ATTCGTCACC TTAGATTAAT CTGTTCTTGT TCAAGAAAAG GTAAAGCGGT

      2740       2750       2760       2770       2780       2790       2800
GAAAAAATGG GGTAATGAGA ATTGGATTGC AGGCAGTTCC ATCTTTAGCG TCCTCACAGG AGAAGGCAAA
CTTTTTTACC CCATTACTCT TAACCTAACG TCCGTCAAGG TAGAAATCGC AGGAGTGTCC TCTTCCGTTT

      2810       2820       2830       2840       2850       2860       2870
GACAGCTATT GAAATGGTGT TACATTTAGA GTCTTTAAAG GACTCACCTT ATGGCACAGA GGATTGGTCA
CTGTCGATAA CTTTACCACA ATGTAAATCT CAGAAATTTC CTGAGTGGAA TACCGTGTCT CCTAACCAGT

      2880       2890       2900       2910       2920       2930       2940
CTTCAAGACA CTAGCAGAGA GCTGTTTTTG GCACCCCCAG CTGGCACCTT CAAGAAGAGT GGCAGCACAC
GAAGTTCTGT GATCGTCTCT CGACAAAAAC CGTGGGGGTC GACCGTGGAA GTTCTTCTCA CCGTCGTGTG

      2950       2960       2970       2980       2990       3000       3010
TTGAGGTTAC CTATGACAAT AACCCTGATA ATCAGACAAG GCACACAATT TGGAATCATG TGTATTATCA
AACTCCAATG GATACTGTTA TTGGGACTAT TAGTCTGTTC CGTGTGTTAA ACCTTAGTAC ACATAATAGT

      3020       3030       3040       3050       3060       3070       3080
AAATGGGGAC GATGTATGGA GAAAAGTATC CAGTGGTGTT GATGCTGTAG GAGTGTACTA TTTAGAACAC
TTTACCCCTG CTACATACCT CTTTTCATAG GTCACCACAA CTACGACATC CTCACATGAT AAATCTTGTG

      3090       3100       3110       3120       3130       3140       3150
GATGGCTATA AAAATTATTA TGTGTTATTT GCTGAGGAGG CCTCTAAGTA CAGCACAACA GGACAATATG
CTACCGATAT TTTTAATAAT ACACAATAAA CGACTCCTCC GGAGATTCAT GTCGTGTTGT CCTGTTATAC

      3160       3170       3180       3190       3200       3210       3220
CTGTAAATTA CAGGGGTAAA AGGTTTACAA ATGTTATGTC TTCCACTAGC TCCCCAAGGG CTGCTGGGGC
GACATTTAAT GTCCCCATTT TCCAAATGTT TACAATACAG AAGGTGATCG AGGGGTTCCC GACGACCCCG

      3230       3240       3250       3260       3270       3280       3290
TCCTGCAGTA CACTCCGACT ACCCAACCCT ATCCGAGAGT GACACCGCCC AGCAATCGAC GTCCATCGAC
AGGACGTCAT GTGAGGCTGA TGGGTTGGGA TAGGCTCTCA CTGTGGCGGG TCGTTAGCTG CAGGTAGCTG

      3300       3310       3320       3330       3340       3350       3360
TACACCGAAC TCCCAGGACA GGGGGAGACC TCGCAGGTCC GACAAAGACA GCAGAAAACA CCTGTACGCA
ATGTGGCTTG AGGGTCCTGT CCCCCTCTGG AGCGTCCAGG CTGTTTCTGT CGTCTTTTGT GGACATGCGT

      3370       3380       3390       3400       3410       3420       3430
GACGGCCTTA CGGACGGCGA AGATCCAGAA GTCCCAGAGG TGGAGGACGA AGAGAAGGAG AATCAACGCC
CTGCCGGAAT GCCTGCCGCT TCTAGGTCTT CAGGGTCTCC ACCTCCTGCT TCTCTTCCTC TTAGTTGCGG

      3440       3450       3460       3470       3480       3490       3500
CTCTAGGACA CCCGGATCTG TCCCTTCTGC GCGAGACGTT GGAAGTATAC ACACAACGCC TCAAAAGGGA
GAGATCCTGT GGGCCTAGAC AGGGAAGACG CGCTCTGCAA CCTTCATATG TGTGTTGCGG AGTTTTCCCT

      3510       3520       3530       3540       3550       3560       3570
CATTCTTCAA GACTTAGACG ACTTCTGCAG GAAGCTTGGG ATCCACCCGT GGTCTGTGTA AAAGGGGGTG
GTAAGAAGTT CTGAATCTGC TGAAGACGTC CTTCGAACCC TAGGTGGGCA CCAGACACAT TTTCCCCCAC

      3580       3590       3600       3610       3620       3630       3640
CCAATCAGCT TAAGTGTCTC AGGTACAGAC TTAAAGCATC TACTCAAGTT GACTTTGACA GCATAAGCAC
GGTTAGTCGA ATTCACAGAG TCCATGTCTG AATTTCGTAG ATGAGTTCAA CTGAAACTGT CGTATTCGTG

      3650       3660       3670       3680       3690       3700       3710
CACATGGCAT TGGACAGATA GAAAAAACAC CGAGAGGATA GGTAGTGCTA GAATGTTAGT AAAGTTTATT
GTGTACCGTA ACCTGTCTAT CTTTTTTGTG GCTCTCCTAT CCATCACGAT CTTACAATCA TTTCAAATAA

      3720       3730       3740       3750       3760       3770       3780
GATGAGGCTC AACGAGAGAA GTTTCTTGAG AGAGTTGCTT TGCCCAGATC AGTGTCTGTG TTTTTGGGAC
CTACTCCGAG TTGCTCTCTT CAAAGAACTC TCTCAACGAA ACGGGTCTAG TCACAGACAC AAAAACCCTG

      3790       3800       3810       3820       3830       3840       3850
AGTTTAATGG GTCTTAAAAT TAATGGAAGT TGATTTTGCT TGGACGTGTG TACATAGTCC CTGTATATAT
TCAAATTACC CAGAATTTTA ATTACCTTCA ACTAAAACGA ACCTGCACAC ATGTATCAGG GACATATATA

      3860       3870       3880       3890       3900       3910       3920
TCCCCTCCTA CCCCCACATA CCTTGAAGCT TGCAACATTG TAACAAATGT ATCGCCTACG TAGAAAACGC
AGGGGAGGAT GGGGGTGTAT GGAACTTCGA ACGTTGTAAC ATTGTTTACA TAGCGGATGC ATCTTTTGCG

      3930       3940       3950       3960       3970       3980       3990
GCTGCCCCCA AAGATATATA CCCCTCATGC AAAATATCAA ACACCTGCCC ACCTGACATT CAAAATAAAA
CGACGGGGGT TTCTATATAT GGGGAGTACG TTTTATAGTT TGTGGACGGG TGGACTGTAA GTTTTATTTT
```

FIGURE 3. (*Continued*)

```
      4000       4010       4020       4030       4040       4050       4060
TTGAGCATAC AACAATTGCT GATAAAATAT TGCAATATGG CAGTCTGGGA GTTTTTTTGG GAGGTTTGGG
AACTCGTATG TTGTTAACGA CTATTTTATA ACGTTATACC GTCAGACCCT CAAAAAAACC CTCCAAACCC

      4070       4080       4090       4100       4110       4120       4130
CATTGGAACA GCCAGAGGCT CTGGAGGAAG AATTGGTTAT ACTCCCCTCG GTGAGGGTGG TGGGGTTAGA
GTAACCTTGT CGGTCTCCGA GACCTCCTTC TTAACCAATA TGAGGGGAGC CACTCCCACC ACCCCAATCT

      4140       4150       4160       4170       4180       4190       4200
GTTGCTACTC GTCCAACTCC AGTAAGGCCT ACAATACCTG TGGAAACAGT AGGCCCCAGT GAAATTTTCC
CAACGATGAG CAGGTTGAGG TCATTCCGGA TGTTATGGAC ACCTTTGTCA TCCGGGGTCA CTTTAAAAGG

      4210       4220       4230       4240       4250       4260       4270
CCATAGATGT TGTAGATCCT ACAGGCCCTG CTGTTATTCC CCTACAAGAT TTAGGTAGAG ACTTCCCAAT
GGTATCTACA ACATCTAGGA TGTCCGGGAC GACAATAAGG GGATGTTCTA AATCCATCTC TGAAGGGTTA

      4280       4290       4300       4310       4320       4330       4340
ACCAACTGTG CAGGTTATTG CAGAAATTCA CCCTATTTCT GACATACCAA ACATTGTTGC ATCTTCAACA
TGGTTGACAC GTCCAATAAC GTCTTTAAGT GGGATAAAGA CTGTATGGTT TGTAACAACG TAGAAGTTGT

      4350       4360       4370       4380       4390       4400       4410
AATGAAGGAG AATCTGCCAT ATTAGATGTG TTACGAGGGA ATGCAACCAT ACGCACTGTT TCAAGAACAC
TTACTTCCTC TTAGACGGTA TAATCTACAC AATGCTCCCT TACGTTGGTA TGCGTGACAA AGTTCTTGTG

      4420       4430       4440       4450       4460       4470       4480
AATACAATAA CCCCTCTTTC ACTGTTGCAT CTACATCTAA TATAAGTGCT GGAGAAGCAT CAACATCAGA
TTATGTTATT GGGGAGAAAG TGACAACGTA GATGTAGATT ATATTCACGA CCTCTTCGTA GTTGTAGTCT

      4490       4500       4510       4520       4530       4540       4550
TATTGTATTT GTTAGCAATG GTTCAGGTGA CAGGGTGGTG GGCGAGGATA TCCCCTTGGT AGAATTAAAC
ATAACATAAA CAATCGTTAC CAAGTCCACT GTCCCACCAC CCGCTCCTAT AGGGGAACCA TCTTAATTTG

      4560       4570       4580       4590       4600       4610       4620
TTAGGCCTTG AAACAGACAC ATCTTCTGTT GTACAAGAAA CAGCATTTTC CAGCAGCACA CCAATTGCTG
AATCCGGAAC TTTGTCTGTG TAGAAGACAA CATGTTCTTT GTCGTAAAAG GTCGTCGTGT GGTTAACGAC

      4630       4640       4650       4660       4670       4680       4690
AAAGACCCTC TTTTAGGCCC TCAAGATTCT ATAATAGGCG TCTATATGAA CAGGTGCAAG TACAAGACCC
TTTCTGGGAG AAAATCCGGG AGTTCTAAGA TATTATCCGC AGATATACTT GTCCACGTTC ATGTTCTGGG

      4700       4710       4720       4730       4740       4750       4760
TAGGTTCGTT GAGCAGCCAC AGTCAATGGT CACTTTTGAT AATCCAGCAT TTGAGCCAGA GCTTGATGAG
ATCCAAGCAA CTCGTCGGTG TCAGTTACCA GTGAAAACTA TTAGGTCGTA AACTCGGTCT CGAACTACTC

      4770       4780       4790       4800       4810       4820       4830
GTGTCTATTA TCTTCCAAAG AGACTTAGAT GCTCTTGCTC AGACACCAGT GCCTGAATTT AGAGATGTAG
CACAGATAAT AGAAGGTTTC TCTGAATCTA CGAGAACGAG TCTGTGGTCA CGGACTTAAA TCTCTACATC

      4840       4850       4860       4870       4880       4890       4900
TTTATCTGAG CAAGCCCACA TTTTCGCGGG AACCAGGGGG ACGGTTAAGG GTTAGCCGCC TTGGCAAAAG
AAATAGACTC GTTCGGGTGT AAAAGCGCCC TTGGTCCCCC TGCCAATTCC CAATCGGCGG AACCGTTTTC

      4910       4920       4930       4940       4950       4960       4970
TTCAACTATT CGTACACGCC TGGGCACAGC AATTGGCGCC AGAACCCACT TTTTCTATGA TTTAAGTTCT
AAGTTGATAA GCATGTGCGG ACCCGTGTCG TTAACCGCGG TCTTGGGTGA AAAAGATACT AAATTCAAGA

      4980       4990       5000       5010       5020       5030       5040
ATTGCTCCAG AAGACTCAAT TGAATTATTG CCTTTAGGTG AGCATAGTCA AACAACAGTC ATTAGTTCCA
TAACGAGGTC TTCTGAGTTA ACTTAATAAC GGAAATCCAC TCGTATCAGT TTGTTGTCAG TAATCAAGGT

      5050       5060       5070       5080       5090       5100       5110
ACTTAGGTGA CACAGCATTT ATACAAGGTG AGACAGCAGA GGATGACTTA GAAGTTATCT CTTTAGAAAC
TGAATCCACT GTGTCGTAAA TATGTTCCAC TCTGTCGTCT CCTACTGAAT CTTCAATAGA GAAATCTTTG

      5120       5130       5140       5150       5160       5170       5180
ACCACAATTA TATTCAGAAG AAGAGCTTTT AGACACAAAC GAAAGTGTGG GCGAAAATTT GCAACTTACT
TGGTGTTAAT ATAAGTCTTC TTCTCGAAAA TCTGTGTTTG CTTTCACACC CGCTTTTAAA CGTTGAATGA

      5190       5200       5210       5220       5230       5240       5250
ATTACTAACT CAGAGGGTGA GGTTTCTATA CTAGATTTAA CACAAAGCAG AGTCAGGCCA CCTTTTGGCA
TAATGATTGA GTCTCCCACT CCAAAGATAT GATCTAAATT GTGTTTCGTC TCAGTCCGGT GGAAAACCGT

      5260       5270       5280       5290       5300       5310       5320
CTGAAGATAC TAGCTTGCAT GTATATTACC CAAATTCTTC TAAAGGGACT CCAATAATTA ATCCTGAAGA
GACTTCTATG ATCGAACGTA CATATAATGG GTTTAAGAAG ATTTCCCTGA GGTTATTAAT TAGGACTTCT
```

FIGURE 3. (*Continued*)

```
      5330       5340       5350       5360       5370       5380       5390
ATCATTTACA CCTTTGGTTA TTATAGCTCT TAACAACTCA ACAGGGGATT TTGAGTTACA TCCTAGTCTT
TAGTAAATGT GGAAACCAAT AATATCGAGA ATTGTTGAGT TGTCCCCTAA AACTCAATGT AGGATCAGAA

      5400       5410       5420       5430       5440       5450       5460
AGAAAGCGTC GTAAAAGAGC TTATGTATAA TGTTTTTCAG ATGGCTGTCT GGTTACCAGC GCAGAATAAG
TCTTTCGCAG CATTTTCTCG AATACATATT ACAAAAAGTC TACCGACAGA CCAATGGTCG CGTCTTATTC

      5470       5480       5490       5500       5510       5520       5530
TTCTATCTTC CTCCCCAGCC CATCACTAGA ATCCTGTCCA CTGATGAATA TGTAACCAGA ACCAATCTCT
AAGATAGAAG GAGGGGTCGG GTAGTGATCT TAGGACAGGT GACTACTTAT ACATTGGTCT TGGTTAGAGA

      5540       5550       5560       5570       5580       5590       5600
TCTACCATGC AACATCTGAA CGTCTACTGC TGGTCGGACA TCCTTTGTTT GAGATCTCCA GTAATCAAAC
AGATGGTACG TTGTAGACTT GCAGATGACG ACCAGCCTGT AGGAAACAAA CTCTAGAGGT CATTAGTTTG

      5610       5620       5630       5640       5650       5660       5670
TGTAACTATA CCAAAAGTGT CACCAAATGC ATTTAGAGTT TTTAGGGTGC GTTTTGCTGA TCCAAATAGA
ACATTGATAT GGTTTTCACA GTGGTTTACG TAAATCTCAA AAATCCCACG CAAAACGACT AGGTTTATCT

      5680       5690       5700       5710       5720       5730       5740
TTTGCATTTG GGGATAAGGC AATTTTTAAT CCAGAAACAG AAAGATTAGT TTGGGGCCTA AGAGGGATAG
AAACGTAAAC CCCTATTCCG TTAAAAATTA GGTCTTTGTC TTTCTAATCA AACCCCGGAT TCTCCCTATC

      5750       5760       5770       5780       5790       5800       5810
AGATAGGTAG AGGCCAGCCT TTAGGTATAG GAATAACGGG CCACCCTCTT TTAAATAAGT TAGATGATGC
TCTATCCATC TCCGGTCGGA AATCCATATC CTTATTGCCC GGTGGGAGAA AATTTATTCA ATCTACTACG

      5820       5830       5840       5850       5860       5870       5880
AGAAAATCCA ACAAATTATA TTAATACTCA TGCAAATGGA GATTCTAGAC AAAATACTGC TTTTGATGCA
TCTTTTAGGT TGTTTAATAT AATTATGAGT ACGTTTACCT CTAAGATCTG TTTTATGACG AAAACTACGT

      5890       5900       5910       5920       5930       5940       5950
AAACAGACAC AAATGTTCCT CGTCGGCTGT ACTCCTGCTT CAGGTGAACA CTGGACAAGT CGTCGTTGCC
TTTGTCTGTG TTTACAAGGA GCAGCCGACA TGAGGACGAA GTCCACTTGT GACCTGTTCA GCAGCAACGG

      5960       5970       5980       5990       6000       6010       6020
CAGGGGAACA AGTGAAACTT GGGGACTGCC CCAGGGTGCA AATGATAGAG TCTGTCATAG AAGATGGTGA
GTCCCCTTGT TCACTTTGAA CCCCTGACGG GGTCCCACGT TTACTATCTC AGACAGTATC TTCTACCACT

      6030       6040       6050       6060       6070       6080       6090
CATGATGGAT ATTGGTTTTG GGGCTATGGA TTTTGCTGCT TTACAGCAAG ACAAGTCTGA TGTCCCTTTA
GTACTACCTA TAACCAAAAC CCCGATACCT AAAACGACGA AATGTCGTTC TGTTCAGACT ACAGGGAAAT

      6100       6110       6120       6130       6140       6150       6160
GATGTTGTTC AAGCAACATG CAAATATCCT GATTATATCA GAATGAACCA TGAAGCCTAT GGCAACTCTA
CTACAACAAG TTCGTTGTAC GTTTATAGGA CTAATATAGT CTTACTTGGT ACTTCGGATA CCGTTGAGAT

      6170       6180       6190       6200       6210       6220       6230
TGTTTTTTTT TGCACGTCGC GAGCAAATGT ATACCAGGCA CTTTTTTACT CGCGGGGGTT CGGTGGGTGA
ACAAAAAAAA ACGTGCAGCG CTCGTTTACA TATGGTCCGT GAAAAAATGA GCGCCCCCAA GCCACCCACT

      6240       6250       6260       6270       6280       6290       6300
TAAGGAGGCA GTCCCACAAA GCCTGTATTT AACAGCAGAT GCTGAACCAA GAACAACTTT AGCAACAACA
ATTCCTCCGT CAGGGTGTTT CGGACATAAA TTGTCGTCTA CGACTTGGTT CTTGTTGAAA TCGTTGTTGT

      6310       6320       6330       6340       6350       6360       6370
AATTATGTAG GCACACCAAG TGGCTCTATG GTTTCATCTG ATGTCCAATT GTTTAATAGA TCTTACTGGC
TTAATACATC CGTGTGGTTC ACCGAGATAC CAAAGTAGAC TACAGGTTAA CAAATTATCT AGAATGACCG

      6380       6390       6400       6410       6420       6430       6440
TTCAGCGATG TCAAGGCCAG AATAATGGCA TTTGCTGGAG AAACCAGTTA TTTATTACAG TTGGAGATAA
AAGTCGCTAC AGTTCCGGTC TTATTACCGT AAACGACCTC TTTGGTCAAT AAATAATGTC AACCTCTATT

      6450       6460       6470       6480       6490       6500       6510
TACCAGAGGA ACAAGTTTAT CTATCAGTAT GAAAAACAAT GCAAGTACTA CATATTCCAA TGCTAATTTT
ATGGTCTCCT TGTTCAAATA GATAGTCATA CTTTTTGTTA CGTTCATGAT GTATAAGGTT ACGATTAAAA

      6520       6530       6540       6550       6560       6570       6580
AATGATTTTC TAAGACATAC TGAAGAATTT GATCTTTCTT TTATAGTTCA GCTTTGTAAA GTAAAGTTAA
TTACTAAAAG ATTCTGTATG ACTTCTTAAA CTAGAAAGAA AATATCAAGT CGAAACATTT CATTTCAATT

      6590       6600       6610       6620       6630       6640       6650
CTCCCGAAAA TCTAGCCTAC ATTCATACAA TGGACCCTAA TATTTTAGAG GATTGGCAAC TATCTGTATC
GAGGGCTTTT AGATCGGATG TAAGTATGTT ACCTGGGATT ATAAAATCTC CTAACCGTTG ATAGACATAG
```

FIGURE 3. (*Continued*)

```
      6660       6670       6680       6690       6700       6710       6720
TCAACCACCT ACCAATCCTC TAGAAGATCA ATATAGGTTT TTAGGGTCTT CCTTGGCAGC AAAATGTCCA
AGTTGGTGGA TGGTTAGGAG ATCTTCTAGT TATATCCAAA AATCCCAGAA GGAACCGTCG TTTTACAGGT

      6730       6740       6750       6760       6770       6780       6790
GAACAGGCGC CTCCTGAGCC CCAGACTGAT CCTTATAGTC AATATAAATT CTGGGAAGTC GATCTCACAG
CTTGTCCGCG GAGGACTCGG GGTCTGACTA GGAATATCAG TTATATTTAA GACCCTTCAG CTAGAGTGTC

      6800       6810       6820       6830       6840       6850       6860
AAAGGATGTC CGAACAATTA GACCAATTTC CACTAGGAAG GAAATTTCTA TATCAAAGTG GCATGACACA
TTTCCTACAG GCTTGTTAAT CTGGTTAAAG GTGATCCTTC CTTTAAAGAT ATAGTTTCAC CGTACTGTGT

      6870       6880       6890       6900       6910       6920       6930
ACGTACTGCT ACTAGTTCCA CCACAAAGCG CAAAACAGTG CGTGTATCTA CGTCAGCCAA GCGCAGGCGT
TGCATGACGA TGATCAAGGT GGTGTTTCGC GTTTTGTCAC GCACATAGAT GCAGTCGGTT CGCGTCCGCA

      6940       6950       6960       6970       6980       6990       7000
AAGGCTTAGT ATATATTATA TATAACTATA TTTATTAGTA GATTATTTAT TATATATTTT TATATTTTTA
TTCCGAATCA TATATAATAT ATATTGATAT AAATAATCAT CTAATAAATA ATATATAAAA ATATAAAAAT

      7010       7020       7030       7040       7050       7060       7070
TACTTTTTAT ACTTGTTTAG TTCTAAATAG ACATGTAAGA TTTACATTAG TATAAGTAGG CATGTATTTA
ATGAAAAATA TGAACAAATC AAGATTTATC TGTACATTCT AAATGTAATC ATATTCATCC GTACATAAAT

      7080       7090       7100       7110       7120       7130       7140
CATAAAATAG TCTTGGAAAC CTTTTATTAG TGAACCATCA TTTACAATAG TGACATCATA GTTCATCTGC
GTATTTTATC AGAACCTTTG GAAAATAATC ACTTGGTAGT AAATGTTATC ACTGTAGTAT CAAGTAGACG

      7150       7160       7170       7180       7190       7200       7210
AATTGCTATT CCATCGTTCT TCACATATTC TACAGTAGTG TTCTCTAGAT TGTATTGCTA TTTTCCTGTT
TTAACGATAA GGTAGCAAGA AGTGTATAAG ATGTCATCAC AAGAGATCTA ACATAACGAT AAAAGGACAA

      7220       7230       7240       7250       7260       7270       7280
AGGCAAACAA CAACATCTGT ACATGGACCA AACAACCCAC TTTCATTTTA TTGTGCTGCA TATATTCCAG
TCCGTTTGTT GTTGTAGACA TGTACCTGGT TTGTTGGGTG AAAGTAAAAT AACACGACGT ATATAAGGTC

      7290       7300       7310       7320       7330       7340       7350
ATTGTTGAGG ATTTATTTGT TTAGACTCCG GTGCATTATA CACAAGTGTG CATTTTTTGT GTTCTCTGAT
TAACAACTCC TAAATAAACA AATCTGAGGC CACGTAATAT GTGTTCACAC GTAAAAAACA CAAGAGACTA

      7360       7370       7380       7390       7400       7410       7420
TGATTGTGTG TTATTTTCCT GCAATATGCA ATAAAAGTGA GCTGTCCTTT CTTTTTGTTA ATCCCTCCCT
ACTAACACAC AATAAAAGGA CGTTATACGT TATTTTCACT CGACAGGAAA GAAAAACAAT TAGGGAGGGA

      7430       7440       7450       7460       7470       7480       7490
ACTCCAATAA AAAATCCCTA CCCCTAAAAT CTGTTTGTGC TGGTTTTATT AATAATTGCG CTCTTTTATA
TGAGGTTATT TTTTAGGGAT GGGGATTTTA GACAAACACG ACCAAAATAA TTATTAACGC GAGAAAATAT

      7500       7510       7520       7530       7540       7550       7560
TAATAAGTAC TATTAACACC GCACCCGTTG TGGCTAATCC CTTATGGTAT TTAAAAGACT ACACCTACAG
ATTATTCATG ATAATTGTGG CGTGGGCAAC ACCGATTAGG GAATACCATA AATTTTCTGA TGTGGATGTC

      7570       7580       7590       7600       7610       7620       7630
GATGTATTGT CTTCATTGTT TATGGTTTAC CGCGCTCCAA AGACGGTTTG CCCAAAGACG GTTTGCCAAC
CTACATAACA GAAGTAACAA ATACCAAATG GCGCGAGGTT TCTGCCAAAC GGGTTTCTGC CAAACGGTTG

      7640       7650       7660       7670       7680       7690       7700
CGCGGTTAGG ACTTGTTTCA ATTTGCTGCC AAACTTATCT GGTCGTGCTC CAACGGGTTT CCTGCCAAGC
GCGCCAATCC TGAACAAAGT TAAACGACGG TTTGAATAGA CCAGCACGAG GTTGCCCAAA GGACGGTTCG

      7710       7720       7730       7740       7750       7760       7770
ACCTAAAACG GTAGGTGTGT ACTCTTTTCA AGAATTAACA AAGGAGATTT CTCCCGCCAA ATTAGTTTCG
TGGATTTTGC CATCCACACA TGAGAAAAGT TCTTAATTGT TTCCTCTAAA GAGGGCGGTT TAATCAAAGC

      7780       7790       7800       7810
AGCGACCGAA TTCGGTCGTA AAAATCTAAA GTGATGATTG TTGTT
TCGCTGGCTT AAGCCAGCAT TTTTAGATTT CACTACTAAC AACAA
```

FIGURE 3. (*Continued*)

```
        10         20         30         40         50         60         70
GTTAATAACA ATCTTGGTTT AAAAAATAGG AGGGACCGAA AACGGTTCAA CCGAAAACGG TTGTATATAA
CAATTATTGT TAGAACCAAA TTTTTTATCC TCCCTGGCTT TTGCCAAGTT GGCTTTTGCC AACATATATT

        80         90        100        110        120        130        140
ACCAGCCCTA AAATTTAGCA AACGAGGCAT TATGGAAAGT GCAAATGCCT CCACGTCTGC AACGACCATA
TGGTCGGGAT TTTAAATCGT TTGCTCCGTA ATACCTTTCA CGTTTACGGA GGTGCAGACG TTGCTGGTAT

       150        160        170        180        190        200        210
GACCAGTTGT GCAAGACGTT TAATCTATCT ATGCATACGT TGCAAATTAA TTGTGTGTTT TGCAAGAATG
CTGGTCAACA CGTTCTGCAA ATTAGATAGA TACGTATGCA ACGTTTAATT AACACACAAA ACGTTCTTAC

       220        230        240        250        260        270        280
CACTGACCAC AGCAGAGATT TATTCATATG CATATAAACA CCTAAAGGTC CTGTTTCGAG GCGGCTATCC
GTGACTGGTG TCGTCTCTAA ATAAGTATAC GTATATTTGT GGATTTCCAG GACAAAGCTC CGCCGATAGG

       290        300        310        320        330        340        350
ATATGCAGCC TGCGCGTGCT GCCTAGAATT TCATGGAAAA ATAAACCAAT ATAGACACTT TGATTATGCT
TATACGTCGG ACGCGCACGA CGGATCTTAA AGTACCTTTT TATTTGGTTA TATCTGTGAA ACTAATACGA

       360        370        380        390        400        410        420
GGATATGCAA CAACAGTTGA AGAAGAAACT AAACAAGACA TCTTAGACGT GCTAATTCGG TGCTACCTGT
CCTATACGTT GTTGTCAACT TCTTCTTTGA TTTGTTCTGT AGAATCTGCA CGATTAAGCC ACGATGGACA

       430        440        450        460        470        480        490
GTCACAAACC GCTGTGTGAA GTAGAAAAGG TAAAACATAT ACTAACCAAG GCGCGGTTCA TAAAGCTAAA
CAGTGTTTGG CGACACACTT CATCTTTTCC ATTTTGTATA TGATTGGTTC CGCGCCAAGT ATTTCGATTT

       500        510        520        530        540        550        560
TTGTACGTGG AAGGGTCGCT GCCTACACTG CTGGACAACA TGCATGGAAG ACATGTTACC CTAAAGGATA
AACATGCACC TTCCCAGCGA CGGATGTGAC GACCTGTTGT ACGTACCTTC TGTACAATGG GATTTCCTAT

       570        580        590        600        610        620        630
TTGTATTAGA CCTGCAACCT CCAGACCCTG TAGGGTTACA TTGCTATGAG CAATTAGTAG ACAGCTCAGA
AACATAATCT GGACGTTGGA GGTCTGGGAC ATCCCAATGT AACGATACTC GTTAATCATC TGTCGAGTCT

       640        650        660        670        680        690        700
AGATGAGGTG GACGAAGTGG ACGGACAAGA TTCACAACCT TTAAAACAAC ATTTCCAAAT AGTGACCTGT
TCTACTCCAC CTGCTTCACC TGCCTGTTCT AAGTGTTGGA AATTTTGTTG TAAAGGTTTA TCACTGGACA

       710        720        730        740        750        760        770
TGCTGTGGAT GTGACAGCAA CGTTCGACTG GTTGTGCAGT GTACAGAAAC AGACATCAGA GAAGTGCAAC
ACGACACCTA CACTGTCGTT GCAAGCTGAC CAACACGTCA CATGTCTTTG TCTGTAGTCT CTTCACGTTG

       780        790        800        810        820        830        840
AGCTTCTGTT GGGAACACTA AACATAGTGT GTCCCATCTG CGCACCGAAG ACCTAACAAC GATGGCGGAC
TCGAAGACAA CCCTTGTGAT TTGTATCACA CAGGGTAGAC GCGTGGCTTC TGGATTGTTG CTACCGCCTG

       850        860        870        880        890        900        910
GATTCAGGTA CAGAAAATGA GGGGTCTGGG TGTACAGGAT GGTTTATGGT AGAAGCTATA GTGCAACACC
CTAAGTCCAT GTCTTTTACT CCCCAGACCC ACATGTCCTA CCAAATACCA TCTTCGATAT CACGTTGTGG

       920        930        940        950        960        970        980
CAACAGGTAC ACAAATATCA GACGATGAGG ATGAGGAGGT GGAGGACAGT GGGTATGACA TGGTGGACTT
GTTGTCCATG TGTTTATAGT CTGCTACTCC TACTCCTCCA CCTCCTGTCA CCCATACTGT ACCACCTGAA

       990       1000       1010       1020       1030       1040       1050
TATTGATGAC AGCAATATTA CACACAATTC ACTGGAAGCA CAGGCATTGT TTAACAGGCA GGAGGCGGAC
ATAACTACTG TCGTTATAAT GTGTGTTAAG TGACCTTCGT GTCCGTAACA AATTGTCCGT CCTCCGCCTG

      1060       1070       1080       1090       1100       1110       1120
ACCCATTATG CGACTGTGCA GGACCTAAAA CGAAAGTATT TAGGTAGTCC ATATGTTAGT CCTATAAACA
TGGGTAATAC GCTGACACGT CCTGGATTTT GCTTTCATAA ATCCATCAGG TATACAATCA GGATATTTGT

      1130       1140       1150       1160       1170       1180       1190
CTATAGCCGA GGCAGTGGAA AGTGAAATAA GTCCACGATT GGACGCCATT AAACTTACAA GACAGCCAAA
GATATCGGCT CCGTCACCTT TCACTTTATT CAGGTGCTAA CCTGCGGTAA TTTGAATGTT CTGTCGGTTT

      1200       1210       1220       1230       1240       1250       1260
AAAGGTAAAG CGACGGCTGT TTCAAACCAG GGAACTAACG GACAGTGGAT ATGGCTATTC TGAAGTGGAA
TTTCCATTTC GCTGCCGACA AAGTTTGGTC CCTTGATTGC CTGTCACCTA TACCGATAAG ACTTCACCTT

      1270       1280       1290       1300       1310       1320       1330
GCTGGAACGG GAACGCAGGT AGAGAAACAT GGCGTACCGG AAAATGGGGG AGATGGTCAG GAAAAGGACA
CGACCTTGCC CTTGCGTCCA TCTCTTTGTA CCGCATGGCC TTTTACCCCC TCTACCAGTC CTTTTCCTGT
```

FIGURE 4. DNA sequence of HPV-6b.

```
      1340       1350       1360       1370       1380       1390       1400
CAGGAAGGGA CATAGAGGGG GAGGAACATA CAGAGGCGGA AGCGCCCACA AACAGTGTAC GGGAGCATGC
GTCCTTCCCT GTATCTCCCC CTCCTTGTAT GTCTCCGCCT TCGCGGGTGT TTGTCACATG CCCTCGTACG

      1410       1420       1430       1440       1450       1460       1470
AGGCACAGCA GGAATATTGG AATTGTTAAA ATGTAAAGAT TTACGGGCAG CATTACTTGG TAAGTTTAAA
TCCGTGTCGT CCTTATAACC TTAACAATTT TACATTTCTA AATGCCCGTC GTAATGAACC ATTCAAATTT

      1480       1490       1500       1510       1520       1530       1540
GAATGCTTTG GGCTGTCTTT TATAGATTTA ATTAGGCCAT TTAAAAGTGA TAAAACAACA TGTTTAGATT
CTTACGAAAC CCGACAGAAA ATATCTAAAT TAATCCGGTA AATTTTCACT ATTTTGTTGT ACAAATCTAA

      1550       1560       1570       1580       1590       1600       1610
GGGTGGTAGC AGGGTTTGGT ATACATCATA GCATATCAGA GGCATTTCAA AAATTAATTG AGCCATTAAG
CCCACCATCG TCCCAAACCA TATGTAGTAT CGTATAGTCT CCGTAAAGTT TTTAATTAAC TCGGTAATTC

      1620       1630       1640       1650       1660       1670       1680
TTTATATGCA CATATACAAT GGCTAACAAA TGCATGGGGA ATGGTATTGT TAGTATTATT AAGATTTAAA
AAATATACGT GTATATGTTA CCGATTGTTT ACGTACCCCT TACCATAACA ATCATAATAA TTCTAAATTT

      1690       1700       1710       1720       1730       1740       1750
GTAAATAAAA GTAGAAGTAC CGTTGCACGT ACACTTGCAA CGCTATTAAA TATACCTGAA AACCAAATGT
CATTTATTTT CATCTTCATG GCAACGTGCA TGTGAACGTT GCGATAATTT ATATGGACTT TTGGTTTACA

      1760       1770       1780       1790       1800       1810       1820
TAATAGAGCC ACCAAAAATA CAAAGTGGTG TTGCAGCCCT GTATTGGTTT CGTACAGGTA TATCAAATGC
ATTATCTCGG TGGTTTTTAT GTTTCACCAC AACGTCGGGA CATAACCAAA GCATGTCCAT ATAGTTTACG

      1830       1840       1850       1860       1870       1880       1890
CAGTACAGTT ATAGGGGAAG CACCAGAATG GATAACACGC CAAACAGTTA TTGAACACGG GTTGGCAGAC
GTCATGTCAA TATCCCCTTC GTGGTCTTAC CTATTGTGCG GTTTGTCAAT AACTTGTGCC CAACCGTCTG

      1900       1910       1920       1930       1940       1950       1960
AGTCAGTTTA AATTAACAGA AATGGTGCAG TGGGCGTATG ATAATGACAT ATGCGAGGAG AGTGAAATTG
TCAGTCAAAT TTAATTGTCT TTACCACGTC ACCCGCATAC TATTACTGTA TACGCTCCTC TCACTTTAAC

      1970       1980       1990       2000       2010       2020       2030
CATTTGAATA TGCACAAAGG GGAGATTTTG ATTCTAATGC ACGAGCATTT TTAAATAGCA ATATGCAGGC
GTAAACTTAT ACGTGTTTCC CCTCTAAAAC TAAGATTACG TGCTCGTAAA AATTTATCGT TATACGTCCG

      2040       2050       2060       2070       2080       2090       2100
AAAATATGTG AAAGATTGTG CAACTATGTG TAGACATTAT AAACATGCAG AAATGAGGAA GATGTCTATA
TTTTATACAC TTTCTAACAC GTTGATACAC ATCTGTAATA TTTGTACGTC TTTACTCCTT CTACAGATAT

      2110       2120       2130       2140       2150       2160       2170
AAACAATGGA TAAAACATAG GGGTTCTAAA ATAGAAGGCA CAGGAAATTG GAAACCAATT GTACAATTCC
TTTGTTACCT ATTTTGTATC CCCAAGATTT TATCTTCCGT GTCCTTTAAC CTTTGGTTAA CATGTTAAGG

      2180       2190       2200       2210       2220       2230       2240
TACGACATCA AAATATAGAA TTCATTCCTT TTTTAACTAA ATTTAAATTA TGGCTGCACG GTACGCCAAA
ATGCTGTAGT TTTATATCTT AAGTAAGGAA AAAATTGATT TAAATTTAAT ACCGACGTGC CATGCGGTTT

      2250       2260       2270       2280       2290       2300       2310
AAAAAACTGC ATAGCCATAG TAGGCCCTCC AGATACTGGG AAATCGTACT TTTGTATGAG TTTAATAAGC
TTTTTTGACG TATCGGTATC ATCCGGGAGG TCTATGACCC TTTAGCATGA AAACATACTC AAATTATTCG

      2320       2330       2340       2350       2360       2370       2380
TTTCTAGGAG GTACAGTTAT TAGTCATGTA AATTCCAGCA GCCATTTTTG GTTGCAACCG TTAGTAGATG
AAAGATCCTC CATGTCAATA ATCAGTACAT TTAAGGTCGT CGGTAAAAAC CAACGTTGGC AATCATCTAC

      2390       2400       2410       2420       2430       2440       2450
CTAAGGTAGC ATTGTTAGAT GATGCAACAC AGCCATGTTG GATATATATG GATACATATA TGAGAAATTT
GATTCCATCG TAACAATCTA CTACGTTGTG TCGGTACAAC CTATATATAC CTATGTATAT ACTCTTTAAA

      2460       2470       2480       2490       2500       2510       2520
GTTAGATGGT AATCCTATGA GTATTGACAG AAAGCATAAA GCATTGACAT TAATTAAATG TCCACCTCTG
CAATCTACCA TTAGGATACT CATAACTGTC TTTCGTATTT CGTAACTGTA ATTAATTTAC AGGTGGAGAC

      2530       2540       2550       2560       2570       2580       2590
CTAGTAACGT CCAACATAGA TATTACTAAA GAAGATAAAT ATAAGTATTT ACATACTAGA GTAACAACAT
GATCATTGCA GGTTGTATCT ATAATGATTT CTTCTATTTA TATTCATAAA TGTATGATCT CATTGTTGTA

      2600       2610       2620       2630       2640       2650       2660
TTACATTTCC AAATCCATTC CCTTTTGACA GAAATGGGAA TGCAGTGTAT GAACTGTCAA ATACAAACTG
AATGTAAAGG TTTAGGTAAG GGAAAACTGT CTTTACCCTT ACGTCACATA CTTGACAGTT TATGTTTGAC
```

FIGURE 4. (*Continued*)

```
      2670       2680       2690       2700       2710       2720       2730
GAAATGTTTT TTTGAAAGAC TGTCGTCAAG CCTAGACATT CAGGATTCTG AGGACGAGGA AGATGGAAGC
CTTTACAAAA AAACTTTCTG ACAGCAGTTC GGATCTGTAA GTCCTAAGAC TCCTGCTCCT TCTACCTTCG

      2740       2750       2760       2770       2780       2790       2800
AATAGCCAAG CGTTTAGATG CGTGCCAGGA ACAGTTGTTA GAACTTTATG AAGAAAACAG TACTGACCTA
TTATCGGTTC GCAAATCTAC GCACGGTCCT TGTCAACAAT CTTGAAATAC TTCTTTTGTC ATGACTGGAT

      2810       2820       2830       2840       2850       2860       2870
CACAAACATG TATTGCATTG GAAATGCATG AGACATGAAA GTGTATTATT ATATAAAGCA AAACAAATGG
GTGTTTGTAC ATAACGTAAC CTTTACGTAC TCTGTACTTT CACATAATAA TATATTTCGT TTTGTTTACC

      2880       2890       2900       2910       2920       2930       2940
GCCTAAGCCA CATAGGAATG CAAGTAGTGC CACCATTAAA GGTGTCCGAA GCAAAAGGAC ATAATGCCAT
CGGATTCGGT GTATCCTTAC GTTCATCACG GTGGTAATTT CCACAGGCTT CGTTTTCCTG TATTACGGTA

      2950       2960       2970       2980       2990       3000       3010
TGAAATGCAA ATGCATTTAG AATCATTATT AAGGACTGAG TATAGTATGG AACCGTGGAC ATTACAAGAA
ACTTTACGTT TACGTAAATC TTAGTAATAA TTCCTGACTC ATATCATACC TTGGCACCTG TAATGTTCTT

      3020       3030       3040       3050       3060       3070       3080
ACAAGTTATG AAATGTGGCA AACACCACCT AAACGCTGTT TTAAAAAACG GGGCAAAACT GTAGAAGTTA
TGTTCAATAC TTTACACCGT TTGTGGTGGA TTTGCGACAA AATTTTTTGC CCCGTTTTGA CATCTTCAAT

      3090       3100       3110       3120       3130       3140       3150
AATTTGATGG CTGTGCAAAC AATACAATGG ATTATGTGGT ATGGACAGAT GTGTATGTGC AGGACAATGA
TTAAACTACC GACACGTTTG TTATGTTACC TAATACACCA TACCTGTCTA CACATACACG TCCTGTTACT

      3160       3170       3180       3190       3200       3210       3220
CACCTGGGTA AAGGTGCATA GTATGGTAGA TGCTAAGGGT ATATATTACA CATGTGGACA ATTTAAAACA
GTGGACCCAT TTCCACGTAT CATACCATCT ACGATTCCCA TATATAATGT GTACACCTGT TAAATTTTGT

      3230       3240       3250       3260       3270       3280       3290
TATTATGTAA ACTTTGTAAA AGAGGCAGAA AAGTATGGGA GCACCAAACA TTGGGAAGTA TGTTATGGCA
ATAATACATT TGAAACATTT TCTCCGTCTT TTCATACCCT CGTGGTTTGT AACCCTTCAT ACAATACCGT

      3300       3310       3320       3330       3340       3350       3360
GCACAGTTAT ATGTTCTCCT GCATCTGTAT CTAGCACTAC ACAAGAAGTA TCCATTCCTG AATCTACTAC
CGTGTCAATA TACAAGAGGA CGTAGACATA GATCGTGATG TGTTCTTCAT AGGTAAGGAC TTAGATGATG

      3370       3380       3390       3400       3410       3420       3430
ATACACCCCC GCACAGACCT CCACCCTTGT GTCCTCAAGC ACCAAGGAAG ACGCAGTGCA AACGCCGCCT
TATGTGGGGG CGTGTCTGGA GGTGGGAACA CAGGAGTTCG TGGTTCCTTC TGCGTCACGT TTGCGGCGGA

      3440       3450       3460       3470       3480       3490       3500
AGGAAACGAG CACGAGGAGT CCAACAGTCC CCTTGCAACG CCTTGTGTGT GGCCCACATT GGACCCGTGG
TCCTTTGCTC GTGCTCCTCA GGTTGTCAGG GGAACGTTGC GGAACACACA CCGGGTGTAA CCTGGGCACC

      3510       3520       3530       3540       3550       3560       3570
ACAGTGGAAA CCACAACCTC ATCACTAACA ATCACGACCA GCACCAAAGA CGGAACAACA GTAACAGTTC
TGTCACCTTT GGTGTTGGAG TAGTGATTGT TAGTGCTGGT CGTGGTTTCT GCCTTGTTGT CATTGTCAAG

      3580       3590       3600       3610       3620       3630       3640
AGCTACGCCT ATAGTGCAAT TTCAAGGTGA ATCCAATTGT TTAAAGTGTT TTAGATATAG GCTAAATGAC
TCGATGCGGA TATCACGTTA AAGTTCCACT TAGGTTAACA AATTTCACAA AATCTATATC CGATTTACTG

      3650       3660       3670       3680       3690       3700       3710
AGACACAGAC ATTTATTTGA TTTAATATCA TCAACGTGGC ACTGGGCCTC CTCAAAGGCA CCACATAAAC
TCTGTGTCTG TAAATAAACT AAATTATAGT AGTTGCACCG TGACCCGGAG GAGTTTCCGT GGTGTATTTG

      3720       3730       3740       3750       3760       3770       3780
ATGCCATTGT AACTGTAACA TATGATAGTG AGGAACAAAG GCAACAGTTT TTAGATGTTG TAAAAATACC
TACGGTAACA TTGACATTGT ATACTATCAC TCCTTGTTTC CGTTGTCAAA AATCTACAAC ATTTTTATGG

      3790       3800       3810       3820       3830       3840       3850
CCCTACCATT AGCCACAAAC TGGGATTTAT GTCACTGCAC CTATTGTAAT TTGTATATAT GTAAATGTGT
GGGATGGTAA TCGGTGTTTG ACCCTAAATA CAGTGACGTG GATAACATTA AACATATATA CATTTACACA

      3860       3870       3880       3890       3900       3910       3920
AAATATATGG TATTGGTGTA ATACAACTGT ACATGTATGG AAGTGGTGCC TGTACAAATA GCTGCAGGAA
TTTATATACC ATAACCACAT TATGTTGACA TGTACATACC TTCACCACGG ACATGTTTAT CGACGTCCTT

      3930       3940       3950       3960       3970       3980       3990
CAACCAGCAC ATTCATACTG CCTGTTATAA TTGCATTTGT TGTATGTTTT GTTAGCATCA TACTTATTGT
GTTGGTCGTG TAAGTATGAC GGACAATATT AACGTAAACA ACATACAAAA CAATCGTAGT ATGAATAACA
```

FIGURE 4. (*Continued*)

```
      4000       4010       4020       4030       4040       4050       4060
ATGGATATCT GAGTTTATTG TGTACACATC TGTGCTAGTA CTAACACTGC TTTTATATTT ACTATTGTGG
TACCTATAGA CTCAAATAAC ACATGTGTAG ACACGATCAT GATTGTGACG AAAATATAAA TGATAACACC

      4070       4080       4090       4100       4110       4120       4130
CTGCTATTAA CAACCCCCTT GCAATTTTTC CTACTAACTC TACTTGTGTG TTACTGTCCC GCATTGTATA
GACGATAATT GTTGGGGGAA CGTTAAAAAG GATGATTGAG ATGAACACAC AATGACAGGG CGTAACATAT

      4140       4150       4160       4170       4180       4190       4200
TACACTACTA TATTGTTACC ACACAGCAAT GATGCTAACA TGTCAATTTA ATGATGGAGA TACCTGGCTG
ATGTGATGAT ATAACAATGG TGTGTCGTTA CTACGATTGT ACAGTTAAAT TACTACCTCT ATGGACCGAC

      4210       4220       4230       4240       4250       4260       4270
GGTTTGTGGT TGTTATGTGC CTTTATTGTA GGGATGTTGG GGTTATTATT GATGCACTAT AGAGCTGTAC
CCAAACACCA ACAATACACG GAAATAACAT CCCTACAACC CCAATAATAA CTACGTGATA TCTCGACATG

      4280       4290       4300       4310       4320       4330       4340
AAGGGGATAA ACACACCAAA TGTAAGAAGT GTAACAAACA CAACTGTAAT GATGATTATG TAACTATGCA
TTCCCCTATT TGTGTGGTTT ACATTCTTCA CATTGTTTGT GTTGACATTA CTACTAATAC ATTGATACGT

      4350       4360       4370       4380       4390       4400       4410
TTATACTACT GATGGTGATT ATATATATAT GAATTAGAGT AAACCGTTTT TTATATTTGT AACAGTGTAT
AATATGATGA CTACCACTAA TATATATATA CTTAATCTCA TTTGGCAAAA AATATAAACA TTGTCACATA

      4420       4430       4440       4450       4460       4470       4480
GCTTTGTATA CCATGGCACA TAGTAGGGCC CGACGACGCA AGCGTGCGTC AGCTACACAG CTATATCAAA
CGAAACATAT GGTACCGTGT ATCATCCCGG GCTGCTGCGT TCGCACGCAG TCGATGTGTC GATATAGTTT

      4490       4500       4510       4520       4530       4540       4550
CATGTAAACT CACTGGAACA TGCCCCCCAG ATGTAATTCC TAAGGTGGAG CACAACACCA TTGCAGATCA
GTACATTTGA GTGACCTTGT ACGGGGGGTC TACATTAAGG ATTCCACCTC GTGTTGTGGT AACGTCTAGT

      4560       4570       4580       4590       4600       4610       4620
AATATTAAAA TGGGGAAGTT TGGGGGTGTT TTTTGGAGGG TTGGGTATAG GCACGGGTTC CGGCACTGGG
TTATAATTTT ACCCCTTCAA ACCCCCACAA AAAACCTCCC AACCCATATC CGTGCCCAAG GCCGTGACCC

      4630       4640       4650       4660       4670       4680       4690
GGTCGTACTG GCTATGTTCC CTTACAAACT TCTGCAAAAC CTTCTATTAC TAGTGGGCCT ATGGCTCGTC
CCAGCATGAC CGATACAAGG GAATGTTTGA AGACGTTTTG GAAGATAATG ATCACCCGGA TACCGAGCAG

      4700       4710       4720       4730       4740       4750       4760
CTCCTGTGGT GGTGGAGCCT GTGGCCCCTT CGGATCCATC TATTGTGTCT TTAATTGAAG AATCGGCAAT
GAGGACACCA CCACCTCGGA CACCGGGGAA GCCTAGGTAG ATAACACAGA AATTAACTTC TTAGCCGTTA

      4770       4780       4790       4800       4810       4820       4830
CATTAACGCA GGGGCGCCTG AAATTGTGCC CCCTGCACAC GGTGGGTTTA CAATTACATC CTCTGAAACA
GTAATTGCGT CCCCGCGGAC TTTAACACGG GGGACGTGTG CCACCCAAAT GTTAATGTAG GAGACTTTGT

      4840       4850       4860       4870       4880       4890       4900
ACTACCCCTG CAATATTGGA TGTATCAGTT ACTAGTCACA CTACTACTAG TATATTTAGA AATCCTGTCT
TGATGGGGAC GTTATAACCT ACATAGTCAA TGATCAGTGT GATGATGATC ATATAAATCT TTAGGACAGA

      4910       4920       4930       4940       4950       4960       4970
TTACAGAACC TTCTGTAACA CAACCCCAAC CACCCGTGGA GGCTAATGGA CATATATTAA TTTCTGCACC
AATGTCTTGG AAGACATTGT GTTGGGGTTG GTGGGCACCT CCGATTACCT GTATATAATT AAAGACGTGG

      4980       4990       5000       5010       5020       5030       5040
CACTGTAACG TCACACCCTA TAGAGGAAAT TCCTTTAGAT ACTTTTGTGG TATCATCTAG TGATAGCGGT
GTGACATTGC AGTGTGGGAT ATCTCCTTTA AGGAAATCTA TGAAAACACC ATAGTAGATC ACTATCGCCA

      5050       5060       5070       5080       5090       5100       5110
CCTACATCCA GTACCCCTGT TCCTGGTACT GCACCTCGGC CTCGTGTGGG CCTATATAGT CGTGCATTGC
GGATGTAGGT CATGGGGACA AGGACCATGA CGTGGAGCCG GAGCACACCC GGATATATCA GCACGTAACG

      5120       5130       5140       5150       5160       5170       5180
ACCAGGTGCA GGTTACAGAC CCTGCATTTC TTTCCACTCC TCAACGCTTA ATTACATATG ATAACCCTGT
TGGTCCACGT CCAATGTCTG GGACGTAAAG AAAGGTGAGG AGTTGCGAAT TAATGTATAC TATTGGGACA

      5190       5200       5210       5220       5230       5240       5250
ATATGAAGGG GAGGATGTTA GTGTACAATT TAGTCATGAT TCTATACACA ATGCACCTGA TGAGGCTTTT
TATACTTCCC CTCCTACAAT CACATGTTAA ATCAGTACTA AGATATGTGT TACGTGGACT ACTCCGAAAA

      5260       5270       5280       5290       5300       5310       5320
ATGGACATAA TTCGTTTGCA CAGACCTGCC ATTGCGTCCC GACGTGGCCT TGTGCGGTAC AGTCGCATTG
TACCTGTATT AAGCAAACGT GTCTGGACGG TAACGCAGGG CTGCACCGGA ACACGCCATG TCAGCGTAAC
```

FIGURE 4. (*Continued*)

```
      5330       5340       5350       5360       5370       5380       5390
GACAACGGGG GTCTATGCAC ACTCGCAGCG GAAAGCACAT AGGGGCCCGC ATTCATTATT TTTATGATAT
CTGTTGCCCC CAGATACGTG TGAGCGTCGC CTTTCGTGTA TCCCCGGGCG TAAGTAATAA AAATACTATA

      5400       5410       5420       5430       5440       5450       5460
TTCACCTATT GCACAGGCTG CAGAAGAAAT AGAAATGCAC CCTCTTGTGG CTGCACAGGA TGATACATTT
AAGTGGATAA CGTGTCCGAC GTCTTCTTTA TCTTTACGTG GGAGAACACC GACGTGTCCT ACTATGTAAA

      5470       5480       5490       5500       5510       5520       5530
GATATTTATG CTGAATCTTT TGAACCTGGC ATTAACCCTA CCCAACACCC TGTTACAAAT ATATCAGATA
CTATAAATAC GACTTAGAAA ACTTGGACCG TAATTGGGAT GGGTTGTGGG ACAATGTTTA TATAGTCTAT

      5540       5550       5560       5570       5580       5590       5600
CATATTTAAC TTCCACACCT AATACAGTTA CACAACCGTG GGGTAACACC ACAGTTCCAT TGTCACTTCC
GTATAAATTG AAGGTGTGGA TTATGTCAAT GTGTTGGCAC CCCATTGTGG TGTCAAGGTA ACAGTGAAGG

      5610       5620       5630       5640       5650       5660       5670
TAATGACCTG TTTTTACAAT CTGGCCCTGA TATAACTTTT CCTACTGCAC CTATGGGAAC ACCCTTTAGT
ATTACTGGAC AAAAATGTTA GACCGGGACT ATATTGAAAA GGATGACGTG GATACCCTTG TGGGAAATCA

      5680       5690       5700       5710       5720       5730       5740
CCTGTAACTC CTGCTTTACC TACAGGCCCT GTTTTCATTA CAGGTTCTGG ATTTTATTTG CATCCTGCAT
GGACATTGAG GACGAAATGG ATGTCCGGGA CAAAAGTAAT GTCCAAGACC TAAAATAAAC GTAGGACGTA

      5750       5760       5770       5780       5790       5800       5810
GGTATTTTGC ACGTAAACGC CGTAAACGTA TTCCCTTATT TTTTTCAGAT GTGGCGGCCT AGCGACAGCA
CCATAAAACG TGCATTTGCG GCATTTGCAT AAGGGAATAA AAAAAGTCTA CACCGCCGGA TCGCTGTCGT

      5820       5830       5840       5850       5860       5870       5880
CAGTATATGT GCCTCCTCCT AACCCTGTAT CCAAAGTTGT TGCCACGGAT GCTTATGTTA CTCGCACCAA
GTCATATACA CGGAGGAGGA TTGGGACATA GGTTTCAACA ACGGTGCCTA CGAATACAAT GAGCGTGGTT

      5890       5900       5910       5920       5930       5940       5950
CATATTTTAT CATGCCAGCA GTTCTAGACT TCTTGCAGTG GGACATCCTT ATTTTTCCAT AAAACGGGCT
GTATAAAATA GTACGGTCGT CAAGATCTGA AGAACGTCAC CCTGTAGGAA TAAAAAGGTA TTTTGCCCGA

      5960       5970       5980       5990       6000       6010       6020
AACAAAACTG TTGTGCCAAA GGTGTCAGGA TATCAATACA GGGTATTTAA GGTGGTGTTA CCAGATCCTA
TTGTTTTGAC AACACGGTTT CCACAGTCCT ATAGTTATGT CCCATAAATT CCACCACAAT GGTCTAGGAT

      6030       6040       6050       6060       6070       6080       6090
ACAAATTTGC ATTGCCTGAC TCGTCTCTTT TCGATCCCAC AACACAACGT TTAGTATGGG CATGCACAGG
TGTTTAAACG TAACGGACTG AGCAGAGAAA AGCTAGGGTG TTGTGTTGCA AATCATACCC GTACGTGTCC

      6100       6110       6120       6130       6140       6150       6160
CCTAGAGGTG GGCAGGGGAC AGCCATTAGG TGTGGGTGTA AGTGGACATC CTTTCCTAAA TAAATATGAT
GGATCTCCAC CCGTCCCCTG TCGGTAATCC ACACCCACAT TCACCTGTAG GAAAGGATTT ATTTATACTA

      6170       6180       6190       6200       6210       6220       6230
GATGTTGAAA ATTCAGGGAG TGGTGGTAAC CCTGGACAGG ATAACAGGGT TAATGTAGGT ATGGATTATA
CTACAACTTT TAAGTCCCTC ACCACCATTG GGACCTGTCC TATTGTCCCA ATTACATCCA TACCTAATAT

      6240       6250       6260       6270       6280       6290       6300
AACAAACACA ATTATGCATG GTTGGATGTG CCCCCCCTTT GGGCGAGCAT TGGGGTAAAG GTAAACAGTG
TTGTTTGTGT TAATACGTAC CAACCTACAC GGGGGGGAAA CCCGCTCGTA ACCCCATTTC CATTTGTCAC

      6310       6320       6330       6340       6350       6360       6370
TACTAATACA CCTGTACAGG CTGGTGACTG CCCGCCCTTA GAACTTATTA CCAGTGTTAT ACAGGATGGC
ATGATTATGT GGACATGTCC GACCACTGAC GGGCGGGAAT CTTGAATAAT GGTCACAATA TGTCCTACCG

      6380       6390       6400       6410       6420       6430       6440
GATATGGTTG ACACAGGCTT TGGTGCTATG AATTTTGCTG ATTTGCAGAC CAATAAATCA GATGTTCCTA
CTATACCAAC TGTGTCCGAA ACCACGATAC TTAAAACGAC TAAACGTCTG GTTATTTAGT CTACAAGGAT

      6450       6460       6470       6480       6490       6500       6510
TTGACATATG TGGCACTACA TGTAAATATC CAGATTATTT ACAAATGGCT GCAGACCCAT ATGGTGATAG
AACTGTATAC ACCGTGATGT ACATTTATAG GTCTAATAAA TGTTTACCGA CGTCTGGGTA TACCACTATC

      6520       6530       6540       6550       6560       6570       6580
ATTATTTTTT TTTCTACGGA AGGAACAAAT GTTTGCCAGA CATTTTTTTA ACAGGGCTGG CGAGGTGGGG
TAATAAAAAA AAAGATGCCT TCCTTGTTTA CAAACGGTCT GTAAAAAAAT TGTCCCGACC GCTCCACCCC

      6590       6600       6610       6620       6630       6640       6650
GAACCTGTGC CTGATACACT TATAATTAAG GGTAGTGGAA ATCGCACGTC TGTAGGGAGT AGTATATATG
CTTGGACACG GACTATGTGA ATATTAATTC CCATCACCTT TAGCGTGCAG ACATCCCTCA TCATATATAC
```

FIGURE 4. (*Continued*)

```
      6660       6670       6680       6690       6700       6710       6720
TTAACACCCC GAGCGGCTCT TTGGTGTCCT CTGAGGCACA ATTGTTTAAT AAGCCATATT GGCTACAAAA
AATTGTGGGG CTCGCCGAGA AACCACAGGA GACTCCGTGT TAACAAATTA TTCGGTATAA CCGATGTTTT

      6730       6740       6750       6760       6770       6780       6790
AGCCCAGGGA CATAACAATG GTATTTGTTG GGGTAATCAA CTGTTTGTTA CTGTGGTAGA TACCACACGC
TCGGGTCCCT GTATTGTTAC CATAAACAAC CCCATTAGTT GACAAACAAT GACACCATCT ATGGTGTGCG

      6800       6810       6820       6830       6840       6850       6860
AGTACCAACA TGACATTATG TGCATCCGTA ACTACATCTT CCACATACAC CAATTCTGAT TATAAAGAGT
TCATGGTTGT ACTGTAATAC ACGTAGGCAT TGATGTAGAA GGTGTATGTG GTTAAGACTA ATATTTCTCA

      6870       6880       6890       6900       6910       6920       6930
ACATGCGTCA TGTGGAAGAG TATGATTTAC AATTTATTTT TCAATTATGT AGCATTACAT TGTCTGCTGA
TGTACGCAGT ACACCTTCTC ATACTAAATG TTAAATAAAA AGTTAATACA TCGTAATGTA ACAGACGACT

      6940       6950       6960       6970       6980       6990       7000
AGTAATGGCC TATATTCACA CAATGAATCC CTCTGTTTTG GAAGACTGGA ACTTTGGGTT ATCGCCTCCC
TCATTACCGG ATATAAGTGT GTTACTTAGG GAGACAAAAC CTTCTGACCT TGAAACCCAA TAGCGGAGGG

      7010       7020       7030       7040       7050       7060       7070
CCAAATGGTA CATTAGAAGA TACCTATAGG TATGTGCAGT CACAGGCCAT TACCTGTCAA AAGCCCACTC
GGTTTACCAT GTAATCTTCT ATGGATATCC ATACACGTCA GTGTCCGGTA ATGGACAGTT TTCGGGTGAG

      7080       7090       7100       7110       7120       7130       7140
CTGAAAAGGA AAAGCCAGAT CCCTATAAGA ACCTTAGTTT TTGGGAGGTT AATTTAAAAG AAAAGTTTTC
GACTTTTCCT TTTCGGTCTA GGGATATTCT TGGAATCAAA AACCCTCCAA TTAAATTTTC TTTTCAAAAG

      7150       7160       7170       7180       7190       7200       7210
TAGTGAATTG GATCAGTATC CTTTGGGACG CAAGTTTTTG TTACAAAGTG GATATAGGGG ACGGTCCTCT
ATCACTTAAC CTAGTCATAG GAAACCCTGC GTTCAAAAAC AATGTTTCAC CTATATCCCC TGCCAGGAGA

      7220       7230       7240       7250       7260       7270       7280
ATTCGTACAG GTGTTAAGCG CCCTGCTGTT TCCAAAGCCT CTGCTGCCCC TAAACGTAAG CGCGCCAAAA
TAAGCATGTC CACAATTCGC GGGACGACAA AGGTTTCGGA GACGACGGGG ATTTGCATTC GCGCGGTTTT

      7290       7300       7310       7320       7330       7340       7350
CTAAAAGGTA ATATATGTGT ATATGTACTG TTATATATAT GTGTGTATGT ACTGTTATGT ATATGTGTGT
GATTTTCCAT TATATACACA TATACATGAC AATATATATA CACACATACA TGACAATACA TATACACACA

      7360       7370       7380       7390       7400       7410       7420
GTGTGTTCTG TGTGTAATGT AAGTTATTTG TGTAATGTGT ATGTGTGTTT ATGTGCAATA AACAATTACC
CACACAAGAC ACACATTACA TTCAATAAAC ACATTACACA TACACACAAA TACACGTTAT TTGTTAATGG

      7430       7440       7450       7460       7470       7480       7490
TCTTGTTACA CCCTGTGACT CAGTGGCTGT TGCACGCGTT TTGGTTTGCA CGCGCCTTAC ACACATAAGT
AGAACAATGT GGGACACTGA GTCACCGACA ACGTGCGCAA AACCAAACGT GCGCGGAATG TGTGTATTCA

      7500       7510       7520       7530       7540       7550       7560
AATATACATG CACAATATAT ATATTTTTGT TTAAAATACT ATACTTTTAT ATTTGCAACC GTTTTCGGTT
TTATATGTAC GTGTTATATA TATAAAAACA AATTTTATGA TATGAAAATA TAAACGTTGG CAAAAGCCAA

      7570       7580       7590       7600       7610       7620       7630
GCCCTTAGCA TACACTTTCC ACCAATTTGT TACAACGTGT TTCCTCTTAA TCCTATATAT TTTGTGCCAG
CGGGAATCGT ATGTGAAAGG TGGTTAAACA ATGTTGCACA AAGGAGAATT AGGATATATA AAACACGGTC

      7640       7650       7660       7670       7680       7690       7700
GTACACATTG CCCTGCCAAG TTGCTTGCCA AGTGCATCAT ATCCTGCCAA CCACACACCT GGCGCCAGGG
CATGTGTAAC GGGACGGTTC AACGAACGGT TCACGTAGTA TAGGACGGTT GGTGTGTGGA CCGCGGTCCC

      7710       7720       7730       7740       7750       7760       7770
TGCGGTATTG CCTTACTCAT AAACCTGTCT TTGTGTTATA CTTTTATGCA CTGTAGCCAA CTCTTAAAAG
ACGCCATAAC GGAATGAGTA TTTGGACAGA AACACAATAT GAAAATACGT GACATCGGTT GAGAATTTTC

      7780       7790       7800       7810       7820       7830       7840
CATTTTTGGC TTGTAGCAGC ACATTTTTTT GCTCTTACTG TTTGGTATAC AATAACATAA AAATGAGTAA
GTAAAAACCG AACATCGTCG TGTAAAAAAA CGAGAATGAC AAACCATATG TTATTGTATT TTTACTCATT

      7850       7860       7870       7880       7890       7900
CCTAAGGTCA CACACCTGCG ACCGGTTTCG GTTATCCACA CCCTACATAT TTCCTTCTTA TA
GGATTCCAGT GTGTGGACGC TGGCCAAAGC CAATAGGTGT GGGATGTATA AAGGAAGAAT AT
```

FIGURE 4. (*Continued*)

```
        10         20         30         40         50         60         70
ACTACAATAA TTCATGTATA AAACTAAGGG CGTAACCGAA ATCGGTTGAA CCGAAACCGG TTAGTATAAA
TGATGTTATT AAGTACATAT TTTGATTCCC GCATTGGCTT TAGCCAACTT GGCTTTGGCC AATCATATTT

        80         90        100        110        120        130        140
AGCAGACATT TTATGCACCA AAAGAGAACT GCAATGTTTC AGGACCCACA GGAGCGACCC AGAAAGTTAC
TCGTCTGTAA AATACGTGGT TTTCTCTTGA CGTTACAAAG TCCTGGGTGT CCTCGCTGGG TCTTTCAATG

       150        160        170        180        190        200        210
CACAGTTATG CACAGAGCTG CAAACAACTA TACATGATAT AATATTAGAA TGTGTGTACT GCAAGCAACA
GTGTCAATAC GTGTCTCGAC GTTTGTTGAT ATGTACTATA TTATAATCTT ACACACATGA CGTTCGTTGT

       220        230        240        250        260        270        280
GTTACTGCGA CGTGAGGTAT ATGACTTTGC TTTTCGGGAT TTATGCATAG TATATAGAGA TGGGAATCCA
CAATGACGCT GCACTCCATA TACTGAAACG AAAAGCCCTA AATACGTATC ATATATCTCT ACCCTTAGGT

       290        300        310        320        330        340        350
TATGCTGTAT GTGATAAATG TTTAAAGTTT TATTCTAAAA TTAGTGAGTA TAGACATTAT TGTTATAGTT
ATACGACATA CACTATTTAC AAATTTCAAA ATAAGATTTT AATCACTCAT ATCTGTAATA ACAATATCAA

       360        370        380        390        400        410        420
TGTATGGAAC AACATTAGAA CAGCAATACA ACAAACCGTT GTGTGATTTG TTAATTAGGT GTATTAACTG
ACATACCTTG TTGTAATCTT GTCGTTATGT TGTTTGGCAA CACACTAAAC AATTAATCCA CATAATTGAC

       430        440        450        460        470        480        490
TCAAAAGCCA CTGTGTCCTG AAGAAAAGCA AAGACATCTG GACAAAAAGC AAAGATTCCA TAATATAAGG
AGTTTTCGGT GACACAGGAC TTCTTTTCGT TTCTGTAGAC CTGTTTTTCG TTTCTAAGGT ATTATATTCC

       500        510        520        530        540        550        560
GGTCGGTGGA CCGGTCGATG TATGTCTTGT TGCAGATCAT CAAGAACACG TAGAGAAACC CAGCTGTAAT
CCAGCCACCT GGCCAGCTAC ATACAGAACA ACGTCTAGTA GTTCTTGTGC ATCTCTTTGG GTCGACATTA

       570        580        590        600        610        620        630
CATGCATGGA GATACACCTA CATTGCATGA ATATATGTTA GATTTGCAAC CAGAGACAAC TGATCTCTAC
GTACGTACCT CTATGTGGAT GTAACGTACT TATATACAAT CTAAACGTTG GTCTCTGTTG ACTAGAGATG

       640        650        660        670        680        690        700
TGTTATGAGC AATTAAATGA CAGCTCAGAG GAGGAGGATG AAATAGATGG TCCAGCTGGA CAAGCAGAAC
ACAATACTCG TTAATTTACT GTCGAGTCTC CTCCTCCTAC TTTATCTACC AGGTCGACCT GTTCGTCTTG

       710        720        730        740        750        760        770
CGGACAGAGC CCATTACAAT ATTGTAACCT TTTGTTGCAA GTGTGACTCT ACGCTTCGGT TGTGCGTACA
GCCTGTCTCG GGTAATGTTA TAACATTGGA AAACAACGTT CACACTGAGA TGCGAAGCCA ACACGCATGT

       780        790        800        810        820        830        840
AAGCACACAC GTAGACATTC GTACTTTGGA AGACCTGTTA ATGGGCACAC TAGGAATTGT GTGCCCCATC
TTCGTGTGTG CATCTGTAAG CATGAAACCT TCTGGACAAT TACCCGTGTG ATCCTTAACA CACGGGGTAG

       850        860        870        880        890        900        910
TGTTCTCAGA AACCATAATC TACCATGGCT GATCCTGCAG GTACCAATGG GGAAGAGGGT ACGGGATGTA
ACAAGAGTCT TTGGTATTAG ATGGTACCGA CTAGGACGTC CATGGTTACC CCTTCTCCCA TGCCCTACAT

       920        930        940        950        960        970        980
ATGGATGGTT TTATGTAGAG GCTGTAGTGG AAAAAAAAAC AGGGGATGCT ATATCAGATG ACGAGAACGA
TACCTACCAA AATACATCTC CGACATCACC TTTTTTTTTG TCCCCTACGA TATAGTCTAC TGCTCTTGCT

       990       1000       1010       1020       1030       1040       1050
AAATGACAGT GATACAGGTG AAGATTTGGT AGATTTTATA GTAAATGATA ATGATTATTT AACACAGGCA
TTTACTGTCA CTATGTCCAC TTCTAAACCA TCTAAAATAT CATTTACTAT TACTAATAAA TTGTGTCCGT

      1060       1070       1080       1090       1100       1110       1120
GAAACAGAGA CAGCACATGC GTTGTTTACT GCACAGGAAG CAAAACAACA TAGAGATGCA GTACAGGTTC
CTTTGTCTCT GTCGTGTACG CAACAAATGA CGTGTCCTTC GTTTTGTTGT ATCTCTACGT CATGTCCAAG

      1130       1140       1150       1160       1170       1180       1190
TAAAACGAAA GTATTTGGTA GTCCACTTAG TGATATTAGT GGATGTGTAG ACAATAATAT TAGTCCTAGA
ATTTTGCTTT CATAAACCAT CAGGTGAATC ACTATAATCA CCTACACATC TGTTATTATA ATCAGGATCT

      1200       1210       1220       1230       1240       1250       1260
TTAAAAGCTA TATGTATAGA AAAACAAAGT AGAGCTGCAA AAAGGAGATT ATTTGAAAGC GAAGACAGCG
AATTTTCGAT ATACATATCT TTTTGTTTCA TCTCGACGTT TTTCCTCTAA TAAACTTTCG CTTCTGTCGC

      1270       1280       1290       1300       1310       1320       1330
GGTATGGCAA TACTGAAGTG GAAACTCAGC AGATGTTACA GGTAGAAGGG CGCCATGAGA CTGAAACACC
CCATACCGTT ATGACTTCAC CTTTGAGTCG TCTACAATGT CCATCTTCCC GCGGTACTCT GACTTTGTGG
```

FIGURE 5. DNA sequence of HPV-16.

```
      1340       1350       1360       1370       1380       1390       1400
ATGTAGTCAG TATAGTGGTG GAAGTGGGGG TGGTTGCAGT CAGTACAGTA GTGGAAGTGG GGGAGAGGGT
TACATCAGTC ATATCACCAC CTTCACCCCC ACCAACGTCA GTCATGTCAT CACCTTCACC CCCTCTCCCA

      1410       1420       1430       1440       1450       1460       1470
GTTAGTGAAA GACACACTAT ATGCCAAACA CCACTTACAA ATATTTTAAA TGTACTAAAA ACTAGTAATG
CAATCACTTT CTGTGTGATA TACGGTTTGT GGTGAATGTT TATAAAATTT ACATGATTTT TGATCATTAC

      1480       1490       1500       1510       1520       1530       1540
CAAAGGCAGC AATGTTAGCA AAATTTAAAG AGTTATACGG GGTGAGTTTT TCAGAATTAG TAAGACCATT
GTTTCCGTCG TTACAATCGT TTTAAATTTC TCAATATGCC CCACTCAAAA AGTCTTAATC ATTCTGGTAA

      1550       1560       1570       1580       1590       1600       1610
TAAAAGTAAT AAATCAACGT GTTGCGATTG GTGTATTGCT GCATTTGGAC TTACACCCAG TATAGCTGAC
ATTTTCATTA TTTAGTTGCA CAACGCTAAC CACATAACGA CGTAAACCTG AATGTGGGTC ATATCGACTG

      1620       1630       1640       1650       1660       1670       1680
AGTATAAAAA CACTATTACA ACAATATTGT TTATATTTAC ACATTCAAAG TTTAGCATGT TCATGGGGAA
TCATATTTTT GTGATAATGT TGTTATAACA AATATAAATG TGTAAGTTTC AAATCGTACA AGTACCCCTT

      1690       1700       1710       1720       1730       1740       1750
TGGTTGTGTT ACTATTAGTA AGATATAAAT GTGGAAAAAA TAGAGAAACA ATTGAAAAAT TGCTGTCTAA
ACCAACACAA TGATAATCAT TCTATATTTA CACCTTTTTT ATCTCTTTGT TAACTTTTTA ACGACAGATT

      1760       1770       1780       1790       1800       1810       1820
ACTATTATGT GTGTCTCCAA TGTGTATGAT GATAGAGCCT CCAAAATTGC GTAGTACAGC AGCAGCATTA
TGATAATACA CACAGAGGTT ACACATACTA CTATCTCGGA GGTTTTAACG CATCATGTCG TCGTCGTAAT

      1830       1840       1850       1860       1870       1880       1890
TATTGGTATA AAACAGGTAT ATCAAATATT AGTGAAGTGT ATGGAGACAC GCCAGAATGG ATACAAAGAC
ATAACCATAT TTTGTCCATA TAGTTTATAA TCACTTCACA TACCTCTGTG CGGTCTTACC TATGTTTCTG

      1900       1910       1920       1930       1940       1950       1960
AAACAGTATT ACAACATAGT TTTAATGATT GTACATTTGA ATTATCACAG ATGGTACAAT GGGCCTACGA
TTTGTCATAA TGTTGTATCA AAATTACTAA CATGTAAACT TAATAGTGTC TACCATGTTA CCCGGATGCT

      1970       1980       1990       2000       2010       2020       2030
TAATGACATA GTAGACGATA GTGAAATTGC ATATAAATAT GCACAATTGG CAGACACTAA TAGTAATGCA
ATTACTGTAT CATCTGCTAT CACTTTAACG TATATTTATA CGTGTTAACC GTCTGTGATT ATCATTACGT

      2040       2050       2060       2070       2080       2090       2100
AGTGCCTTTC TAAAAAGTAA TTCACAGGCA AAAATTGTAA AGGATTGTGC AACAATGTGT AGACATTATA
TCACGGAAAG ATTTTTCATT AAGTGTCCGT TTTTAACATT TCCTAACACG TTGTTACACA TCTGTAATAT

      2110       2120       2130       2140       2150       2160       2170
AACGAGCAGA AAAAAAACAA ATGAGTATGA GTCAATGGAT AAAATATAGA TGTGATAGGG TAGATGATGG
TTGCTCGTCT TTTTTTTGTT TACTCATACT CAGTTACCTA TTTTATATCT ACACTATCCC ATCTACTACC

      2180       2190       2200       2210       2220       2230       2240
AGGTGATTGG AAGCAAATTG TTATGTTTTT AAGGTATCAA GGTGTAGAGT TTATGTCATT TTTAACTGCA
TCCACTAACC TTCGTTTAAC AATACAAAAA TTCCATAGTT CCACATCTCA AATACAGTAA AAATTGACGT

      2250       2260       2270       2280       2290       2300       2310
TTAAAAAGAT TTTTGCAAGG CATACCTAAA AAAAATTGCA TATTACTATA TGGTGCAGCT AACACAGGTA
AATTTTTCTA AAAACGTTCC GTATGGATTT TTTTTAACGT ATAATGATAT ACCACGTCGA TTGTGTCCAT

      2320       2330       2340       2350       2360       2370       2380
AATCATTATT TGGTATGAGT TTAATGAAAT TTCTGCAAGG GTCTGTAATA TGTTTTGTAA ATTCTAAAAG
TTAGTAATAA ACCATACTCA AATTACTTTA AAGACGTTCC CAGACATTAT ACAAAACATT TAAGATTTTC

      2390       2400       2410       2420       2430       2440       2450
CCATTTTTGG TTACAACCAT TAGCAGATGC CAAAATAGGT ATGTTAGATG ATGCTACAGT GCCCTGTTGG
GGTAAAAACC AATGTTGGTA ATCGTCTACG GTTTTATCCA TACAATCTAC TACGATGTCA CGGGACAACC

      2460       2470       2480       2490       2500       2510       2520
AACTACATAG ATGACAATTT AAGAAATGCA TTGGATGGAA ATTTAGTTTC TATGGATGTA AAGCATAGAC
TTGATGTATC TACTGTTAAA TTCTTTACGT AACCTACCTT TAAATCAAAG ATACCTACAT TTCGTATCTG

      2530       2540       2550       2560       2570       2580       2590
CATTGGTACA ACTAAAATGC CCTCCATTAT TAATTACATC TAACATTAAT GCTGGTACAG ATTCTAGGTG
GTAACCATGT TGATTTTACG GGAGGTAATA ATTAATGTAG ATTGTAATTA CGACCATGTC TAAGATCCAC

      2600       2610       2620       2630       2640       2650       2660
GCCTTATTTA CATAATAGAT TGGTGGTGTT TACATTTCCT AATGAGTTTC CATTTGACGA AAACGGAAAT
CGGAATAAAT GTATTATCTA ACCACCACAA ATGTAAAGGA TTACTCAAAG GTAAACTGCT TTTGCCTTTA
```

FIGURE 5. (*Continued*)

```
      2670       2680       2690       2700       2710       2720       2730
CCAGTGTATG AGCTTAATGA TAAGAACTGG AAATCCTTTT TCTCAAGGAC GTGGTCCAGA TTAAGTTTGC
GGTCACATAC TCGAATTACT ATTCTTGACC TTTAGGAAAA AGAGTTCCTG CACCAGGTCT AATTCAAACG

      2740       2750       2760       2770       2780       2790       2800
ACGAGGACGA GGACAAGGAA AACGATGGAG ACTCTTTGCC AACGTTTAAA TGTGTGTCAG GACAAAATAC
TGCTCCTGCT CCTGTTCCTT TTGCTACCTC TGAGAAACGG TTGCAAATTT ACACACAGTC CTGTTTTATG

      2810       2820       2830       2840       2850       2860       2870
TAACACATTA TGAAAATGAT AGTACAGACC TACGTGACCA TATAGACTAT TGGAAACACA TGCGCCTAGA
ATTGTGTAAT ACTTTTACTA TCATGTCTGG ATGCACTGGT ATATCTGATA ACCTTTGTGT ACGCGGATCT

      2880       2890       2900       2910       2920       2930       2940
ATGTGCTATT TATTACAAGG CCAGAGAAAT GGGATTTAAA CATATTAACC ACCAAGTGGT GCCAACACTG
TACACGATAA ATAATGTTCC GGTCTCTTTA CCCTAAATTT GTATAATTGG TGGTTCACCA CGGTTGTGAC

      2950       2960       2970       2980       2990       3000       3010
GCTGTATCAA AGAATAAAGC ATTACAAGCA ATTGAACTGC AACTAACGTT AGAAACAATA TATAACTCAC
CGACATAGTT TCTTATTTCG TAATGTTCGT TAACTTGACG TTGATTGCAA TCTTTGTTAT ATATTGAGTG

      3020       3030       3040       3050       3060       3070       3080
AATATAGTAA TGAAAAGTGG ACATTACAAG ACGTTAGCCT TGAAGTGTAT TTAACTGCAC CAACAGGATG
TTATATCATT ACTTTTCACC TGTAATGTTC TGCAATCGGA ACTTCACATA AATTGACGTG GTTGTCCTAC

      3090       3100       3110       3120       3130       3140       3150
TATAAAAAAA CATGGATATA CAGTGGAAGT GCAGTTTGAT GGAGACATAT GCAATACAAT GCATTATACA
ATATTTTTTT GTACCTATAT GTCACCTTCA CGTCAAACTA CCTCTGTATA CGTTATGTTA CGTAATATGT

      3160       3170       3180       3190       3200       3210       3220
AACTGGACAC ATATATATAT TTGTGAAGAA GCATCAGTAA CTGTGGTAGA GGGTCAAGTT GACTATTATG
TTGACCTGTG TATATATATA AACACTTCTT CGTAGTCATT GACACCATCT CCCAGTTCAA CTGATAATAC

      3230       3240       3250       3260       3270       3280       3290
GTTTATATTA TGTTCATGAA GGAATACGAA CATATTTTGT GCAGTTTAAA GATGATGCAG AAAAATATAG
CAAATATAAT ACAAGTACTT CCTTATGCTT GTATAAAACA CGTCAAATTT CTACTACGTC TTTTTATATC

      3300       3310       3320       3330       3340       3350       3360
TAAAAATAAA GTATGGGAAG TTCATGCGGG TGGTCAGGTA ATATTATGTC CTACATCTGT GTTTAGCAGC
ATTTTTATTT CATACCCTTC AAGTACGCCC ACCAGTCCAT TATAATACAG GATGTAGACA CAAATCGTCG

      3370       3380       3390       3400       3410       3420       3430
AACGAAGTAT CCTCTCCTGA AATTATTAGG CAGCACTTGG CCAACCACCC CGCCGCGACC CATACCAAAG
TTGCTTCATA GGAGAGGACT TTAATAATCC GTCGTGAACC GGTTGGTGGG GCGGCGCTGG GTATGGTTTC

      3440       3450       3460       3470       3480       3490       3500
CCGTCGCCTT GGGCACCGAA GAAACACAGA CGACTATCCA GCGACCAAGA TCAGAGCCAG ACACCGGAAA
GGCAGCGGAA CCCGTGGCTT CTTTGTGTCT GCTGATAGGT CGCTGGTTCT AGTCTCGGTC TGTGGCCTTT

      3510       3520       3530       3540       3550       3560       3570
CCCCTGCCAC ACCACTAAGT TGTTGCACAG AGACTCAGTG GACAGTGCTC CAATCCTCAC TGCATTTAAC
GGGGACGGTG TGGTGATTCA ACAACGTGTC TCTGAGTCAC CTGTCACGAG GTTAGGAGTG ACGTAAATTG

      3580       3590       3600       3610       3620       3630       3640
AGCTCACACA AAGGACGGAT TAACTGTAAT AGTAACACTA CACCCATAGT ACATTTAAAA GGTGATGCTA
TCGAGTGTGT TTCCTGCCTA ATTGACATTA TCATTGTGAT GTGGGTATCA TGTAAATTTT CCACTACGAT

      3650       3660       3670       3680       3690       3700       3710
ATACTTTAAA ATGTTTAAGA TATAGATTTA AAAAGCATTG TACATTGTAT ACTGCAGTGT CGTCTACATG
TATGAAATTT TACAAATTCT ATATCTAAAT TTTTCGTAAC ATGTAACATA TGACGTCACA GCAGATGTAC

      3720       3730       3740       3750       3760       3770       3780
GCATTGGACA GGACATAATG TAAAACATAA AAGTGCAATT GTTACACTTA CATATGATAG TGAATGGCAA
CGTAACCTGT CCTGTATTAC ATTTTGTATT TTCACGTTAA CAATGTGAAT GTATACTATC ACTTACCGTT

      3790       3800       3810       3820       3830       3840       3850
CGTGACCAAT TTTTGTCTCA AGTTAAAATA CCAAAAACTA TTACAGTGTC TACTGGATTT ATGTCTATAT
GCACTGGTTA AAAACAGAGT TCAATTTTAT GGTTTTTGAT AATGTCACAG ATGACCTAAA TACAGATATA

      3860       3870       3880       3890       3900       3910       3920
GACAAATCTT GATACTGCAT CCACAACATT ACTGGCGTGC TTTTTGCTTT GCTTTGTGTG CTTTTGTGTG
CTGTTTAGAA CTATGACGTA GGTGTTGTAA TGACCGCACG AAAAACGAAA CGAAACACAC GAAAACACAC

      3930       3940       3950       3960       3970       3980       3990
TCTGCCTATT AATACGTCCG CTGCTTTTGT CTGTGTCTAC ATACACATCA TTAATAATAT TGGTATTACT
AGACGGATAA TTATGCAGGC GACGAAAACA GACACAGATG TATGTGTAGT AATTATTATA ACCATAATGA
```

FIGURE 5. (*Continued*)

```
      4000       4010       4020       4030       4040       4050       4060
ATTGTGGATA ACAGCAGCCT CTGCGTTTAG GTGTTTTATT GTATATATTA TATTTGTTTA TATACCATTA
TAACACCTAT TGTCGTCGGA GACGCAAATC CACAAAATAA CATATATAAT ATAAACAAAT ATATGGTAAT

      4070       4080       4090       4100       4110       4120       4130
TTTTTAATAC ATACACATGC ACGCTTTTTA ATTACATAAT GTATATGTAC ATAATGTAAT TGTTACATAT
AAAAATTATG TATGTGTACG TGCGAAAAAT TAATGTATTA CATATACATG TATTACATTA ACAATGTATA

      4140       4150       4160       4170       4180       4190       4200
AATTGTTGTA TACCATAACT TACTATTTTT TCTTTTTTAT TTTCATATAT AATTTTTTTT TTTGTTTGTT
TTAACAACAT ATGGTATTGA ATGATAAAAA AGAAAAAATA AAAGTATATA TTAAAAAAAA AAACAAACAA

      4210       4220       4230       4240       4250       4260       4270
TGTTTGTTTT TTAATAAACT GTTATTACTT AACAATGCGA CACAAACGTT CTGCAAAACG CACAAAACGT
ACAAACAAAA AATTATTTGA CAATAATGAA TTGTTACGCT GTGTTTGCAA GACGTTTTGC GTGTTTTGCA

      4280       4290       4300       4310       4320       4330       4340
GCATCGGCTA CCCAACTTTA TAAAACATGC AAACAGGCAG GTACATGTCC ACCTGACATT ATACCTAAGG
CGTAGCCGAT GGGTTGAAAT ATTTTGTACG TTTGTCCGTC CATGTACAGG TGGACTGTAA TATGGATTCC

      4350       4360       4370       4380       4390       4400       4410
TTGAAGGCAA AACTATTGCT GAACAAATAT TACAATATGG AAGTATGGGT GTATTTTTTG GTGGGTTAGG
AACTTCCGTT TTGATAACGA CTTGTTTATA ATGTTATACC TTCATACCCA CATAAAAAAC CACCCAATCC

      4420       4430       4440       4450       4460       4470       4480
AATTGGAACA GGGTCGGGTA CAGGCGGACG CACTGGGTAT ATTCCATTGG GAACAAGGCC TCCCACAGCT
TTAACCTTGT CCCAGCCCAT GTCCGCCTGC GTGACCCATA TAAGGTAACC CTTGTTCCGG AGGGTGTCGA

      4490       4500       4510       4520       4530       4540       4550
ACAGATACAC TTGCTCCTGT AAGACCCCCT TTAACAGTAG ATCCTGTGGG CCCTTCTGAT CCTTCTATAG
TGTCTATGTG AACGAGGACA TTCTGGGGGA AATTGTCATC TAGGACACCC GGGAAGACTA GGAAGATATC

      4560       4570       4580       4590       4600       4610       4620
TTTCTTTAGT GGAAGAAACT AGTTTTATTG ATGCTGGTGC ACCAACATCT GTACCTTCCA TTCCCCCAGA
AAAGAAATCA CCTTCTTTGA TCAAAATAAC TACGACCACG TGGTTGTAGA CATGGAAGGT AAGGGGGTCT

      4630       4640       4650       4660       4670       4680       4690
TGTATCAGGA TTTAGTATTA CTACTTCAAC TGATACCACA CCTGCTATAT TAGATATTAA TAATACTGTT
ACATAGTCCT AAATCATAAT GATGAAGTTG ACTATGGTGT GGACGATATA ATCTATAATT ATTATGACAA

      4700       4710       4720       4730       4740       4750       4760
ACTACTGTTA CTACACATAA TAATCCCACT TTCACTGACC CATCTGTATT GCAGCCTCCA ACACCTGCAG
TGATGACAAT GATGTGTATT ATTAGGGTGA AAGTGACTGG GTAGACATAA CGTCGGAGGT TGTGGACGTC

      4770       4780       4790       4800       4810       4820       4830
AAACTGGAGG GCATTTTACA CTTTCATCAT CCACTATTAG TACACATAAT TATGAAGAAA TTCCTATGGA
TTTGACCTCC CGTAAAATGT GAAAGTAGTA GGTGATAATC ATGTGTATTA ATACTTCTTT AAGGATACCT

      4840       4850       4860       4870       4880       4890       4900
TACATTTATT GTTAGCACAA ACCCTAACAC AGTAACTAGT AGCACACCCA TACCAGGGTC TCGCCCAGTG
ATGTAAATAA CAATCGTGTT TGGGATTGTG TCATTGATCA TCGTGTGGGT ATGGTCCCAG AGCGGGTCAC

      4910       4920       4930       4940       4950       4960       4970
GCACGCCTAG GATTATATAG TCGCACAACA CAACAGGTTA AAGTTGTAGA CCCTGCTTTT GTAACCACTC
CGTGCGGATC CTAATATATC AGCGTGTTGT GTTGTCCAAT TTCAACATCT GGGACGAAAA CATTGGTGAG

      4980       4990       5000       5010       5020       5030       5040
CCACTAAACT TATTACATAT GATAATCCTG CATATGAAGG TATAGATGTG GATAATACAT TATATTTTTC
GGTGATTTGA ATAATGTATA CTATTAGGAC GTATACTTCC ATATCTACAC CTATTATGTA ATATAAAAAG

      5050       5060       5070       5080       5090       5100       5110
TAGTAATGAT AATAGTATTA ATATAGCTCC AGATCCTGAC TTTTTGGATA TAGTTGCTTT ACATAGGCCA
ATCATTACTA TTATCATAAT TATATCGAGG TCTAGGACTG AAAAACCTAT ATCAACGAAA TGTATCCGGT

      5120       5130       5140       5150       5160       5170       5180
GCATTAACCT CTAGGCGTAC TGGCATTAGG TACAGTAGAA TTGGTAATAA ACAAACACTA CGTACTCGTA
CGTAATTGGA GATCCGCATG ACCGTAATCC ATGTCATCTT AACCATTATT TGTTTGTGAT GCATGAGCAT

      5190       5200       5210       5220       5230       5240       5250
GTGGAAAATC TATAGGTGCT AAGGTACATT ATTATTATGA TTTAAGTACT ATTGATCCTG CAGAAGAAAT
CACCTTTTAG ATATCCACGA TTCCATGTAA TAATAATACT AAATTCATGA TAACTAGGAC GTCTTCTTTA

      5260       5270       5280       5290       5300       5310       5320
AGAATTACAA ACTATAACAC CTTCTACATA TACTACCACT TCACATGCAG CCTCACCTAC TTCTATTAAT
TCTTAATGTT TGATATTGTG GAAGATGTAT ATGATGGTGA AGTGTACGTC GGAGTGGATG AAGATAATTA
```

FIGURE 5. (*Continued*)

```
      5330       5340       5350       5360       5370       5380       5390
AATGGATTAT ATGATATTTA TGCAGATGAC TTTATTACAG ATACTTCTAC AACCCCGGTA CCATCTGTAC
TTACCTAATA TACTATAAAT ACGTCTACTG AAATAATGTC TATGAAGATG TTGGGGCCAT GGTAGACATG

      5400       5410       5420       5430       5440       5450       5460
CCTCTACATC TTTATCAGGT TATATTCCTG CAAATACAAC AATTCCTTTT GGTGGTGCAT ACAATATTCC
GGAGATGTAG AAATAGTCCA ATATAAGGAC GTTTATGTTG TTAAGGAAAA CCACCACGTA TGTTATAAGG

      5470       5480       5490       5500       5510       5520       5530
TTTAGTATCA GGTCCTGATA TACCCATTAA TATAACTGAC CAAGCTCCTT CATTAATTCC TATAGTTCCA
AAATCATAGT CCAGGACTAT ATGGGTAATT ATATTGACTG GTTCGAGGAA GTAATTAAGG ATATCAAGGT

      5540       5550       5560       5570       5580       5590       5600
GGGTCTCCAC AATATACAAT TATTGCTGAT GCAGGTGACT TTTATTTACA TCCTAGTTAT TACATGTTAC
CCCAGAGGTG TTATATGTTA ATAACGACTA CGTCCACTGA AAATAAATGT AGGATCAATA ATGTACAATG

      5610       5620       5630       5640       5650       5660       5670
GAAAACGACG TAAACGTTTA CCATATTTTT TTTCAGATGT CTCTTTGGCT GCCTAGTGAG GCCACTGTCT
CTTTTGCTGC ATTTGCAAAT GGTATAAAAA AAAGTCTACA GAGAAACCGA CGGATCACTC CGGTGACAGA

      5680       5690       5700       5710       5720       5730       5740
ACTTGCCTCC TGTCCCAGTA TCTAAGGTTG TAAGCACGGA TGAATATGTT GCACGCACAA ACATATATTA
TGAACGGAGG ACAGGGTCAT AGATTCCAAC ATTCGTGCCT ACTTATACAA CGTGCGTGTT TGTATATAAT

      5750       5760       5770       5780       5790       5800       5810
TCATGCAGGA ACATCCAGAC TACTTGCAGT TGGACATCCC TATTTTCCTA TTAAAAAACC TAACAATAAC
AGTACGTCCT TGTAGGTCTG ATGAACGTCA ACCTGTAGGG ATAAAAGGAT AATTTTTTGG ATTGTTATTG

      5820       5830       5840       5850       5860       5870       5880
AAAATATTAG TTCCTAAAGT ATCAGGATTA CAATACAGGG TATTTAGAAT ACATTTACCT GACCCCAATA
TTTTATAATC AAGGATTTCA TAGTCCTAAT GTTATGTCCC ATAAATCTTA TGTAAATGGA CTGGGGTTAT

      5890       5900       5910       5920       5930       5940       5950
AGTTTGGTTT TCCTGACACC TCATTTTATA ATCCAGATAC ACAGCGGCTG GTTTGGGCCT GTGTAGGTGT
TCAAACCAAA AGGACTGTGG AGTAAAATAT TAGGTCTATG TGTCGCCGAC CAAACCCGGA CACATCCACA

      5960       5970       5980       5990       6000       6010       6020
TGAGGTAGGT CGTGGTCAGC CATTAGGTGT GGGCATTAGT GGCCATCCTT TATTAAATAA ATTGGATGAC
ACTCCATCCA GCACCAGTCG GTAATCCACA CCCGTAATCA CCGGTAGGAA ATAATTTATT TAACCTACTG

      6030       6040       6050       6060       6070       6080       6090
ACAGAAAATG CTAGTGCTTA TGCAGCAAAT GCAGGTGTGG ATAATAGAGA ATGTATATCT ATGGATTACA
TGTCTTTTAC GATCACGAAT ACGTCGTTTA CGTCCACACC TATTATCTCT TACATATAGA TACCTAATGT

      6100       6110       6120       6130       6140       6150       6160
AACAAACACA ATTGTGTTTA ATTGGTTGCA AACCACCTAT AGGGGAACAC TGGGGCAAAG GATCCCCATG
TTGTTTGTGT TAACACAAAT TAACCAACGT TTGGTGGATA TCCCCTTGTG ACCCCGTTTC CTAGGGGTAC

      6170       6180       6190       6200       6210       6220       6230
TACCAATGTT GCAGTAAATC CAGGTGATTG TCCACCATTA GAGTTAATAA ACACAGTTAT TCAGGATGGT
ATGGTTACAA CGTCATTTAG GTCCACTAAC AGGTGGTAAT CTCAATTATT TGTGTCAATA AGTCCTACCA

      6240       6250       6260       6270       6280       6290       6300
GATATGGTTC ATACTGGCTT TGGTGCTATG GACTTTACTA CATTACAGGC TAACAAAAGT GAAGTTCCAC
CTATACCAAG TATGACCGAA ACCACGATAC CTGAAATGAT GTAATGTCCG ATTGTTTTCA CTTCAAGGTG

      6310       6320       6330       6340       6350       6360       6370
TGGATATTTG TACATCTATT TGCAAATATC CAGATTATAT TAAAATGGTG TCAGAACCAT ATGGCGACAG
ACCTATAAAC ATGTAGATAA ACGTTTATAG GTCTAATATA ATTTTACCAC AGTCTTGGTA TACCGCTGTC

      6380       6390       6400       6410       6420       6430       6440
CTTATTTTTT TATTTACGAA GGGAACAAAT GTTTGTTAGA CATTTATTTA ATAGGGCTGG TACTGTTGGT
GAATAAAAAA ATAAATGCTT CCCTTGTTTA CAAACAATCT GTAAATAAAT TATCCCGACC ATGACAACCA

      6450       6460       6470       6480       6490       6500       6510
GAAAATGTAC CAGACGATTT ATACATTAAA GGCTCTGGGT CTACTGCAAA TTTAGCCAGT TCAAATTATT
CTTTTACATG GTCTGCTAAA TATGTAATTT CCGAGACCCA GATGACGTTT AAATCGGTCA AGTTTAATAA

      6520       6530       6540       6550       6560       6570       6580
TTCCTACACC TAGTGGTTCT ATGGTTACCT CTGATGCCCA AATATTCAAT AAACCTTATT GGTTACAACG
AAGGATGTGG ATCACCAAGA TACCAATGGA GACTACGGGT TTATAAGTTA TTTGGAATAA CCAATGTTGC

      6590       6600       6610       6620       6630       6640       6650
AGCACAGGGC CACAATAATG GCATTTGTTG GGGTAACCAA CTATTTGTTA CTGTTGTTGA TACTACACGC
TCGTGTCCCG GTGTTATTAC CGTAAACAAC CCCATTGGTT GATAAACAAT GACAACAACT ATGATGTGCG
```

FIGURE 5. (*Continued*)

```
      6660       6670       6680       6690       6700       6710       6720
AGTACAAATA TGTCATTATG TGCTGCCATA TCTACTTCAG AAACTACATA TAAAAATACT AACTTTAAGG
TCATGTTTAT ACAGTAATAC ACGACGGTAT AGATGAAGTC TTTGATGTAT ATTTTTATGA TTGAAATTCC

      6730       6740       6750       6760       6770       6780       6790
AGTACCTACG ACATGGGGAG GAATATGATT TACAGTTTAT TTTTCAACTG TGCAAAATAA CCTTAACTGC
TCATGGATGC TGTACCCCTC CTTATACTAA ATGTCAAATA AAAAGTTGAC ACGTTTTATT GGAATTGACG

      6800       6810       6820       6830       6840       6850       6860
AGACGTTATG ACATACATAC ATTCTATGAA TTCCACTATT TTGGAGGACT GGAATTTTGG TCTACAACCT
TCTGCAATAC TGTATGTATG TAAGATACTT AAGGTGATAA AACCTCCTGA CCTTAAAACC AGATGTTGGA

      6870       6880       6890       6900       6910       6920       6930
CCCCCAGGAG GCACACTAGA AGATACTTAT AGGTTTGTAA CCCAGGCAAT TGCTTGTCAA AAACATACAC
GGGGGTCCTC CGTGTGATCT TCTATGAATA TCCAAACATT GGGTCCGTTA ACGAACAGTT TTTGTATGTG

      6940       6950       6960       6970       6980       6990       7000
CTCCAGCACC TAAAGAAGAT GATCCCCTTA AAAAATACAC TTTTTGGGAA GTAAATTTAA AGGAAAAGTT
GAGGTCGTGG ATTTCTTCTA CTAGGGGAAT TTTTTATGTG AAAAACCCTT CATTTAAATT TCCTTTTCAA

      7010       7020       7030       7040       7050       7060       7070
TTCTGCAGAC CTAGATCAGT TTCCTTTAGG ACGCAAATTT TTACTACAAG CAGGATTGAA GGCCAAACCA
AAGACGTCTG GATCTAGTCA AAGGAAATCC TGCGTTTAAA AATGATGTTC GTCCTAACTT CCGGTTTGGT

      7080       7090       7100       7110       7120       7130       7140
AAATTTACAT TAGGAAAACG AAAAGCTACA CCCACCACCT CATCTACCTC TACAACTGCT AAACGCAAAA
TTTAAATGTA ATCCTTTTGC TTTTCGATGT GGGTGGTGGA GTAGATGGAG ATGTTGACGA TTTGCGTTTT

      7150       7160       7170       7180       7190       7200       7210
AACGTAAGCT GTAAGTATTG TATGTATGTT GAATTAGTGT TGTTTGTTGT GTATATGTTT GTATGTGCTT
TTGCATTCGA CATTCATAAC ATACATACAA CTTAATCACA ACAAACAACA CATATACAAA CATACACGAA

      7220       7230       7240       7250       7260       7270       7280
GTATGTGCTT GTAAATATTA AGTTGTATGT GTGTTTGTAT GTATGGTATA ATAAACACGT GTGTATGTGT
CATACACGAA CATTTATAAT TCAACATACA CACAAACATA CATACCATAT TATTTGTGCA CACATACACA

      7290       7300       7310       7320       7330       7340       7350
TTTTAAATGC TTGTGTAACT ATTGTGTCAT GCAACATAAA TAAACTTATT GTTTCAACAC CTACTAATTG
AAAATTTACG AACACATTGA TAACACAGTA CGTTGTATTT ATTTGAATAA CAAAGTTGTG GATGATTAAC

      7360       7370       7380       7390       7400       7410       7420
TGTTGTGGTT ATTCATTGTA TATAAACTAT ATTTGCTACA TCCTGTTTTT GTTTTATATA TACTATATTT
ACAACACCAA TAAGTAACAT ATATTTGATA TAAACGATGT AGGACAAAAA CAAAATATAT ATGATATAAA

      7430       7440       7450       7460       7470       7480       7490
TGTAGCGCCA GGCCCATTTT GTAGCTTCAA CCGAATTCGG TTGCATGCTT TTTGGCACAA AATGTGTTTT
ACATCGCGGT CCGGGTAAAA CATCGAAGTT GGCTTAAGCC AACGTACGAA AAACCGTGTT TTACACAAAA

      7500       7510       7520       7530       7540       7550       7560
TTTAAATAGT TCTATGTCAG CAACTATGGT TTAAACTTGT ACGTTTCCTG CTTGCCATGC GTGCCAAATC
AAATTTATCA AGATACAGTC GTTGATACCA AATTTGAACA TGCAAAGGAC GAACGGTACG CACGGTTTAG

      7570       7580       7590       7600       7610       7620       7630
CCTGTTTTCC TGACCTGCAC TGCTTGCCAA CCATTCCATT GTTTTTTACA CTGCACTATG TGCAACTACT
GGACAAAAGG ACTGGACGTG ACGAACGGTT GGTAAGGTAA CAAAAAATGT GACGTGATAC ACGTTGATGA

      7640       7650       7660       7670       7680       7690       7700
GAATCACTAT GTACATTGTG TCATATAAAA TAAATCACTA TGCGCCAACG CCTTACATAC CGCTGTTAGG
CTTAGTGATA CATGTAACAC AGTATATTTT ATTTAGTGAT ACGCGGTTGC GGAATGTATG GCGACAATCC

      7710       7720       7730       7740       7750       7760       7770
CACATATTTT TGGCTTGTTT TAACTAACCT AATTGCATAT TTGGCATAAG GTTTAAACTT CTAAGGCCAA
GTGTATAAAA ACCGAACAAA ATTGATTGGA TTAACGTATA AACCGTATTC CAAATTTGAA GATTCCGGTT

      7780       7790       7800       7810       7820       7830       7840
CTAAATGTCA CCCTAGTTCA TACATGAACT GTGTAAAGGT TAGTCATACA TTGTTCATTT GTAAAACTGC
GATTTACAGT GGGATCAAGT ATGTACTTGA CACATTTCCA ATCAGTATGT AACAAGTAAA CATTTTGACG

      7850       7860       7870       7880       7890       7900
ACATGGGTGT GTGCAAACCG ATTTTGGGTT ACACATTTAC AAGCAACTTA TATAATAATA CTAA
TGTACCCACA CACGTTTGGC TAAAACCCAA TGTGTAAATG TTCGTTGAAT ATATTATTAT GATT
```

FIGURE 5. (*Continued*)

II. RESTRICTION SITE ANALYSIS

The DNA sequences of BPV-1, CRPV, HPV-1a, HPV-6b, and HPV-16 have been analyzed for restriction sites using the programs of Mount and Conrad (Conrad and Mount, 1982; Mount and Conrad, 1984). These results are shown in Figs. 6–10, respectively. Numbers refer to nucleotide positions of the first nucleotide of the recognition site. Ambiguous bases are defined at the bottom of each figure. Restriction enzymes marked with the symbol "#" indicate asymmetric recognition sequences and will therefore be listed twice for each sequence. Caution should therefore be exercised when an enzyme with an asymmetric recognition sequence is listed only once under "Enzymes which do not cut," since this indicates that that enzyme does cut the DNA.

Aat II (GACGTC)
769

Acc I (GTMKAC)
3004, 3966, 4366, 6686

Aha II (GPCGYC)
769, 2944, 6143, 7275

Alu I (AGCT)
51, 65, 264, 312,
438, 741, 764, 1208,
1303, 1336, 1368, 1565,
1643, 1667, 1761, 1910,
1958, 1997, 2008, 2028,
2098, 2267, 2390, 2899,
3044, 3826, 4171, 4459,
4707, 4785, 5044, 5139,
5634, 6671, 6714, 6732,
6754, 6960, 7098, 7677,
7709, 7728, 7890

Ava I (CYCGPG)
945, 3408, 4685, 4840

Ava II (GGWCC)
93, 488, 527, 641,
2904, 3259, 3337, 3373,
4425, 4473, 4682, 4720,
4890, 5057, 5132, 6378,
7457, 7819

BamH I (GGATCC)
4451

Ban I (GGYPCC)
2944, 3456, 4975, 5952,
6143, 7275, 7282

Ban II (GPGCYC)
1341, 3073, 3414, 6388,
7649

Ban III (ATCGAT)
5474, 6835, 7477

Bbv I# (GCAGC)
578, 1129, 1301, 1637,
1665, 2373, 2483, 3042,
3136, 4175, 4783, 5926,
6215, 6820, 7874

Bbv I# (GCTGC)
66, 359, 458, 470,
1634, 1860, 2268, 2740,
3918, 3932, 4172, 4382,
4708, 4786, 6317

Bcl I (TGATCA)
1811, 3737, 3838

Bgl I (GCCNNNNNGGC)
612, 2812, 6526, 7826

Bgl II (AGATCT)
1515, 6250, 6276, 6946

Bsm I# (GCATTC)
4497, 5225

Bsp 1286 (GDGCHC)
1341, 3073, 3414, 3450,
3986, 4191, 4252, 4976,
5513, 5951, 6388, 6543,
7630, 7649, 7790

BstE II (GGTNACC)
2405

BstN I (CCWGG)
567, 992, 1040, 1052,
1232, 2402, 2948, 3615,
3683, 3796, 4538, 4717,
6547, 7279

BstX I (CCANNNNNNTGG)
3882

Cfo I (GCGC)
1663, 1740, 2638, 2945,
3024, 3540, 5725, 6144,
7276, 7699

FIGURE 6. Restriction enzyme sites in BPV-1 (nucleotide positions). # = asymmetric site; N = A,G,C,T; P = A,G; Y = C,T; W = A,T; S = G,C; M = A,C; K = G,T; D = A,G,T; H = A,T,C.

```
Cla I      (ATCGAT)
    5474,   6835,   7477

Dde I      (CTNAG)
     126,    802,    907,   1659,
    1726,   1825,   1912,   1953,
    2322,   2720,   2861,   4151,
    4339,   4677,   5078,   5307,
    5364,   5784,   6048,   6309,
    6640,   6973,   7100,   7300,
    7337

Dpn I      (GATC)
     978,   1516,   1812,   1867,
    2689,   3175,   3226,   3738,
    3749,   3839,   3975,   4392,
    4452,   4988,   5831,   6128,
    6251,   6277,   6947,   6977,
    7746,   7838

Dra I      (TTTAAA)
     796,   1088,   1347,   1680,
    2093,   4942,   5813,   6424

EcoR I     (GAATTC)
    2113

EcoR II    (CCWGG)
     567,    992,   1040,   1052,
    1232,   2402,   2948,   3615,
    3683,   3796,   4538,   4717,
    6547,   7279

Fnu4H I    (GCNGC)
      66,    359,    458,    470,
     578,   1129,   1301,   1634,
    1637,   1665,   1860,   2268,
    2373,   2483,   2740,   3042,
    3136,   3918,   3932,   4172,
    4175,   4382,   4708,   4783,
    4786,   5884,   5926,   6215,
    6317,   6820,   7369,   7874

FnuD II    (CGCG)
     839,   7353

Fok I#     (GGATG)
     511,    888,    995,   2336,
    2606,   2871,   6442,   7022,
    7610,   7670

Fok I#     (CATCC)
    3203,   3251,   5004,   5552,
    5744,   7218

Fsp I      (TGCGCA)
    3023

Hae II     (PGCGCY)
    2944,   5724,   6143,   7275

Hae III    (GGCC)
     620,    690,   1440,   1805,
    2145,   2820,   3093,   3352,
    4528,   4570,   5385,   5627,
    5770,   6038,   6525,   7587

Hga I#     (GACGC)
     455,   3104,   6314,   7351

Hga I#     (GCGTC)
     844,   4534,   7535,   7557

HgiA I     (GWGCWC)
    3986,   4191,   7630,   7790

Hha I      (GCGC)
    1663,   1740,   2638,   2945,
    3024,   3540,   5725,   6144,
    7276,   7699

Hinc II    (GTYPAC)
       1,   1008,   7143

Hind III   (AAGCTT)
    6959

Hinf I     (GANTC)
     515,    833,    876,    972,
    1589,   1657,   1830,   2499,
    3239,   3505,   3528,   3557,
    4607,   4640,   5076,   5450,
    6061,   6866,   7288

HinP I     (GCGC)
    1663,   1740,   2638,   2945,
    3024,   3540,   5725,   6144,
    7276,   7699

Hpa I      (GTTAAC)
       1

Hpa II     (CCGG)
      80,    669,    946,   1586,
    3095,   3442,   3460,   3812,
    7210,   7366,   7415,   7766,
    7902

Hph I#     (GGTGA)
     304,   1074,   2505,   2757,
    3127,   3639,   7718

Hph I#     (TCACC)
      20,    518,   1037,   1695,
    3740,   4394,   5411,   6003,
    7893

Kpn I      (GGTACC)
    3456

Mbo I      (GATC)
     978,   1516,   1812,   1867,
    2689,   3175,   3226,   3738,
    3749,   3839,   3975,   4392,
    4452,   4988,   5831,   6128,
    6251,   6277,   6947,   6977,
    7746,   7838

Mbo II#    (GAAGA)
     275,    757,   1186,   1620,
    2601,   3484,   3487,   3517,
    4985,   5174,   5243,   5346,
    6182,   6704,   6911,   6993,
    6996,   7044

Mbo II#    (TCTTC)
    1048,   3211,   3310,   3430,
    3858,   5318,   7784

Mlu I      (ACGCGT)
    7352

Mnl I#     (CCTC)
     157,    653,    663,    814,
    1229,   2148,   2321,   2399,
    3209,   3354,   3376,   3434,
    3793,   4657,   4761,   4778,
    4923,   5306,   5316,   5353,
    5457,   5556,   5645,   5930,
    6152,   6155,   6257,   6298,
    6639,   6789,   6824,   6831,
    7846

Mnl I#     (GAGG)
     307,    452,    639,    909,
     939,    999,   1044,   1077,
    1245,   1251,   1256,   1272,
    1385,   1449,   1461,   1528,
    1654,   1827,   2303,   2433,
```

FIGURE 6. (*Continued*)

2586, 2589, 2592, 2604,
2959, 3031, 3133, 3490,
3514, 3578, 4344, 4550,
4568, 4679, 4688, 4836,
4860, 5126, 5261, 5566,
5922, 5937, 6311, 6384,
6845, 7575, 7721

Mst II (CCTNAGG)
4676, 5783

Nar I (GGCGCC)
2944, 6143, 7275

Nci I (CCSGG)
79, 668, 945, 946,
3441, 7209, 7414

Nco I (CCATGG)
2878, 3089

Nde I (CATATG)
6698

Nhe I (GCTAGC)
5429, 6715

Nla III (CATG)
90, 186, 202, 236,
325, 403, 478, 549,
848, 1374, 1521, 1713,
1917, 1979, 2292, 2488,
2618, 2632, 2879, 3021,
3090, 3584, 3781, 4239,
4257, 4362, 5144, 5171,
5240, 5588, 5711, 6117,
6331, 6553, 6570, 7272

Nla IV (GGNNCC)
641, 941, 1134, 2903,
2944, 3322, 3337, 3456,
4424, 4451, 4472, 4473,
4526, 4570, 4681, 4682,
4889, 4890, 4975, 5056,
5131, 5768, 5952, 6143,
6647, 7200, 7275, 7282,
7456, 7507, 7650, 7819,
7859

Nru I (TCGCGA)
838

Nsi I (ATGCAT)
1015, 2329

Pst I (CTGCAG)
471, 522, 576, 1299,
1500, 1635, 1861, 2775,
3275, 3401, 4173, 4383,
4602, 4781

Pvu II (CAGCTG)
4784, 5043

Rsa I (GTAC)
239, 1236, 2306, 2447,
2700, 2782, 2997, 3018,
3149, 3364, 3457, 3784,
3874, 3994, 4231, 4402,
4635, 4963, 5152, 5163,
5231, 5342, 5530, 5940,
6078, 6286, 6615, 7213,
7390, 7618, 7758

Sau3A I (GATC)
978, 1516, 1812, 1867,
2689, 3175, 3226, 3738,
3749, 3839, 3975, 4392,
4452, 4988, 5831, 6128,
6251, 6277, 6947, 6977,
7746, 7838

Sau96 I (GGNCC)
93, 488, 527, 641,
689, 1804, 2145, 2904,
3259, 3337, 3373, 4425,
4473, 4527, 4570, 4682,
4720, 4890, 5057, 5132,
5769, 6378, 6525, 7457,
7586, 7819

Sca I (AGTACT)
5162, 7389

ScrF I (CCNGG)
79, 567, 668, 945,
946, 992, 1040, 1052,
1232, 2402, 2948, 3441,
3615, 3683, 3796, 4538,
4717, 6547, 7209, 7279,
7414

SfaN I# (GCATC)
1017, 1143, 3250, 3472,
4442, 5551, 7341

SfaN I# (GATGC)
172, 356, 587, 889,
1505, 1631, 2283, 2529,
4592, 4619, 5978, 6044,
6443, 7378

Sma I (CCCGGG)
945

Spe I (ACTAGT)
3911, 6493

Sph I (GCATGC)
1978, 2617

Ssp I (AATATT)
2064, 2415, 2520, 4823,
6096

Stu I (AGGCCT)
3351, 6037

Sty I (CCWWGG)
2766, 2878, 3089, 3535,
4397, 6987, 7656

Taq I (TCGA)
697, 767, 1471, 1652,
1780, 5475, 6836, 7405,
7478, 7794

Tha I (CGCG)
839, 7353

Tth111 I (GACNNNGTC)
3235, 5076

Tth111 II# (CAAPCA)
1600, 1968, 2152, 2814,
2981, 3724, 4243, 4902,
5009, 5249, 6011, 6032,
6172, 6888

Tth111 II# (TGYTTG)
777, 1435, 1626, 2294,
2888, 4101, 4113, 5973,
6281, 6795

Xba I (TCTAGA)
6133

FIGURE 6. (*Continued*)

Xho II (PGATCY)
1515, 1866, 4451, 4987,
6250, 6276, 6946

Xma I (CCCGGG)
945

Xma III (CGGCCG)
619

Restriction enzyme sites not present in BPV-1:

Apa I (GGGCCC)
Bal I (TGGCCA)
Bsm I# (GAATGCN)
BssH II (GCGCGC)
EcoR V (GATATC)
Nae I (GCCGGC)
Not I (GCGGCCGC)
PaeR 71 (CTCGAG)
Pvu I (CGATCG)
Sac I (GAGCTC)
Sac II (CCGCGG)
Sal I (GTCGAC)
Sfi I (GGCCNNNNNGGCC)
SnaB I (TACGTA)
Xho I (CTCGAG)
Xmn I (GAANNNNTTC)

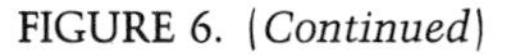

FIGURE 6. (*Continued*)

Aat II (GACGTC)
2928, 4822

Acc I (GTMKAC)
4571, 5842

Aha II (GPCGYC)
2766, 2928, 4822

Alu I (AGCT)
185, 436, 501, 796,
850, 900, 1094, 1103,
1264, 1459, 1658, 1724,
1856, 1869, 1879, 1965,
2214, 2402, 2599, 2773,
3197, 3453, 3712, 4007,
4020, 4064, 4135, 4318,
4455, 4686, 4794, 4982,
5100, 5322, 5417, 5710,
6291, 6784, 6975, 7066,
7206, 7218, 7444

Apa I (GGGCCC)
3802

Ava I (CYCGPG)
609, 3982, 5440

Ava II (GGWCC)
326, 657, 817, 889,
1448, 2220, 2303, 2864,
3070, 3231, 3362, 4230,
4699, 4995, 5437, 5653,
6451, 7699

Bal I (TGGCCA)
27, 7624

BamH I (GGATCC)
810, 6358

Ban I (GGYPCC)
2563, 2879, 5993, 6120,
7671

Ban II (GPGCYC)
1878, 3802, 4783, 6333

Bbv I# (GCAGC)
189, 2243, 4206, 4316,
4514, 4618, 4621, 4777,
6289

Bbv I# (GCTGC)
186, 259, 626, 901,
1003, 1725, 2045, 2048,
2096, 2750, 3652, 3910,
4405

Bcl I (TGATCA)
4899, 7213

Bgl II (AGATCT)
2123, 5074, 7660

Bsm I# (GAATGCN)
2384

Bsp 1286 (GDGCHC)
1280, 1338, 1878, 2897,
3177, 3734, 3778, 3802,
4054, 4152, 4783, 6333,
6968

FIGURE 7. Restriction enzyme sites in CRPV (nucleotide position). # = asymmetric site; N = A,G,C,T; P = A,G; Y = C,T; W = A,T; S = G,C; M = A,C; K = G,T; D = A,G,T; H = A,T,C.

```
BstE II     (GGTNACC)
    3842

BstN I      (CCWGG)
     115,    706,    748,    814,
    1332,   1523,   1613,   3847,
    4693,   5156,   5277,   5346,
    5547,   5632,   5695,   6849,
    7795

BstX I      (CCANNNNNNTGG)
    2854,   2917,   3326,   5156

Cfo I     (GCGC)
     412,    996,   1129,   1245,
    2094,   2778,   3129,   3690,
    4176,   4209,   4775

Dde I       (CTNAG)
      64,     96,    144,   1091,
    1096,   1132,   1403,   1552,
    1738,   1980,   2055,   2413,
    2613,   2829,   3077,   3084,
    3121,   3132,   3157,   3217,
    3358,   4202,   4222,   4679,
    5002,   5218,   5871,   6125,
    6336,   6935,   7068,   7200,
    7541,   7683

Dpn I       (GATC)
      68,    100,    364,    682,
     811,    925,   1186,   1503,
    2010,   2124,   3261,   4027,
    4690,   4900,   5075,   5087,
    5239,   5266,   6095,   6359,
    6545,   6729,   7214,   7223,
    7549,   7562,   7661

Dra I       (TTTAAA)
    1808,   2061,   3057,   7404

EcoR I      (GAATTC)
    1063

EcoR II     (CCWGG)
     115,    706,    748,    814,
    1332,   1523,   1613,   3847,
    4693,   5156,   5277,   5346,
    5547,   5632,   5695,   6849,
    7795

EcoR V      (GATATC)
      39,   5111

Fnu4H I     (GCNGC)
     186,    189,    259,    626,
     901,   1003,   1725,   2045,
    2048,   2096,   2243,   2561,
    2750,   3652,   3723,   3784,
    3886,   3910,   4206,   4316,
    4405,   4514,   4618,   4621,
    4777,   5342,   5866,   6289

FnuD II     (CGCG)
    2801,   3639,   3815,   4003,
    4210,   5335

Fok I#      (GGATG)
    2333,   2513,   2518,   2786,
    4770,   4905,   4914,   5205,
    5280,   5899,   6305,   7358

Fok I#      (CATCC)
    2797,   4838,   5042,   5451,
    5457,   5937,   5966,   7522

Fsp I     (TGCGCA)
     995

Hae II      (PGCGCY)
     411,   1128,   1244,   2093,
    2777,   3128

Hae III     (GGCC)
      28,    317,    709,    746,
    1217,   2606,   2653,   3762,
    3803,   3863,   3885,   4660,
    5153,   5300,   5385,   5393,
    6149,   6469,   6649,   7262,
    7625

Hga I#      (GACGC)
    2766,   3637,   5355,   5801

Hga I#      (GCGTC)
     650,   1274,   5336,   6309

HgiA I      (GWGCWC)
    1280,   1878,   3177,   4152,
    6968

Hha I     (GCGC)
     412,    996,   1129,   1245,
    2094,   2778,   3129,   3690,
    4176,   4209,   4775

Hinc II     (GTYPAC)
     736,   2256,   4141,   4571,
    6991

Hind III     (AAGCTT)
    1093,   1657,   5709

Hinf I      (GANTC)
     612,   1305,   1643,   1740,
    1755,   1828,   2452,   3136,
    3190,   3595,   3904,   4198,
    4544,   4651,   5464,   6042,
    6162,   6211,   6557,   6746,
    7820

HinP I      (GCGC)
     412,    996,   1129,   1245,
    2094,   2778,   3129,   3690,
    4176,   4209,   4775

Hpa I     (GTTAAC)
    2256,   6991

Hpa II      (CCGG)
     617,    655,    676,    727,
     860,    892,   1342,   3093,
    3865,   3892,   3914,   3983,
    4556,   4841,   4895,   5947,
    7462,   7531,   7697,   7703,
    7716,   7748,   7780

Hph I#      (GGTGA)
     589,    868,   1114,   2594,
    3105,   3478,   4897,   6077

Hph I#      (TCACC)
     375,   1882,   3707,   3844,
    5731

Mbo I     (GATC)
      68,    100,    364,    682,
     811,    925,   1186,   1503,
    2010,   2124,   3261,   4027,
    4690,   4900,   5075,   5087,
    5239,   5266,   6095,   6359,
    6545,   6729,   7214,   7223,
    7549,   7562,   7661

Mbo II#     (GAAGA)
     464,    467,    763,    766,
    3096,   3742,   3757,   3953,
    3971,   5108,   6413,   6579,
    6640,   6898,   7088
```

FIGURE 7. (*Continued*)

```
Mbo II#     (TCTTC)
   3294,  3697,  3700,  6606,
   6926

Mlu I     (ACGCGT)
   3814

Mnl I#     (CCTC)
   2145,  2690,  2828,  2901,
   2906,  3302,  3424,  3693,
   3979,  4103,  4636,  4663,
   5001,  5014,  5223,  5454,
   5785,  5870,  6155,  7142,
   7374

Mnl I#     (GAGG)
    214,   403,   552,   688,
    772,   775,   802,   805,
    808,  1180,  1183,  1443,
   1446,  1545,  2617,  3102,
   3115,  3148,  3463,  3485,
   3578,  3760,  3940,  4535,
   4559,  4598,  4681,  4882,
   5023,  5352,  5600,  5749,
   5779,  6129,  6137,  6147,
   6356,  6482,  6621,  6680,
   6947,  7511

Mst II     (CCTNAGG)
   4678

Nae I     (GCCGGC)
   3864

Nci I     (CCSGG)
    616,   654,  1341,  3092,
   3891,  3914,  3982,  3983,
   7461,  7696,  7702,  7715,
   7747,  7779

Nco I     (CCATGG)
   3382,  6904

Nde I     (CATATG)
   2314,  4742

Nhe I     (GCTAGC)
   2774,  4021,  5101

Nla III     (CATG)
     20,   444,  1238,  1542,
   1947,  2300,  2444,  2486,
   2494,  2591,  2639,  2684,
   3383,  3687,  4157,  4179,
   4296,  4377,  4430,  4720,
   4801,  4921,  5620,  5998,
   6236,  6265,  6424,  6793,
   6905,  6941,  7024,  7469,
   7804,  7831

Nla IV     (GGNNCC)
    325,   709,   718,   810,
    817,   888,  1526,  2563,
   2879,  3362,  3762,  3802,
   4229,  4230,  4698,  4884,
   4995,  5437,  5652,  5781,
   5993,  6120,  6358,  6484,
   6774,  7671,  7699

Nru I     (TCGCGA)
   4002

Nsi I     (ATGCAT)
    294,  4293,  5934,  7414

Pst I     (CTGCAG)
    187,   260,   330,   902,
    916,  4504,  4730,  7106

Pvu II     (CAGCTG)
   4454,  4793,  6290

Rsa I     (GTAC)
    452,  1049,  1372,  1708,
   2183,  3342,  3818,  4343,
   4581,  4835,  5122,  5623,
   6855,  6917,  7096,  7278,
   7338,  7632

Sac I     (GAGCTC)
   1878

Sal I     (GTCGAC)
   4571

Sau3A I     (GATC)
     68,   100,   364,   682,
    811,   925,  1186,  1503,
   2010,  2124,  3261,  4027,
   4690,  4900,  5075,  5087,
   5239,  5266,  6095,  6359,
   6545,  6729,  7214,  7223,
   7549,  7562,  7661

Sau96 I     (GGNCC)
    326,   657,   709,   745,
    817,   889,  1217,  1448,
   2220,  2303,  2652,  2864,
   3070,  3231,  3362,  3762,
   3802,  3803,  4230,  4660,
   4699,  4995,  5153,  5392,
   5437,  5653,  6451,  6648,
   7699

ScrF I     (CCNGG)
    115,   616,   654,   706,
    748,   814,  1332,  1341,
   1523,  1613,  3092,  3847,
   3891,  3914,  3982,  3983,
   4693,  5156,  5277,  5346,
   5547,  5632,  5695,  6849,
   7461,  7696,  7702,  7715,
   7747,  7779,  7795

SfaN I#     (GCATC)
     72,  1321,  1538,  1574,
   5936,  6714

SfaN I#     (GATGC)
   1467,  2790,  3036,  5491,
   6306,  6670,  7413

Sma I     (CCCGGG)
   3982

Spe I     (ACTAGT)
   6394,  7420

Sph I     (GCATGC)
   4178,  7023,  7830

Ssp I     (AATATT)
    195,   539,  5662,  6859

Stu I     (AGGCCT)
   2605,  5299,  5384

Sty I     (CCWWGG)
    585,   741,  3382,  4518,
   6882,  6904

Taq I     (TCGA)
    800,  1061,  1308,  1594,
   1621,  3416,  4112,  4572,
   4880,  5174,  6145,  6207,
   6273,  7738,  7770
```

FIGURE 7. (*Continued*)

```
Tha I      (CGCG)
     2801,  3639,  3815,  4003,
     4210,  5335

Tth111 I       (GACNNNGTC)
     2483,  3745

Tth111 II#       (CAAPCA)
     1832,  2728,  4095,  4751,
     4960,  5054,  5368,  5400,
     5423,  6082,  6280

Tth111 II#       (TGYTTG)
      251,   788,  1462,  4042,
     6000,  6300

Xba I      (TCTAGA)
     4665,  5241,  5775

Xho II       (PGATCY)
      363,   681,   810,   924,
     2123,  5074,  5238,  5265,
     6358,  6728,  7660

Xma I      (CCCGGG)
     3982

Xmn I      (GAANNNNTTC)
     1527,  2603
```

Restriction enzyme sites not present in CRPV:

```
Ban III    (ATCGAT)
Bgl I      (GCCNNNNNGGC)
Bsm I#     (GCATTC)
BssH II    (GCGCGC)
Cla I      (ATCGAT)
Kpn I      (GGTACC)
Nar I      (GGCGCC)
Not I      (GCGGCCGC)
PaeR 71    (CTCGAG)
Pvu I      (CGATCG)
Sac II     (CCGCGG)
Sca I      (AGTACT)
Sfi I      (GGCCNNNNNGGCC)
SnaB I     (TACGTA)
Xho I      (CTCGAG)
Xma III    (CGGCCG)
```

FIGURE 7. (*Continued*)

```
Aat II      (GACGTC)
     3278

Acc I      (GTMKAC)
     3475,  5552,  6189

Aha II       (GPCGYC)
      949,  1343,  3278,  4657,
     4935,  6726

Alu I      (AGCT)
      132,   216,   743,  1182,
     1294,  1604,  1767,  1880,
     2063,  2356,  2452,  2632,
     2804,  2890,  2909,  3198,
     3533,  3577,  3877,  4750,
     5134,  5262,  5345,  5408,
     6560,  7390

Ava II      (GGWCC)
       97,   120,   555,  2129,
     3326,  6612,  7235

BamH I      (GGATCC)
     3539

Ban I      (GGYPCC)
      343,   615,   949,  2900,
     2913,  3567,  4935,  6726

Ban II      (GPGCYC)
     2355,  3217,  6736

Bbv I#      (GCAGC)
     2230,  2630,  2932,  4603,
     4703,  6706

Bbv I#       (GCTGC)
      217,  1189,  1643,  1646,
     3210,  3921,  6055,  7265,
     7655

Bcl I      (TGATCA)
      226

Bgl I      (GCCNNNNNGGC)
     4885

Bgl II      (AGATCT)
      594,  1667,  5582,  6358

Bsm I#      (GAATGCN)
     4379

Bsp 1286       (GDGCHC)
      780,   789,  1914,  2355,
     3217,  4922,  6736,  7675

BstE II      (GGTNACC)
      896,  2945,  5441
```

FIGURE 8. Restriction enzyme sites in HPV-1a (nucleotide position). # = asymmetric site; N = A,G,C,T; P = A,G; Y = C,T; W = A,T; S = G,C; M = A,C; K = G,T; D = A,G,T; H = A,T,C.

```
BstN I    (CCWGG)
     558,   961,   973,  2498,
    3303,  4863,  4919,  5950,
    5981,  6194

BstX I    (CCANNNNNNTGG)
    6387

Cfo I    (GCGC)
     731,   950,  1065,  1308,
    3459,  3919,  4936,  5449,
    6727,  6888,  6921,  7478,
    7592

Dde I    (CTNAG)
     210,   419,   653,  1053,
    1161,  1617,  1877,  1944,
    2247,  2601,  2612,  3112,
    3124,  3513,  3588,  4550,
    4784,  4798,  4836,  5042,
    5087,  5189,  5388,  5728,
    6520,  6734,  6935

Dpn I    (GATC)
     227,   595,   724,   764,
    1235,  1668,  2495,  2673,
    3382,  3445,  3540,  3757,
    4215,  5583,  5659,  6359,
    6541,  6676,  6748,  6781

Dra I    (TTTAAA)
    1027,  1328,  1538,  2078,
    2344,  2456,  2480,  2834,
    5790,  7540

EcoR I    (GAATTC)
     967,  7778

EcoR II    (CCWGG)
     558,   961,   973,  2498,
    3303,  4863,  4919,  5950,
    5981,  6194

EcoR V    (GATATC)
    4527

Fnu4H I    (GCNGC)
     217,   952,  1189,  1271,
    1310,  1643,  1646,  2230,
    2630,  2932,  3210,  3921,
    4603,  4703,  4885,  6055,
    6706,  7265,  7655

FnuD II    (CGCG)
     720,  1066,  1307,  1365,
    3460,  3918,  4855,  6178,
    6211,  7591,  7631

Fok I#    (GGATG)
     272,  1165,  1198,  1855,
    2320,  2427,  5081,  6794,
    7560

Fok I#    (CATCC)
    5379,  5569

Hae II    (PGCGCY)
     730,   949,  4935,  6726

Hae III    (GGCC)
    1270,  1789,  3119,  3364,
    4156,  4182,  4224,  4554,
    4636,  5236,  5725,  5752,
    5779,  6385

Hga I#    (GACGC)
    1363

Hga I#    (GCGTC)
     581,  1344,  2778,  4658,
    5396

HgiA I    (GWGCWC)
     789,  1914,  2355,  7675

Hha I    (GCGC)
     731,   950,  1065,  1308,
    3459,  3919,  4936,  5449,
    6727,  6888,  6921,  7478,
    7592

Hinc II    (GTYPAC)
       1,  3618,  6576

Hind III    (AAGCTT)
    2062,  3532,  3876

Hinf I    (GANTC)
      86,   332,   995,  1220,
    1748,  2690,  2829,  2841,
    2993,  3420,  4350,  4645,
    4983,  5230,  5297,  5319,
    5489,  5851,  5998,  7304

HinP I    (GCGC)
     731,   950,  1065,  1308,
    3459,  3919,  4936,  5449,
    6727,  6888,  6921,  7478,
    7592

Hpa I    (GTTAAC)
       1,  6576

Hpa II    (CCGG)
     118,   347,  3442,  7308

Hph I#    (GGTGA)
     896,  1279,  2492,  4110,
    4506,  5007,  5046,  5067,
    5196,  5923,  6016,  6226

Hph I#    (TCACC)
    2844,  4298,  5620

Kpn I    (GGTACC)
     343

Mbo I    (GATC)
     227,   595,   724,   764,
    1235,  1668,  2495,  2673,
    3382,  3445,  3540,  3757,
    4215,  5583,  5659,  6359,
    6541,  6676,  6748,  6781

Mbo II#    (GAAGA)
     356,  1811,  2576,  2670,
    2924,  3379,  3409,  4087,
    4980,  5127,  5130,  5253,
    5316,  6010,  6532,  6673

Mbo II#    (TCTTC)
     564,  1588,  1676,  3189,
    3504,  4332,  4572,  4771,
    5286,  5466,  5528,  6697,
    7158,  7570

Mnl I#    (CCTC)
     145,   619,   652,   717,
     888,  1052,  1622,  2132,
    2234,  2360,  2600,  2782,
    3121,  3319,  3430,  3489,
    3854,  3943,  4106,  4423,
    4627,  4639,  5470,  5785,
    5898,  6667,  6730,  7414
```

FIGURE 8. (*Continued*)

```
Mnl I#     (GAGG)
     341,   440,   610,   613,
     634,   637,   748,   854,
     869,   881,   947,  1853,
    2184,  2582,  2859,  2943,
    3114,  3117,  3397,  3403,
    3674,  3714,  4051,  4075,
    4084,  4113,  4375,  4524,
    4758,  5079,  5193,  5199,
    5732,  5750,  6235,  6446,
    6628,  7287

Mst II     (CCTNAGG)
    1052

Nar I      (GGCGCC)
     949,  4935,  6726

Nci I      (CCSGG)
    3441

Nhe I      (GCTAGC)
    1085,  1601

Nla III    (CATG)
     405,  1493,  1692,  2280,
    2468,  2997,  3643,  3946,
    5268,  5536,  5839,  6021,
    6107,  6139,  6852,  7032,
    7061,  7232

Nla IV     (GGNNCC)
     343,   615,   949,  1268,
    2900,  2913,  3218,  3539,
    3567,  4182,  4859,  4935,
    5723,  6612,  6726

Nru I      (TCGCGA)
    6177

Nsi I      (ATGCAT)
     406,  2313,  5627

Pst I      (CTGCAG)
    3223,  3525

Pvu II     (CAGCTG)
     742,  2908

Rsa I      (GTAC)
     344,   828,  1372,  1532,
    1698,  2160,  3065,  3128,
    3228,  3354,  3593,  3830,
    4581,  4680,  4912,  5909,
    6485,  6863,  7229,  7497,
    7719

Sac I      (GAGCTC)
    2355

Sac II     (CCGCGG)
    7630

Sau3A I    (GATC)
     227,   595,   724,   764,
    1235,  1668,  2495,  2673,
    3382,  3445,  3540,  3757,
    4215,  5583,  5659,  6359,
    6541,  6676,  6748,  6781

Sau96 I    (GGNCC)
      97,   120,   555,  1269,
    1788,  2129,  3326,  4182,
    4224,  4636,  5724,  5778,
    6612,  7235

Sca I      (AGTACT)
    6484,  7496

ScrF I     (CCNGG)
     558,   961,   973,  2498,
    3303,  3441,  4863,  4919,
    5950,  5981,  6194

SfaN I#    (GCATC)
     241,  3606,  4329,  4437,
    4467

SfaN I#    (GATGC)
     273,  1598,  2243,  2267,
    3051,  4788,  5806,  5875,
    6268

SnaB I     (TACGTA)
    3907

Spe I      (ACTAGT)
    2393,  6871

Ssp I      (AATATT)
    1211,  2122,  2399,  4016,
    6619

Stu I      (AGGCCT)
    3118,  4155,  4553

Sty I      (CCWWGG)
      40,  3204,  4534,  4689,
    4889,  6701

Taq I      (TCGA)
     705,  1144,  1223,  3276,
    3286,  6779,  7768

Tha I      (CGCG)
     720,  1066,  1307,  1365,
    3460,  3918,  4855,  6178,
    6211,  7591,  7631

Tth111 I   (GACNNNGTC)
    1363

Tth111 II#   (CAAPCA)
     988,  3958,  4318,  5019,
    6100,  7214,  7239,  7696

Tth111 II#   (TGYTTG)
     267,  2114,  3817,  5576,
    7452

Xba I      (TCTAGA)
    5854,  6669,  7184

Xho II     (PGATCY)
     594,  1667,  3381,  3444,
    3539,  4214,  5582,  6358

Xmn I      (GAANNNNTTC)
     751,  2527,  2706,  5454
```

Restriction enzyme sites not present in HPV-1a:

```
Apa I      (GGGCCC)
Ava I      (CYCGPG)
Bal I      (TGGCCA)
Ban III    (ATCGAT)
Bsm I#     (GCATTC)
BssH II    (GCGCGC)
Cla I      (ATCGAT)
Fsp I      (TGCGCA)
Mlu I      (ACGCGT)
Nae I      (GCCGGC)
Nco I      (CCATGG)
Nde I      (CATATG)
```

FIGURE 8. (*Continued*)

```
Not I       (GCGGCCGC)
PaeR 71     (CTCGAG)
Pvu I       (CGATCG)
Sal I       (GTCGAC)
Sfi I       (GGCCNNNNNGGCC)
Sma I       (CCCGGG)
Sph I       (GCATGC)
Xho I       (CTCGAG)
Xma I       (CCCGGG)
Xma III     (CGGCCG)
```

FIGURE 8. (*Continued*)

```
Acc I      (GTMKAC)
     617,   1559,   2060,   4416,
    7815

Aha II     (GPCGYC)
    1162,   4773,   7691

Alu I      (AGCT)
     484,    623,    771,    894,
    1260,   2308,   3571,   3910,
    4263,   4461,   4469

Apa I      (GGGCCC)
    4436,   5363

Ava I      (CYCGPG)
    6658

Ava II     (GGWCC)
      33,    257,   1071,   3491,
    5038,   7203

BamH I     (GGATCC)
    4722

Ban I      (GGYPCC)
    3697,   3895,   4773,   7691

Ban II     (GPGCYC)
    4436,   5363

Bbv I#     (GCAGC)
     285,   1447,   1783,   2348,
    3288,   5345,   7786

Bbv I#     (GCTGC)
     298,    508,   2223,   3911,
    4060,   5407,   5440,   6488,
    7253

Bgl I      (GCCNNNNNGGC)
    6703

Bsm I#     (GAATGCN)
     206,   1471,   2628,   2886

Bsm I#     (GCATTC)
    5369

Bsp 1286   (GDGCHC)
    3259,   3438,   4436,   4528,
    4786,   5363,   6258

BssH II    (GCGCGC)
    7270

BstE II    (GGTNACC)
    6185

BstN I     (CCWGG)
    1217,   2755,   3153,   4193,
    5062,   5112,   5485,   6191,
    6724,   7627,   7688,   7695

BstX I     (CCANNNNNNTGG)
    7681

Cfo I      (GCGC)
     292,    471,    810,   1372,
    4774,   7228,   7270,   7272,
    7472,   7692

Dde I      (CTNAG)
     392,    625,   2381,   2708,
    2873,   2976,   3183,   3999,
    4520,   6337,   6681,   7103,
    7439,   7564,   7842

Dpn I      (GATC)
    4546,   4723,   6014,   6053,
    7088,   7151

Dra I      (TTTAAA)
      18,    670,   1465,   1510,
    1675,   1897,   2010,   2212,
    3050,   3212,   3610,   7123,
    7520

EcoR I     (GAATTC)
    2188

EcoR II    (CCWGG)
    1217,   2755,   3153,   4193,
    5062,   5112,   5485,   6191,
    6724,   7627,   7688,   7695

EcoR V     (GATATC)
    3994,   5979

Fnu4H I    (GCNGC)
     271,    285,    298,    508,
    1447,   1783,   2223,   2348,
    3288,   3424,   3911,   4060,
    5345,   5407,   5440,   5794,
    6488,   6663,   7253,   7786
```

FIGURE 9. Restriction enzyme sites in HPV-6b (nucleotide position). # = asymmetric site; N = A,G,C,T; P = A,G; Y = C,T; W = A,T; S = G,C; M = A,C; K = G,T; D = A,G,T; H = A,T,C.

```
FnuD II      (CGCG)
    293,     472,    7271,    7455,
   7471

Fok I#      (GGATG)
    707,     877,     939,    4232,
   4848,    5193,    5448,    5857,
   6254,    6364

Fok I#      (CATCC)
   4817,    5045,    5731,    5924,
   6137,    6813

Fsp I    (TGCGCA)
    809

Hae II      (PGCGCY)
   1371,    4773,    7227,    7691

Hae III      (GGCC)
   1505,    2263,    2870,    3481,
   3685,    4437,    4676,    4713,
   5078,    5089,    5296,    5364,
   5623,    5695,    5796,    6089,
   6937,    7045

Hga I#      (GACGC)
   1162,    3410,    4445,    7167

Hga I#      (GCGTC)
   4456,    5284,    6865
HgiA I      (GWGCWC)
   3259,    3438,    4528

Hha I     (GCGC)
    292,     471,     810,    1372,
   4774,    7228,    7270,    7272,
   7472,    7692

Hinc II      (GTYPAC)
   6377,    6650

Hind III      (AAGCTT)
   2307

Hinf I      (GANTC)
    659,     841,    1990,    2704,
   2960,    3350,    3447,    3599,
   4750,    5218,    5473,    6038,
   6955,    7437

HinP I      (GCGC)
    292,     471,     810,    1372,
   4774,    7228,    7270,    7272,
   7472,    7692

Hpa I    (GTTAAC)
   6650

Hpa II      (CCGG)
   1297,    4610,    7862

Hph I#      (GGTGA)
   3596,    4354,    6323,    6503

Hph I#      (TCACC)
   5392

Mbo I     (GATC)
   4546,    4723,    6014,    6053,
   7088,    7151

Mbo II#      (GAAGA)
    369,     372,     537,     629,
    817,    2088,    2551,    2719,
   2780,    3407,    4747,    5413,
   6875,    6971,    7016

Mbo II#      (TCTTC)
   6827

Mlu I    (ACGCGT)
   7454

Mnl I#      (CCTC)
    118,     578,    2266,    2515,
   3378,    3393,    3517,    3687,
   3690,    4690,    4820,    5074,
   5080,    5149,    5431,    5822,
   5825,    6678,    6960,    6995,
   7206,    7248,    7419,    7603

Mnl I#      (GAGG)
     30,      94,     268,     635,
    859,     937,     943,     946,
    952,    1042,    1129,    1345,
   1351,    1363,    1579,    1945,
   2085,    2318,    2710,    2716,
   3242,    3444,    3740,    4586,
   4939,    4993,    5191,    5242,
   6095,    6572,    6683,    7115

Mst II      (CCTNAGG)
   4519,    7841

Nar I      (GGCGCC)
   4773,    7691

Nco I      (CCATGG)
   4421

Nde I      (CATATG)
    235,     280,    1100,    1938,
   3729,    5165,    6445,    6498

Nla III      (CATG)
    312,     529,     533,     542,
    969,    1288,    1396,    1529,
   1643,    2074,    2335,    2414,
   2807,    2827,    2834,    3201,
   3710,    3882,    4169,    4422,
   4481,    4499,    5215,    5738,
   5891,    6081,    6247,    6459,
   6799,    6862,    6869,    7497

Nla IV      (GGNNCC)
     32,    2989,    3491,    3697,
   3895,    4436,    4606,    4704,
   4713,    4722,    4773,    5362,
   5363,    6580,    7691

Nsi I      (ATGCAT)
    171,     238,     530,    1640,
   2824,    2951,    4336,    6244

Pst I      (CTGCAG)
   3912,    5408,    6489

Rsa I      (GTAC)
    493,     741,     848,     872,
    917,    1294,    1387,    1697,
   1709,    1802,    1823,    2161,
   2231,    2286,    2321,    2790,
   3879,    3902,    4012,    4028,
   4267,    4625,    5051,    5066,
   5203,    5307,    6300,    6314,
   6792,    6859,    7008,    7215,
   7305,    7329,    7631

Sau3A I      (GATC)
   4546,    4723,    6014,    6053,
   7088,    7151
```

FIGURE 9. (*Continued*)

```
Sau96 I     (GGNCC)
       33,    257,   1071,   2263,
     2869,   3481,   3491,   3684,
     4436,   4437,   4675,   4713,
     5038,   5088,   5363,   5364,
     5623,   5695,   7203

Sca I     (AGTACT)
     2789,   4027

ScrF I     (CCNGG)
     1217,   2755,   3153,   4193,
     5062,   5112,   5485,   6191,
     6724,   7627,   7688,   7695

SfaN I#     (GCATC)
     3311,   3975,   5730,   6812,
     7664

SfaN I#     (GATGC)
     2377,   2401,   2747,   3179,
     4161,   4251,   5858

Spe I     (ACTAGT)
     4669,   4861,   4876

Sph I     (GCATGC)
     1395,   6080

Ssp I     (AATATT)
      994,   1413,   4551,   4842

Stu I     (AGGCCT)
     6088

Sty I     (CCWWGG)
      466,   3402,   3428,   4421

Taq I     (TCGA)
      266,    724,   6051

Tha I     (CGCG)
      293,    472,   7271,   7455,
     7471

Tth111 II#     (CAAPCA)
     1379,   1861,   2803,   3029,
     3096,   3265,   3396,   4305,
     4477,   6233

Tth111 II#     (TGYTTG)
     6540,   6762,   7652,   7809

Xba I     (TCTAGA)
     5903

Xho II     (PGATCY)
     4722,   6013,   7087

Xmn I     (GAANNNNTTC)
     2188
```

Restriction enzyme sites not present in HPV-6b:

```
Aat II        (GACGTC)
Bal I         (TGGCCA)
Ban III       (ATCGAT)
Bcl I         (TGATCA)
Bgl II        (AGATCT)
Cla I         (ATCGAT)
Kpn I         (GGTACC)
Nae I         (GCCGGC)
Nci I         (CCSGG)
Nhe I         (GCTAGC)
Not I         (GCGGCCGC)
Nru I         (TCGCGA)
PaeR 71       (CTCGAG)
Pvu I         (CGATCG)
Pvu II        (CAGCTG)
Sac I         (GAGCTC)
Sac II        (CCGCGG)
Sal I         (GTCGAC)
Sfi I         (GGCCNNNNNGGCC)
Sma I         (CCCGGG)
SnaB I        (TACGTA)
Tth111 I      (GACNNNGTC)
Xho I         (CTCGAG)
Xma I         (CCCGGG)
Xma III       (CGGCCG)
```

FIGURE 9. (*Continued*)

```
Acc I     (GTMKAC)
      781,   1167,   1971,   2089,
     3687,   3702,   3828,   3955,
     4138,   4946,   5667,   6479,
     6850

Aha II     (GPCGYC)
     1309

Alu I     (AGCT)
      156,    552,    652,    684,
     1196,   1223,   1604,   2297,
     2671,   3571,   4477,   5065,
     5503,   6369,   7094,   7147,
     7443

Apa I     (GGGCCC)
     4528

Ava II     (GGWCC)
      112,    498,    679,   2713,
     5471

Bal I     (TGGCCA)
     3398,   5990

BamH I     (GGATCC)
     6150

Ban I     (GGYPCC)
      880,   1309,   2928,   3442,
     5377
```

FIGURE 10. Restriction enzyme sites in HPV-16 (nucleotide position). # = asymmetric site; N = A,G,C,T; P = A,G; Y = C,T; W = A,T; S = G,C; M = A,C; K = G,T; D = A,G,T; H = A,T,C.

```
Ban II      (GPGCYC)
     707,    4528

Bbv I#      (GCAGC)
    1476,    1809,    1812,    2295,
    3356,    3390,    4004,    4741,
    5297,    6042

Bbv I#      (GCTGC)
     157,    1224,    1578,    3940,
    5648,    6672

Bsp 1286     (GDGCHC)
     707,     813,     831,    2439,
    3441,    3545,    4528,    4587,
    6580

BstE II     (GGTNACC)
    6533,    6612

BstN I      (CCWGG)
    4883,    5528,    6180,    6864,
    6902,    7428

BstX I      (CCANNNNNNTGG)
    2891,    6590,    6823

Cfo I       (GCGC)
    1310,    2862,    7425,    7672

Dde I       (CTNAG)
      24,     654,     845,    1146,
    1285,    3515,    3534,    4335,
    5199,    5692,    7761

Dpn I       (GATC)
     525,     622,     871,    3479,
    4520,    4538,    5072,    5234,
    6151,    6951,    7014

Dra I       (TTTAAA)
     301,    1445,    1494,    1539,
    2775,    2905,    3265,    3624,
    3645,    3667,    6986,    7282,
    7491,    7520,    7752

EcoR I      (GAATTC)
    6818,    7453

EcoR II     (CCWGG)
    4883,    5528,    6180,    6864,
    6902,    7428

Fnu4H I     (GCNGC)
     157,    1224,    1476,    1578,
    1809,    1812,    2295,    3356,
    3390,    3412,    3940,    4004,
    4741,    5297,    5648,    5924,
    6042,    6672

FnuD II     (CGCG)
    3414

Fok I#      (GGATG)
     666,     904,     913,     954,
    1161,    2483,    2504,    3076,
    5708,    6014,    6224

Fok I#      (CATCC)
    3868,    4788,    5579,    5752,
    5775,    5994,    7389

Hae II      (PGCGCY)
    1309,    7424

Hae III     (GGCC)
    1952,    2590,    2889,    3399,
    4467,    4529,    5106,    5660,
    5936,    5991,    6588,    7061,
    7431,    7765

Hga I#      (GACGC)
    4437,    7030

HgiA I      (GWGCWC)
    3545,    4587,    6580

Hha I       (GCGC)
    1310,    2862,    7425,    7672

Hinc II     (GTYPAC)
    3208

Hinf I      (GANTC)
     274,     474,     745,    2129,
    2580,    2760,    3532,    7631

HinP I      (GCGC)
    1310,    2862,    7425,    7672

Hpa II      (CCGG)
      57,     501,     700,    3494,
    5375

Hph I#      (GGTGA)
     997,    1511,    2172,    3631,
    5564,    6183,    6228,    6438

Hph I#      (TCACC)
    5303,    7778

Kpn I       (GGTACC)
     880,    5377

Mbo I       (GATC)
     525,     622,     871,    3479,
    4520,    4538,    5072,    5234,
    6151,    6951,    7014

Mbo II#     (GAAGA)
     440,     799,     892,    1000,
    1251,    3175,    3448,    4562,
    4814,    5243,    6879,    6945

Mnl I#      (CCTC)
    1788,    2541,    3371,    3555,
    4008,    4469,    4745,    5118,
    5301,    5391,    5676,    5899,
    6538,    6858,    6930,    7108,
    7117

Mnl I#      (GAGG)
     224,     658,     661,     664,
     895,     928,    1395,    2170,
    2733,    2739,    3199,    4767,
    5658,    5952,    6738,    6834,
    6868

Mst II      (CCTNAGG)
    4334

Nar I       (GGCGCC)
    1309

Nci I       (CCSGG)
    5374

Nco I       (CCATGG)
     863

Nde I       (CATATG)
     279,    3126,    3761,    4986,
    5001,    6358
```

FIGURE 10. (*Continued*)

```
Nla III     (CATG)
       13,    173,    561,    565,
      586,    864,   1066,   1314,
     1330,   1666,   1672,   2859,
     3091,   3235,   3313,   3707,
     4076,   4296,   4314,   5294,
     5593,   5742,   6157,   6732,
     7308,   7464,   7546,   7793,
     7842

Nla IV     (GGNNCC)
      112,    880,   1309,   2928,
     3442,   4528,   5377,   6150

Nsi I     (ATGCAT)
      253,    562,   2476,   3139

Pst I     (CTGCAG)
      875,   3692,   4755,   5238,
     6787,   7003

Pvu II     (CAGCTG)
      551,    683

Rsa I     (GTAC)
      196,    766,    791,    881,
      899,   1111,   1373,   1452,
     1804,   1921,   1944,   2526,
     2575,   2822,   3619,   3680,
     4107,   4311,   4428,   4601,
     4800,   5127,   5140,   5172,
     5204,   5226,   5378,   5387,
     6160,   6310,   6430,   6447,
     6652,   6722,   7529,   7641

Sau3A I     (GATC)
      525,    622,    871,   3479,
     4520,   4538,   5072,   5234,
     6151,   6951,   7014

Sau96 I     (GGNCC)
      112,    498,    679,   1951,
     2713,   4528,   4529,   5471,
     5935,   6587,   7431

Sca I     (AGTACT)
     5225

ScrF I     (CCNGG)
     4883,   5374,   5528,   6180,
     6864,   6902,   7428

SfaN I#     (GCATC)
     3181,   3867,   4271

SfaN I#     (GATGC)
      955,   1105,   2406,   2430,
     3274,   3634,   4580,   5558,
     6543

SnaB I     (TACGTA)
     5169

Spe I     (ACTAGT)
     1461,   4568,   4865

Sph I     (GCATGC)
     7463

Ssp I     (AATATT)
      181,    718,   1176,   1440,
     1633,   1845,   3330,   3976,
     4366,   5453,   5813,   6551,
     7224

Stu I     (AGGCCT)
     4466

Sty I     (CCWWGG)
      863,   3437,   4906

Taq I     (TCGA)
      505

Tha I     (CGCG)
     3414

Tth111 I     (GACNNNGTC)
     2708

Tth111 II#     (CAAPCA)
      161,    202,    691,   1425,
     1890,   2965,   4300,   5162,
     5728,   6089,   6093,   7047,
     7880

Tth111 II#     (TGYTTG)
     4193,   4197,   4201,   6400,
     6911,   7181,   7196,   7206,
     7216,   7242,   7288,   7539,
     7581

Xho II     (PGATCY)
     4519,   5071,   6150

Xmn I     (GAANNNNTTC)
     4814
```

FIGURE 10. (*Continued*)

```
Restriction enzyme sites not present in HPV-16:

Aat II      (GACGTC)
Ava I       (CYCGPG)
Ban III     (ATCGAT)
Bcl I       (TGATCA)
Bgl I       (GCCNNNNNGGC)
Bgl II      (AGATCT)
Bsm I#      (GAATGCN)
Bsm I#      (GCATTC)
BssH II     (GCGCGC)
Cla I       (ATCGAT)
EcoR V      (GATATC)
Fsp I       (TGCGCA)
Hga I#      (GCGTC)
Hind III    (AAGCTT)
Hpa I       (GTTAAC)
Mbo II#     (TCTTC)
Mlu I       (ACGCGT)
Nae I       (GCCGGC)
Nhe I       (GCTAGC)
Not I       (GCGGCCGC)
Nru I       (TCGCGA)
PaeR 71     (CTCGAG)
Pvu I       (CGATCG)
Sac I       (GAGCTC)
Sac II      (CCGCGG)
Sal I       (GTCGAC)
Sfi I       (GGCCNNNNNGGCC)
Sma I       (CCCGGG)
Xba I       (TCTAGA)
Xho I       (CTCGAG)
Xma I       (CCCGGG)
Xma III     (CGGCCG)
```

FIGURE 10. (*Continued*)

III. GENOME ORGANIZATION

Papillomaviruses are unusual in their genomic organization in that only one strand appears to be expressed as mRNA and protein (Danos *et al.*, 1983). The coding strands for BPV-1, CRPV, HPV-1a, HPV-6b, and HPV-16 were analyzed for all major open reading frames (ORFs) using the programs of Mount and Conrad. These results are shown in Tables I–V, respectively. Nomenclature of the ORFs follows that first used for BPV-1 (Danos *et al.*, 1983) where "E" refers to the "early" or transforming region of BPV-1 and "L" refers to the "late" or wart-specific region. The same nomenclature is used for other papillomavirus ORFs which are either homologous at the protein level or are located in the same region of the genome. Note that ORFs analogous to E3 and E8 of BPV-1 are not found in all viruses.

The genomic organization of the papillomaviruses is shown schematically in Fig. 11. Transcription is from left to right. The ORFs are shown as open boxes. Dashed vertical lines within the ORFs indicate the position of the first methionine codon. The location of mRNA start sites (and therefore putative transcriptional promoters), where known, are indicated by "P." For BPV-1 the transformed cell mRNA start sites (P_T) are located at n. 89, approximately n. 2440, approximately n. 3100 (Yang *et al.*, 1985a; Stenlund *et al.*, 1985), and n. 7185 (Baker and Howley, personal communication) and the major productive infection mRNA start site (P_L)

at approximately n. 7250 (Baker and Howley, personal communication). mRNA start sites for CPRV in VX2 carcinomas (P_E) are found at n. 87 and n. 158 (2-kb mRNA) and at n. 905 and n. 975 (1.25-kb mRNA) (Danos *et al.*, 1985) and for CRPV in papillomas (P_L) at n. 7523 (Felix Wettstein, personal communication). HPV-1a mRNA 5′ termini have been mapped by electron microscopy heteroduplex analysis to the middle of the E7 ORF (major promoter) and just upstream of the E6 ORF (Louis Chow, Tom Broker, Lorne Taichman, and Sheilla Reilly, personal communication). The major start site for HPV-16 mRNA in CaSki cells and cervical tumors is at n. 97 (Smotkin and Wettstein, 1986).

The location of the polyadenylation signal AATAAA is indicated in Fig. 11 by a solid triangle (▲). The polyadenylation signals actually used in BPV-1 are at n. 4180 for the transforming region mRNAs and at n. 7156 for the late mRNAs. There is no AATAAA in HPV-6b at the end of the early region. The alternate polyadenylation signal ATTAAA (■) is present at n. 4554 and probably functions as the early region polyadenlylation signal.

IV. PROTEIN COMPARISONS

The putative protein sequences predicted from the DNA sequences of each virus (Figs. 1–5) and ORF data in Tables I–V were compared for

TABLE I. Open Reading Frames of BPV-1

Open Reading Frame	Frame	First Nucleotide	Methionines	Nucleotide Preceding Stop Codon
E1	3	813	849, 1506, 1596, 1632, 1794, 1938, 1980, 2169, 2619	2663
E2	1	2581	2608, 2653, 2830, 2914, 3022, 3091, 3802	3837
E3	3	3267		3551
E4	2	3173	3191	3526
E5	3	3714	3879, 3924	4010
E6	1	49	91, 187	501
E7	2	449	479	859
E8	1	1099	1204, 1270, 1321, 1375, 1399	1479
L1	2	5597	5609, 6194, 6197, 6305, 6332, 6554, 6701, 6785	7093
L2	2	4172	4187	5593

TABLE II. Open Reading Frames of CRPV

Open Reading Frame	Frame	First Nucleotide	Methionines	Nucleotide Preceding Stop Codon
E1	3	1356	1362, 1911, 2196, 2226, 2268, 2301, 2364, 2445, 2472, 2481, 2487, 2520, 2592, 2640, 2685	3167
E2	1	3088	3112, 3334, 3688, 4180	4281
E4	2	3332	3377, 3473, 3476, 3497, 3539, 3590, 3611, 3620, 3659, 3668	4012
E5	2	4265	4274, 4511	4576
E6	1	100	154, 445	972
E7	1	979	1075	1356
E8	2	182	323	472
L1	2	5780	5828, 5999, 6266, 6299, 6425, 6449, 6536, 6563, 6590, 6671, 6794, 7025	7342
L2	1	4303	4378, 4891, 4993, 5365	5853

TABLE III. Open Reading Frames of HPV-1a

Open Reading Frame	Frame	First Nucleotide	Methionines	Nucleotide Preceding Stop Codon
E1	2	773	812, 1673, 1778, 1922, 1949, 1958, 1964, 2069, 2153, 2294, 2306, 2429	2647
E2	3	2568	2592, 2637, 2745, 2814, 3186, 3693	3794
E4	1	3157	3181	3558
E5	1	3877	4027	4152
E6	2	86	104	523
E7	1	502	529, 541	807
E8	3	45	165, 207, 339	353
L1	1	5347	5413, 5431, 5893, 5992, 6022, 6025, 6046, 6133, 6160, 6187, 6328, 6469, 6610, 6796, 6853	6936
L2	3	3894	3897, 4716	5417

TABLE IV. Open Reading Frames of HPV-6b

Open Reading Frame	Frame	First Nucleotide	Methionines	Nucleotide Preceding Stop Codon
E1	1	715	832, 886, 970, 1651, 1747, 1912, 2023, 2056, 2083, 2092, 2296, 2428, 2440, 2467	2778
E2	2	2696	2723, 2828, 2867, 2888, 2945, 2951, 2987, 3023, 3107, 3173 3809	3826
E4	3	3240	3255, 3285	3581
E5a	2	3872	3887	4159
E5b	1	4003	4159, 4162, 4234, 4252, 4336, 4369	4374
E6	3	30	102, 171, 534, 543	551
E7	2	440	530	823
L1	2	5678	5789, 6221, 6248, 6374, 6398, 6485, 6539, 6800, 6863, 6935, 6953	7288
L2	1	4378	4423, 4681, 5251, 5335, 5425, 5653	5799

TABLE V. Open Reading Frames of HPV-16

Open Reading Frame	Frame	First Nucleotide	Methionines	Nucleotide Preceding Stop Codon
E1a	1	859	865	1167
E1b	3	1104	1293, 1482, 1680, 1770, 1776, 1779, 1941, 2085, 2121, 2127, 2193, 2223, 2325, 2334, 2421, 2502	2810
E2	1	2725	2755, 2860, 2899, 3139, 3841	3849
E4	2	3332		3616
E5	2	3863		4096
E6	2	65	83, 104, 512	556
E7	1	544	562, 595, 811	855
L1	3	5526	5559, 5637, 6081, 6234, 6258, 6345, 6399, 6531, 6660, 6798, 6816	7151
L2	2	4133	4235, 4385, 4826, 5594	5653

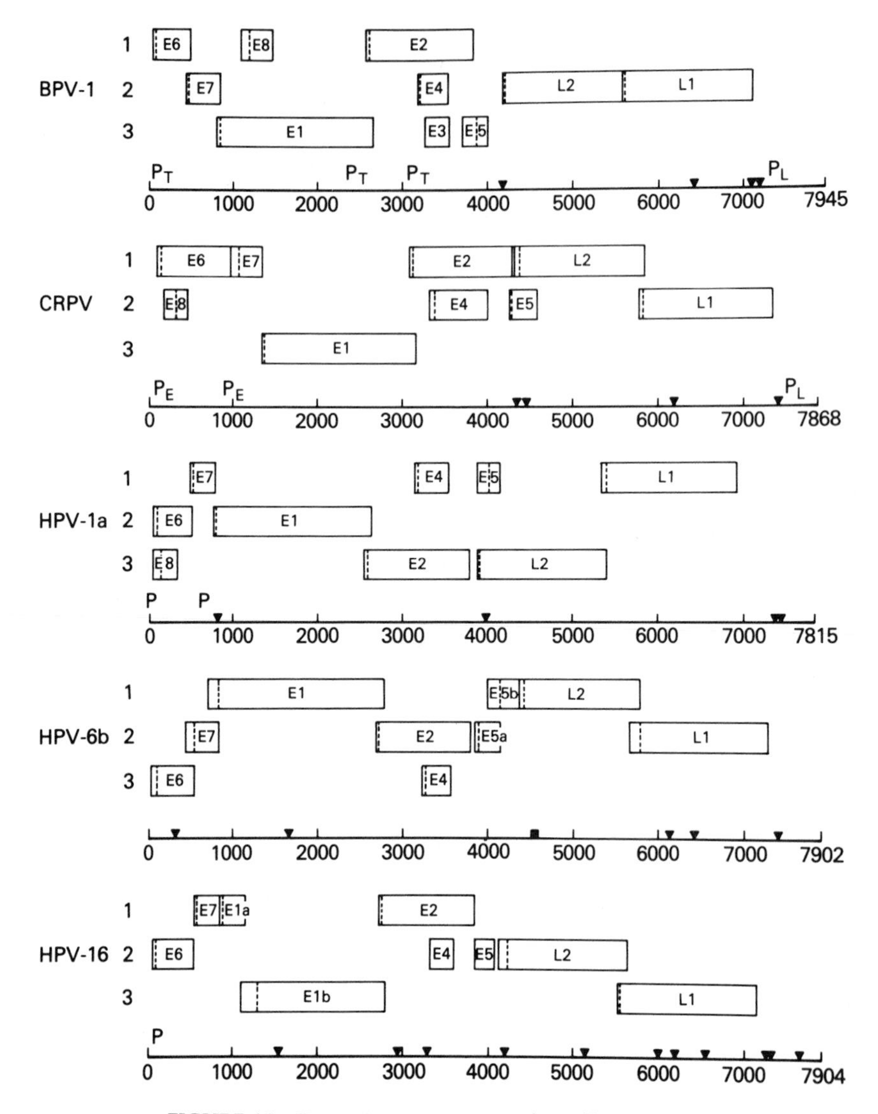

FIGURE 11. Genomic organization of papillomaviruses.

each ORF. The protein sequences for a single ORF were first compared to each other using the FASTP program of Lipman and Pearson (1985), resulting in similarity scores for each pair of proteins. This score was used to arrange the proteins such that adjacent pairs were usually most similar. Each protein pair was then aligned using the PRTALN program of Wilbur and Lipman (1983). A gap penalty of 1 was used for the E1 and L1 proteins and a penalty of 4 for the other ORFs to minimize fragmentation in areas of low homology and thus facilitate alignment of multiple proteins. These aligned pairs of proteins were then manually aligned into a multiple protein comparison by extension of gaps. The statistical significance (z value) of protein similarity was computed with the RDF program of Lipman and Pearson (1985) where:

$$z = \frac{(\text{similarity score} - \text{mean of random scores})}{(\text{standard deviation of random scores})}$$

and z values are interpreted as follows:

$z > 3$	possibly significant
$z > 6$	probably significant
$z > 10$	significant

Similarity scores and z values for the following comparisons are given in Table VI. In the following figures, ":" indicates identity between adjacent pairs of proteins at that position. A solid square (■) above a comparison indicates that all the proteins are identical at that position.

A. E1 ORF

The putative E1 proteins are very highly conserved over the COOH-terminal approximately 60% of the proteins (maximum aligned z value of 212 for BPV-1 versus CRPV), but are much less conserved over the NH_2-terminal portion (Fig. 12). Approximately 23% of the residues are identical in all five papillomaviruses compared. The E1 ORF is split into two ORFs (E1a, E1b) in HPV-16. Sequence analysis of the HPV-16 genome present in the SiHa cervical carcinoma cell line has shown that the E1 ORF is not split in this cell line (Baker *et al.*, 1987), indicating that the published sequence (Seedorf *et al.*, 1985) may be that of a mutant genome. A search of the NBRF (National Biomedical Research Foundation) protein data bank using the FASTP program (Lipman and Pearson, 1985) indicates regions of local homology between E1 and SV40 and polyoma T antigen (Py T) (Fig. 12) (Clertant and Seif, 1984). In Fig. 12 a short region of homology with Py T is shown which corresponds to the ATPase active site of T antigen. A ":" indicates homology with at least one papillomavirus E1 protein; "." indicates a conservative amino acid change (Schulz and Schirmer, 1979). The best alignment is with BPV-1 (initial z value = 13.22; aligned z value = 6.9).

B. E2 ORF

The putative E2 proteins are strongly homologous in their NH_2- and COOH-terminal domains (maximum z value of 94 for HPV-1a versus HPV-6b), but are very divergent in the middle domain which overlaps the

TABLE VI. Similarity Scores and Z Values for Protein Comparisons

VIRUS PAIR	ORF	INITIAL SCORE	INITIAL Z	ALIGNED SCORE	ALIGNED Z	MEAN SCORE	MEAN S.D.	MEAN ALIGNED SCORE	MEAN ALIGNED S.D.
BPV-1/CRPV	E1	903	151.16	1223	211.77	34.1	5.75	36.7	5.60
CRPV/HPV-1a	E1	1182	181.50	1516	112.96	34.0	6.33	34.7	13.11
HPV-1a/HPV-16	E1a	51	5.95	58	6.97	24.9	4.39	28.0	4.30
HPV-1a/HPV16	E1b	1001	216.16	1410	92.66	33.1	4.48	31.7	14.87
HPV-16/HPV-6b	E1a	145	14.17	222	15.43	28.3	8.23	26.0	12.71
HPV-16/HPV-6b	E1b	1498	296.35	1632	176.17	29.3	4.96	31.8	9.08
Py T/BPV1	E1	101	13.22	101	6.90	34.7	5.02	34.9	9.57
Py T/CRPV	E1	86	7.68	89	3.46	37.2	6.36	31.1	16.75
Py T/HPV-1a	E1	73	5.29	86	2.99	37.5	6.72	30.5	18.56
Py T/HPV-16	E1b	72	4.95	77	3.66	35.6	7.34	37.6	10.76
Py T/HPV-6b	E1	72	6.34	78	3.93	33.2	6.13	35.8	10.72
BPV-1/CRPV	E2	391	67.90	447	69.80	28.5	5.34	33.5	5.92
CRPV/HPV-1a	E2	591	104.44	606	62.14	30.9	5.36	32.8	9.22
HPV-1a/HPV-6b	E2	350	56.48	478	94.27	27.3	5.71	33.4	4.72
HPV-6b/HPV-16	E2	661	101.06	833	66.83	30.2	6.24	30.8	12.00
c-mos/BPV-1	E2	29	-0.35	31	-0.18	30.7	4.80	32.8	10.34
c-mos/CRPV	E2	54	3.07	66	3.13	32.9	6.86	34.2	10.18
c-mos/HPV-1a	E2	42	1.48	57	2.20	32.8	6.20	34.2	10.34
c-mos/HPV-6b	E2	48	3.33	48	1.46	28.6	5.82	32.8	10.41
c-mos/HPV-16	E2	34	0.59	34	0.34	30.5	5.96	30.0	11.79
BPV-1/CRPV	E4	35	0.97	35	0.60	29.0	6.17	29.6	9.08
CRPV/HPV-6b	E4	44	3.17	52	2.45	27.3	5.26	30.4	8.85
HPV-6b/HPV-16	E4	51	2.44	136	7.62	32.7	7.48	27.9	14.19
HPV-16/HPV-1a	E4	31	0.16	55	2.70	29.9	7.02	30.1	9.24
DPV/BPV-1	E5	165	19.41	171	9.49	34.1	6.74	29.1	14.95
BPV-1/BPV-2	E5	285	59.71	285	10.15	43.7	4.04	15.3	26.56
BPV-2/HPV-6b	E5a	61	4.67	87	3.08	39.3	4.65	19.7	21.83
HPV-6b/HPV-16	E5a	47	1.31	91	5.05	41.2	4.44	7.4	16.55
CRPV/HPV-1a	E5	119	18.80	122	17.67	24.3	5.04	28.2	5.31
CRPV/HPV-1a	E6	190	25.06	198	29.52	33.8	6.23	36.5	5.47
HPV-1a/HPV-16	E6	201	30.60	210	18.38	32.2	5.51	35.2	9.51
HPV-16/HPV-6b	E6	415	55.37	421	29.30	31.2	6.93	31.4	13.30
HPV-6b/BPV-1	E6	146	17.17	187	9.67	34.6	6.49	27.9	16.45
BPV-1/CRPV	E7	112	19.21	112	16.46	27.2	4.41	30.3	4.96
CRPV/HPV-6b	E7	99	9.71	172	21.02	28.0	7.31	30.5	6.73
HPV-6b/HPV-16	E7	239	43.22	294	31.57	27.5	4.89	28.6	8.41
HPV-16/HPV-1a	E7	139	16.64	169	22.99	29.0	6.61	33.1	5.91
BPV-1/CRPV	L1	410	57.64	1407	267.14	32.7	6.55	36.3	5.13
CRPV/HPV-1a	L1	557	85.97	1683	166.12	32.9	6.10	33.1	9.93
HPV-1a/HPV-16	L1	610	87.75	1615	154.04	34.2	6.56	35.2	10.26
HPV-16/HPV-6b	L1	903	185.04	1979	137.58	33.1	4.70	30.3	14.16
DPV/BPV-1	L2	275	43.11	597	30.28	34.5	5.58	24.8	18.90
BPV-1/CRPV	L2	191	27.62	213	13.10	34.0	5.68	36.3	13.49
CRPV/HPV-1a	L2	309	43.82	408	21.08	35.0	6.25	32.8	17.80
HPV-1a/HPV-11	L2	271	33.76	302	13.96	34.3	7.01	26.2	19.75
HPV-11/HPV-6b	L2	1513	225.95	1932	95.27	35.4	6.54	25.6	20.01
HPV-6b/HPV-16	L2	442	74.84	1010	49.22	36.0	5.42	25.9	19.99

E4 ORF (Fig. 13). Approximately 12% of the residues are identical for all five viruses compared. It has been reported that the COOH-terminal domain of E2 is homologous to a portion of the *c-mos* oncogene (Fig. 13) (Danos and Yaniv, 1984). The significance of this homology is questionable by the RDF test (maximum z value of 3.33 for HPV-6b).

C. E4 ORF

The putative E4 proteins are only weakly homologous at the amino acid sequence level. Therefore, the E4 alignment is not shown. The maximum aligned z value is 7.62 for HPV-6b versus HPV-16 (Table VI).

```
BPV-1                                TSCVHVVNLASVMANDKGSNWDSGLGCSY--LLTE  33
                                                 ::       : : :     : ::
CRPV                                           LKMAEGTDPLDDCG-G----FLDTE  20
                                                :::       :        ::
HPV-1a                              TSCAHCAPYSDSKMADNKGTEND-------WFLVEA  29
                                               : :::  ::          :: :::
HPV-16                                         STMADPAGTNGEEGTGCNGWFYVEA  25
E1a                                             ::::  ::  : : :: ::: :::
HPV-6b    QQRSTGCAVYRNRHQRSATASVGNTKHSVSHLRTEDLTTMADDSGTENE-GSGCTGWFMVEA  61

BPV-1     AECESDKENEEPGAGVELSVESDRYDSQDEDFVDNASVFQGNHLE------VFQALEKKAGE  89
          : :      ::                      :   : ::: ::       ::  :
CRPV      ADCLDCDNLEEDLTELFDADTVS-----S-LLDDTDQV-QGNSLE------PFQHHEATETL  69
            ::    :::  : : : : ::     : :::   :  :::::        :   :  :
HPV-1a    TDCEET--LEE--TSLGDLDNVSCVSDLSDLLDEAPQS-QGNSLE------LFHKQESLESE  80
                :       :     : :                      :     ::  ::
HPV-16    VVEKKTGDAI----SDDENENDSD-TGEDLVDFIVNDNDYLTQAETETAHALFTAQEAKQHR  82
E1a         :   :: :    :::: :   :  : : ::::   :       :  : :::  :::   :
HPV-6b    IVQHPTGTQI----SDDEDEEVED-SGYDMVDFIDDSNITHNSLE---AQALFNRQEADTHY  115

BPV-1     EQILNLKRKVLGSSQNSSGSEASETP---VKRR---------KSGAKRR-LFAEN-------  131
            :  ::::   :   : :   :          :        :   : : ::
CRPV      KSIEHLKRKYVDSPDKSLGIDNSVNA---LSPRLQAFSLSGQKKAVKKR-LFGTD-------  120
               ::::   ::                ::::   :   : :    : ::   :
HPV-1a    QELNALKRKLLYSPQARSADETDIAS---ISPRLETISITKQDKKRYRRQLFSQD---DSGL  136
               :::: : ::               :::::  :    :     :: ::      :::
HPV-16    DAVQVLKRKYLVVHLV-----ILVDV                                      103
E1a         :: ::::::    :     :   :
HPV-16    RCSTGSKTKVFGSPLSD----ISGCVDNNISPRLKAICLEKQSRAAKRR-LF---ESEDSGY  54
E1b             : :  :::       :   :   ::::: ::    :    ::: ::   :  ::::
HPV-6b    ATVQDLKRKYLGSPYVSPINTIAEAVESEISPRLDAIKLTRQPKKVKRR-LFQTRELTDSGY  176

BPV-1     -------EANRVLTPLQVQGE-GEGRQELNEEQA--------------------ISHLHL--  163
                 ::    ::: :  : : : :                            :
CRPV      -----GDEAASGAESLQV--ESGFGSQQS-------------------------VSDTPV--  148
                ::   ::  ::                                          :
HPV-1a    ELSLLQDETENIDESTQVDQQQK-------------------------------EHTGEVGA  167
                         :::                                     ::
HPV-16    GNTEVETQQM-----LQVEGRHETETPCSQYSGGSGGGCSQYSS-GSGGEGVSERHTICQTP  110
E1b       :  :::          :::     :           ::  :    :   ::  : ::    :
HPV-6b    GYSEVEAGTG-----TQVEKHGVPEN----------GGDGQEKDTGRDIEG--EEHTEAEAP  221

BPV-1     --------------QLVKSKN-ATVFKLGLFKSLFLCSFHDITRLFKNDKTTNQQWVLAVFG  210
                              : : :   ::       ::   :: :  :::    ::    :
CRPV      --------------TDILNANTARVKHLLLFRQAHSVSFSELTRTFQSDKTMSWDWVG---G  193
                          :: :   :   :  :     :::  :::   : ::   :::
HPV-1a    A------------GVNILKASNIRAALLSRFKDTAGVSFTDLTRSYKSNKTCCGDWV-----  212
                         : :: ::  :: :  ::   ::::  : :  ::::  : ::
HPV-16    LTN----------ILNVLKTSNAKAAMLAKFKELYGVSFSELVRPFKSNKSTCCDWC-----  157
E1b        ::          ::  ::     :: : ::::  : ::  : ::::: : :: ::
HPV-6b    -TNSVREHAGTAGILELLKCKDLRAALLGKFKECFGLSFIDLIRPFKSDKTTCLDWV-----  277
```

FIGURE 12. Comparison of putative E1 proteins.

```
BPV-1    LAE--VFFEASFELLKKQCSFLQMQKRSH-------------EGGTCAVYLIC---FNTAK  253
         :: :             ::  :::               : : : :      :   :
CRPV     LADIHV------SVLES----LQTSLRSHCVY--VQ-YDLNFAE--TNASSLLLLLRFKAQK  240
         :: :      :: :    :::    ::::   :          :       : :: ::::::
HPV-1a   LAVWGVRENLIDSVKE----LLQT----HCVY--IQ-LEHAVTE---KNRFLFLLVRFKAQK  260
          :  :      :: :    :::       : :  :: :  :          : :::: :  :
HPV-16   IAAFGLTPSIADSIKT----LLQ----QYCLYLHIQSL--ACSW---GMVVL-LLVRYKCGK  205
 Elb      : ::   ::          :         :: ::: :     :     ::: :  : : :  :
HPV-6b   VAGFGIHHSISEAFQK----LIE----PLSLYAHIQWL--TNAW---GMVLL-VLLRFKVNK  325

BPV-1    SRETVRNLMANTLNVREECLMLQPAKIRGLSAALFWFKSSLSPATLKHGALPEWIRAQTTLN  315
            :  :  :      : :      : : : :     :::: :           :  ::::  ::  :
CRPV     CRDGVKALLSQLLGVQDLKVLLEPPKTRSVAVALFWYKRAMVSGVFSYGPMPEWITQQTNVN  302
            :  :  :      : :    : :::: :::: ::::::: : : ::  :     ::: :::   :
HPV-1a   SRETVIKLITTILPVDASYILSEPPKSRSVAAALFWYKRSMSSTVFTWGTTLEWIAQQTLIN  322
          :::  ::      : :        :::: :: :::: :::      :       : : :::  ::
HPV-16   NRETIEKLLSKLLCVSPMCMMIEPPKLRSTAAALYWYKTGISNISEVYGDTPEWIQRQTVLQ  267
 Elb      : :     :  ::        : :::::  :   :::::    ::::: : : :   :::: ::::
HPV-6b   SRSTVARTLATLLNIPENQMLIEPPKIQSGVAALYWFRTGISNASTVIGEAPEWITRQTVIE  387

BPV-1    ESL-QTEK-FDFGTMVQWAYDHKYAEESKIAYEYALAAGSDSNARAFLATNSQAKHVKDCAT  375
                :: :     :::::::     :: ::: ::   :   : ::::::: ::::: : ::
CRPV     HQM-LQEKPFQLSVMVQWAYDNHLQDESSIAYKYAMLAETDENARAFLASNSQAKYVRDCCN  363
         ::     : :: :   ::::::::     :   ::: :: ::    :::::::: :::: ::: ::
HPV-1a   HQL-DSESPFELCKMVQWAYDNGHTEECKIAYYYAVLADEDENARAFLSSNSQEKYVVKDCAQ  383
         :     :     :::   ::::::::         ::: :: :::     :: ::: :::: : :::::
HPV-16   HSFNDCT--FELSQMVQWAYDNDIVDDSEIAYKYAQLADTNSNASAFLKSNSQAKIVKDCAT  327
 Elb     :     :       : :   ::::::::::    ::::   :::   :    ::: ::: :: ::: ::::::
HPV-6b   HGLADSQ--FKLTEMVQWAYDNDICEESEIAFEYAQRGDFDSNARAFLNSNMQAKYVKDCAT  447

BPV-1    MVRHYLRAETQALSMPAYIKARCKLATGEGSWKSILTFFNYQNIELITFINALKLWLKGIPK  437
         ::: :::::        : :  :  :         : ::      :     :   :  :  :    ::   :  : ::
CRPV     MVRLYLRAEMRQMTMSAWINYRLDGMNDDGDWKVVVHFLRHQRVEFIPFMVKLKAFLRGTPK  425
         ::: :::::: :: :: ::     ::        : ::    : ::: : :::: ::     :   : : ::
HPV-1a   MVRHYLRAEMAQMSMSEWIFRKLDNVEGSGNWKEIVRFLRFQEVEFISFMIAFKDLLCGKPK  445
         : ::: :::   ::::: ::      : :     : :: :: ::: : ::: ::    :  :   : : ::
HPV-16   MCRHYKRAEKKQMSMSQWIKYRCDRVDDGGDWKQIVMFLRYQGVEFMSFLTALKRFLQGIPK  389
 Elb     :::::: ::    ::  :::: :            : :: :: ::: :  ::   :::  :    : : ::
HPV-6b   MCRHYKHAEMRKMSIKQWIKHRGSKIEGTGNWKPIVQFLRHQNIEFIPFLTKFKLWLHGTPK  509
                                                             :::: ...::
Py T Ag                                                LFKMLKLLTENVPK  565

BPV-1    KNCLAFIGPPNTGKSMLCNSLIHFLGGSVLSFANHKSHFWLASLADTRAALVDDATHACWRY  499
         :::  : :::: :::  : :::  : : ::::::  :::::  :::     ::::::: :::
CRPV     KNCMVFYGPPNSGKSYFCMSLIRLLAGRVLSFANSRSHFWLQPLADAKLALVDDATSACWDF  487
         :::        :::: ::: :: ::  :: : : :   :: : :::::::::::  : ::::  :::
HPV-1a   KNCLLIFGPPNTGKSMFCTSLLKLLGGKVISYCNSKSQFWLQPLADAKIGLLDDATKPCWDY  507
         ::: :  :  ::::: :  :: ::: : : ::    :::: ::::::::::: ::::: ::: :
HPV-16   KNCILLYGAANTGKSLFGMSLMKFLQGSVICFVNSKSHFWLQPLADAKIGMLDDATVPCWNY  451
 Elb     ::::     :     :::: : :::  :: : ::  ::: :::::::: :::    ::::: ::: :
HPV-6b   KNCIAIVGPPDTGKSYFCMSLISFLGGTVISHVNSSSHFWLQPLVDAKVALLDDATQPCWIY  571
         .  ::: :: ::::. :...:::::::: :.
Py T Ag  RRNILFRGPVNSGKTGLAAALISLLGGKSLN                                 596
```

FIGURE 12. *(Continued)*

```
BPV-1    FDTYLRNALDGYPVSIDRKHKAAVQIKAPPLLVTSNIDVQAEDRYLYLHSRVQTFRFEQPCT  561
         ::::::::: : : : ::::   :: :::: : : ::    ::  :: ::   : : :
CRPV     IDTYLRNALDGNPISVDLKHKAPIEIKCPPLLITTNVDVKSDDRWRYLFSRICVFNFLQELP  549
         : : ::::::: :  :::: ::  ::::::::: : :::::  : :: :::  : :  : :
HPV-1a   MDIYMRNALDGNTICIDLKHRAPQQIKCPPLLITSNIDVKSDTCWMYLHSRISAFKFAHEFP  569
         :   :::::::      : :::    : :::::::::::      :  : ::: :   : :  :::
HPV-16   IDDNLRNALDGNLVSMDVKHRPLVQLKCPPLLITSNINAGTDSRWPYLHNRLVVFTFPNEFP  513
 E1b     :   :: ::::  : : : ::  :   :::::: ::::           ::: :   ::::: ::
HPV-6b   MDTYMRNLLDGNPMSIDRKHKALTLIKCPPLLVTSNIDITKEDKYKYLHTRVTTFTFPNPFP  633


BPV-1    DE-SGEQPFNITDADWKSFFVRLWGRLDLIDEEEDSEEDGDSMRTFTCSARNTNAVD         617
            :         :: ::::: : :  : : : : :   :: :             :
CRPV     IR-NGTPVYELNDANWKSFFKRFWSTLELSDPE-DEGDDGGSQPALRLHTGGTSQSL         604
            :: :   : : :::::: :::  ::::: : ::: ::  :  :::
HPV-1a   FKDNGDPGFSLTDENWKSFFERFWQQLELSDQE-DEGNDGKPQQSLRLTARAANEPI         625
         :  :: :   : : :::::: : :   : :   : :  :::
HPV-16   FDENGNPVYELNDKNWKSFFSRTWSRLSLHEDE-DKENDGDSLPTFKCVSGQNTNTL         569
 E1b     :: ::: ::::    ::: :: :   : :      : : : ::     : :: :    ::
HPV-6b   FDRNGNAVYELSNTNWKCFFERLSSSLDIQDSE-DEE-DGSNSQAFRCVPGTVVRTL         688
```

FIGURE 12. (*Continued*)

```
BPV-1     LTRRRIVKRMETACERLHVAQETQMQLIEKSSDKLQDHILYWTAVRTENTLLYAARKKGVT   61
            :     ::   ::   ::    : :: :  :       :    : :  ::    : :
CRPV       ATRKTRVTMEALSQRLDSIQEELLSLYEKESTSLESQLQHWNLLRKEQVLLHFCKKHGIR   60
           : ::: ::: :: :::   :: :   :::  :    : :    ::: : :::: ::   : :
HPV-1a     VTKKTRETMENLSSRLDLLQEQLMNLYEQDSKLIEDQIKQWNLIRQEQVLFHFARKNGVM   60
                 ::     :::  ::::  ::: :             :    : : ::   :    :
HPV-6b    TFRILRTRKMEAIAKRLDACQEQLLELYEENSTDLHKHVLHWKCMRHESVLLYKAKQMGLS   61
            :     ::     ::  ::     :  ::  ::::  :     :: :: :     ::: ::
HPV-16   VCTRTRTRKTMETLCQRLNVCQDKILTHYENDSTDLRDHIDYWKHMRLECAIYYKAREMGFK   62


BPV-1    VLGHCRVPHSVVCQERAKQAIEMQLSLQELSKTEFGDEPWSLLDTSWDRYMSEPKRCFKKGA  123
         ::    ::     :: ::::::: :      :         :::: : :::  :  :  :   :::
CRPV     QLGYTPVPSLLTSQECAKQAIEMVLYIESLLRSPYSDEPWTLQDTSRERFESPPQKTFKKNP  122
           :    ::::  ::: :: ::::::   :::  :::     : : ::::::: :   ::  ::::
HPV-1a   RIGLQAVPSLASSQEKAKTAIEMVLHLESLKDSPYGTEDWSLQDTSRELFLAPPAGTFKKSG  122
         :: : :: :   :   :     ::::   :::::        :    : : :: :: :     ::   ::: :
HPV-6b   HIGMQVVPPLKVSEAKGHNAIEMQMHLESLLRTEYSMEPWTLQETSYEMWQTPPKRCFKKRG  123
         ::  :::: : ::   :     ::: :  ::          :: : ::::  : :     : :  : :: :
HPV-16   HINHQVVPTLAVSKNKALQAIELQLTLETIYNSQYSNEKWTLQDVSLEVYLTAPTGCIKKHG  124


BPV-1    RVVEVEFDGNASNTNWYTVVYSNLYMRTEDGWQLAKAGADGTGLYYCTMAGAGRIYYSRFGDE  185
           :::  ::   : : ::              ::                  :              :: :
CRPV     AIVEVYYDGDRGNNNEYTLWGIFIIGNADGEWVKTESGVDYRGIYYVDSEGNYVYYVDFSTD  184
           :: ::     :       : :         : :   : :   ::::  : ::     :     ::: :
HPV-1a   STLEVTYDNNPDNQTRHTIWNHVYYQNGDDVWRKVSSGVDAVGVYYLEHDGYKNYYVLFAEE  184
          : ::  :      :          :  :: :  :    : :: : ::: : ::       : ::: :  :
HPV-6b   KTVEVKFDGCANNTMDYVVWTDVYVQDNDT-WVKVHSMVDAKGIYYTCGQF-KTYYVNFVKE  183
         :::: :::      ::: :  ::  :                :   ::   : ::          :: : :
HPV-16   YTVEVQFDGDICNTMHYTNWTHIYICEEAS-VTVVEGQVDYYGLYYVHEGI-RTYFVQFKDD  184
```

FIGURE 13. Comparison of putative E2 proteins.

```
BPV-1   AARFSTTGHYSVRDQDRVYAGVSSTSSDFRDRP---DGVWVASEGPEGDPAGKEAEPAQPVS   244
CRPV    AGRFAANGHYDVVFQNMRLSSSVTSSPQPLVSA---PEDTVPEEAPDSAVPAAQKKTGPKTT   243
HPV-1a  ASKYSTTGQYAVNYRGKRFTNVMSSTSSPRAAG---APAVHSDYPTLSESDTAQQSTSIDYT   243
HPV-6b  AEKYGSTKHWEVCYGSTVICSPASVSSTTQEVS---IPESTTYTPA----------------   226
HPV-16  AEKYSKNKVWEVHAGGQVILCPTSVFSSNEVSSPEIIRQHLANHPA----------------   230

BPV-1   SLLG-------SPACGPIRAGLGWVRDGPRSHPYNFPAGSGGSILRSSSTPVQGTVPVDLAS   299
CRPV    RTLG-------RRRSRSPGVQRRPAKQRKQAAPDEADSAAGDI-------------------   279
HPV-1a  ELPGQGETSQVRQRQQKTPVRRRPYGRRRSRSPRGGGRREGES------------------   286
HPV-6b  --------------QTSTLVSSSTKEDAVQTPPRKRARGVQQS------------------   255
HPV-16  --------------ATHTKAVALGTEETQTTIQRPRSEPDTGN------------------   259

BPV-1   RQEEEEQSPDSTEEEPVTLP----RRTTNDGFHLLKAGGSCFALISGTANQV-KCYRFRVKKN  357
CRPV    RPPAPEDVGRRTTTVGRTPP----GRNRRLRELITEASDPPVICLKGGHNQL-KCLRYRLKSK  337
HPV-1a  TPSRTPGSVPSARDVGSIHTTPQKGHSSRLRRLLQEAWDPPVVCVKGGANQL-KCLRYRLKAS  348
HPV-6b  PCNALCVAHIGPVDSGN-HNLITNNHDQHQRRNNSNSSATPIVQFQGESNCL-KCFRYRLNDR  316
HPV-16  PCHTTKLLHRDSVDSAP-ILTAFNSS--HKGRINCNSNTTPIVHLKGDANTL-KCLRYRFKK-  317
H c-mos                                        EPH-CRTGGQLSLGKCLKYSLDVV  186

BPV-1   HRHRYENCTTTWFTVADNGAERQGQAQILITFGSPSQRQDFLKHVPLPPGMNISGFTASLDF   419
CRPV    HSSLFDCISTTWSWVDTTSTCRLGSGRMLIKFADSEQRDKFLSRVPLPSTTQVFLGNFYGL   398
HPV-1a  TQVDFDSISTTWHWTDRKNTERIGSARMLVKFIDEAQREKFLERVALPRSVSVFLGQFNGS   409
HPV-6b  HRHLFDLISSTWHWASSKAPHKH--AIVTVTYDSEEQRQQFLDVVKIPPTISHKLGFMSLHLL  377
HPV-16  HCTLYTAVSSTWHWTGHNVKHKS--AIVTLTYDSEWQRDQFLSQVKIPKTITVSTGFMSI   375
H c-mos NGLLF-LHSQSIVHLDLKPANILISEQDVCKISDFGCSEK-LEDLLCFQTPSYPLGGTYTHRA  247
```

FIGURE 13. (*Continued*)

D. E5 ORF

The putative E5 proteins for BPV-1, DPV (Groff and Lancaster, 1985), and BPV-2 (D. E. Groff, R. Mitra, and W. D. Lancaster, manuscript in preparation) are very homologous after the first methionine (48% of the residues are identical for all three viruses) (Fig. 14A). The E5 gene product is known to have a transforming function (Yang *et al.*, 1985b; Schiller *et al.*, 1986). Since DPV, BPV-1, and BPV-2 are unique in that they all produce fibropapillomas, it is probable that the E5 gene product of DPV, BPV-1 and BPV-2 is a fibroblast-specific transforming protein. The putative E5 proteins of CRPV and HPV-1a (both produce cutaneous lesions) show some homology with each other (Fig. 14B) but not with the other E5 proteins.

E. E6 ORF

The putative E6 proteins show limited overall homology at the amino acid sequence level (less than 10% of the positions are identical for all five viruses) (Fig. 15). Of interest are the identically positioned CXXC sequences found in all five E6 proteins. Similar cysteine-rich repeats are also found in polyoma and SV40 small T antigen (Ghosh *et al.*, 1979).

F. E7 ORF

The putative E7 proteins show limited, but significant (maximum aligned *z* value of 31 for HPV-6b versus HPV-16) homology (Fig. 16, Table VI). Two CXXC repeats similar to those present in E6 are also observed.

A

```
DPV       MNHPGLFLFLGLVFGVQLLLLVFILFFFFVWWDQFGCKCENFHM   44
          :      ::::::   ::::: : : :: : :: : : :
BPV-1     MPNLWFLLFLGLVAAMQLLLLLFLLLFFLVYWDHFECSCTGLPF   44
          :::::::::::::::::::::::::: :::::::::::::::::
BPV-2     MPNLWFLLFLGLVAAMQLLLLLFLLLFLLVYWDHFECSCTGLPF   44
```

B

```
CRPV      GIFMGFSDVYACNPFPSAAFVTQRFFVPINLAHTQKVSWLHGHENAGLHHKTFIQHAKLLAI   62
                                            :            :  :    :::
HPV-1a                                    SLQHCNKCIAYVENALPPKIYTPHAKYQTP   30

CRPV      AQLTYRINLKTKQLQIKFCSMAALVFSLEDLESVVPEVLGVD                      104
          : ::  : :   :: :: : ::   :  :      :: :
HPV-1a    AHLTFKIKLSIQQLLIKYCNMAVWEFFWEVWALEQPEALEEELVILPSVRVVGLELLLVQLQ   92
```

FIGURE 14. Comparison of putative E5 proteins.

```
CRPV                DRNALPGISAYKTRNLEQMENCLPRSLEKLQQILQISLEDLPFGCIFCGKLL  52
                                            :    :   : :   :    : ::   :
HPV-1a                           DSAQDQMATPIRTVRQLSESLCIPYIDVLLPCNFCNYFL  39
                                            :   ::   :     :  :  :  :   :
HPV-16                YKSRHFMHQKRTAMFQDPQERPRKLPQLCTELQTTIHDIILECVYCKQQL  50
                      ::                        :::       :     ::::   :
HPV-6b    EGPKTVQPKTVVYKPALKFSKRGIMESANASTSATTIDQLCKTFNLSMHTLQINCVFCKNAL  62
                                                   : :          :  :   :
BPV-1                             PANYKKLLTDPGFHMDLKPFARTNPFSGLDCLWCREPL  38
                                                                :  :
                                                                CXXC

CRPV      GAAEKQLFKCTGLCIVWHKGWPYGTCRDCTVLSCALDLYCHLALTAPALEAEALVGGQEISSW 114
            ::: ::    : ::        : : :      :              : :
HPV-1a    SNAEKLLFDHFDLHLVWRDNLVFGCCQGCARTVSLLEFVLYYQESYEVPEIEEILDRPLLQI 101
             :   :   :: : ::        : :    :       :  :       :    ::
HPV-16    LRREVYDFAFRDLCIVYRDGNPYAVCDKCLKFYSKISEYRHYCYSLYGTTLEQQYNKPLCDL 112
             : :  :   :     :::::   : :: : :: ::  :::  : : ::  :        :
HPV-6b    TTAEIYSYAYKHLKVLFRGGYPYAACACCLEFHGKINQYRHFDYAGYATTVEEETKQDILDV 124
          :         :   :  : :  : : : : :::                     :
BPV-1     TEVDAFRCMVKDFHVVIREGCRYGACTICLENCLATERRLWQGVPVTGEEAELLHGKTLDRL 100
                                    : :
                                   CXXC

CRPV      FMRCTVCGRRLTIPEKIELRARNCTLCCIDKGQYFQWRGHCSSCKLSDQGDLGGYPPSPGSR 176
            ::  :   :   :: :                        :: : :
HPV-1a    ELRCVTCIKKLSVAEKLEVVSNGERVHRVRNRLK----AKCSLCRLYAI              146
            ::  : : :   ::        : :  : :        :  :
HPV-16    LIRCINCQKPLCPEEKQRHLDKKQRFHNIRGRWT----GRCMSCCRSSRTRRETQL       164
          ::::  : ::::  ::  :   : ::      :     :::  :
HPV-6b    LIRCYLCHKPLCEVEKVKHILTKARFIKLNCTWK----GRCLHCWTTCMEDMLP         174
           :::  :   :   ::  : :    : :          :::  :         :
BPV-1     CIRCCYCGGKLTKNEKHRHVLFNEPFCKTRANII---RGRCYDCCRHGSRSKYP         151
             :  :                                 :  :
             CXXC                                 CXXC

CRPV      CGECLTPVDLEELGLYPGPEGTYPDLVDLGPGVFGEEDEEGGGLFDSFEEEDPGPNQCGCFFC 239
          :  :                                                        :  :
          CXXC                                                        CXXC

CRPV      TSYPSGTGDTDINQGPAGAAGIALQSDPVCFCENCINFTEFR                     291
                                         :  :
                                         CXXC
```

FIGURE 15. Comparison of putative E6 proteins.

G. L1 ORF

The L1 ORF encodes the major capsid protein of the papillomaviruses (Engel *et al.*, 1983; Pilacinski *et al.*, 1984). The L1 proteins are the most highly conserved of the papillomavirus proteins (Fig. 17, Table VI). Approximately 29% of the residues are identical for all five viruses.

H. L2 ORF

The putative L2 proteins of DPV (Groff and Lancaster, 1985), BPV-1, CRPV, HPV-1a, HPV-6b, HPV-11 (Dartman *et al.*, 1986), and HPV-16 were compared (Fig. 18, Table VI) and shown to be well conserved only

```
BPV-1     LEDAATTAADMVQGPNTHRNLDDSPAGPLLILSPCAGTPTRSPAAPDAPDFRLPCHFGRPTR  62
                                  :      ::        :
CRPV                          YFCYPVRRAASLSFFFIYYCFSFCTGFIEFCKMIGRTPKLSE  42
                                                              : ::   :
HPV-6b                          SRKGKTYTNQGAVHKAKLYVEGSLPTLLDNMHGRHVTLKD  40
                                                              :::   ::
HPV-16                                                  RNPAVIMHGDTPTLHE  16
                                                              : :  : :
HPV-1a                                               FVSLVCYITMVGEMPALKD  19


                                                           █  █
BPV-1     KRGPTTPPLSSPGKLCATGPR------------------RVYSVTVC--CGNCGKELTFAVK  104
               :    :     :                      : : : :   :  :     :
CRPV      LVLGETAEALSLHCDEALENLSDDDEEDHQDRQVFI--ERPYAVSVP--CKRCRQTISFVCV  100
           ::                  :  :: :  : :                :  :      :
HPV-6b    IVLDLQPPDPVGLHCYEQLVDSSEDEVDEVDGQDSQ--PLKQHFQIVTCCCGCDSNVRLVVQ  100
            :::::     : ::::: :::: : :: ::   :  :   :  ::: :: :::  :: ::
HPV-16    YMLDLQPETTD-LYCYEQLNDSSEEE-DEIDGPAGQAEPDRAHYNIVTFCCKCDSTLRLCVQ   76
            : : :   : :  :        ::    :            :  :  :  :    :: :
HPV-1a    LVLQLEPSVLD-LDLYCY------EE-VPPDDIEEELVSPQQPYAVVASCAYCEKLVRLTVL   73
                                                           :  :
                                                           CXXC


                     ██   █   ██ █
BPV-1     TSSTSLLGFEHLLNSDLDLLCPRCESRERHGKR                               137
                     ::   : : :: :
CRPV      CAPEAIRTLNRLLSASLSLVCPECCN                                      126
          :    ::    ::   :  ::: :
HPV-6b    CTETDIREVQQLLLGTLNIVCPICAPKT                                    128
           :  :::    :: ::: ::::::  :
HPV-16    STHVDIRTLEDLLMGTLGIVCPICSQKP                                    104
            :  :: :: ::   : :::: :
HPV-1a    ADHSAIRQLEELLLRSLNIVCPLCTLQRQ                                   102
                              :  :
                              CXXC
```

FIGURE 16. Comparison of putative E7 proteins.

```
BPV-1   MALWQQGQ-KLYLPPT-PVSKVLCSETYVQRKSIFYHAETERLLTIGHPYYPVSI---   53
CRPV    MAVWLSTQNKFYLPPQ-PVTKIPSTDEYVTRTNVFYYASSDRLLTVGHPYYEIRDK--   55
HPV-1a  MAVWLPAQNKFYLPPQ-PITRILSTDEYVTRTNLFYHATSERLLLVGHPLFEISSNQT   57
HPV-16  MSLWLPSEATVYLPPV-PVSKVVSTDEYVARTNIYYHAGTSRLLAVGHPYFPIKKPNN   57
HPV-6b  M--WRPSDSTVYVPPPNPVSKVVATDAYVTRTNIFYHASSSRLLAVGHPYFSIKRAN-   55

GAKTVPKVSANQYRVFKIQLPDPNQFALPDRTVHNPSKERLVWAVIGVQVSRGQPLGGTVTGHPTF  119
GTMLVPKVSPNQYRVFRIKLPDPNKFAFGDKQLYDPEKERLVWCLRGIEVNRGQPLGVSVTGNPIF  121
VTI--PKVSPNAFRVFRVRFADPNRFAFGDKAIFNPETERLVWGLRGIEIGRGQPLGIGITGHPLL  121
NKILVPKVSGLQYRVFRIHLPDPNKFGFPDTSFYNPDTQRLVWACVGVEVGRGQPLGVGISGHPLL  123
-KTVVPKVSGYQYRVFKVVLPDPNKFALPDSSLFDPTTQRLVWACTGLEVGRGQPLGVGVSGHPFL  120

NALLDAENV--NRKVTTQTTDDRKQTGLDAKQQQILLLGCTPAEGEYWTTARPCVTDRLENGACPP  183
NKFDDVENPTKYYNNHADQQDYRKSMAFDPKQVQLLMLGCVPATGEHWAQAKQCAEDPPQQTDCPP  187
NKLDDAENPTNYINTHANG-DSRQNTAFDAKQTQMFLVGCTPASGEHWT-SRRCPGEQVKLGDCPR  185
NKLDDTENASAYAAN-AGV-DNRECISMDYKQTQLCLIGCKPPIGEHWGKGSPCTNVAVNPGDCPP  187
NKYDDVEN-SGSGGN-PGQ-DNRVNVGMDYKQTQLCMVGCAPPLGEHWGKGKQCTNTPVQAGDCPP  183

LELKNKHIEDGDMMEIGFGAANFKEINASKSDLPLDIQNEICLYPDYLKMAEDAAGNSMFFFARKE  249
IELVNTVIEDGDMCEIGFGAMDHKTLQASLSEVPLELAQSISKYPDYLKMQKDQFGDSMFFYARRE  253
VQMIESVIEDGDMMDIGFGAMDFAALQQDKSDVPLDVVQATCKYPDYIRMNHEAYGNSMFFFARRE  251
LELINTVIQDGDMVHTGFGAMDFTTLQANKSEVPLDICTSICKYPDYIKMVSEPYGDSLFFYLRRE  253
LELITSVIQDGDMVDTGFGAMNFADLQTNKSDVPIDICGTTCKYPDYLQMAADPYGDRLFFFLRKE  249

QVYVRHIWTRGG--SEKEAPTTDFYLKN---NKGDATLKIPSVHFGSPSGSLVSTDNQIFNRPYWL  310
QMYARHFFSRAG--GDKENVKSRAYIKRTQMQGEANANIATDNYCITPSGSLVSSDSQVFNRAYWL  317
QMYTRHFFTRGGSVGDKEAVPQSLYL---TADAEPRTTLATTNYVGTPSGSMVSSDVQLFNRSYWL  314
QMFVRHLFNRAGTVG--ENVPDDLYIKGSGSTAN----LASSNYFPTPSGSMVTSDAQIFNKPYWL  313
QMFARHFFNRAGEVG--EPVPDTLIIKGSGNRTS----VGSSIYVNTPSGSLVSSEAQLFNKPYWL  309
```

FIGURE 17. Comparison of putative L1 proteins.

```
FRAQGMNNGIAWNNLLFLTVGDNTRGTNLTISVASD-GTPLTEYDSSKFNVYHRHMEEYKLAFILE  375
 :::::::  : :  : :: :::::: :     :      :     :  : :: ::: : : :
QKAQGMNNGVCWDNQIFVTVVDNTRGTILSLVTKSKEQIKKTHGKTVHFSSYLRHVEEYELQFVLQ  383
:  :: ::: :: :: : :: :::::: ::   :       :      :   ::: ::  : :  :
QRCQGQNNGICWRNQLFITVGDNTRGTSLSISMKNNAS---TTYSNANFNDFLRHTEEFDLSFIVQ  377
:: :: :::::: :::: :: : :: :  :       :   ::: : ::   ::: :: :: :: :
QRAQGHNNGICWGNQLFVTVVDTTRSTNMSLCAAISTS--ETTYKNTNFKEYLRHGEEYDLQFIFQ  377
: :::::::::::::::::::::::::: :::   ::   : : :   ::: :: ::::::::::
QKAQGHNNGICWGNQLFVTVVDTTRSTNMTLCASVTTS--ST-YTNSDYKEYMRHVEEYDLQFIFQ  372

LCSVEITAQTVSHLQGLMPSVLENWEIGVQPPTSSILEDTYRYIE-SP-AT-KCASNVIPAKE-DP  437
:: :  :    : :    :    ::   :   :  ::: :::   :  :: ::    :    ::
LCKVKLTPENLSYLHSMHPTIIDNWQLSVSAQPSGTLEDQYRYLQ-SI-AT-KCPPPEPPKENTDP  446
:::::::::: : : : : :   ::::::  :   :::::: :  :  :  :::   ::   :::
LCKVKLTPENLAYIHTMDPNILEDWQLSVSQPPTNPLEDQYRFLG-SSLAA-KCPEQAPPEPQTDP  441
:::  ::     ::: :   :::::      ::   ::: :::      :  :   :: :    ::
LCKITLTADVMTYIHSMNSTILEDWNFGLQPPPGGTLEDTYRFVT-QAIACQKHTPPAPKED--DP  440
:: ::: : :: ::: ::   :::::::: ::: :::::::: :  ::: ::: ::   :    ::
LCSITLSAEVMAYIHTMNPSVLEDWNFGLSPPPNGTLEDTYRYVQSQAITCQKPTPEKEK---PDP  435

YAGFKFWNIDLKEKLSLDLDQFPLGRRFLAQQGAGCSTVRKRRISQKT--------SSKPAKKKKK  495
   :::  :: ::::  ::: :::: :: : :                        :::
YKNYKFWEVDLSEKLSDQLDQYPLGRKFLNQSGLQRIGTKRPAPAPVSIVK-----SSKRKRRT    505
:  :::::::: :  : :::: ::::::: :::                       : :: :
YSQYKFWEVDLTERMSEQLDQFPLGRKFLYQSGMTQRTATSSTTKRKTVRV---STSAKRRRKA    502
   : :::: : :  :  ::::::::::: : :                :       :: :::
LKKYTFWEVNLKEKFSADLDQFPLGRKFLLQAGLKAKPKFTLGKRKATPTTSSTSTTAKRKKRKL  505
 :   :::::::::::  ::: ::::::::: :            :          :   :: : :
YKNLSFWEVNLKEKFSSELDQYPLGRKFLLQSGYRGRSSIRTGVKRPAVSKASAAPKRKRAKTKR  500
```

FIGURE 17. (*Continued*)

over the NH_2-terminal domain (22% identical for an approximately 73-amino-acid stretch). The L2 proteins clearly fall into three homologous groups: fibropapilloma-producing (DPV, BPV-1), cutaneous (CRPV, HPV-1a), and mucosal (HPV-6b, HPV-11, and HPV-16). It is thus likely that the L2 protein is involved in host cell range.

```
DPV                     ...YNKCTMPPLKRVKRANPYDLYRTCKRWK-C  137
                                 :   ::::::  :::::::    :
BPV-1                      AAAIKMSARKRVKRASAYDLYRTCKQAGTC   30
                              :   : :: :  ::   : : ::: :: :
CRPV    PIPISSFCNATFFCTNKSCTYAKGVMVARSRKRRAAPQDIYPTCKIAGNC   50
                                  : :  :::: :::: :::    :
HPV-1a                               QMYRLRRKRAAPKDIYPSCKISNTC   25
                                       :::::      :  ::    ::
HPV-11                          LYIMKPRARRRKRASATQLYQTCKATGTC   29
                                    ::::::::::::::::::: ::::
```

FIGURE 18. Comparison of putative L2 proteins.

```
HPV-6b                              SKPFFIFVTVYALYTMAHSRARRRKRASATQLYQTCKLTGTC   42
                                              : :       : : ::::::::: :::  :::
HPV-16  LLYTITYYFFFFIFIYNFFFCLFVCFLINCYYLTMRHKRSAKRTKRASATQLYKTCKQAGTC   62


DPV     PLMSFLKVEGKTVADKYCSMGSMGVYWRLALHGSGRPTQGG------YVPLRGGGSSTSLSS  193
        :     :::: : :::     :     :        :            : :::  ::  :: :
BPV-1   PPDVIPKVEGDTIADKILKFGGLAIYLGGLGIGTWSTGRVAAGGSPRYTPLRTAGSTSSLAS   92
        : :    : :  :::::::  : :     :::::               ::   ::::    :       :
CRPV    PADIQNKFENKTIADKILQYGSLGVFFGGLGISSAGGS----GGRLGYTPLSGGGGRVIAAA  108
        : ::::: :   ::::::::::::::: :::::   : ::         ::: :::::   :::    :
HPV-1a  PPDIQNKIEHTTIADKILQYGSLGVFLGGLGIGTARGS----GGRIGYTPLGEGGGVRVATR   83
        :::     : ::::::: ::  :::::: :::::::   ::         ::: :: :::
HPV-11  PPDVIPKVEHTTIADQILKWGSLGVFFGGLGIGTGAGS----GGRAGYIPLGSSPKPAITGG   87
        ::::::::: ::::::::::::::::::::::::: :         ::: :: ::   : :: :: :
HPV-6b  PPDVIPKVEHNTIADQILKWGSLGVFFGGLGIGTGSGT----GGRTGYVPLQTSAKPSITSG  100
        ::: :::::  ::: :::   :: ::::::::::::::         :::::: :: :
HPV-16  PPDIIPKVEGKTIAEQILQYGSMGVFFGGLGIGTGSGT----GGRTGYIPLGTRPPTATDTL  120


DPV     R--GSGSSTSISRPFAG----------GIPLETLETVGAFRPGIIEEVAPTLEGVLPDAPAV  243
           ::   :   ::  :           :::: :::: :: :::   :           ::: :::
BPV-1   I--GSRAVTAGTRPSIGA---------GIPLDTLETLGALRPGVYEDT------VLPEAPAI  137
                :                  :   :   :   :   :   :   :            :      :
CRPV    P--VRPPITTESVGPLDIVPE-VADPGGPTLVSLHELPAETPYVSSTN------VTGDGAAE  161
        :  ::: :   : :::   : :   : :: ::      :   :      :
HPV-1a  PTPVRPTIPVETVGPSEIFPIDVVDPTGPAVIPLQDLGRDFPIPTVQV------IAEIHPIS  139
        :    ::    :: : ::                       :         :            :    :
HPV-11  PAA-RPPVLVEPVAPS------------------DPSIVSLIEESAI------INAGAPEV  123
        : : :::: :::::::                  :::::::::::::      :::::::
HPV-6b  PMA-RPPVVVEPVAPS------------------DPSIVSLIEESAI------INAGAPEI  136
            :::  :  :: ::                 ::::::: ::         : ::::
HPV-16  APV-RPPLTVDPVGPS------------------DPSIVSLVEETSF------IDAGAPTS  156


DPV     VT---PEAV------PVDQGLSGLDV--AREVTQESLITFLQPEGPDDIAVLELRPTEHDQT  294
        ::   : ::      : : :: :            : ::: : :::: ::::::: :      :
BPV-1   VT---PDAV------PADSGLDALSI--GTDSSTETLITLLEPEGPEDIAVLELQPLDRP-T  187
             :              :  :                :  :        :  :
CRPV    PL---PAGH------GGSQISDVTSG--TSGTVSRTHINNPVFEAPMTGDQDVSDVHVFA-H  211
             :         : : : ::  :   :  :::::  ::: :         :
HPV-1a  DI---PNIVASSTNEGESAILDVLRGNATIRTVSRTQYNNPSFTVASTSNISAGEASTSD-I  197
             :            ::      :     :    :          :              :::
HPV-11  VP---PTQGGFTITSSESTTPAILDVSVTNHTTTSVFQNPLFTEPSVIQPQPPVEASGHI-L  181
        ::   :  ::::::::: ::::::::::: :::::  : ::  ::::::  :::::::: ::: :
HPV-6b  VP---PAHGGFTITSSETTTPAILDVSVTSHTTTSIFRNPVFTEPSVTQPQPPVEANGHI-L  194
        ::   :      :: :: :  :::::::      :  :  :    :: :: ::: ::  :  :  ::
HPV-16  VPSIPPDVSGFSITTSTDTTPAILDINNTVTTVTTHN-NPTFTDPSVLQPPTPAETGGHF-T  216


DPV     HLISTSTHPNPLFHAPIQQSSIIAETSGSENIFVGGGGVGSTT----GEEIELTLFGQPKTS  352
           :    :     ::: :   : :::::: ::::::: : : :      :: :::: :: : ::
BPV-1   WQVSNAVHQSSAYHAPLQLQSSIAETSGLENIFVGGSGLGDTG----GENIELTYFGSPRTS  245
           :      :       :             :    :    :    :                   :  :
CRPV    SESSITINQTENTGGELIEMVPLRHPPRSEGDFRETSFSTSTPI--------------PDRS  259
           :                   :              :: :: ::::                    :
HPV-1a  VFVSNGSGDRVVGEDIPLVELNLGLETDTSSVVQETAFSSSTPI--------------AERP  245
                  :                    :::         ::                    :
```

FIGURE 18. *(Continued)*

```
HPV-11  ISAPTITSQHVEDIPLDTFVVSSSDSGPTSSTPLPRAFPRPRVGLYSRALQQVQVTDPAFLS  243
        ::::: ::   : :::::::::::::::::::: :   :::::::::::: :::::::::::
HPV-6b  ISAPTVTSHPIEEIPLDTFVVSSSDSGPTSSTPVPGTAPRPRVGLYSRALHQVQVTDPAFLS  256
         :  :   :  :::: ::: ::      ::::: ::  :  : :::::   :: : ::::
HPV-16  LSSSTISTHNYEEIPMDTFIVSTNPNTVTSSTPIPGSRPVARLGLYSRTTQQVKVVDPAFVT  278


DPV     TPEGPINRGRG-IFNWFNRTYYTQVPVEDPDEIAAAGSYVFENALYDSKAYKHEQQPWLSRP  413
        ::       :: : :::   :::::: :::
BPV-1   TPRSIASKSRG-ILNWFSKRYYTQVPTEDPEVFSS---------------------------  279
          :::   ::                      :
CRPV    ALRSINVASRR-YQQVQVENPAFLNRPRELVQFENTFDNP----------------------  298
          :      :: : ::::  : :   :   : :      :
HPV-1a  SFRPSRFYNRRLYEQVQVQDPRFVEQPQSMVTFDNPAFEP----------------------  285
               :    ::   :      :
HPV-11  TPQRLVTYDNPVYEGEDVSLQFTHESIHN---------------------------------  272
        ::::: ::::::::::::: :: : ::::
HPV-6b  TPQRLITYDNPVYEGEDVSVQFSHDSIHN---------------------------------  285
        ::  ::::::: ::: ::       :  :
HPV-16  TPTKLITYDNPAYEGIDVDNTLYFSSNDN---------------------------------  307


DPV     QDAPEFDFQDAVRLLQGPSGRVGWSRIIRPTSIGTRSGVRVGPLYHLRQSFSTIDEPETIEL  475
        :             : ::::::: :    :    ::::  :::  : : : ::: :    :
BPV-1   QTFANPLYEAEPAVLKGPSGRVGLSQVYKPDTLTTRSGTEVGPQLHVRYSLSTIHEDV--EA  339
                  :   :              :
CRPV    -----AFVDDEQLSLLFEQDLDTVVATPDPAFQDVVRLSRPSFTQSRAGRVRVSRLGR--TL  353
                     :  :  :::    :: : : ::: :: : :     :: ::::::
HPV-1a  --------ELDEVSIIFQRDLDALAQTPVPEFRDVVYLSKPTFSREPGGRLRVSRLGKSS--  337
                :     :   :         :   :     :  : :    :         :
HPV-11  -----APDEAFMDIIRLHRPAITSRRGLVRFSRIGQRGSMYTRSGQHIGARIHYFQDISP--  327
             ::::::::::::::::: ::::::: ::::::::: :::: ::::::::: ::::
HPV-6b  -----APDEAFMDIIRLHRPAIASRRGLVRYSRIGQRGSMHTRSGKHIGARIHYFYDISP--  340
             :::  : ::  :::::  :::   ::::::      ::::: :::  :: :: :
HPV-16  -SINIAPDPDFLDIVALHRPALTSRRTGIRYSRIGNKQTLRTRSGKSIGAKVHYYYDLST--  366


DPV     IPSTVDEEEVLTGVPESAEGPDAEYSDIDLQSIGSDEPLLGT---GIIYPLVGGGQIFLCMH  534
        :: ::::           :                   :          :   :    :
BPV-1   IPYTVDENTQGLAFVPLHEEQAGFEEIELDDFSETHRLLPQN---TSSTPVGSGVRRSLIPT  398
           :      : :     :     :  : :               :  :
CRPV    TMQTRSGKAFGPAKHFYYELSSIAEGPEPDILIPESEQETSF---TDATSKDTQQEAEVYAD  412
        :  :: : : :   :: : :::::       :       :           ::       :
HPV-1a  TIRTRLGTAIGARTHFFYDLSSIAPEDSIELLPLGEHSQTTV---ISSNLGDTAFIQGETAE  396
                          :   :  :          :: :            :
HPV-11  VTQAAEEIELHPLVAAENDTFDIYAEPFDPIPDPVQHSVTQS---YLTSTPNTLSQS-----  381
          ::::::: ::::::  :::::::: : :   : :: ::     ::::::::  :
HPV-6b  IAQAAEEIEMHPLVAAQDDTFDIYAESFEPGINPTQHPVTNISDTYLTSTPNTVTQP-----  397
        :  : :             :                     : ::  :  :
HPV-16  IDPAEEIELQTITPSTYTTTSHAASPTSINNGLYDIYADDFITDTSTTPVPSVPSTSLSGYI  428


DPV     RAPVGW-----------------------------SSGTYINHEGQSRDDGEYVIDNGGQSN  567
                                                             : :    :
BPV-1   -----QEFSAT------------------------RPTGVVTYGSPDTYSASPVTDPDSTSP  431
                  :                        :  :             ::   ::
CRPV    -----GSTLETDTSADENL--------TLVFSDRGRGQGSHVPIPGKSTIGGPVNIGDSKYY  461
                                                           : : : :
```

FIGURE 18. (*Continued*)

```
HPV-1a    -----DDLEVISLETPQLY--------SEEELLDTNESVGENLQLTITNSEGEVSILDLTQS  445
                 :   :    :          :                  :: :
HPV-11    WGNTTVPLSIPSDWFVQSGPDITFPTASMGTPFSPVTPALPTGPVFITGSDFYLHPTWYFAR  443
          ::::::::: : : : ::::::::::: :::::::::::::::::::::: ::::: :::::
HPV-6b    WGNTTVPLSLPNDLFLQSGPDITFPTAPMGTPFSPVTPALPTGPVFITGSGFYLHPAWYFAR  459
            ::: :         : :::::            :  :  :          :::::  :  :
HPV-16    PANTTIPFGGAYNIPLVSGPDIPINITDQAPSLIPIVPGSPQYTIIADAGDFYLHPSYYMLR  490


DPV       I------TPT-------------------VVIDGSIALSLEYFRHYYLHPSLLRRKRKRNPI  604
                 : :                     :         :   : :::::::   ::
BPV-1     SLVIDDTTTT-------------------PIIIIDGHTVDLYSSNYTLHPSLLRKRKKRKHA  474
           :    ::                         :            :         : ::
CRPV      TLNPGETTSF-------------------EADVISPVFIFEGNADGTYYLEEPLRKKRRKSI  504
             :   :                     :      : :   :  :   :   ::: :
HPV-1a    RVRPPFGTEDTSLHVYYPNSSKGTPIINPEESFTPLVIIALNNSTGDFELHPSLRKRRKRAY  507
          : :       :
HPV-11    RRRKRIPLFFTDVAA                                                 458
           ::::::::: ::::
HPV-6b    KRRKRIPLFFSDVAA                                                 474
          ::::: : :::::
HPV-16    KRRKRLPYFFSDVSLAA                                               507


DPV       FI                                                              606

CRPV      FLLADGSVAVYAE                                                   517

HPV-1a    V                                                               508
```

FIGURE 18. (*Continued*)

REFERENCES

Baker, C. C., Phelps, W. C., Lindgren, V., Braun, M. J., Gonda, M.A., and Howley, P. M., 1987, A structural and transcriptional analysis of HPV-16 sequences in cervical carcinoma cell lines, *J. Virol* (in press).

Chen, E. Y., Howley, P. M., Levinson, A. D., and Seeburg, P. H., 1982, The primary structure and genetic organization of the bovine papillomavirus type 1 genome, *Nature* **299:**529–534.

Clad, A., Gissmann, L., Meier, B., Freese, U. K., and Schwarz, E., 1982, Molecular cloning and partial nucleotide sequence of human papillomavirus type 1a DNA, *Virology* **118:**254–259.

Clertant, P., and Seif, I., 1984, A common function for polyoma virus large-T and papillomavirus E1 proteins, *Nature* **311:**276–279.

Conrad, B., and Mount, D., 1982, Microcomputer programs for DNA sequence analysis, *Nucleic Acids Res.* **10:**31–38.

Danos, O., and Yaniv, M., 1984, An homologous domain between the *c-mos* gene product and a papillomavirus polypeptide with putative role in cellular transformation, in *Cancer Cells*, Part II: *Oncogenes and Viral Genes* (G. F. Vande Woude, A. J. Levine, W. C. Topp, and J. D. Watson, eds.), pp. 291–294, Cold Spring Harbor Laboratory, Cold Spring Harbor, N.Y.

Danos, O., Katinka, M., and Yaniv, M., 1982, Human papillomavirus 1a complete DNA sequence: A novel type of genome organization among papovaviridae, *EMBO J.* **1:**231–236.

Danos, O., Engel, L. W., Chen, E. Y., Yaniv, M., and Howley, P. M., 1983, A comparative analysis of the human type 1a and bovine type 1 papillomavirus genomes, *J. Virol.* **46:**557–566.

Danos, O., Engel, L. W., Chen, E. Y., Yaniv, M., and Howley, P. M., 1983, A comparative analysis of the human type 1a and bovine type 1 papillomavirus genomes, *J. Virol.* **46:**557–566.

Dartmann, K., Schwarz, E., Gissmann, L., and zur Hausen, H., 1986, The nucleotide sequence and genome organization of human papilloma virus type 11, *Virology* **151:**124–130.

Engel, L., Heilman, D., and Howley, P., 1983, Trancriptional organization of bovine papillomavirus type 1, *J. Virol* **47:**516–528.

Fuchs, P. G., Iftner, T., Weninger, J., and Pfister, H., 1986, Epidermodysplasia verruciformis-associated human papillomavirus 8: Genomic sequence and comparative analysis, *J. Virol.* **58:**626–634.

Ghosh, P., Piatak, M., Reddy, V., Swinscoe, J., Lebowitz, P., and Weissman, S., 1979, Transcription of the SV40 genome in virus transformed cells and early lytic infection, *Cold Spring Harbor Symp. Quant. Biol.* **44:**31–39.

Giri, I., Danos, O., and Yaniv, M., 1985, Genomic structure of the cottontail rabbit (Shope) papillomavirus, *Proc. Natl. Acad. Sci. USA* **82:**1580–1584.

Groff, D. E., and Lancaster, W. D., 1985, Molecular cloning and nucleotide sequence of deer papillomavirus, *J. Virol.* **56:**85–91.

Lipman, D., and Pearson, W., 1985, Rapid and sensitive protein similarity searches, *Science* **227:**1435–1441.

Mount, D., and Conrad, B., 1984, Microcomputer programs for backtranslation of protein to DNA sequences and analysis of ambiguous DNA sequences, *Nucleic Acids Res.* **12:**819–824.

Pilacinski, W. P., Glassman, D. L., Krzyzek, R. A., Sadowski, P. L., and Robbins, A. K., 1984, Cloning and expression in *Escherichia coli* of the bovine papillomavirus L1 and L2 open reading frames, *Biotechnology* **1:**356–360.

Schiller, J., Vousden, K., Vass, W., and Lowy, D., 1986, The E5 open reading frame of bovine papillomavirus type 1 encodes a transforming gene, *J. Virol.* **57:**1–6.

Schulz, G., and Schirmer, R., 1979, *Principles of Protein Structure*, pp. 14–16, Springer-Verlag, Berlin.

Schwarz, E., Dürst, M., Demankowski, C., Lattermann, O., Zech, R., Wolfsperger, E., Suhai, S., and zur Hausen, H., 1983, DNA sequence and genome organization of genital human papillomavirus type 6b, *EMBO J.* **2:**2341–2348.

Seedorf, K., Krammer, G., Dürst, M., Suhai, S., and Rowekamp, W., 1985, Human papillomavirus type 16 DNA sequence, *Virology* **145:**181–185.

Smotkin, D., and Wettstein, F. O., 1986, Transcription of human papillomavirus type 16 early genes in a cervical cancer and a cancer-derived cell line and identification of the E7 protein, *Proc. Natl. Acad. Sci. USA* **83:**4680–4684.

Stenlund, A., Zabielski, J., Ahola, H., Moreno-Lopez, J., and Pettersson, U., 1985, Messenger RNAs from the transforming region of bovine papilloma virus type 1, *J. Mol. Biol.* **182:**541–554.

Wilbur, W., and Lipman, D., 1983, Rapid similarity searches of nucleic acid and protein data banks, *Proc. Natl. Acad. Sci. USA* **80:**726–730.

Yang, Y., Okayama, H., and Howley, P. M., 1985a, Bovine papillomavirus contains multiple transforming genes, *Proc. Natl. Acad. Sci. USA* **82:**1030–1034.

Yang, Y.-C., Spalholz, B., Rabson, M., and Howley, P., 1985b, Dissociation of transforming and transactivation functions for bovine papillomavirus type 1, *Nature* **318:**575–577.

Index